Arthritis Research

METHODS IN MOLECULAR MEDICINE™

John M. Walker, SERIES EDITOR

139. **Vascular Biology Protocols,** edited by *Nair Sreejayan and Jun Ren, 2007*
138. **Allergy Methods and Protocols,** edited by *Meinir G. Jones and Penny Lympany, 2007*
137. **Microtubule Protocols,** edited by Jun Zhou, 2007
136. **Arthritis Research:** *Methods and Protocols, Vol. 2,* edited by *Andrew P. Cope, 2007*
135. **Arthritis Research:** *Methods and Protocols, Vol. 1,* edited by *Andrew P. Cope, 2007*
134. **Bone Marrow and Stem Cell Transplantation,** edited by *Meral Beksac, 2007*
133. **Cancer Radiotherapy,** edited by *Robert A. Huddart and Vedang Murthy, 2007*
132. **Single Cell Diagnostics:** *Methods and Protocols,* edited by *Alan Thornhill, 2007*
131. **Adenovirus Methods and Protocols, Second Edition, Vol. 2:** *Ad Proteins, RNA, Lifecycle, Host Interactions, and Phylogenetics,* edited by *William S. M. Wold and Ann E. Tollefson, 2007*
130. **Adenovirus Methods and Protocols, Second Edition, Vol. 1:** *Adenoviruses, Ad Vectors, Quantitation, and Animal Models,* edited by *William S. M. Wold and Ann E. Tollefson, 2007*
129. **Cardiovascular Disease:** *Methods and Protocols, Volume 2: Molecular Medicine,* edited by *Qing K.Wang, 2006*
128. **Cardiovascular Disease:** *Methods and Protocols, Volume 1: Genetics,* edited by *Qing K. Wang, 2006*
127. **DNA Vaccines:** *Methods and Protocols, Second Edition,* edited by *Mark W. Saltzman, Hong Shen, and Janet L. Brandsma, 2006*
126. **Congenital Heart Disease:** *Molecular Diagnostics,* edited by *Mary Kearns-Jonker, 2006*
125. **Myeloid Leukemia:** *Methods and Protocols,* edited by *Harry Iland, Mark Hertzberg, and Paula Marlton, 2006*
124. **Magnetic Resonance Imaging:** *Methods and Biologic Applications,* edited by *Pottumarthi V. Prasad, 2006*
123. **Marijuana and Cannabinoid Research:** *Methods and Protocols,* edited by *Emmanuel S. Onaivi, 2006*
122. **Placenta Research Methods and Protocols:** *Volume 2,* edited by *Michael J. Soares and Joan S. Hunt, 2006*
121. **Placenta Research Methods and Protocols:** *Volume 1,* edited by *Michael J. Soares and Joan S. Hunt, 2006*
120. **Breast Cancer Research Protocols,** edited by *Susan A. Brooks and Adrian Harris, 2006*
119. **Human Papillomaviruses:** *Methods and Protocols,* edited by *Clare Davy and John Doorbar, 2005*
118. **Antifungal Agents:** *Methods and Protocols,* edited by *Erika J. Ernst and P. David Rogers, 2005*
117. **Fibrosis Research:** *Methods and Protocols,* edited by *John Varga, David A. Brenner, and Sem H. Phan, 2005*
116. **Inteferon Methods and Protocols,** edited by *Daniel J. J. Carr, 2005*
115. **Lymphoma:** *Methods and Protocols,* edited by *Timothy Illidge and Peter W. M. Johnson, 2005*
114. **Microarrays in Clinical Diagnostics,** edited by *Thomas O. Joos and Paolo Fortina, 2005*
113. **Multiple Myeloma:** *Methods and Protocols,* edited by *Ross D. Brown and P. Joy Ho, 2005*
112. **Molecular Cardiology:** *Methods and Protocols,* edited by *Zhongjie Sun, 2005*
111. **Chemosensitivity:** *Volume 2, In Vivo Models, Imaging, and Molecular Regulators,* edited by *Rosalyn D. Blumethal, 2005*
110. **Chemosensitivity:** *Volume 1, In Vitro Assays,* edited by *Rosalyn D. Blumethal, 2005*
109. **Adoptive Immunotherapy:** *Methods and Protocols,* edited by *Burkhard Ludewig and Matthias W. Hoffman, 2005*
108. **Hypertension:** *Methods and Protocols,* edited by *Jérôme P. Fennell and Andrew H. Baker, 2005*
107. **Human Cell Culture Protocols,** *Second Edition,* edited by *Joanna Picot, 2005*
106. **Antisense Therapeutics,** *Second Edition,* edited by *M. Ian Phillips, 2005*
105. **Developmental Hematopoiesis:** *Methods and Protocols,* edited by *Margaret H. Baron, 2005*
104. **Stroke Genomics:** *Methods and Reviews,* edited by *Simon J. Read and David Virley, 2004*
103. **Pancreatic Cancer:** *Methods and Protocols,* edited by *Gloria H. Su, 2004*
102. **Autoimmunity:** *Methods and Protocols,* edited by *Andras Perl, 2004*
101. **Cartilage and Osteoarthritis:** *Volume 2, Structure and In Vivo Analysis,* edited by *Frédéric De Ceuninck, Massimo Sabatini, and Philippe Pastoureau, 2004*
100. **Cartilage and Osteoarthritis:** *Volume 1, Cellular and Molecular Tools,* edited by *Massimo Sabatini, Philippe Pastoureau, and Frédéric De Ceuninck, 2004*
99. **Pain Research:** *Methods and Protocols,* edited by *David Z. Luo, 2004*
98. **Tumor Necrosis Factor:** *Methods and Protocols,* edited by *Angelo Corti and Pietro Ghezzi, 2004*
97. **Molecular Diagnosis of Cancer:** *Methods and Protocols, Second Edition,* edited by *Joseph E. Roulston and John M. S. Bartlett, 2004*
96. **Hepatitis B and D Protocols:** *Volume 2, Immunology, Model Systems, and Clinical Studies,* edited by *Robert K. Hamatake and Johnson Y. N. Lau, 2004*

METHODS IN MOLECULAR MEDICINE™

Arthritis Research

Methods and Protocols

Volume 1

Edited by

Andrew P. Cope

*The Kennedy Institute of Rheumatology,
Imperial College London, London UK*

HUMANA PRESS ✸ TOTOWA, NEW JERSEY

© 2007 Humana Press Inc.
999 Riverview Drive, Suite 208
Totowa, New Jersey 07512

www.humanapress.com

All rights reserved. No part of this book may be reproduced, stored in a retrieval system, or transmitted in any form or by any means, electronic, mechanical, photocopying, microfilming, recording, or otherwise without written permission from the Publisher. Methods in Molecular Biology™ is a trademark of The Humana Press Inc.

All papers, comments, opinions, conclusions, or recommendations are those of the author(s), and do not necessarily reflect the views of the publisher.

This publication is printed on acid-free paper. ∞
ANSI Z39.48-1984 (American Standards Institute)
Permanence of Paper for Printed Library Materials.

Cover design by Nancy K. Fallatt.

Cover illustration: *(Background)* Figure 1 from Chapter 24. *(Inset)* Figure 2 from Chapter 15.

For additional copies, pricing for bulk purchases, and/or information about other Humana titles, contact Humana at the above address or at any of the following numbers: Tel.: 973-256-1699; Fax: 973-256-8341; E-mail: orders@humanapr.com; or visit our Website: www.humanapress.com

Photocopy Authorization Policy:
Authorization to photocopy items for internal or personal use, or the internal or personal use of specific clients, is granted by Humana Press Inc., provided that the base fee of US $30.00 per copy is paid directly to the Copyright Clearance Center at 222 Rosewood Drive, Danvers, MA 01923. For those organizations that have been granted a photocopy license from the CCC, a separate system of payment has been arranged and is acceptable to Humana Press Inc. The fee code for users of the Transactional Reporting Service is: [978-1-58829-344-2/07 $30.00].

Printed in the United States of America. 10 9 8 7 6 5 4 3 2 1

eISBN: 978-1-59745-401-8

Library of Congress Cataloging in Publication Data
Arthritis research : methods and protocols / edited by Andrew P. Cope.
    p. ; cm. — (Methods in molecular biology ; v. 135-136)
  Includes bibliographical references and index.
  ISBN 1-58829-344-0 (v. 1 : alk. paper) — ISBN 1-58829-918-X (v. 2 : alk. paper)
  1. Arthritis—Laboratory manuals. 2. Arthritis—Molecular aspects. I. Cope, Andrew P. II. Series: Methods in molecular biology (Clifton, N.J.) ; v. 135-136.
  [DNLM: 1. Arthritis—Laboratory Manuals. 2. Laboratory Techniques and Procedures—Laboratory Manuals. W1 ME9616J v.135-136 2007 / WE 25 A787 2007]
RC933.A665245 2007
616.7'220072—dc22   2006019975

Preface

> "................. do not go where the path may lead,
> go instead where there is no path and leave a trail"
>
> Ralph Waldo Emerson

The postgenomic era is upon us and with it comes a growing need to understand the function of every gene and its contribution to physiological and pathological processes. Such advances will underpin our understanding of the molecular basis of common chronic inflammatory and degenerative diseases and inspire the development of targeted therapy. Any postgenomic approach for exploring gene function must necessarily address gene expression and regulation, localization of gene products in diseased tissue, manipulation of expression by transgenesis or knockdown technology, and combine these studies with appropriate manipulations in relevant in vivo models. To validate potential therapeutic targets in any depth requires a growing repertoire of assays and disease models that underpin key pathogenic pathways. The same repertoire of tools must be employed to rigorously evaluate process specific biomarkers, which may be of diagnostic and prognostic value. Indeed, measuring the impact of our interventions remains a major challenge for the future.

The rheumatic diseases encompass prototypic chronic inflammatory and degenerative diseases. It would be true to say that experimental procedures adapted for investigating the pathogenesis of diseases such as rheumatoid arthritis have contributed greatly to recent advances in biological therapy. *Arthritis Research: Methods and Protocols* seeks to crystallize methods and protocols that have contributed to such advances in molecular medicine. These volumes are timely because the tools are now accessible to most laboratories. Also included are newer technologies, some of which are still evolving and whose impact are yet to be realized. It is important to note that in these volumes there is something for everyone—basic scientists, clinician scientists, and clinicians alike—with contributions from leaders in their field covering imaging and immunohistochemistry, analysis of cartilage, and bone catabolism, as well as leukocyte trafficking and migration. Combine volumes 1 and 2 and the end product is a concise set of protocols condensing decades of experience and expertise. From the outset of this project it was always the intention that this compendium should provide a unique resource at the bench that would be used in ways that will facilitate the endeavors of clinicians at the bedside in the future.

Acknowledgments

I wish to thank many friends and colleagues for their enthusiasm, support and invaluable contributions toward this project. I am also very grateful to Mandy Wilcox for her dedicated secretarial assistance in compiling the finished product. The research carried out by the Editor's laboratory at the Kennedy Institute is supported by grants from the Wellcome Trust and the Arthritis Research Campaign, UK.

Andrew P. Cope

Contents

Preface v
Ackowledgments vii
Contributors xviii
Color Plate xvii
Contents for Volume 2 xix

PART I SYNOVIAL JOINT MORPHOLOGY, HISTOPATHOLOGY, AND IMMUNOHISTOCHEMISTRY

1 Imaging Inflamed Synovial Joints
   Ai Lyn Tan, Helen I. Keen, Paul Emery, and Dennis McGonagle 3

2 Arthroscopy as a Research Tool: *A Review*
   Richard J. Reece 27

3 Immunohistochemistry of the Inflamed Synovium
   Martina Gogarty and Oliver FitzGerald 47

4 *In Situ* Hybridization of Synovial Tissue
   *Stefan Kuchen, Christian A. Seemayer, Michel Neidhart,
   Renate E. Gay, and Steffen Gay* 65

5 Subtractive Hybridization
   *Jörg H. W. Distler, Oliver Distler, Michel Neidhart,
   and Steffen Gay* 77

6 Laser Capture as a Tool for Analysis of Gene Expression
   in Inflamed Synovium
   *Ulf Müller-Ladner, Martin Judex, Elena Neumann,
   and Steffen Gay* 91

7 Preparation of Mononuclear Cells from Synovial Tissue
   Jonathan T. Beech and Fionula M. Brennan 105

8 Quantitative Image Analysis of Synovial Tissue
   Pascal O. van der Hall, Maarten C. Kraan, and Paul Peter Tak 121

PART II CARTILAGE MATRIX AND BONE BIOLOGY

9 Cartilage Histomorphometry
   Ernst B. Hunziker 147

10 Image Analysis of Aggrecan Degradation in Articular
   Cartilage with Formalin-Fixed Samples
   *Barbara Osborn, Yun Bai, Anna H. K. Plaas,
   and John D. Sandy* 167

11	*In Situ* Detection of Cell Death in Articular Cartilage **Samantha N. Redman, Ilyas M. Khan, Simon R. Tew, and Charles W. Archer**	183
12	Measurement of Glycosaminoglycan Release from Cartilage Explants **John S. Mort and Peter J. Roughley**	201
13	Assessment of Collagenase Activity in Cartilage **Tim E. Cawston and Tanya G. Morgan**	211
14	Assessment of Gelatinase Expression and Activity in Articular Cartilage **Rosalind M. Hembry, Susan J. Atkinson, and Gillian Murphy**	227
15	Analysis of MT1-MMP Activity in Cells **Richard D. Evans and Yoshifumi Itoh**	239
16	Analysis of TIMP Expression and Activity **Linda Troeberg and Hideaki Nagase**	251
17	Bone Histomorphology in Arthritis Models **Georg Schett and Birgit Tuerk**	269
18	Generation of Osteoclasts In Vitro, and Assay of Osteoclast Activity **Naoyuki Takahashi, Nobuyuki Udagawa, Yasuhiro Kobayashi, and Tatsuo Suda**	285

PART III CELL TRAFFICKING, MIGRATION, AND INVASION

19	Isolation and Analysis of Large and Small Vessel Endothelial Cells **Justin C. Mason, Elaine A. Lidington, and Helen Yarwood**	305
20	Analysis of Flow-Based Adhesion In Vitro **Oliver Florey and Dorian O. Haskard**	323
21	Analysis of Leukocyte Recruitment in Synovial Microcirculation by Intravital Microscopy **Gabriela Constantin**	333
22	Angiogenesis in Arthritis: *Methodological and Analytical Details* **Ursula Fearon and Douglas J. Veale**	343
23	Analysis of Inflammatory Leukocyte and Endothelial Chemotactic Activity **Zoltán Szekanecz and Alisa Koch**	359
24	Acquisition, Culture, and Phenotyping of Synovial Fibroblasts **Sanna Rosengren, David L. Boyle, and Gary S. Firestein**	365
25	Genotyping of Synovial Fibroblasts: *cDNA Array in Combination with RAP-PCR in Arthritis* **Elena Neumann, Martin Judex, Steffen Gay, and Ulf Müller-Ladner**	377

26 Gene Transfer to Synovial Fibroblast: *Methods and Evaluation in the SCID Mouse Model*
   Ingmar Meinecke, Edita Rutkauskaite, Antje Cinski, Ulf Müller-Ladner, Steffen Gay, and Thomas Pap *393*

27 In Vitro Matrigel Fibroblast Invasion Assay
   Tanja C.A. Tolboom and Tom W.J. Huizinga *413*

28 Culture and Analysis of Circulating Fibrocytes
   Timothy E. Quan and Richard Bucala *423*

Index *435*

Contributors

CHARLES W. ARCHER • *Cardiff School of Biosciences, Cardiff, Wales*
SUSAN J. ATKINSON • *Department of Oncology, University of Cambridge, Cambridge Institute for Medical Research, Cambridge, UK*
YUN BAI • *Shriners Hospital for Children, Tampa, FL*
JONATHAN T. BEECH • *Kennedy Institute of Rheumatology Division, Faculty of Medicine, Imperial College London, London, UK*
DAVID L. BOYLE • *Division of Rheumatology, Allergy, and Immunology, School of Medicine, University of California at San Diego, La Jolla, CA*
FIONULA M. BRENNAN • *Kennedy Institute of Rheumatology Division, Faculty of Medicine, Imperial College London, London, UK*
RICHARD BUCALA • *Yale University School of Medicine, Department of Medicine, Section of Rheumatology, New Haven, CT*
TIM E. CAWSTON • *Musculoskeletal Research Group, School of Clinical Medical Sciences, Newcastle Upon Tyne, UK*
ANTJE CINSKI • *Division of Molecular Medicine of Musculoskeletal Tissue, Department of Orthopaedics, University Hospital Munster, Germany*
GABRIELA CONSTANTIN • *Department of Pathology, Division of General Pathology, University of Verona, Verona, Italy*
ANDREW COPE • *The Kennedy Institute of Rheumatology Division, Imperial College London, London, UK*
JORG H. W. DISTLER • *Centre of Experimental Rheumatology, Department of Rheumatology, University of Zurich, Zurich, Switzerland*
OLIVER DISTLER • *Centre of Experimental Rheumatology, Department of Rheumatology, University of Zurich, Zurich, Switzerland*
PAUL EMERY • *Academic Unit of Musculoskeletal Disease, Chapel Allerton Hospital, Leeds, UK*
RICHARD D. EVANS • *Kennedy Institute of Rheumatology Division, Faculty of Medicine, Imperial College London, London, UK*
URSULA FEARON • *St. Vincent's University Hospital, Elm Park, Dublin, Ireland*
GARY S FIRESTEIN • *Division of Rheumatology, Allergy and Immunology, School of Medicine, University of California at San Diego, La Jolla, CA*
OLIVER FITZGERALD • *Rheumatology Department, St. Vincent's University Hospital, Dublin, Ireland*
OLIVER FLOREY • *The Eric Bywaters Centre for Vascular Inflammation, Imperial College London, Hammersmith Hospital, London, UK*

RENATE E. GAY • *Centre of Experimental Rheumatology, Department of Rheumatology, University Hospital Zurich, Zurich, Switzerland*

STEFFEN GAY • *Centre of Experimental Rheumatology, Department of Rheumatology, University Hospital Zurich, Zurich, Switzerland*

MARTINA GOGARTY • *Rheumatology Department, St. Vincent's University Hospital, Dublin, Ireland*

DORIAN O. HASKARD • *The Eric Bywaters Centre for Vascular Inflammation, Imperial College London, Hammersmith Hospital, London, UK*

ROSALIND M. HEMBRY • *School of Biological Sciences, University of East Anglia, Norwich, UK*

TOM W. J. HUIZINGA • *Leiden University Medical Centre, Department of Rheumatology, RC Leiden, The Netherlands*

ERNST B. HUNZIKER • *ITI Research Institute for Dental and Skeletal Biology, University of Bern, Bern, Switzerland*

YOSHIFUMI ITOH • *Kennedy Institute of Rheumatology Division, Faculty of Medicine, Imperial College London, London, UK*

MARTIN JUDEX • *Sidney Kimmel Cancer Centre, San Diego, CA*

HELEN I. KEEN • *Academic Unit of Musculoskeletal Disease, Chapel Allerton Hospital, Leeds, UK*

ILYAS M. KHAN • *Cardiff School of Biosciences, Cardiff, UK*

YASUHIRO KOBAYASHI • *Institute for Oral Science, Matsumoto Dental University, Nagano, Japan*

ALISA E. KOCH • *University of Michigan Health System, Internal Medicine/ Rheumatology and Veterans' Administration, Ann Arbor, MI*

MAARTEN C. KRAAN • *Division of Clinical Immunology and Rheumatology, Academic Medical Centre, University of Amsterdam, Amsterdam, The Netherlands*

STEFAN KUCHEN • *Centre of Experimental Rheumatology, Department of Rheumatology, University Hospital Zurich, Zurich, Switzerland*

ELAINE A. LIDINGTON • *BHF Cardiovascular Medicine Unit, The Eric Bywaters Centre, Imperial College London, Hammersmith Hospital, London, UK*

JUSTIN C. MASON • *Cardiovascular Medicine Unit, The Eric Bywaters Centre, Imperial College London, Hammersmith Hospital, London, UK*

DENNIS MCGONAGLE • *Academic Unit of Musculoskeletal Disease, Chapel Allerton Hospital, Leeds, UK*

INGMAR MEINECKE • *Department of Traumatology and Division of Molecular Medicine of Musculoskeletal Tissue, University Hospital, Munster, Germany*

TANYA G. MORGAN • *Musculoskeletal Research Group, School of Clinical Medical Sciences, University of Newcastle, Newcastle Upon Tyne, UK*

Contributors

JOHN S. MORT • *Joint Diseases Laboratory, Shriners Hospital for Children, Montreal, Quebec, Canada*

ULF MÜLLER-LADNER • *Division of Internal Medicine I, Division of Rheumatology and Clinical Immunology, University of Regensburg, Regensburg, Germany*

GILLIAN MURPHY • *Department of Oncology, University of Cambridge, Cambridge Institute for Medical Research, Cambridge, UK*

HIDEAKI NAGASE • *Kennedy Institute of Rheumatology Division, Faculty of Medicine, Imperial College London, London, UK*

MICHEL NEIDHART • *Centre of Experimental Rheumatology, Department of Rheumatology, University of Zurich, Zurich, Switzerland*

ELENA NEUMANN • *Division of Internal Medicine I, Division of Rheumatology and Clinical Immunology, University of Regensburg, Regensburg, Germany*

BARBARA OSBORN • *Department of Internal Medicine, Division of Rheumatology, University of South Florida, Tampa, FL*

THOMAS PAP • *Division of Molecular Medicine and Musculoskeletal Tissue, University Hospital Munster, Munster, Germany*

ANNA H. K. PLAAS • *Department of Internal Medicine, Division of Rheumatology, University of South Florida, Tampa, FL*

TIMOTHY E. QUAN • *Yale University School of Medicine, Department of Medicine, Section of Rheumatology, New Haven, CT*

SAMANTHA N. REDMAN • *Cardiff School of Biosciences, Cardiff, UK*

RICHARD J. REECE • *Academic Unit of Musculokeletal Medicine, Clinical Pharmacology Unit (Rheumatism Research), University of Leeds, Chapel Allerton Hospital, Leeds, UK*

SANNA ROSENGREN • *Division of Rheumatology, Allergy and Immunology, School of Medicine, University of California at San Diego, La Jolla, CA*

PETER J. ROUGHLEY • *Genetics Unit, Shriners Hospital for Children and Department of Surgery and Human Genetics, McGill University, Montreal, Quebec, Canada*

EDITA RUTKAUSKAITE • *Division of Molecular Medicine of Musculoskeletal Tissue, Department of Orthopaedics, University Hospital Munster, Germany*

JOHN D. SANDY • *Shriners Hospital for Children, Tampa, FL*

GEORG SCHETT • *Department of Internal Medicine 3 and Institute of Clinical Immunology, University of Erlangen-Nuremberg, Erlangen, Germany*

CHRISTIAN A. SEEMAYER • *Centre of Experimental Rheumatology and WHO Collaborating Centre for Molecular Biology and Novel Therapeutic Strategies for Rheumatic Diseases, University Hospital, Zurich, Switzerland*

TATSUO SUDA • *Research Centre for Genomic Medicine, Saitama Medical School, Saitama, Japan*

ZOLTÁN SZEKANECZ • *Rheumatology Division, Third Department of Medicine, University of Debrecan Medical and Health Sciences Centre, Debrecen, Hungary*

PAUL PETER TAK • *Division of Clinical Immunology and Rheumatology, Academic Medical Centre, University of Amsterdam, Amsterdam, The Netherlands*

NAOYUKI TAKAHASHI • *Institute for Oral Science, Matsumoto Dental University, Nagano, Japan*

AI LYN TAN • *Academic Unit of Musculoskeletal Disease, Chapel Allerton Hospital, Leeds, UK*

SIMON R. TEW • *UK Centre for Tissue Engineering, Faculty of Life Sciences, The University of Manchester, Manchester, UK*

TANJA C. A. TOLBOOM • *Department of Rheumatology, Leiden University Medical Centre, Leiden, The Netherlands*

LINDA TROEBERG • *Kennedy Institute of Rheumatology Division, Faculty of Medicine, Imperial College London, London, UK*

BIRGIT TUERK • *Department of Internal Medicine III, Division of Rheumatology, University of Vienna, Vienna, Austria*

NOBUYUKI UDAGAWA • *Department of Biochemistry, Matsumoto Dental University, Nagano, Japan*

PASCAL O. VAN DER HALL • *Division of Clinical Immunology and Rheumatology, Academic Medical Centre, University of Amsterdam, Amsterdam, The Netherlands*

DOUGLAS J. VEALE • *St. Vincent's University Hospital, Elm Park, Dublin, Ireland*

HELEN YARWOOD • *Department of Biology, Imperial College London, London, UK*

Color Plate

The following color illustrations are printed in the insert that follows p. 268.

Chapter 1
    Fig. 4: Ultrasonography of a metacarpophalangeal joint.

Chapter 10
    Fig. 1: Chondrocyte-mediated processing of ADAMTS-4.
    Fig. 2: IHC with non-immune IgG and anti-NITEGE of freshly excised bovine articular cartilage.
    Fig. 3: IHC with non-immune IgG and anti-NITEGE of cartilage after explant culture in RA.
    Fig. 4: IHC of freshly excised bovine articular cartilage using anti-NITEGE, anti-MT4MMP, and anti-ADAMTS4
    Fig. 5: IHC of bovine articular cartilage after 4 d of serum-free culture.
    Fig. 6: IHC of bovine articular cartilage after 4 d of serum-free culture with 3 μM RA.

Chapter 11
    Fig. 4: Use of DAPI to detect apoptotic nuclei.
    Fig. 5: Use of Live/Dea kit to detect live and dead cells.

Chapter 18
    Fig. 1: Enzyme histochemistry for TRAP and ALP in mouse bone marrow cultures.
    Fig. 2: Purified TRAP-positive osteoclasts.

Chapter 22
    Fig. 1: Demonstration of a 2.7-mm-diameter arthroscope.
    Fig. 2: Arthroscopic images of chondropathy and synovitis in RA.
    Fig. 3: Arthroscopic images of straight and tortuoous vessels.
    Fig. 4: CD31 staining in early PsA synovial tissue section.

Chapter 25
    Fig. 4: Comparison views using different normalization settings of two cDNA arrays.

Contents for Volume 2

Preface
Acknowledgment
Contributors
Color Plate

PART I IMMUNOBIOLOGY

1. Phenotypic Analysis of B Cells and Plasma Cells
   Henrik E. Mei, Taketoshi Yoshida, Gwendowlin Muehlinghaus, Falk Hiepe Thomas Dorner, Andreas Radbruch and Bimba F. Hoyer
2. Detection of Antigen Specific B Cells in Tissues
   Marie Wahren-Herlenius and Stina Salomonsson
3. Single Cell Analysis of Synovial Tissue B Cells
   Hye-Jung Kim and Claudia Berek
4. Tracking Antigen Specific CD4$^+$ T Cells with Soluble MHC Molecules
   John A. Gebe and William W. Kwok
5. Analysis of Antigen Reactive T Cells
   Clare Alexander, Richard C. Duggleby, Jane C. Goodall, Malgosia K. Matyszak, Natasha Telyatnikova, and J. S. Hill Gaston
6. Identification and Manipulation of Antigen Specific T Cells with Artificial Antigen Presenting Cells
   Eva Koffeman, Elissa Keogh, Mark Klein, Berent Prakken, and Salvatore Albani
7. Analysis of Th1/Th2 T Cell Subsets
   Alla Skapenko and Hendrik Schulze-Koops
8. Analysis of the T Cell Receptor Repertoire of Synovial T Cells
   Lucy Wedderburn and Douglas J. King
9. The Assessment of T Cell Apoptosis in Synovial Fluid
   Karim Razza, Dagmar Scheel-Toellner, Janet M. Lord, Arne N. Akbar, Christopher D. Buckley, and Mike Salmon
10. Assay of T Cell Contact Dependent Monocyte-Macrophage Functions
    Danielle Burger and Jean-Michel Dayer
11. Phenotypic and Functional Analysis of Synovial Natural Killer Cells
    Nicola Dalbeth and Margaret F. C. Callan
12. Identification and Isolation and Synovial Dendritic Cells
    Allison R Pettit, Lois Cavanagh, Amanda Boyce, Jagadish Padmanabha, Judy Peng, and Ranjeny Thomas

Part II Animal Models of Arthritis

13 The Use of Animal Models for Rheumatoid Arthritis
   Rikard Holmdahl
14 Collagen Induced Arthritis in Mice
   Richard O. Williams
15 Collagen Induced Arthritis in Rats
   Marie M. Griffiths, Grant W. Cannon, Tim Corsi, Van Reese, and Kandie Kunzler
16 Collagen Antibody Induced Arthritis (CAIA)
   Kutty Selva Nandakumar and Rikard Holmdahl
17 Arthritis Induced with Minor Cartilage Proteins
   Stefan Carlsen, Shemin Lu, and Rikard Holmdahl
18 Murine Antigen-Induced Arthritis
   Wim van den Berg, Leo A. B. Joosten, and Peter L.E.M. van Lent
19 Pristane Induced Arthritis in the Rat
   Peter Olofsson and Rikard Holmdahl
20 The K/BxN Mouse Model of Inflammatory Arthritis: *Theory and Practice*
   Paul Monach, Kimie Hattori, Haochu Huang, Elzbieta Hyatt, Jody Morse[1], Linh Nguyen, Adriana Ortiz-Lopez, Hsin-Jung Wu, Diane Mathis, and Christophe Benoist
21 Analysis of Arthritic Lesions in the Del1 Mouse: *A Model for OA*
   Anna-Marja Säämänen, Mika Hyttinen, and Eero Vuorio

Part III Application of New Technologies to Define Novel Therapeutic Targets

22 Gene Expression Profiling in Rheumatology
   Tineke C.M.T. van der Pauw Kraan, Lisa G.M. van Baarsen, Francois Rustenburg, Belinda Baltus, Mike Fero, and Cornelis L. Verweij
23 Differential Display Reverse Transcription-Polymerase Chain Reaction (DDRT-PCR) to Identify Novel Biomolecules in Arthritis Research
   Manir Ali and John D. Isaacs
24 Two-Dimensional Electrophoresis of Proteins Secreted from Articular Cartilage
   Monika Hermansson, Jeremy Saklatvala, and Robin Wait
25 Mapping Lymphocyte Plasma Membrane Proteins: *A Proteomic Approach*
   Matthew J. Peirce, Jeremy Saklatvala, Robin Wait, and Andrew P. Cope
26 In Vivo Phage Display Selection in the Human/SCID Mouse Chimera Model for Defining Synovial Specific Determinants
   Lewis Lee, Toby Garrood, and Costantino Pitzalis
27 Adenoviral Targeting of Signal Transduction Pathways in Synovial Cell Cultures
   Alison Davis, Corinne Taylor, Catherine Willetts, Clive Smith, and Brian M.J. Foxwell

Index

I

SYNOVIAL JOINT MORPHOLOGY, HISTOPATHOLOGY, AND IMMUNOHISTOCHEMISTRY

1

Imaging Inflamed Synovial Joints

Ai Lyn Tan, Helen I. Keen, Paul Emery, and Dennis McGonagle

Summary

As a research tool, medical imaging has considerably increased our understanding of joint inflammation in vivo. Although conventional radiography remains important for diagnosis, other imaging modalities such as magnetic resonance imaging (MRI) and ultrasonography show considerable potential for the study of joint inflammation in the research setting. The advances in MRI and ultrasound in the last decade has lead to significant conceptual advances into the microanatomical basis for arthritis. The purpose of this chapter is to describe how MRI and ultrasound can be used in the research environment. Other imaging techniques such as computed tomography and scintigraphy have little use in studying inflamed synovial joints due to their relative inability to demonstrate inflammation, or poor resolution or both. However, molecular imaging using high-resolution positron emission tomography or single photon emission computed tomography is being increasingly used to study inflammation in the experimental setting and also in man.

Key Words: Magnetic resonance imaging; ultrasonography; synovial joints; imaging; inflammation; synovitis; enthesitis; erosions.

1. Introduction

Imaging has played an important role in rheumatology both clinically and in research. Over the past few decades there have been significant advances in the technique and application of various imaging modalities in studying arthritis. Imaging techniques have become more sensitive and sophisticated and more widely utilized in rheumatology than ever before. As the evidence for very early diagnosis and treatment of various rheumatological conditions increases, so has the importance of imaging. Whereas the diagnostic value of modern imaging techniques is being increasingly realized the purpose of this chapter is to focus on imaging as a research adjunct in man.

All joint structures may be involved in arthritis including the synovial lining, the synovial fluid, the joint capsule, the ligaments and tendons, and of course cartilage. Conventional radiography (CR) may be the first choice imaging investigation in clinical practice; however, because of its relative inability to image soft tissue structures and therefore its inability to image inflammation in joints, its use is limited. Nevertheless, CR is often used to document the sequelae of joint inflammation including bone erosions, new bone formation or loss of joint space; it is still the mainstay of research into the potential for a new therapy to prevent bone erosion. Various radiographic scoring methods have been devised and have been used as outcome measures in many studies *(1,2)*.

Magnetic resonance imaging (MRI) and ultrasonography (US) do not involve ionizing radiation, making them good tools for serial evaluation of subjects. Furthermore, MRI and US are particularly good at depicting synovitis and other soft tissue structures in joints. Both MRI and US have the capability to monitor the progression of arthritis, and can therefore be used to evaluate therapy *(3–9)*. MRI can be used to identify prebone erosion abnormalities such as bone oedema *(10)*, and both modalities can be used to define synovitis and enthesitis *(11–14)*. US is perhaps preferable at demonstrating tendon pathology than MRI *(13)*. This chapter concentrates on the use of MRI and US in imaging inflamed synovial joints, both of which show tremendous potential in rheumatology.

2. Magnetic Resonance Imaging

Unlike US, most rheumatologists generally do not tend to perform MRI themselves. Magnetic resonance imaging is a highly technical skill that requires a qualified radiographer to perform. Nevertheless, it is important to understand the technique of MRI and how to request an appropriate MRI investigation in order to achieve optimum image acquisition for a particular purpose. There have been significant developments in MRI of the synovial joints of the musculoskeletal system, particularly the development of more sophisticated scanners, novel and improved MRI sequences with better image processing, and strategies for more superior resolution *(15)*. More recently, portable MRI scanners have been developed which are more straightforward to operate and can be managed by rheumatologists and other researchers, but their resolution is rather less, resulting in more limited research potential.

2.1. MRI Equipment

There are a few key components in MRI that are controlled by a central computer. Most advances in the development of these components are in fine-tuning the equipment and to stretch their varying capabilities in order to achieve better quality images.

2.1.1. MRI Scanner

Magnetic resonance imaging, as the name suggests, has the magnet as its main component. The type of MRI scanner is often denoted by the strength of the magnet, designated by the term tesla (T). Conventionally, most musculoskeletal imaging is performed on a 1.5T MRI scanner. More recently magnetic fields as high as 3T and as low as 0.2T have been used to image the musculoskeletal system.

One of the main advantages of a 3T MRI scanner compared to a 1.5T scanner is higher signal-to-noise ratio (SNR), thereby producing higher resolution images and improved imaging speed *(16)*. In contrast the low field MRI scanners compromise on the image quality and scanning time. However, because of the lower field strength employed, it tends to be more portable and therefore more likely to be a dedicated extremity scanner. The low field scanners are open scanners, and are less claustrophobic not requiring shielding and are more comfortable for patients. These scanners are also generally cheaper than the higher field scanners. It has been found to be comparable to a conventional whole body scanner in detecting bone erosions and synovitis *(16,17)*, although it is less sensitive for identifying bone marrow oedema *(17)*.

2.1.2. MRI Coils

For MRI to work, the magnetic field has to be transmitted from and/or received by a conductor in the form of MRI coils. There are many types of MRI coils for imaging the different anatomy, mainly comprising of volume coils, surface coils, and gradient coils of different shapes and sizes. In essence the smaller the coil, and the closer the coil is to the object of interest, the smaller the volume and the higher the SNR, resulting in an image of higher resolution. The recent development of microscopy MRI coils has opened up the challenge to study anatomical structures in vivo at the greatest possible resolution to date. Resolution as high as 100 µm have been achieved in viewing small joints like the fingers *(19)*. Phased array coils utilises a number of small coils to achieve a high SNR and still maintaining the field of view *(20)*. These arrays of coils have been exploited by parallel imaging techniques such as SMASH (SiMultaneous Acquisition of Spatial Harmonics) *(21)* and SENSE (SENSitivity Encoding) *(22)* which speed up MRI scanning time. Recent developments utilize whole body surface coil technology whereby as many as 32 phased-array surface coils are used simultaneously, thereby reducing scanning time significantly *(23)*.

2.2. MRI Display

MRI images can either be printed on an acetate film or viewed electronically. However most imaging units are slowly phasing out films and replacing them with Picture Archiving and Communications Systems (PACS) *(24)*. The benefits of using PACS over hard copy films are numerous, and have shown to be promising clinically, academically, as well as in research *(25)*. The advantage of viewing the MRI electronically is that the displayed image can be modified and other imaging software can be used on the images. MRI images are normally stored as Digital Imaging and Communications in Medicine (DICOM) files.

2.3. MRI Sequences

Unlike CR where only one type of image is possible the number of MRI sequences available is large and increasing all the time. Depending on what structural component an investigator wishes to study, and depending on the water or fat content of the anatomy, then different sequences need to be used. The most commonly used pulse sequence is known as a spin echo (SE) sequence, where T1-weighted and T2-weighted sequences are produced, so-called as a result of the time constant that describes how the magnetization responds. T1-weighted images are characterized by short repetition time (TR) and echo time (TE), where fat appear bright and water dark. T1-weighted sequences produce good anatomical details, whereas T2-weighted imaging highlights presence of water or synovial fluid within joints with fat appearing dark. However, the brightness of fat can mask the appearance of any pathological fluid observed; a technique of fat-suppression is therefore required to overcome this. Fat suppression techniques have proven particularly useful in rheumatology for imaging the joints in rheumatoid arthritis (RA) *(26,27)*, spondyloarthropathy *(22,29,30)* and osteoarthritis (OA). Fat suppression is often achieved by Short Tau Inversion Recovery (STIR) pulse sequence. An alternative to fat suppression is water excitation which has been reported to produce better images *(31,32)*. We are presently in an age of rapid progress in the development of MRI sequences, and a recent development has been ultrashort echo time (UTE) which has been developed for the better visualisation of joint fibrocartilage and tendons *(11,33,34)*.

2.3.1. Contrast Agents

Contrast agents can be administered during MRI to enhance certain abnormalities in arthritis. The most commonly used contrast agent in musculoskeletal examination is gadolinium diethylene triamine tetrapentaacetic acid (Gd-DTPA). The purpose of using contrast agents is to enhance and delineate inflammation in the synovial joints. It can be used to differentiate between fluid and synovitis, both appearing as high signal on T2-weighted images. On T1-weighted post contrast MRI, inflamed synovium appears bright, whereas fluid in joints does not (*see* **Figs. 1–3**). Inflammation within synovial joints can

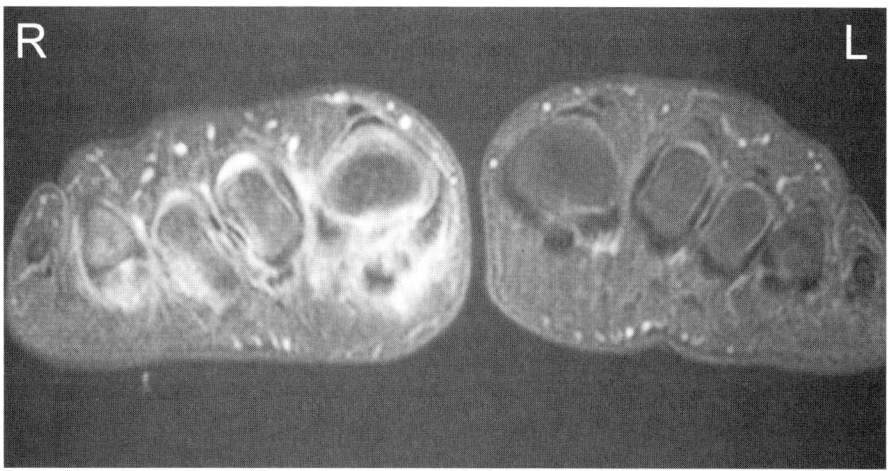

Fig. 1. T1-weighted, fat suppressed post-gadolinium axial magnetic resonance images of the metatarsal heads of both feet of a patient with inflammatory arthritis. When compared to the normal left foot (L), there was enhancement around the metatarsal heads on the right foot (R) indicating inflammation or synovitis

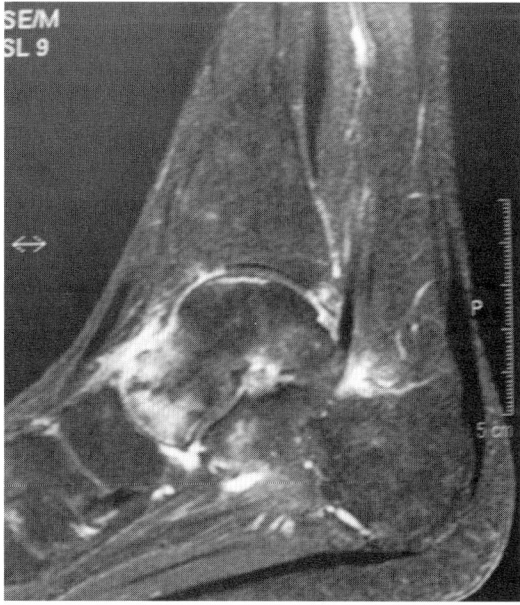

Fig. 2. T1-weighted fat suppressed post-gadolinium sagittal magnetic resonance images of a painful ankle showing patchy enhancement in the talus and the surrounding soft tissues indicating widespread inflammation in the ankle joint.

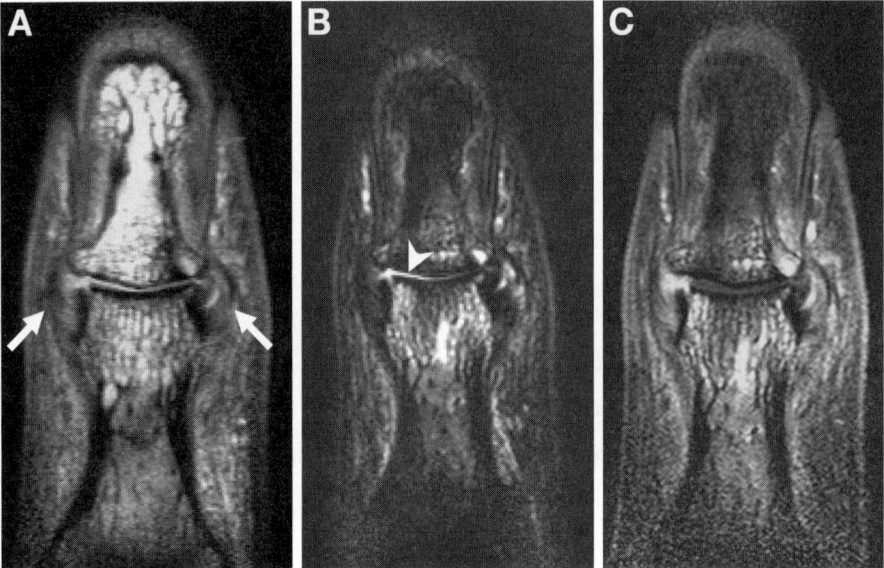

Fig. 3. A distal interphalangeal joint of a patient with psoriatic arthritis affecting the fingers displayed in three MRI sequences to demonstrate the different characteristics of the joint. (**A**) is a proton density-weighted 2D spin echo image that highlights the structure of the abnormally thickened collateral ligaments (arrows). (**B**) is a T2-weighted, fat-suppressed spin echo image and (**C**) a T1-weighted fat suppressed, spin echo post-gadolinium image of the same joint. These latter two MRI protocols show high signal in the areas of the joint that are inflamed which included the bones (bone oedema), collateral ligaments, soft tissues, and the nail bed; however synovial fluid in the joint space (arrow head) that shows up as high signal on the T2-weighted protocol does not enhance post contrast, thereby differentiating between fluid and oedema in the joint.

further be characterized using dynamic contrast enhanced MRI (DEMRI) by measuring capillary permeability parameters such as initial rate of contrast enhancement and maximal enhancement *(35–40)*.

2.4. Contraindications to MRI

Although most modern MRI scanners are quite safe, there are potentially serious problems if certain screening criteria are not adhered to *(41,42)*. There are three main areas of contraindication to MRI: (1) the effect of the magnetic field, (2) the contrast material used, and (3) the confined space of whole body scanners. Most material that succumbs to magnetic field has potential adverse effects either to the patients and personnel involved, or to the devices themselves. Patients should be screened for the presence of cardiac pacemakers or implantable cardioverter defibrillators, intracranial aneurysm clips, and certain drug delivery patches that contain metallic foils. An allergy to contrast materials and pregnancy precludes the use of the contrast material during MRI. Claustrophobia may prevent a successful scanning session, and may require appropriate sedation during the scan.

2.5. MRI Features of Inflamed Synovial Joint

The common features of an inflamed synovial joint such as synovitis, enthesitis, bone erosions and bone oedema, and cartilage damage are readily demonstrated on MRI. There are certain MRI sequences that can optimize these observations. Not all of these features can be best demonstrated on one sequence alone, as visibility of some features is compromised whereas the anatomy of interest is being highlighted.

2.5.1. Synovitis

Synovitis appears as high signal on T2-weighted sequences often alongside fluid that also appears as high signal. To differentiate between these and to confirm if the high signal on T2-weighted signal is indeed synovitis, a T1-weighted post contrast with gadolinium often shows up synovitis as bright whereas fluid remains dark (*see* **Figs. 1–3**). Volume of synovitis can be obtained using processed DEMRI, and disease activity can be measured using the rate of gadolinium enhancement. Studies have used this method to investigate the mechanism of inflammatory arthritis, to monitor response to therapy, and to differentiate between the different arthropathies *(36–40,43,44)*. Computer software has been applied to dynamic MRI to quantify and qualify the degree of synovitis in various synovial joints by measuring the rate of enhancement of gadolinium.

2.5.2. Enthesitis

The same MRI protocol for synovitis can be applied to investigate enthesitis which is the inflammation of the attachment of tendons, ligaments, or joint capsule to bones *(12)*. In addition, as inflamed entheses may appear thickened with an irregular appearance, other sequences can exploit these characteristics to better visualize these structures. Because of the multiplanar capability of MRI, both the cross-section and the longitudinal images of these structures can and should be demonstrated. T1-weighted SE sequences generally depict any entheseal abnormality quite clearly. These structures however tend to have low signal; this property can be utilized by proton-weighted density SE sequence, where the entheses appear as dark bands because these structures contain low density of protons or hydrogen atoms (*see* **Fig. 3A**). More recently, because the entheses have a very short T2 component and therefore very low signal on T1-weighted sequences, ultrashort TE (UTE) have been explored to detect signals from these entheseal structures with short T2s *(11,33,34)*.

2.5.3. Bone Erosions

MRI is the most sensitive imaging modality that can detect and monitor bone erosions. MRI can also illustrate pre-erosive changes such as bone oedema and synovitis *(10)*. As T1-weighted sequences demonstrate anatomical details very well, it is the most appropriate MRI sequence for revealing bone erosions. Bone erosions on MRI are defined as bone defects with sharp margins visible on T1-weighted images in two planes (coronal and axial) with a cortical break seen in at least one plane according to Outcome Measures in Rheumatology Clinical Trials (OMERACT) group recommendation *(45)*. It is important to note that MRI changes resembling mild synovitis or small bone erosions are sometimes found in the joints of healthy subjects, although the absence of other signs such as bone oedema may help distinguish between this and pathological changes *(46)*.

2.5.4. Bone Oedema

MRI has been shown to be a sensitive imaging modality in detecting bone marrow oedema which may precede bone erosions in RA *(47)*. As such, the ability to detect bone oedema early on in disease has both diagnostic and therapeutic implications *(48)*. In OA bone oedema has been shown to be a risk factor for structural damage in the knee joint and pain *(49,50)*. Bone oedema can be detected on T1-weighted sequences as areas of low signal, but is better seen on other sequences such as STIR, turbo SE (TSE) STIR, and T2-weighted fat suppressed TSE, as high signal *(51,52)*. The lesions tend to appear clearer on T1-weighted post contrast sequence with fat suppression *(53)*, but has been found to add little to the STIR sequence in identifying bone oedema in spinal lesions of ankylosing spondylitis patients *(54,55)*

2.5.5. Cartilage

The cartilage is an important structure in the synovial joint which can be affected in various arthropathies, but most commonly studied in relation to OA. MRI of cartilage is well reviewed in a recent article by Verstraete et al. *(56)*. The fat-saturated intermediate-weighted fast SE and spoiled gradient echo sequences are said to be the most appropriate sequences for studying cartilage abnormalities *(57)*, in particular the three-dimensional (3D) spoiled gradient-echo (SPGR) *(58)* or 3D fast low-angle shot (FLASH) sequences with fat saturation and 3D double-echo steady-state (DESS) sequences *(60,61)*. MRI of cartilage is further enhanced using a 3T scanner instead of a 1.5T scanner *(62)*.

One of the most common indications for obtaining an MRI of cartilage is to determine the volume of cartilage *(63)*. Understanding the structural properties of cartilage may also be important in studying OA; more recently, a method of diffusion tensor imaging (DTI) commonly used to study other anatomical structures, have been shown to be a potential imaging tool in assessing the cartilage in this manner *(64,65)*. DTI allows the measurement of the direction and magnitude of the diffusion of water in tissues. The fine structures of the cartilage, such as the collagenase fibers, can be analyzed using DTI, thus allowing the study of early changes in the cartilage in arthritis *(64)*. Other sequences such as T2 mapping, T1rho mapping, and delayed gadolinium-enhanced MRI of cartilage have been developed which are complementary techniques and look at the composition of the cartilage matrix *(66–69)*.

2.6. MRI of Animal Models

Most models of adjuvant induced arthritis are small animals. High field MRI up to 7T can be used in these animal models as they can be anaesthetized, and can therefore be scanned for longer periods. The advantage therefore is that better resolution of the synovial joints can be achieved.

The most recent innovation in MRI of animal models with arthritis is in molecular imaging, with the ability to image macrophage activity that has an important role in arthritis *(70)*. Several studies have used MRI to study macrophage activity following the intravenous administration of iron oxide particles *(71,72)*. Although macrophage labeling with ultra-small superparamagnetic iron oxide particles have been used in the MRI evaluation of various inflammatory conditions, its use in studying arthritis is limited *(73,74)*.

Other uses of MRI in animal models of arthritis are to study the composition and characteristics of cartilage in OA in vivo, particularly in the early stages of disease, with the potential to monitor therapy *(75,76)*.

2.7. Recent Developments in MRI

Future developments in MRI include molecular imaging which may offer insights into understanding the pathogenesis of disease. MRI of cells such as mesenchymal fibroblasts labeled with superparamagnetic iron oxide (SPIO) and ultra-small superparamagnetic iron oxide (USPIO) *(77)* allows the potential to study synovial joints at a cellular or molecular level. A different MRI technique for studying molecular structures is magnetic resonance spectroscopy (MRS). This technique has been used to study cartilage in vitro that can be applied to monitor treatment for OA *(78)*.

Another novel technique is using a blood-oxygenation-level-dependent (BOLD) imaging sequence that measures tissue oxygenation for perfusion, and is currently used mainly in examination of the brain *(79)*. This technique can be developed to study the synovial joints, since hypoxia has been implicated in arthritic joints like RA *(80)*.

Higher contrast and improved contrast-to-noise ratios between tissues can be made possible by the use of different doses and concentrations of gadolinium-based contrast agents, but application of this has yet to be studied in relation to the synovial joints *(81–84)*.

3. Ultrasonography

Musculoskeletal ultrasonography (MUS) is increasingly used by rheumatologists in clinical practice *(85)* and is now part of core rheumatology training in some European countries. This has seen an increase in research publications providing mounting evidence to support an ongoing role in the management of rheumatology patients. The increasing interest and scientific exploration into the utility of MUS in rheumatology has prompted many rheumatologists to learn the technique and apply it in their clinical practice. The process of developing and sustaining a MUS service, whether it be for clinical or research purposes, requires both knowledge acquisition and experience along with the appropriate facilities to perform the service.

3.1. Acquisition of Skills and Knowledge

A recent study of worldwide experts in MUS reports a wide variation in the type and amount of training undertaken and that formal training regimes or competency assessment is rare *(86)*. Currently there remains a lack of universal recommendations regarding training curricula or agreement on competency for rheumatologists wishing to develop expertise in MUS. Currently work is underway to develop such a curricula and a consensus agreement amongst MUS experts considering appropriate indications, anatomical areas, knowledge, and skills was recently published as the first step in establishing best practice guidelines for rheumatologists performing ultrasonography *(87)*. The European League Against Rheumatism (EULAR) have established a working group to develop guidelines regarding equipment specification and image acquisition in response to the increasing practice of MUS by rheumatologists *(88)*. Introductory teaching courses that aim to introduce basic concepts and aid in understanding the utility of MUS (rather than provide formal and practical training) are now commonplace at rheumatology conferences. However, there remains little consensus on what constitutes appropriate training, assessment of competency and appropriate maintenance of skills and knowledge. At the 3rd British Society for Rheumatology MUS course, suggested first steps included:

1. *Attendance at an introductory MUS course*: Various rheumatology organizations, such as The British Society for Rheumatology and the EULAR, run short courses

which generally last a couple of days. Courses take place at regular intervals throughout the year and at various sites across the world.
2. *Develop knowledge of the literature*: It is essential to understand the principles behind the techniques in order to appropriately understand and interpret results. It is also important to appreciate the evidence base reflected in research publications, in particular the extent to which MUS has been scientifically tested and validated in order to understand the indications and limitations of this technique in clinical practice. This knowledge will also guide further research questions and projects.

3.2. Ultrasound Equipment and Facilities

3.2.1. The Ultrasound Machine

The ultrasound machine is likely to be the largest financial outlay when setting up a MUS practice. However, technology is rapidly improving and costs of equipment are falling. Current machines range from about approx £5,000 to £150 000, and ongoing servicing may be between 5 and 7% of the capital per annum. Given this, it is important to fully research and consider precisely what you require prior to purchasing a machine.

Specifications to consider include the probe type, resolution, array of modes (e.g., 2D, 3D, Doppler) and frequency range. The desirable specifications will depend on what type of imaging you are going to be undertaking and it may be appropriate to take advice form the manufacturers about what would best suit your requirements. Higher frequency probes provide the best resolution at the expense of reduced depth of penetration, so are probably best for imaging relatively superficial joints such as the hands and feet, whereas lower frequencies are more suitable for deeper joints such as the hip. The size, shape and weight of the ultrasound probe may also be an important consideration.

Portability of the machine is one of the advantages of US over other imaging techniques such as conventional MRI and radiography. However increased portability is generally associated with poorer image quality and it is worth considering whether it is easier to move the patient rather than the machine. There may be many machines that meet your requirements, so ease of use may be the deciding factor. An ideal situation would to use a machine on loan for a period before committing to the purchase.

3.2.2. Software and Storage Systems

Software and storage systems for archiving images need to be compatible with local needs and medico-legal requirements. The machines themselves only store a limited number of images so an additional computer storage facility providing an image database may be required. Consider what you will be archiving, volumes of data, who should have access , and how easy it needs to be to access stored images. All machines are usually sold with a basic package of features but have options for additional software. These may be designed to improve image quality or offer additional image processing features such as image reconstruction or 3D capability. Again it is important to consider what use such additional features will be to your practice before committing to their purchase.

3.2.3. The Scanning Room

A sufficiently large room, in to which patients can be appropriately scanned will avoid the issue of needing to move the machine. The room then needs to be appropriately private and isolated such that exposed areas can be scanned while maintaining patient confidentiality and privacy.

Ensuring adequate space is a key issue. The room needs to contain not only the scanning machine, but also a bed or chair for the patient, a chair for the ultrasonographer, storage space for manuals and other accessories, such as equipment to carry out procedures including aspiration and injection, as well as space for observers or people accompanying the patient. Consider whether bed or chair bound patients can be manoeuvred around the room, through the door or into the corridor.

The machines tend to generate a lot of heat when on, so a room with air conditioning or good ventilation is essential to maximize the patient and ultrasonographer's comfort. A dimmer switch is ideal to allow control over the amount of light in the room, ensuring the room can be appropriately darkened to allow optimal viewing of images. A wash basin will allow the ultrasonographer to wash hands between examinations, as required by most infection control policies. In addition, patients may need access to cleaning facilities to remove excess gel. This is best positioned outside the scanning room in order to least disrupt the schedule of examinations.

3.3. Principles behind US techniques

Ultrasound utilizes the principle of sound waves reflecting back off matter to the source in order to produce an image. Sound waves travel at differing speeds through different matter, and is reflected (an echo) from the interface of materials with differing densities. "B mode" ultrasonography produces 2D images in gray scale, which displays echoes returning to the source as pixels in varying shades of brightness to produce a picture. The pixel brightness is in proportion to the intensity of the echo.

The Doppler technique can be added to gray scale to provide information about the vascularity of the tissue being imaged. Doppler ultrasonography utilizes the principle that the echo frequency is altered when reflected off moving objects. This principle can be applied to MUS as color Doppler or power Doppler. Power Doppler displays the amplitude of the Doppler signal as a color spectrum, providing information only about the power of the signal, but is sensitive to low flow. Color Doppler displays the range of frequencies reflected as color, encoding both velocity and directional information. In musculoskeletal imaging, the direction and velocity of vascular flow is often less important than the amplitude of the flow, so power Doppler is the more commonly applied technique.

3.4. MUS Features of the Inflamed Synovial Joint

MUS can accurately depict features of inflammation in joints and periarticular structures such as synovitis, effusion, tenosynovitis and tendonopathy, enthesitis, and cortical bone erosions. Doppler techniques offer the ability to further quantify inflammation by providing a measure of vascularity. Unlike MRI, MUS is unable to provide much information about subcortical bone and is unable to visualize bone marrow oedema as sound waves are largely reflected off the surface of bone and are unable to penetrate its dense substance. In addition, MUS provides only limited visualization of some structures, as a result of the intrinsic principles of MUS and the structure of some joints. For example, superficial structures may be visualized with greater clarity than deep structures, as high frequency sound waves produce the best structural definition, but do not penetrate deeply. In addition, structures closely opposed or where there may be limited access for the ultrasound probe can be difficult to visualize, such as the cortical bone between metacarpal joints 3 and 4. Nevertheless, MUS has advantages over MRI of portability, the lack of ionizing radiation or contrast agents, feasibility and accessibility, and can be performed by an appropriately trained rheumatologist at the same time as conducting a clinical assessment. It has been suggested that MUS is better able to define tendon pathology than MRI *(13)*. MUS therefore represents an excellent clinical tool in the assessment of inflammatory arthritis and lends itself to temporal use to monitor changes.

3.4.1. Synovitis

The OMERACT MUS group has proposed definitions of synovial hypertrophy and synovial fluid *(89)*, which together are composites of synovitis. "Synovial hypertrophy" is defined as abnormal hypoechoic intra-articular tissue that is nondisplaceable and poorly compressible and which may exhibit Doppler signal. On a gray scale image, synovial hypertrophy would appear darker than surrounding fat. "Synovial fluid" is defined as abnormal hypoechoic or anechoic intra-articular material that is displaceable and compressible but does not exhibit Doppler signal. On grey scale, fluid would appear black compared to surrounding fat (*see* **Fig. 4A,B**).

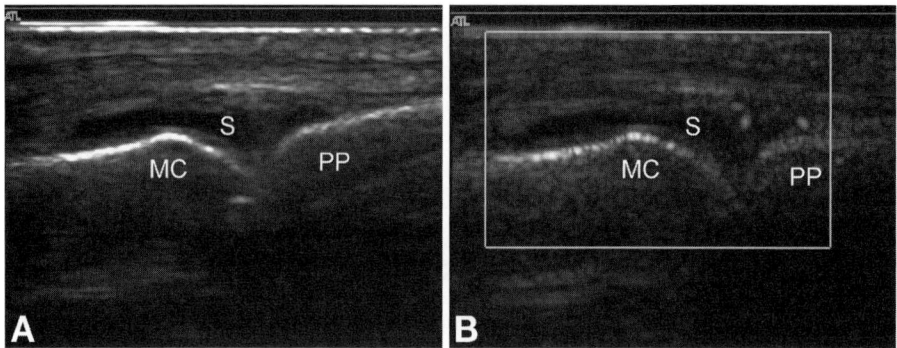

Fig. 4. Ultrasonography of a metacarpophalangeal joint of a patient with rheumatoid arthritis in (**A**) (gray scale) and (**B**) (power Doppler) showing synovitis (S) in longitudinal plane. MC. metacarpal head; PP, proximal phalanx. (Figure appears in color insert following p. 268.)

MUS has been demonstrated to be sensitive in the detection/assessment of synovitis, and validated against many other imaging techniques, including MRI and arthroscopy *(14,90–93)*. MUS has consistently been demonstrated to be superior to clinical examination in detecting synovitis *(94–96)*.

The addition of Doppler techniques to gray scale imaging has been utilized in research to assess and demonstrate vascularity of synovium *(97,98)*, and is thought to reflect the relative activity of inflammation. While Doppler techniques remain a research tool, studies have demonstrated temporal changes which correlate with improvement resulting from therapy *(7–9)*.

3.4.2. Enthesopathy

"Enthesopathy" is defined by OMERACT as "abnormally hypoechoic (loss of fibrillar architecture) and/or thickened tendon or ligament at its bony attachment seen in two perpendicular planes which may exhibit Doppler signal and/ or bony changes including enthesophytes, erosions or irregularity" *(89)*. Typical findings on gray scale include increased tendon diameter and decreased echogenicity or a heterogeneous darkening of the tendon near the point of insertion. Once again, MUS has been shown to be more sensitive to the presence of enthesitis than clinical examination *(99)*.

3.4.4. Tenosynovitis

"Tenosynovitis" is defined as a "hypo or anechoic thickened tissue with or without fluid within the tendon sheath which is seen in two perpendicular planes and which may exhibit Doppler signal" *(89)*. When healthy tendons typically appear as well demarcated structures characterized by a fibrillar pat-

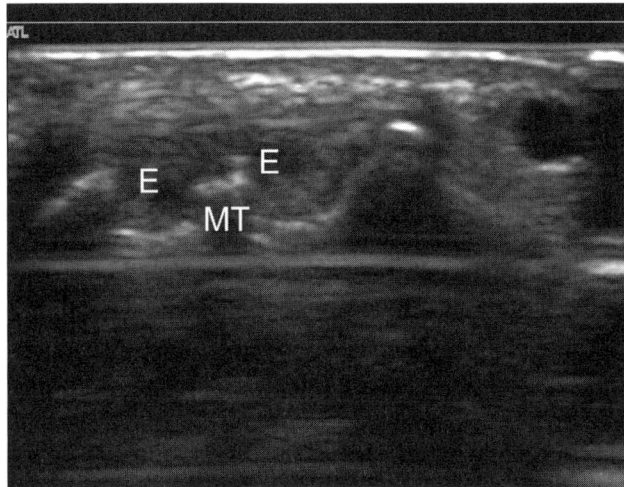

Fig. 5. Ultrasonography of a metatarsal joint showing erosions (E) in transverse plane. MT, metatarsal head.

tern of echos (appearing as tightly packed lines). Tenosynovitis on gray scale can appear as a dark halo around the tendon as the tendon sheath widens. In addition the tendon itself may lose its fibrillar pattern, with diffuse or focal thickening of the tendon, diffuse or focal dark lesions within the tendon, irregular margins, or even complete. Associated peritendonous oedema can also occur. Tendons are well visualized by MUS as a result of their superficial position and it has been said that MUS is the gold standard for imaging tendons because of the increased sensitivity over MRI although studies comparing the two modalities demonstrate conflicting results depending on the definition of tendon damage used *(13)*.

3.4.5. Erosions

"Bone erosion" is defined by OMERACT as the "intra-articular discontinuity of the bone surface which is visible in two perpendicular planes" *(89)* (*see* **Fig. 5**). Erosions detected by MUS have been validated against MRI, confirming accuracy and demonstrating high levels of reproducibility. It has also been demonstrated that the MUS is more sensitive to the presence of erosions than conventional radiography (CR) *(90,100–103)*.

3.5. Future Developments in MUS

Ultrasound is still undergoing validation as a research tool. For the researcher wishing to study arthritis and taking into consideration the factors mentioned above, then the use of ultrasound to discriminate inflammation from fibrosis or fatty tissue deposition using Doppler techniques or ultrasound contrast agents is a potentially exciting development for both diagnosis and for monitoring therapy. Technology is also advancing such that 3- and 4-dimensional images will soon be reproducible with ultrasonographonic techniques. This is another exciting area of future investigation.

4. Other Imaging Modalities

Radiography, although still the first line investigation in many arthritis, is unable to study inflammation of the joints. The advantage of computed tomography (CT) over CR is its tomographic ability, which permits the visualization of bone erosions that may be missed on CR. However, for the same reason as CR, CT is not used to study inflamed synovial joints as it demonstrates joint-based inflammation very poorly or not at all. As mentioned above, US and MRI have various features which allow inflamed synovial joints to be studied in detail. Another imaging modality that has potential for further developments in studying inflammation in joints is scintigraphy.

4.1. Scintigraphy

Scintigraphy, including single-photon emission computed tomography (SPECT) and positron emission tomography (PET) have roles in orthopaedic practice but little clinical application in rheumatology at the present time. Like radiography, CT and scintigraphy involve ionizing radiation and are not widely used in the rheumatology setting. Scintigraphy has some applications in rheumatic diseases including inflamed synovial joints *(104)*. The physiological changes in bones in arthritis can be determined in vivo using scintigraphy, which cannot be established from the other imaging modalities mentioned. The physiologic aspect of inflamed synovial joints is important as it often predates structural damage *(91,105)*. PET and SPECT permit molecular imaging and may have potential for further developments in studying joint inflammation because of both the higher resolution of commercially available scanners and the potential for development of new molecular probes specific for inflammation *(106)*. The techniques may permit imaging of structural abnormalities before visually evidenced on other imaging modalities *(91,105)*. The disease activity of RA can be measured by determining the metabolic activity of synovitis using 18F-fluorodeoxyglucose (18F-FDG) PET *(107)*.

5. Summary

This review outlines how the two important imaging modalities, MRI and US, can be applied to study inflamed synovial joints. They are both versatile techniques that often complement each other. They allow us to further our understanding of pathophysiology of inflamed synovial joints in vivo, and to investigate how inflammatory arthritis may respond to therapy. These techniques continue to be developed and improved, which will no doubt enhance their potential in imaging inflamed synovial joints.

References

1. van der Heijde, D. (2004) Quantification of radiological damage in inflammatory arthritis: rheumatoid arthritis, psoriatic arthritis and ankylosing spondylitis. *Best Pract. Res. Clin. Rheumatol.* **18,** 847–860.
2. van der Heijde, D., Sharp, J., Wassenberg, S., and Gladman, D. D. (2005) Psoriatic arthritis imaging: a review of scoring methods. *Ann. Rheum. Dis.* **64,** ii61–ii64.
3. Lee, J., Lee, S. K., Suh, J. S., Yoon, M., Song, J. H., and Lee, C. H. (1997) Magnetic resonance imaging of the wrist in defining remission of rheumatoid arthritis. *J. Rheumatol.* **24,** 1303–1308.
4. Kalden-Nemeth, D., Grebmeier, J., Antoni, C., Manger, B., Wolf, F., and Kalden, J. R. (1997) NMR monitoring of rheumatoid arthritis patients receiving anti-TNF-alpha monoclonal antibody therapy. *Rheumatol. Int.* **16,** 249–255.
5. Ostergaard, M., Stoltenberg, M., Henriksen, O., and Lorenzen, I. (1996) Quantitative assessment of synovial inflammation by dynamic gadolinium-enhanced magnetic resonance imaging. A study of the effect of intra-articular methylprednisolone on the rate of early synovial enhancement. *Br. J. Rheumatol.* **35,** 50–59.
6. Sugimoto, H., Takeda, A., and Kano, S. (1998) Assessment of disease activity in rheumatoid arthritis using magnetic resonance imaging: quantification of pannus volume in the hands. *Br. J. Rheumatol.* **37,** 854–861.
7. Terslev, L., Torp-Pedersen, S., Qvistgaard, E., Danneskiold-Samsoe, B., and Bliddal, H. (2003) Estimation of inflammation by Doppler ultrasound: quantitative changes after intra-articular treatment in rheumatoid arthritis. *Ann. Rheum. Dis.* **62,** 1049–1053.
8. Hau, M., Kneitz, C., Tony, H. P., Keberle, M., Jahns, R., and Jenett, M. (2002) High resolution ultrasound detects a decrease in pannus vascularisation of small finger joints in patients with rheumatoid arthritis receiving treatment with soluble tumour necrosis factor alpha receptor (etanercept). *Ann. Rheum. Dis.* **61,** 55–58.
9. Newman, J. S., Laing, T. J., McCarthy, C. J., and Adler, R. S. (1996) Power Doppler sonography of synovitis: assessment of therapeutic response—preliminary observations. *Radiology* **198,** 582–584.
10. Stewart, N. R., Crabbe, J. P., and McQueen, F. M. (2004) Magnetic resonance imaging of the wrist in rheumatoid arthritis: demonstration of progression between 1 and 6 years. *Skeletal Radiol.* **33,** 704–711. Epub 2004 Oct 15.
11. Robson, M. D., Benjamin, M., Gishen, P., and Bydder, G. M. (2004) Magnetic resonance imaging of the Achilles tendon using ultrashort TE (UTE) pulse sequences. *Clin. Radiol.* **59,** 727–735.
12. McGonagle, D., Gibbon, W., and Emery, P. (1998) Classification of inflammatory arthritis by enthesitis. *Lancet* **352,** 1137–1140.
13. Grassi, W., Filippucci, E., Farina, A., and Cervini, C. (2000) Sonographic imaging of tendons. *Arthritis Rheum.* **43,** 969–976.

14. Karim, Z., Wakefield, R. J., Quinn, M., et al. (2004) Validation and reproducibility of ultrasonography in the detection of synovitis in the knee: a comparison with arthroscopy and clinical examination. *Arthritis Rheum.* **50,** 387–394.
15. Young, I. R. and Bydder, G. M. (2003) Magnetic resonance: new approaches to imaging of the musculoskeletal system. *Physiol. Meas.* **24,** R1–23.
16. Gold, G. E., Suh, B., Sawyer-Glover, A., and Beaulieu, C. (2004) Musculoskeletal MRI at 3.0 T: initial clinical experience. *Am. J. Roentgenol.* **183,** 1479–1486.
17. Ejbjerg, B., Narvestad, E., Jacobsen, S., Thomsen, H. S., and Ostergaard, M. (2005) Optimised, low cost, low field dedicated extremity MRI is highly specific and sensitive for synovitis and bone erosions in rheumatoid arthritis wrist and finger joints: a comparison with conventional high-field MRI and radiography. *Ann. Rheum. Dis.* **13,** 13.
18. Taouli, B., Zaim, S., Peterfy, C. G., et al. (2004) Rheumatoid arthritis of the hand and wrist: comparison of three imaging techniques. *Am. J. Roentgenol.* **182,** 937–943.
19. Tan, A. L., Grainger, A. J., Tanner, S. F., et al. (2005) High-resolution magnetic resonance imaging for the assessment of hand osteoarthritis. *Arthritis Rheum.* **52,** 2355–2365.
20. Brown, R., Mareyam, A., Reid, E., and Wang, Y. (2004) Novel RF coil geometry for lower extremity imaging. *Magn. Reson. Med.* **51,** 635–639.
21. Sodickson, D. K., and Manning, W. J. (1997) Simultaneous acquisition of spatial harmonics (SMASH): fast imaging with radiofrequency coil arrays. *Magn. Reson. Med.* **38,** 591–603.
22. Pruessmann, K. P., Weiger, M., Scheidegger, M. B., and Boesiger, P. (1999) SENSE: sensitivity encoding for fast MRI. *Magn. Reson. Med.* **42,** 952–962.
23. Fenchel, M., Scheule, A. M., Stauder, N. I., et al. (2005) Atherosclerotic Disease: Whole-Body Cardiovascular Imaging with MR System with 32 Receiver Channels and Total-Body Surface Coil Technology—Initial Clinical Results. *Radiology* **22,** 22.
24. Lemke, H. U. (2003) PACS developments in Europe. *Comput. Med. Imaging Graph.* **27,** 111–120.
25. Ozsunar, Y., Keceli, M., Koseoglu, K., Coskun, G., and Karaman, C. (2003) PACS utilization in radiologic research. *Acad Radiol.* **10,** 32–36.
26. Sugimoto, H., Takeda, A., Masuyama, J., and Furuse, M. (1996) Early-stage rheumatoid arthritis: diagnostic accuracy of MR imaging. *Radiology* **198,** 185–192.
27. McGonagle, D., Conaghan, P. G., O'Connor, P., et al. (1999) The relationship between synovitis and bone changes in early untreated rheumatoid arthritis: a controlled magnetic resonance imaging study. *Arthritis Rheum.* **42,** 1706–1711.
28. McGonagle, D., Gibbon, W., O'Connor, P., Green, M., Pease, C., and Emery, P. (1998) Characteristic magnetic resonance imaging entheseal changes of knee synovitis in spondylarthropathy. *Arthritis Rheum.* **41,** 694–700.
29. Marzo-Ortega, H., McGonagle, D., O'Connor, P., and Emery, P. (2001) Efficacy of etanercept in the treatment of the entheseal pathology in resistant

spondylarthropathy: a clinical and magnetic resonance imaging study. *Arthritis Rheum.* **44**, 2112–2117.
30. Tan, A. L., Marzo-Ortega, H., O'Connor, P., Fraser, A., Emery, P., and McGonagle, D. (2004) Efficacy of anakinra in active ankylosing spondylitis: a clinical and magnetic resonance imaging study. *Ann. Rheum. Dis.* **63**, 1041–1045. Epub 2004 Apr 5.
31. Hauger, O., Dumont, E., Chateil, J. F., Moinard, M., and Diard, F. (2002) Water excitation as an alternative to fat saturation in MR imaging: preliminary results in musculoskeletal imaging. *Radiology* **224**, 657–663.
32. Thomasson, D., Purdy, D., and Finn, J. P. (1996) Phase-modulated binomial RF pulses for fast spectrally-selective musculoskeletal imaging. *Magn. Reson. Med.* **35**, 563–568.
33. Robson, M. D., Gatehouse, P. D., Bydder, M., and Bydder, G. M. (2003) Magnetic resonance: an introduction to ultrashort TE (UTE) imaging. *J. Comput. Assist. Tomogr.* **27**, 825–846.
34. Gatehouse, P. D., and Bydder, G. M. (2003) Magnetic resonance imaging of short T2 components in tissue. *Clin. Radiol.* **58**, 1–19.
35. Rhodes, L. A., Conaghan, P. G., Radjenovic, A., Grainger, A. J., Emery, P., and McGonagle, D. (2005) Further evidence that a cartilage-pannus junction synovitis predilection is not a specific feature of rheumatoid arthritis. *Ann. Rheum. Dis.* **64**, 1347–1349.
36. Cimmino, M. A., Parodi, M., Innocenti, S., et al. (2005) Dynamic magnetic resonance of the wrist in psoriatic arthritis reveals imaging patterns similar to those of rheumatoid arthritis. *Arthritis Res. Ther.* **7**, R725–731. Epub 2005 Apr 1.
37. Cimmino, M. A., Innocenti, S., Livrone, F., Magnaguagno, F., Silvestri, E., and Garlaschi, G. (2003) Dynamic gadolinium-enhanced magnetic resonance imaging of the wrist in patients with rheumatoid arthritis can discriminate active from inactive disease. *Arthritis Rheum.* **48**, 1207–1213.
38. Tan, A. L., Tanner, S. F., Conaghan, P. G., et al. (2003) Role of metacarpophalangeal joint anatomic factors in the distribution of synovitis and bone erosion in early rheumatoid arthritis. *Arthritis Rheum.* **48**, 1214–1222.
39. Rhodes, L. A., Tan, A. L., Tanner, S. F., et al. (2004) Regional variation and differential response to therapy for knee synovitis adjacent to the cartilage-pannus junction and suprapatellar pouch in inflammatory arthritis: implications for pathogenesis and treatment. *Arthritis Rheum.* **50**, 2428–2432.
40. Reece, R. J., Kraan, M. C., Radjenovic, A., et al. (2002) Comparative assessment of leflunomide and methotrexate for the treatment of rheumatoid arthritis, by dynamic enhanced magnetic resonance imaging. *Arthritis Rheum.* **46**, 366–372.
41. Kanal, E., Borgstede, J. P., Barkovich, A. J., et al. (2002) American College of Radiology White Paper on MR Safety. *Am. J. Roentgenol.* **178**, 1335–1347.
42. Kanal, E., Borgstede, J. P., Barkovich, A. J., et al. (2004) American College of Radiology White Paper on MR Safety: 2004 update and revisions. *Am. J. Roentgenol.* **182**, 1111–1114.

43. Ostergaard, M., Lorenzen, I., and Henriksen, O. (1994) Dynamic gadolinium-enhanced MR imaging in active and inactive immunoinflammatory gonarthritis. *Acta. Radiol.* **35,** 275–281.
44. Palosaari, K., Vuotila, J., Takalo, R., et al. (2004) Contrast-enhanced dynamic and static MRI correlates with quantitative 99Tcm-labelled nanocolloid scintigraphy. Study of early rheumatoid arthritis patients. *Rheumatology (Oxford)* **43,** 1364-73. Epub 2004 Jul 6.
45. Conaghan, P., Edmonds, J., Emery, P., et al. (2001) Magnetic resonance imaging in rheumatoid arthritis: summary of OMERACT activities, current status, and plans. *J. Rheumatol.* **28,** 1158–1162.
46. Ejbjerg, B., Narvestad, E., Rostrup, E., Szkudlarek, M., Jacobsen, S., Thomsen, H. S., and Ostergaard, M. (2004) Magnetic resonance imaging of wrist and finger joints in healthy subjects occasionally shows changes resembling erosions and synovitis as seen in rheumatoid arthritis. *Arthritis Rheum.* **50,** 1097–1106.
47. McQueen, F. M., Stewart, N., Crabbe, J., et al. (1999) Magnetic resonance imaging of the wrist in early rheumatoid arthritis reveals progression of erosions despite clinical improvement. *Ann. Rheum. Dis.* **58,** 156–163.
48. Benton, N., Stewart, N., Crabbe, J., Robinson, E., Yeoman, S., and McQueen, F. M. (2004) MRI of the wrist in early rheumatoid arthritis can be used to predict functional outcome at 6 years. *Ann. Rheum. Dis.* **63,** 555–561.
49. Felson, D. T., McLaughlin, S., Goggins, J., et al. (2003) Bone marrow edema and its relation to progression of knee osteoarthritis. *Ann. Intern. Med.* **139,** 330–336.
50. Felson, D. T., Chaisson, C. E., Hill, C. L., et al. (2001) The association of bone marrow lesions with pain in knee osteoarthritis. *Ann. Intern. Med.* **134,** 541–549.
51. Jones, K. M., Unger, E. C., Granstrom, P., Seeger, J. F., Carmody, R. F., and Yoshino, M. (1992) Bone marrow imaging using STIR at 0.5 and 1.5 T. *Magn. Reson. Imaging* **10,** 169–176.
52. Arndt, W. F., 3rd, Truax, A. L., Barnett, F. M., Simmons, G. E., and Brown, D. C. (1996) MR diagnosis of bone contusions of the knee: comparison of coronal T2-weighted fast spin-echo with fat saturation and fast spin-echo STIR images with conventional STIR images. *Am. J. Roentgenol.* **166,** 119–124.
53. Nakahara, N., Uetani, M., Hayashi, K., Kawahara, Y., Matsumoto, T., and Oda, J. (1996) Gadolinium-enhanced MR imaging of the wrist in rheumatoid arthritis: value of fat suppression pulse sequences. *Skeletal Radiol*. **25,** 639–647.
54. Maksymowych, W. P., Inman, R. D., Salonen, D., et al. (2005) Spondyloarthritis Research Consortium of Canada magnetic resonance imaging index for assessment of spinal inflammation in ankylosing spondylitis. *Arthritis Rheum.* **53,** 502–509.
55. Braun, J., Baraliakos, X., Golder, W., et al. (2003) Magnetic resonance imaging examinations of the spine in patients with ankylosing spondylitis, before and after successful therapy with infliximab: evaluation of a new scoring system. *Arthritis Rheum.* **48,** 1126–1136.
56. Verstraete, K. L., Almqvist, F., Verdonk, P., et al. (2004) Magnetic resonance imaging of cartilage and cartilage repair. *Clin. Radiol*. **59,** 674–689.

57. Trattnig, S., Mlynarik, V., Huber, M., Ba-Ssalamah, A., Puig, S., and Imhof, H. (2000) Magnetic resonance imaging of articular cartilage and evaluation of cartilage disease. *Invest. Radiol* .**35,** 595–601.
58. Disler, D. G., McCauley, T. R., Kelman, C. G., et al. (1996) Fat-suppressed three-dimensional spoiled gradient-echo MR imaging of hyaline cartilage defects in the knee: comparison with standard MR imaging and arthroscopy. *Am. J. Roentgenol.* **167,** 127–132.
59. Recht, M. P., Piraino, D. W., Paletta, G. A., Schils, J. P., and Belhobek, G. H. (1996) Accuracy of fat-suppressed three-dimensional spoiled gradient-echo FLASH MR imaging in the detection of patellofemoral articular cartilage abnormalities. *Radiology* **198,** 209–212.
60. Hardy, P. A., Recht, M. P., Piraino, D., and Thomasson, D. (1996) Optimization of a dual echo in the steady state (DESS) free-precession sequence for imaging cartilage. *J. Magn. Reson. Imaging* **6,** 329–335.
61. Ruehm, S., Zanetti, M., Romero, J., and Hodler, J. (1998) MRI of patellar articular cartilage: evaluation of an optimized gradient echo sequence (3D-DESS). *J. Magn. Reson. Imaging* **8,** 1246–1251.
62. Masi, J. N., Sell, C. A., Phan, C., et al. (2005) Cartilage MR imaging at 3.0 versus that at 1.5 T: preliminary results in a porcine model. *Radiology* **236,** 140–150.
63. Gandy, S. J., Brett, A. D., Dieppe, P. A., et al. (2005) Measurement of cartilage volumes in rheumatoid arthritis using MRI. *Br. J. Radiol.* **78,** 39–45.
64. Filidoro, L., Dietrich, O., Weber, J., et al. (2005) High-resolution diffusion tensor imaging of human patellar cartilage: feasibility and preliminary findings. *Magn. Reson. Med.* **53,** 993–998.
65. Glaser, C. (2005) New techniques for cartilage imaging: T2 relaxation time and diffusion-weighted MR imaging. *Radiol. Clin. North Am.* **43,** 641–653, vii.
66. Gold, G. E., and Beaulieu, C. F. (2001) Future of MR imaging of articular cartilage. *Semin. Musculoskelet. Radiol.* **5,** 313–327.
67. Dardzinski, B. J., Mosher, T. J., Li, S., Van Slyke, M. A., and Smith, M. B. (1997) Spatial variation of T2 in human articular cartilage. *Radiology* **205,** 546–550.
68. Duvvuri, U., Kudchodkar, S., Reddy, R., and Leigh, J. S. (2002) T(1rho) relaxation can assess longitudinal proteoglycan loss from articular cartilage in vitro. *Osteoarthritis Cartilage* **10,** 838–844.
69. Bashir, A., Gray, M. L., Boutin, R. D., and Burstein, D. (1997) Glycosaminoglycan in articular cartilage: in vivo assessment with delayed Gd(DTPA)(2-)-enhanced MR imaging. *Radiology* **205,** 551–558.
70. Burmester, G. R., Stuhlmuller, B., Keyszer, G., and Kinne, R. W. (1997) Mononuclear phagocytes and rheumatoid synovitis. Mastermind or workhorse in arthritis? *Arthritis Rheum.* **40,** 5–8.
71. Weissleder, R., Elizondo, G., Wittenberg, J., Rabito, C. A., Bengele, H. H., and Josephson, L. (1990) Ultrasmall superparamagnetic iron oxide: characterization of a new class of contrast agents for MR imaging. *Radiology* **175,** 489–493.
72. Bulte, J. W., and Kraitchman, D. L. (2004) Iron oxide MR contrast agents for molecular and cellular imaging. *NMR Biomed.* **17,** 484–499.

73. Lutz, A. M., Seemayer, C., Corot, C., et al. (2004) Detection of synovial macrophages in an experimental rabbit model of antigen-induced arthritis: ultrasmall superparamagnetic iron oxide-enhanced MR imaging. *Radiology* **233**, 149–157.
74. Beckmann, N., Falk, R., Zurbrugg, S., Dawson, J., and Engelhardt, P. (2003) Macrophage infiltration into the rat knee detected by MRI in a model of antigen-induced arthritis. *Magn. Reson. Med.* **49**, 1047–1055.
75. Wheaton, A. J., Borthakur, A., Dodge, G. R., Kneeland, J. B., Schumacher, H. R., and Reddy, R. (2004) Sodium magnetic resonance imaging of proteoglycan depletion in an in vivo model of osteoarthritis. *Acad. Radiol.* **11**, 21–28.
76. Nissi, M. J., Toyras, J., Laasanen, M. S., et al. (2004) Proteoglycan and collagen sensitive MRI evaluation of normal and degenerated articular cartilage. *J. Orthop. Res.* **22**, 557–564.
77. Sun, R., Dittrich, J., Le-Huu, M., et al. (2005) Physical and biological characterization of superparamagnetic iron oxide- and ultrasmall superparamagnetic iron oxide-labeled cells: a comparison. *Invest. Radiol.* **40**, 504–513.
78. Shapiro, E. M., Borthakur, A., Dandora, R., Kriss, A., Leigh, J. S., and Reddy, R. (2000) Sodium visibility and quantitation in intact bovine articular cartilage using high field (23)Na MRI and MRS. *J. Magn. Reson.* **142**, 24–31.
79. Turner, R., Howseman, A., Rees, G. E., Josephs, O., and Friston, K. (1998). Functional magnetic resonance imaging of the human brain: data acquisition and analysis. *Exp. Brain Res.* **123**, 5–12.
80. Taylor, P. C., and Sivakumar, B. (2005) Hypoxia and angiogenesis in rheumatoid arthritis. *Curr. Opin. Rheumatol.* **17**, 293–298.
81. Tombach, B., Benner, T., Reimer, P., et al. (2003) Do highly concentrated gadolinium chelates improve MR brain perfusion imaging? Intraindividually controlled randomized crossover concentration comparison study of 0.5 versus 1.0 mol/L gadobutrol. *Radiology* **226**, 880–888.
82. Essig, M., Lodemann, K. P., LeHuu, M., Schonberg, S. O., Hubener, M., and Van Kaick, G. (2002) [Comparison of MultiHance and Gadovist for cerebral MR perfusion imaging in healthy volunteers]. *Radiologe* **42**, 909–915.
83. Goyen, M., Lauenstein, T. C., Herborn, C. U., Debatin, J. F., Bosk, S., and Ruehm, S. G. (2001) 0.5 M Gd chelate (Magnevist) versus 1.0 M Gd chelate (Gadovist): dose-independent effect on image quality of pelvic three-dimensional MR-angiography. *J. Magn. Reson. Imaging* **14**, 602–607.
84. Thilmann, O., Larsson, E. M., Bjorkman-Burtscher, I. M., Stahlberg, F., and Wirestam, R. (2005) Comparison of contrast agents with high molarity and with weak protein binding in cerebral perfusion imaging at 3 T. *J. Magn. Reson. Imaging* **30**, 30.
85. Wakefield, R. J., Goh, E., Conaghan, P. G., Karim, Z., and Emery, P. (2003) Musculoskeletal ultrasonography in Europe: results of a rheumatologist-based survey at a EULAR meeting. *Rheumatology (Oxford)* **42**, 1251–1253.
86. Brown, A. K., O'Connor, P. J., Wakefield, R. J., Roberts, T. E., Karim, Z., and Emery, P. (2004) Practice, training, and assessment among experts performing musculoskeletal ultrasonography: toward the development of an international

consensus of educational standards for ultrasonography for rheumatologists. *Arthritis Rheum.* **51,** 1018–1022.
87. Brown, A. K., O'Connor P, J., Roberts, T. E., Wakefield, R. J., Karim, Z., and Emery, P. (2005) Recommendations for musculoskeletal ultrasonography by rheumatologists: setting global standards for best practice by expert consensus. *Arthritis Rheum.* **53,** 83–92.
88. Backhaus, M., Burmester, G. R., Gerber, T., et al. (2001). Guidelines for musculoskeletal ultrasound in rheumatology. *Ann. Rheum. Dis.* **60,** 641–649.
89. Wakefield, R. J., Balint, P. V., Szkudlarek, M., et al. (2005) Preliminary definitions of Rheumatoid Arthritis pathology. *J. Rheumatol.* **32,** 2485–2487.
90. Backhaus, M., Burmester, G. R., Sandrock, D., et al. (2002) Prospective two year follow up study comparing novel and conventional imaging procedures in patients with arthritic finger joints. *Ann. Rheum. Dis.* **61,** 895–904.
91. Backhaus, M., Kamradt, T., Sandrock, D., et al. (1999) Arthritis of the finger joints: a comprehensive approach comparing conventional radiography, scintigraphy, ultrasound, and contrast-enhanced magnetic resonance imaging. *Arthritis Rheum.* **42,** 1232–1245.
92. Conaghan, P., Wakefield, R. J., and O'Connor P, J. (1998) MCPJ assessment in early RA: a comparison between x-ray, MRI, high-resolution ultrasound and clinical examination. *Arthritis Rheum.* **41,** S246.
93. van Holsbeeck, M., van Holsbeeck, K., Gevers, G., et al. (1988) Staging and follow-up of rheumatoid arthritis of the knee. Comparison of sonography, thermography, and clinical assessment. *J. Ultrasound. Med.* **7,** 561–566.
94. Wakefield, R. J., Karim, Z., and Conaghan, P. (1999) Sonography is more sensitive than clinical examination at detecting synovitis in the metatarsophalangeal joints than clinical examination. *Arthritis Rheum.* **42,** S352.
95. Wakefield, R. J., Green, M. J., Marzo-Ortega, H., et al. (2004) Should oligoarthritis be reclassified? Ultrasound reveals a high prevalence of subclinical disease. *Ann. Rheum. Dis.* **63,** 382–385.
96. Kane, D., Balint, P. V., and Sturrock, R. D. (2003) Ultrasonography is superior to clinical examination in the detection and localization of knee joint effusion in rheumatoid arthritis. *J. Rheumatol.* **30,** 966–971.
97. Walther, M., Harms, H., Krenn, V., Radke, S., Faehndrich, T. P., and Gohlke, F. (2001) Correlation of power Doppler sonography with vascularity of the synovial tissue of the knee joint in patients with osteoarthritis and rheumatoid arthritis. *Arthritis Rheum.* **44,** 331–338.
98. Schmidt, W. A., Volker, L., Zacher, J., Schlafke, M., Ruhnke, M., and Gromnica-Ihle, E. (2000) Colour Doppler ultrasonography to detect pannus in knee joint synovitis. *Clin. Exp. Rheumatol.* **18,** 439–444.
99. Balint, P. V., Kane, D., Wilson, H., McInnes, I. B., and Sturrock, R. D. (2002) Ultrasonography of entheseal insertions in the lower limb in spondyloarthropathy. *Ann. Rheum. Dis.* **61,** 905–910.
100. Klocke, R., Glew, D., Cox, N., and Blake, D. R. (2001) Sonographic erosions of the rheumatoid little toe. *Ann. Rheum. Dis.* **60,** 896–897.

101. Weidekamm, C., Koller, M., Weber, M., and Kainberger, F. (2003) Diagnostic value of high-resolution B-mode and doppler sonography for imaging of hand and finger joints in rheumatoid arthritis. *Arthritis Rheum.* **48,** 325–333.
102. Grassi, W., Filippucci, E., Farina, A., Salaffi, F., and Cervini, C. (2001) Ultrasonography in the evaluation of bone erosions. *Ann. Rheum. Dis.* **60,** 98–103.
103. Wakefield, R. J., Gibbon, W. W., Conaghan, P. G., et al. (2000) The value of sonography in the detection of bone erosions in patients with rheumatoid arthritis: a comparison with conventional radiography. *Arthritis Rheum.* **43,** 2762–2770.
104. Colamussi, P., Prandini, N., Cittanti, C., Feggi, L., and Giganti, M. (2004) Scintigraphy in rheumatic diseases. *Best Pract. Res. Clin. Rheumatol.* **18,** 909–926.
105. Mottonen, T. T., Hannonen, P., Toivanen, J., Rekonen, A., and Oka, M. (1988) Value of joint scintigraphy in the prediction of erosiveness in early rheumatoid arthritis. *Ann. Rheum. Dis.* **47,** 183–189.
106. Marshall, D., and Haskard, D. O. (2003) Quantifying inflammation in vivo using radiolabeled antibodies and leukocytes. *Methods Mol. Biol.* **225,** 273–282.
107. Beckers, C., Ribbens, C., Andre, B., et al. (2004) Assessment of disease activity in rheumatoid arthritis with (18)F-FDG PET. *J. Nucl. Med.* **45,** 956–964.

2

Arthroscopy as a Research Tool
A Review

Richard J. Reece

Summary

Arthroscopy continues to experience a growth in interest from the rheumatology community reflecting a common desire to gain better understanding of the underlying processes in inflammatory and degenerative joint diseases. Arthroscopy provides the ability to assess the internal appearances of a joint in a well tolerated and repeatable manner, to obtain tissue samples from the principle site of pathology within the joint and thus confers on it the role of "gold standard" amongst currently available imaging techniques. The evolution of arthroscopy is reviewed together with an overview of the evidence obtained from its research application in the rheumatology. Methodology for the conduct of arthroscopy and synovial biopsy is described.

Key Words: Arthroscopy; synovitis; synovial biopsy; chondroscopy; chondropathy; magnetic resonance imaging; ultrasound; history; diagnosis; rheumatoid arthritis; osteoarthritis.

1. Introduction

It was in 1877 that the cystoscope was developed by Max Nitze. A 7.3-mm diameter cystoscope was subsequently used in 1918 by a Japanese group, led by Kenji Takagi, to make the first examination of a cadaveric knee joint utilizing saline distention and irrigation of the joint *(1)*. From 1920 onward, Takagi worked to develop a practical arthroscope. However, it was not until 1931 that he was able to apply the technique successfully in patients, utilizing the so-called "No. 1 arthroscope." This was 3.5 mm in diameter, made of stainless steel, contained a small bulb for illumination, had a narrow field of view, and required the operator to look directly through the eyepiece. A separate sheath was needed for the acquisition of punch biopsies of the synovium. Successful black and white photography of arthroscopic appearances followed in 1932, with the capture of color pictures and movie sequences by 1936.

In 1921 in Europe, Bircher reported on successful knee arthroscopy in a patient for the purpose of meniscal examination by means of a laparoscope and gas insufflation of the joint space *(2)*. His arthroscopy career was short and stopped in 1922. From the United States Kreuscher reported on the use of arthroscopy to examine early meniscal lesions in 1925 *(3)*, followed in 1931 by Mayer and Finkelstein's report of successful synovial punch biopsy *(4)*. Thereafter the description of arthroscopy in cadaveric joints other than the knee by Burman *(5)* and a report of clinical experience with arthroscopy in 1934 from Burman, Mayer and Finkelstein followed *(6)*. Their work with arthroscopy did not continue, but in Germany workers continued to describe the utility of arthroscopy in the diagnosis of meniscal lesions until 1955 *(7)*.

Throughout the 1930s Takagi and coworkers developed and refined the rudimentary no. 1 arthroscope. They reported on their experience in 57 procedures with the first 12 designs in 1938 at the annual meeting of the Japanese Orthopaedic Association, when arthroscopy was adopted as a major theme *(1)*. Between 1933 and 1939 Iino compared arthroscopic findings in cadaveric knees with those on formal dissection of the joints *(8)*. This work formed the basic principles of arthroscopic diagnosis in the knee still employed today. In 1941 Miki concluded that the optimum medium for joint cavity distention and irrigation was normal saline prewarmed to 35°C delivering a water column pressure of 50 to 70 cm *(9)*.

In the aftermath of World War II, arthroscopy development was revitalized in Japan by pupils of Takagi. Mizumachi, Sakakibara, and Watanabe undertook clinical assessments that resulted in, among other things, the first description of the medial synovial plica *(10–12)*. Watanabe's primary aim was to develop a durable, stainless steel arthroscope with an eyepiece complete with adjustable focus, through which the operator could view the joint contents. The minimum focal length he considered necessary was 1 mm. He reasoned that comprehensive arthroscopic examination would be optimized if the facility existed to alter the arthroscopic angle of view, from 0° (straight ahead) to 90° (side-viewing), by switching lenses during the procedure. Other attributes he desired were minimizing the arthroscope's diameter for practicality while achieving sufficient illumination from a bare bulb situated within the arthroscope itself, coupled with robustness of construction permitting repeated cycles of cleaning and sterilization. Lavage of the joint space was essential and to be delivered by means of a side arm with tap that he incorporated within the design of the outer arthroscopic sheath. Eventually, in 1958, he was able to perfect the no. 21 arthroscope which encompassed all the above features within a sheath of maximum 4.9-mm diameter *(13)*. This model formed the prototype upon which all modern arthroscopes are now based. Using the no. 21 arthroscope, Watanabe was able to collate the images incorporated within his *Atlas of Arthroscopy (7)*. In 1969 the no. 21 arthroscope was modified by replacing the 90° angle of view lens with a 30° one.

Through the 1960s, Watanabe and his colleague Takeda used this instrument in over 800 arthroscopies. They further refined the principles of arthroscopic knee diagnosis, particularly of internal knee derangements and published the second edition of their atlas in 1969 *(14,7)*. Interest in arthroscopy by both orthopaedic surgeons and rheumatologists grew and the technique spread to the United States, Canada, and several European countries once more. Coincident was the collaboration between Watanabe and a Japanese optical glass manufacturer of Selfoc® glass rod lenses. These lenses were modified and incorporated into the available arthroscopes to enhance image quality, permitting development of narrower arthroscopes such as the no. 24 model, just 1.7 mm in diameter *(7)*. Practical accessibility to smaller joints thus became a realistic prospect *(15)*.

The experience of arthroscopy in the hands of rheumatologists in the United Kingdom and United States was reported in the late 1960s and early 1970s *(16,17)*. Their findings were consistent with those previously described. Synovitis appeared macroscopically to manifest and be most severe adjacent to cartilage surfaces such as the patella. In the mid-1970s attempts were made to correlate macroscopic appearances at arthroscopy and synovial histology against standard clinical measures in cohorts of patients with rheumatoid arthritis (RA) *(18,19)*. Reliable conclusions could not be drawn but the investigators were struck by the heterogeneity of synovitis.

Synovitis can be recognized macroscopically by an increase in vascular markings of the synovium, capillary hyperaemia, and proliferation of the depth of the synovial lining and sublining layers. This can range from oedema through a "cobble-stone like" appearance (granularity) to true villous formation. The wide variation in the appearance of villi is well documented *(20)*. Other variable features of synovitis include the presence of fibrin, rice-bodies, and patchy areas of pigmentation reflecting areas of previous haemorrhage. Pannus is considered to occur when the synovial tissue is observed to encroach across the surface of the cartilage becoming adherent and invasive. The synovial-cartilage interface has therefore also attracted the rheumatologist's interest *(21–23)*.

Through the 1980s and into the early 1990s the use of arthroscopy by rheumatologists largely died out, workers preferring the bedside closed needle biopsy technique to obtain synovial tissue despite its limitations *(25,26)*. With this closed needle method the first concerted efforts were made to address questions including the role of immunocompetent cells in the pathogenesis of RA *(27)*, the optimal location for biopsy *(28)*, the subclinical extent of knee joint synovitis *(29)*, the determination of tissue markers of disease severity *(30)*, and to assess the effects of therapeutic interventions *(31)*. These questions would be further addressed with reawakened interest in rheumatological arthroscopy *(32)*. Meanwhile, Lindblad and coworkers began a series of arthroscopic studies initially looking at the specificity of synovial histopathology for different diseases *(33)*. It was found that features, such as the inflammatory cellular infiltrate considered specific to RA, were in fact common to other conditions in this study of 22 patients that included psoriatic arthritis, reactive arthritis, juvenile idiopathic arthritis, and gout. Subsequently, in a cohort of 10 osteoarthritis (OA) patients studied, knee arthroscopy revealed macroscopic signs of inflammation most pronounced adjacent to cartilage surfaces. Analysis of the synovial biopsies obtained confirmed the presence of an inflammatory cellular infiltrate assessed to be most intense from sites of greatest macroscopic change *(34)*.

From the same center a study of the normal knee joints of healthy volunteers was also undertaken *(35)*. This confirmed that healthy joint synovium has a white, shiny surface comprising a thin translucent membrane either directly overlying the fibrous joint capsule, or having a variable depth of connective tissue beneath it, that may contain considerable fatty deposits. A dispersed vascular network of fine caliber blood vessels is visible within the sublining layer that can extend into the slender villi. Microscopically the synovial lining layer was found to be 1 to 2 cells in depth. T-lymphocytes are present in close proximity to the blood vessels of the sublining layer, together with occasional more diffusely scattered cells of macrophage-fibroblast lineage.

Perhaps the greatest practical impact arose from the semiquantitative synovitis scoring system proposed by Linblad and Hedfors when addressing the issue of intra-articular variation of synovitis *(36)*. They recognized the primacy of two macroscopic features considered to be ubiquitous with synovitis. The first one is that of membrane vascularity of which there are two components each individually assigned a score of 0 or 1 depending on their absence or presence at the biopsy site:

1. Capillary hyperaemia,
2. Increased vascular markings.

The second ubiquitous feature is that of synovial membrane proliferation which has three components. The predominant proliferative appearance at the biopsy site is scored:

1. Normal/no proliferation = 0,
2. Granularity = 1,
3. Villous hypertrophy = 2.

Thus a hypervascular, hyperaemic synovium with villous hypertrophy scores a maximum of 4. This scoring system has been adapted in standard arthroscopy practice to grade synovitis in anatomically defined areas (e.g. the medial gutter of the knee in conjunction with a 100 mm visual analogue scale [VAS] for global features of vascularity and membrane proliferation). Other synovitis scoring systems have also been proposed *(37–39)* but have not been widely adopted into practice however *(40)*. The morphology of vascular markings and synovial proliferation and may be of value in distinguishing between RA and psoriatic or reactive arthritis of less than 12 mo clinical duration *(41)*. Furthermore high inter- and intraobserver coefficients of correlation were demonstrated in this study (using the Linblad and Hedfors semiquantitative scoring system) for two experienced arthroscopists, with somewhat lower coefficients for a relatively inexperienced trained arthroscopist. Work to refine and standardize the assessment of macroscopic synovitis is currently underway (personal communication, Professor Oliver FitzGerald, St. Vincent's Hospital, Dublin).

In the last decade arthroscopes have slimmed to as little as 1 mm in diameter but remain comparable with larger diameter arthroscopes *(42–45)*. The practice of undertaking small joint arthroscopy and synovial biopsy has become more widespread *(46–49)*. Congruity has been demonstrated in synovial biopsies obtained arthroscopically both from MCPs/wrists and from the knee in RA patients *(50)*. The question regarding the extent of subclinical synovitis in RA was revisited by Kraan and colleagues. They confirmed the presence of rheumatoid synovitis in clinically normal knees *(51)*. Sequential knee arthroscopy and synovial biopsy has been successful in patients after time intervals as short as 48 h in the assessment of therapeutic interventions *(52)*.

The three recognized indications for undertaking arthroscopy by rheumatologists are:

1. Diagnostic,
2. Therapeutic,
3. Research.

However, macroscopic assessment by arthroscopy rarely provides diagnostic information *(24)*. Notable exceptions include the synovial slough of Behcet's disease, the intense haemosiderin staining of pigmented villo-nodular synovitis, the blackened cartilage in ochronosis, and the crystalline deposits of uric acid or calcium pyrophosphate. However, precision synovial biopsy from operator selected areas under direct vision is without equal when histological confirmation is required.

Arthroscopy has the potential to be employed in multicenter studies assessing the effects of disease modifying drugs in OA. It is prerequisite that the quantification of chondropathy be simple to use, reliable, valid, and capable of detecting change. The chondropathy of OA is initially recognized by fine fibrillation of the chondral surface, advancing through superficial and deep fissuring to progressive thinning and eventual complete loss of cartilage with exposure of the subchondral boney surface. Two chondroscopic methods, among several described in the literature, have been critically evaluated. The inter- and intraobserver reliabilities for both proposed systems were found to be high and statistically significant *(53–56)*.

The first one, proposed by Noyes and Stabler (later adopted by the American College of Rheumatology) took into account depth, size, and location (documented on a knee diagram) of chondral lesions according to six anatomically defined cartilage surfaces (patella, trochlear, medial and lateral femoral condyles, and the medial and lateral tibial plateaus). An overall score (0–350 points) was then assigned. The second method was proposed and validated by Ayral on behalf of the Société Française d'Arthroscopie (SFA) and has since been adopted for use by European rheumatologists. The SFA system uses a composite score based upon two component parts:

1. SFA grade
   0 = normal cartilage
   I = chondromalacia (softening/fine fibrillation of the surface)
   II = superficial fissuring
   III = deep fissuring down to but not exposing subchondral bone
   IV = exposure of subchondral bone +/– erosion
2. SFA score
   This is a continuous variable with a value between 0 and 100 which is obtained using the following mathematical formula:
   Size (%) of grade I lesions × 0.14, plus
   Size (%) of grade II lesions × 0.34, plus
   Size (%) of grade III lesions × 0.65, plus
   Size (%) of grade IV lesions × 1.00
   (% = average percentage of chondral surface assessed to be affected)

The SFA grade and score is obtained for each of the six anatomical chondral surfaces as defined above. This system has been successfully applied in a multicenter study comparing Tenidap and Piroxicam on the progression of medial tibio-femoral joint OA by chondroscopy and standard radiological progression in 665 patients over a 12-mo follow-up period *(57)*. The chondroscopic system demonstrated success in detecting statistically significant disease progression and appears to be more sensitive than does radiological progression in so doing. Ayral has proposed and validated a system for grading synovitis in OA also *(58)*.

Currently, a number of manufacturers produce robust, reusable, precision engineered arthroscopes and ancillary equipment. The Hopkins® glass rod lens system has superseded the Selfoc® system. Interchangeable arthroscopes in a range of lengths and diameters can be easily sourced, having viewing angles of 0°, 20 to 30°, and 90° and complete with high quality lens coatings. Arthroscopic tools, such as grasping biopsy forceps, are also available in a range of sizes with corresponding sheaths and introducers. Light sources have become externalized from the arthroscope itself and now attach by means of a flexible cable. Tungsten and halogen light sources are being superseded by xenon. No longer does the operator have to view the joint directly through the eye-piece. Instead a small camera head attachment locks into position over the eyepiece and conveys images via a flexible cable to the 1 or 3 chip digital camera unit and then projected on to a conveniently placed monitor or "head-up" display. Images captured can be conveniently undertaken on to video-printer, CD, or Zip disc in preference to super-VHS tape.

Consequently, the opportunity to undertake arthroscopy safely and reliably on a day-case basis requiring local anaesthesia alone has beckoned strongly to the rheumatology community and will continue to do so *(59,60)*. It has become common practice in rheumatological arthroscopy to utilize arthroscopes measuring 2.7 to 4.0 mm in diameter with a 30° angle of view in knees, whereas 1.0 to 2.4 mm diameter arthroscopes (0 or 30° angle of view) are to be preferred in small and medium sized joints. The tolerability and patient safety of arthroscopy and synovial sampling has been surveyed worldwide across 32 units currently undertaking the technique *(40)*. Recognized complications include wound complications (0.1%), intra-articular sepsis (0.1%), and deep venous thrombosis (0.2%). The highest reported complication was haemarthrosis (0.9%), which only occurred in relation to cartilage rather than synovial biopsy however. Reported complications rates in orthopaedic arthroscopy are higher, possibly because of the use of a tourniquet, a limb-holding device during arthroscopic procedures and the complex nature of some techniques undertaken *(61)*. Furthermore, general anaesthesia carries risks not shared with the local anaesthetic techniques employed in rheumatological arthroscopy.

It has been stated previously that arthroscopy is the current "gold standard" for the assessment of synovitis. In recent years we have witnessed the application of high-resolution ultra-sound scanning (HRUS) and magnetic resonance imaging (MRI) for synovitis assessment too. They have been directly compared with each other and standard radiographs to determine synovitis extent and the presence of erosions *(62)*. HRUS is cheaper, portable, and repeatable but is more subject to operator bias than is the more expensive and time consuming MRI. Arthroscopic and HRUS findings in knees are highly comparable for synovial membrane proliferation *(63)* but HRUS is limited to the supra-patellar pouch and medial and lateral gutters due to the acoustic shadow that is cast by the patella, and its inability to access the intercondylar notch well. The adoption of power Doppler for the assessment of vascularity and blood flow compared to arthroscopy and biopsy findings such as vascular endothelial growth factor (VEGF) expression is an exciting prospect worthy of further study *(64)*.

An integrated imaging approach is an attractive proposition in seeking to achieve better understanding of disease pathogenesis and the effects of therapeutic interventions. A small number of clinical trials have been undertaken where arthroscopic synovitis assessment and synovial biopsy analysis have been undertaken in conjunction with dynamic enhanced MRI and HRUS *(65–67)*. When compared with standard clinical outcome measures differential treatment effects can be demonstrated in modest sized patient cohorts.

The complexity of the processes within an inflamed or degenerate joint is gradually becoming better understood. The advances in immunohistochemistry and digital image analysis as well as the new methods of proteomic and genomic assay will take our understanding further. This is well recognized not only within our scientific and clinical community but also within the pharmaceutical industry. It is therefore certain that the quest for knowledge through expanded investigation of the synovium will continue. Arthroscopy remains unique in being able to directly image the synovium, cartilage, and other joint contents and to permit tissue sampling in a safe, repeatable, and reliable manner. The expansion of arthroscopy in rheumatology is certain; however this demand must be satisfied by the availability of adequately trained and competent operators.

2. Materials

1. Operating theatre or suitable procedure room and equipment sterilization facilities.
2. Arthroscopy light source (Xenon or Halogen) and light cable.
3. Digital (3-chip or 1-chip) camera (PAL/NTSC), arthroscopic camera head attachment and lead.
4. Image recording medium (CD/Zip disc digital image recorder/Super VHS video recorder/Video Colour Printer) and consumables.
5. 50-cm color medical monitor.
6. Equipment stacking trolley (for **steps 2–5** above).
7. Procedure trolley.
8. Sterile drapes, gowns, and gloves.
9. Arthroscope (knees: 2.7- to 4.0-mm diameter, 30° angle of view; : medium joints: 1.6 to 2.4-mm diameter, 30° angle of view; small joints: 1.0 to 1.9-mm diameter, 0 or 30° angle of view), blunt and sharp introducers and arthroscopic sheath.

10. Drainage/biopsy cannula (1.6 to 4.0-mm diameter) and biopsy forceps.
11. Irrigation tubing (intravenous administration sets).
12. 1000 m: unit doses 0.9% saline prewarmed to 37°C for irrigation.
13. No. 11 disposable knife blade.
14. Syringes (60 mL, 20 mL, 10 mL) and needles.
15. Local anaesthesia: 20 mL 1% xylocaine with adrenaline for large/medium joints; 10 mL 1% Lidocaine for small joints).
16. Wound closure material (Steristrips®, wound clips, or 3/0 nylon suture on curved needle).
17. Occlusive dressings, crepe, and wool bandaging.
18. Benzodiazepine (oral or intravenous administration) for patient sedation.
19. Formalin filled pot for biopsy (if to undergo histological analysis).
20. Aluminium foil and Nunc cryo pen.
21. OCT embedding medium and moulds (minimum two per patient).
22. Cryopreservation tubes for storage of samples for polymerase chain reaction (PCR) (minimum two per patient).
23. Universal container for storage of OCT blocks.
24. Pastettes.
25. Scalpel blades—for biopsy handling.

3. Methods

The methods described below outline: (1) the setting-up process for an arthroscopic procedure, (2) the conduct of an arthroscopic knee examination, (3) biopsy sampling from operator selected sites, and (4) immediate preparation of synovial biopsies prior to storage at –80°C.

3.1. Setting Up for an Arthroscopy Procedure

3.1.1. Preparing the Operating Theatre/Procedure Room

This would be undertaken by nursing and other theatre staff involved and should include a brief inspection of imaging equipment on the stacking trolley to ensure it is fully functional, that all leads are correctly attached and image recording consumables are present.

3.1.2. Patient Preparation and Consent

Informed consent must be obtained from the patient undergoing the procedure in accordance with standard practice within that institution. Written information explaining the procedure should be made available to the patient. Post-procedure information detailing what is to be expected, what is abnormal, and what is potentially serious should also be available. Sedation must be administered in accordance with standard protocols at the institution concerned and should only be undertaken if facilities and skills for intubation and resuscitation are available.

3.1.3. Skin Disinfection and Draping

The skin of all the surfaces of the joint to be arthroscoped should be thoroughly disinfected with proprietary skin preparation solutions (e.g. chlorhexidine), prior to draping the operating field. Adherence to aseptic precautions should be followed at all times.

3.1.4. Portal Selection and Local Anaesthesia

Standard anatomically defined arthroscopy portals are well described in standard orthopaedic texts. Local anaesthesia (not exceeding maximum recommended doses) should be administered at the chosen portals by infiltration in to the skin and deeper tissues, down to and including the joint capsule, while avoiding instillation into the joint cavity (adrenaline may affect vascularity). Plain local anaesthesia may be instilled directly into the joint space without detriment to the synovial macroscopic appearances or affecting the biopsies obtained. Alternative local anaesthetic approaches such as regional nerve blocks are acceptable. A topical anesthetic cream can be applied over the portals and left *in situ* for at least 45 min prior to the local anesthetic being infiltrated to minimise patient discomfort.

3.1.5. White Balance Setting

Always perform white balance setting on the digital camera unit, according to equipment manufacturer's instructions before cannulation of the joint space. Select standby mode.

3.2. Conducting a Knee Arthroscopy

3.2.1. Joint Cannulation

The knee joint is initially distended by injecting a volume of 50 to 60 mL of prewarmed saline taken from the unit dose for irrigation. The knife is then used to make a single stab incision at the first portal only, through the skin, and is extended down to and through the capsule. A small amount of the instilled fluid will be seen to leak from the wound on removal of the knife. The arthoscopic sheath with the blunt introducer locked in position is introduced to the joint through this incision. When the operator is confident that the sheath is within the joint space the blunt introducer can be unlocked and removed from the sheath. With care the arthroscope is then inserted and locked into position. The lavage inflow tubing, light cable and camera head attachment may then be attached securely.

3.2.2. Establishing Lavage of the Joint

The container of lavage fluid (pre-warmed to 37°C) is suspended 70 cm above the top of the theater table/procedure couch and connected to the arthroscope sheath by means of intravenous drip tubing with a tap for regulating flow. The lavage is commenced and the joint is allowed to become distended once more. The cannulation process can then be repeated at the second chosen portal for the biopsy cannula to be inserted. The drainage tubing is connected to this cannula. The lavage fluid may drain away safely under gravity to a suitable receptacle at floor level. Single or twin Luer locks with taps for controlling flow are fitted as standard to arthroscopic sheaths and drainage/biopsy cannulae. Ensure that the lavage is kept constantly flowing throughout the procedure. Typically, a 1000 mL volume is required for a knee procedure taking 15 to 20 min. Additional fluid should be prewarmed and available for immediate use. The lavage should be continued after completion of the joint examination and biopsy sampling ensuring that the fluid is as clear as possible before removing the instruments from the joint.

3.2.3. Getting Orientated Within the Joint

Turn the digital camera unit on from standby mode. Orientate the arthroscope within the joint. Always remember that the light cable attaches to the arthroscope directly opposite to the angle of view (except when using a forward looking [0°] instrument). The cable leading from the camera head attachment should always be kept pointing directly at the floor. This ensures that "up is up" on the projected image. As the arthroscope is manouvered within the joint, the camera head attachment can be loosened, rotated, and repositioned to keep the cable pointing down to the floor at all times. Ensure the image is focused clearly on the monitor screen. The in-flowing lavage fluid will carry debris and proliferative synovium away from the field of view. Identify the anatomical starting point for the examination (e.g. the trochlear of the femur in the case of the knee) and ensure the arthroscope is correctly orientated.

3.2.4. Systematic Examination of the Synovium and Chondral Surfaces

Using a steady sweeping motion at all times begin the inspection of the floor of the patello-femoral joint space moving progressively in a cranial direction. At the proximal reflection, invert the scope and repeat the sweeping action looking at the roof of the joint space until reaching the proximal pole of the patella. Continue while inspecting the patella surfaces until the original starting point is reached. Invert the arthroscope once more. From the starting point now move medially along the trochlear to enter the medial gutter and inspect its walls. Return to the starting point and repeat, moving laterally to access the lateral gutter in a likewise fashion. Once again return to the starting point. To inspect the tibio-femoral joint (TFJ) space it will be necessary to gently position the patient's leg over the side/end of the table/couch while holding the heel of that foot firmly in the nonoperating hand. Manouver the arthroscope medially along the trochlear once again. The arthroscope will then enter the medial tibio-femoral joint space. Follow the medial femoral condyle down until the medial meniscus and tibal plateau can clearly be seen. Placing gentle valgus strain on the knee will reveal the posterior horn of the meniscus. Moving laterally the arthroscope will enter the intercondylar space. Withdraw the arthroscope slightly to inspect the anterior cruciate ligament and synovium. The arthroscope should now be returned to its start point once again. The leg can simultaneously be placed back upon the couch. Formal inspection of the lateral compartment is difficult under local anaesthesia through the infero-lateral portal. To inspect this compartment it is necessary to re-enter the joint through an infero-medial portal (anaesthetized as above) and to repeat the systematic inspection as for medial TFJ space starting from the trochlear reference start point. The inspection can be conducted while applying gentle varus strain to the knee. Alternatively, this compartment may be inspected through the infero-lateral portal with accessibility assisted by flexion of the knee combined with external rotation of the hip to place the mid-shin across the contralateral leg for support. The comfort of the patient must be observed at all times and in our experience the quality of information obtained and synovial biopsies is not impaired by omitting routine inspection of the lateral TFJ. The findings in each area should be recorded and if necessary a scoring measure can be utilized.

3.2.5. Synovial Biopsy

The sites selected for biopsy by the operator should be chosen to be representative of the synovial appearances throughout the joint.

3.2.6. Completion of Procedure

The instruments should be removed in turn. Initially turn the digital camera unit back to stand-by mode. Disconnect the light cable and camera head attachment while continuing the lavage. Unlock and carefully remove the arthroscope from its sheath. Turn off the lavage inflow at the Luer lock and tap on the infusion tubing prior to disconnection from the sheath. Remove the sheath from the joint. Apply firm but gentle pressure to the joint to encourage as much lavage fluid to exit the joint via the biopsy/drainage cannula prior to disconnecting the drainage tubing from the Luer lock. Remove this cannula and return to procedure trolley along with all other equipment for disposal or cleaning and subsequent sterilization as appropriate. Dry the skin and close the wounds with chosen method (e.g. sutures or wound clips). Up to 10 mL 0.5% bupivacaine (and 40–80 mg DepoMedrone® or equivalent at operator's discretion) may also be instilled at this time by direct injection through one of the portals used. Synovial fluid replacement (e.g. Viscoseal®) could be instilled in place of the above. An occlusive dressing is placed over the wound. Finally, a crepe and wool bandage may be wrapped firmly around the joint to remain *in situ* for 24 h only. The dressing may be removed after 3 d but the wound closures remain in place for 5 to 7 d. The procedure should be documented and recorded in the case record.

3.3. Synovial Sampling Under Direct Vision

3.3.1. Biopsy Forceps

Straight dismantling spoon or grasping forceps are recommended. Punch forceps may crush the tissue samples obtained.

3.3.2. Insertion and Triangulation

The biopsy forceps should be carefully inserted through the end of the drainage/biopsy cannulae. These come fitted with rubber bung and/or hinged flap to prevent leakage of lavage fluid. Ensure the end of the drainage/biopsy tube can be seen within the supra-patellar pouch on the monitor (if not visible within the joint consider repeating the cannulation process). Advance the biopsy forceps into the joint space and towards the area selected for biopsy until the jaws of the forceps contact the synovium (triangulation). With the arthroscope held in the other hand or by the scrubbed assistant attempt to keep the biopsy forceps in view at all times.

3.3.3. Selection of Biopsy Sites

Biopsies may be taken from the area immediately adjacent to the patella, immediately proximal to the trochlear and from the medial gutter walls. Other sites of specific interest can also be selected however. The number and location of the biopsies should be recorded and immediately processed by a laboratory technician present at the procedure. Care should be exercised with areas of scarring resulting from previous biopsy or areas with particularly large blood vessels (may result in uncontrollable blood loss and haemarthrosis). Fibrin deposits and other debris within the joint can easily be removed with the biopsy forceps to gain better access to synovium of interest.

3.3.4. Taking Biopsies

Carefully open the jaws fully apply gentle pressure to the synovial membrane and then firmly close the jaws. Gently pull the forceps away from the synovium with a half twist to the left or right. Keep the jaws firmly shut and withdraw the forceps from the joint. Open the forceps jaws only when outside the joint to place biopsy on to damp gauze. Repeat as often as required.

3.4. Biopsy Specimen Handling

1. Prefill the moulds with OCT and remove bubbles with a pastette. Label aluminium foil specimen wrappings with Nunc cryo pen. Label formalin pot and cryo-preservative tubes clearly.
2. Assess for quality selecting the larger ones for freezing into OCT (two if possible), one into theformalin pot for routine histology, and the rest divided between the cryo-preservative tubes.
3. Using a scalpel blade place a biopsy into the OCT, ensuring all surfaces are well covered.
4. Using long tongs lower the mould, whilst keeping it horizontal, into the flask of liquid nitrogen. Remove when frozen and express the OCT block from the mould. Wrap it in the labelled foil and replace into liquid nitrogen.
5. Divide the biopsies for PCR between the two cryo-preservation tubes and seal.
6. Carefully place the cryo-preservative tubes into the liquid nitrogen.
7. Place the biopsy for histology into the labelled formalin filled pot and seal.
8. Place frozen OCT blocks into a labelled universal container. Store with PCR samples at $-80°C$.

4. Notes

1. Training to competence in the technique is a vital prerequisite for anyone wishing to undertake this technique. Training guidelines are available on the ILAR website (*www.ILAR.org*). Arthroscopy training courses for rheumatologists are held at various locations. Further information for surgical training can be sought from the Royal College of Surgeons in London (*www.rcseng.ac.uk*) or from rheumatological and orthopaedic specialist societies. Prolonged attachments to specialist centres undertaking the technique already would be strongly advised.
2. Postoperative analgesia is not usually required in addition to that the patient is usually prescribed. Encourage early mobilization of the joint (after 24 h) and rapid return to normal activities. Perioperative use of non-steroidal anti-inflammatory drugs (NSAIDs) has been demonstrated to be efficacious and is not associated with increased incidence of postoperative bleeding complications. Patients should be advised to avoid aspirin (other than for low dose CVS prophylaxis) in first wk preceding the arthroscopy. Warfarinized patients can omit the two doses prior to the procedure and restart their usual dose the day after the arthroscopy. Diabetic patients are usually placed as the first case on the list.
3. Patient sedation is rarely necessary (<10% for knees). The equivalent of 4 to 6 mg Midazolam by slow intravenous injection is usually sufficient. Rapid reversal of the sedation and full resuscitation facilities should be immediately available.
4. It is sometimes difficult to establish and maintain an adequate flow of lavage fluid. This may result from debris washed from the joint obstructing the cannulae or tubing. This can easily be dislodged and the flow restored. If this is not the cause check that the drainage/biopsy cannula is visible and within the joint space. Joints that have been chronically swollen may contain turbid synovial fluid that

should be thoroughly lavaged from the joint before attempting systematic inspection.
5. Always maintain the field of view on the monitor as a circle. Excessive force on the arthroscope within its sheath will distort the image. This force should be removed immediately and the arthroscope repositioned within the joint. Failure to do so may result in patient discomfort and breakage of the arthroscope.
6. Post-procedure complications include wound infection or leakage, joint effusion, septic arthritis, and haemarthrosis. Wound complications should be managed by careful attention to wound toilet, swabbing of microbiological investigation of the causative organism, and a short course of appropriate antibiotics if indicated clinically. Occasionally, wounds can leak synovial fluid for up to 2 wk following the procedure. Applying a small absorbent dressing regularly and administering prophylactic antibiotics until the flow has stemmed is indicated. A postprocedure joint effusion is not unusual and subsides within a few days but should be differentiated from septic arthritis if there is clinical concern. Signs of intra-articular joint sepsis include pyrexia, flu-like symptoms, joint swelling, and overlying skin redness. This should be investigated and treated as high priority.

Acknowledgments

The author wishes to thank many collaborators for sharing their arthroscopy experiences. The methods here are those established within Professor Paul Emery's Unit in Leeds. I would especially like to thank clinical and laboratory colleagues in Leeds, present and past, as well as those with whom I have collaborated at other centers. None of this would have been possible without the help and efforts of the Day Case Theatre and Ward staff at Chapel Allerton Hospital, Leeds. For the synovial biopsy methodology I am grateful to Anne English, Karen Henshaw and Diane Carscadden. Staffan Linblad deserves special thanks for introducing me to arthroscopy and inspiring me to begin to learn the craft.

References

1. Watanabe, M. (1986) Memories of the Early Days of Arthroscopy. Arthroscopy **2,** 209–214.
2. Bircher, E. (1921) Die arthroendoskopie. Zentralblatt für Chirurgie **48,** 1460–1461
3. Kreuscher, PH. (1931) Semilunar cartilage disease: a plea for early recognition by means of the arthroscope and the early treatment of this condition. *IMJ* **47,** 290–292.
4. Finkelstein, H. and Mayer, L. (1931) The arthroscope, a new method of examining joints. *J. Bone Joint Surg.* **13,** 583–588.
5. Burman, M. S. (1931) Arthroscopy or the direct visualization of joints: an experimental cadaver study. *J. Bone Joint Surg.* **13,** 669–695.
6. Burman, M. S., Finkelstein, H., and Mayer, L. (1934) Arthroscopy of the knee joint. *J. Bone Joint Surg.* **16,** 255–268.

7. Watanabe, M., Takeda, S., and Ikeuchi, H. (1957, 1969 (2nd ed), 1978 (3rd ed.)) Atlas of Arthroscopy. Igaku-Shoin, Tokyo, New York.
8. Iino, S. (1939) Normal arthroscopic findings of the knee joints in adults. *J. Jap. Orthop. Ass.* **14,** 467–523.
9. Miki, M. (1941) Influences of the temperature and pressure of the medium on the arthroscopic findings of the blood vessels of the synovial membrane. *J. Jap. Orthop. Ass.* **16,** 405–409.
10. Mizumachi, S., Kawashima, W., and Okamura, T. (1948) So called synovial shelf in the knee joint. *J. Jap. Orthop. Ass.* **22,** 1–5.
11. Sakakibara, J. (1976) Arthroscopic study on Iino's band (plica synovialis mediopatellaris). *J. Jap. Orthop. Ass.* **50,** 513–522.
12. Watanabe, M. and Takeda, S. (1953) On the popularization of arthroscopy. *J. Jap. Orthop. Ass.* **27,** 258–265.
13. Watanabe, M. and Takeda, S. (1960) The No. 21 arthroscope. *J. Jap. Orthop. Ass.* **34,** 1041.
14. Watanabe, M. (1968) Arthroscopic diagnosis of the internal derangements of the knee joint. *J. Jap. Orthop. Ass.* **42,** 993–1002.
15. Watanabe, M. (1972) Arthroscope, present and future. *Surg. Ther.* **26,** 73–77.
16. Palmer, D. G. (1967) Synovial villi: an examination of these structures within the anterior compartment of the knee and metacarpophalangeal joints. *Arthritis Rheum.* **10,** 451–458.
17. Jayson, M. I. V. and St. J. Dixon, A. (1968) Arthroscopy of the Knee in Rheumatic Diseases. *Ann. Rheum. Dis.* **27,** 503–511.
18. Yates, D. B. and Scott, J. T. (1975) Rheumatoid synovitis and joint disease. *Ann. Rheum. Dis.* **34,** 1–6.
19. Henderson, D. R. F., Jayson, M. I. V., and Tribe, C. R. (1975) Lack of correlation of synovial histology with joint damage in RA. *Ann. Rheum. Dis.* **34,** 7–15.
20. Kurosaka, M., Ohnu, O., and Hirohata, K. (1991) Arthroscopic evaluation of synovitis in the knee joints. *Arthroscopy* **7,** 162–170.
21. Salisbury, R.B., and Nottage, W.M. (1985) A new evaluation of gross pathological changes and concepts of rheumatoid articular degeneration. *Clin. Orthop. Rel. Res.* **199,** 242–247.
22. Allard, S. A., Bayliss, M. T., and Maini, R. N. (1990) The synovium-cartilage junction of the normal human knee. Implications for joint destruction and repair. *Arthritis Rheum.* **33,** 1170–1179.
23. Allard, S. A., Muirden, K. D., Maini, R. N. (1991) Correlations of histopathological features of pannus with patterns of damage in different joints in rheumatoid arthritis. *Ann. Rheum. Dis.* **50,** 278–283.
24. O'Rourke, K. S. and Ike, R. W. (1994) Diagnostic arthroscopy in the arthritis patient. *Rheum. Dis. Clin. North. Am.* **20,** 321–342.
25. Gibson, T., Fagg, N., Highton, J., Wilton, M., and Dyson, M. (1985) The diagnostic value of synovial biopsy in patients with arthritis of unknown cause. *Br. J. Rheumatol.* **24,** 232–241.

26. Gallagher, P. J., Blake, D. R., and Lever, J. V. (1985) Audit of closed synovial biopsy in patients with arthritis of unknown cause. *Scand. J. Rheumatol.* **14,** 307–314.
27. Rooney, M., Condell, D. Quinlan, W., et al. (1988) Analysis of the histological variation of synovitis in rheumatoid arthritis. *Arthritis. Rheum.* **31,** 956–963.
28. Hutton, C. W., Hinton, C., and Dieppe, P. A. (1987) Intra-articular variation of synovial changes in knee arthritis: biopsy study comparing changes in patello-femoral synovium and the medial tibiofemoral synovium. *Br. J. Rheumatol.* **26,** 5–8.
29. Soden, M., Rooney, M., Cullen, A., Whelan, A., Feighery, C., and Bresnihan, B. (1989) Immunohistological features in the synovium obtained from clinically uninvolved knee joints of patients with rheumatoid arthritis. *Br. J. Rheumatol.* **28,** 287–292.
30. Rooney, M., Whelan, A., Feighery, C., and Bresnihan, B. (1989) Changes in lymphocyte infiltration of the synovial membrane and the clinical course of rheumatoid arthritis. *Arthritis Rheum.* **32,** 361–369.
31. Kirkham, B., Portek, I., Soon Lee, C., et al. (1997) Intra-articular variability of synovial membrane histology, immunohistochemistry, and cytokine mRNA expression in patients with rheumatoid arthritis. *J. Rheumatol.* **26,** 777–784.
32. Tak, P. P. and Bresnihan, B. (2000) The pathogenesis and prevention of joint damage in rheumatoid arthritis. Advances from synovial biopsy and tissue analysis. *Arthritis Rheum.* **43,** 2619–2633.
33. Lindblad, S. Klareskog, L., Hedfors, E., Forsum, U., and Sündström, C. (1983) Phenotypic characterization of synovial tissue cells in situ in different types of synovitis. *Arthritis Rheum.* **26,** 1321–1332.
34. Lindblad, S. and Hedfors, E. (1987) Arthroscopic and immunohistologic characterization of knee joint synovitis in osteoarthritis. *Arthritis Rheum.* **30,** 1081–1088.
35. Lindblad, S. and Hedfors, E. (1987) The synovial membrane of healthy individuals – immunohistochemical overlap with synovitis. *Clin. Exp. Immunol.* **69,** 41–47.
36. Lindblad, S. and Hedfors, E. Intra-articular variations in synovitis. Local macroscopic and microscopic signs of inflammatory activity are significantly correlated. *Arthritis Rheum.* **28,** 977–986.
37. Zschäbitz, A., Neurath, M., Grevenstein, J., Koepp, H., and Stofft, E. (1992) Correlative histologic and arthroscopic evaluation in rheumatoid knee joints. *Surg. Endosc.* **6,** 277–282.
38. Paus, A. C. and Pahle, J.A. (1990) Arthroscopic evaluation of the synovial lining before and after open surgery of the knee joint in patients with chronic inflammatory joint diseases. *Scand. J. Rheum.* **19,** 193–201.
39. Ike, R. W. (1995) Arthroscopy: an outcome measure of synovitis? *Rheum. Eur.* **(suppl 2),** 134–138.
40. Kane, D., Veale, D. J., FitzGerald, O., and Reece, R. (2002) Survey of arthroscopy performed by rheumatologists. *Rheumatol.* **41,** 210–215.
41. Reece, R. J., Canete, J. D., Parsons, W. J., Emery, P., and Veale, D.J. (1999) Distinct vascular patterns of early synovitis in psoriatic, reactive and rheumatoid arthritis. *Arthritis Rheum.* **42,** 1481–1484

42. Halbrecht, J. L. and Jackson D. W. (1992) Office Arthroscopy: A Diagnostic Alternative. *Arthroscopy* **8**, 320–326.
43. Wei, N., Delauter, S. K., and Erlichman, M. S. (1993) Office knee arthroscopy: the first 100 cases. *Rheumatol. Rev.* **2**, 151–158.
44. Ike, R.W., and O'Rourke, K.S. (1993) Detection of intra-articular abnormalities in osteoarthritis of the knee: a pilot study comparing needle arthroscopy with standard arthroscopy. *Arthritis Rheum.* **36**, 1353–1363.
45. Gramas, D. A., Antounian, F. S., Peterfy, C. G., Genant, H. K., and Lane, N. E. (1995) Assessment of needle arthroscopy, standard arthroscopy, physical examination and magnetic resonance imaging in knee pain: a pilot study. *J. Clin. Rheumatol.* **1**, 26–34.
46. Sekiya, I., Kobayahi, M., Takeda, Y., and Matsui, N. (2002) Arthroscopy of the proximal interphalangeal and metacarpophalangeal joints in rheumatoid hands. *Arthroscopy* **18**, 292–297.
47. Rozmaryn, L. M. and Wei, N. (1999) Metacarpophalangeal arthroscopy. *Arthroscopy* **15**, 333–337.
48. Wei, N., Delauter, S. K., Erlichman, M. S., Rozmaryn, L. M., Beard, S. J., and Henry, D. L. (1999) Arthroscopic synovectomy of the metacarpophalangeal joint in refractory rheumatoid arthritis: a technique. *Arthroscopy* **15**, 265–268.
49. Ostendorf, B., Dann, P., Wedekind, F., et al. (1999) Miniarthroscopy of metacarpophalangeal joints in rheumatoid arthritis. Rating diagnostic value in synovitis staging and efficiency of synovial biopsy. *J. Rheumatol.* **26**, 1901–1908.
50. Kraan, M. C. Reece, R. J., Smeets, T. J., Veale, D. J., Emery, P., and Tak, P. P. (2002) Comparison of synovial tissues from the knee joints and the small joints of rheumatoid arthritis patients: Implications for pathogenesis and evaluation of treatment. *Arthritis Rheum.* **46**, 2034–2038.
51. Tak, P. P., Smeets, T. J., Daha, M. R., et al. (1997) Analysis of the synovial cell infiltrate in early rheumatoid synovial tissue in relation to local disease activity. *Arthritis Rheum.* **40**, 217–225.
52. Smeets, T. J., Kraan, M. C., van Loon, M. E., and Tak, P. P. (2003) Tumor necrosis factor alpha blockade reduces the synovial cell infiltrate early after initiation of treatment, but apparently not by induction of apoptosis in synovial tissue. *Arthritis Rheum.* **48**, 2155–2162.
53. Ayral, X., Gueguen, A., Ike, R. W., et al. (1998) Inter-observer reliability of the arthroscopic quantification of chondropathy of the knee. *Osteoarth. Cart.* **6**, 160–166.
54. Noyes, F. and Stabler, C. L. (1989) A system for grading articular cartilage lesions at arthroscopy. *Am. J. Sports Med.* **17**, 1505–1513.
55. Ayral, X. and Dougados, M. (1995) Viability of chondroscopy as a means of cartilage assessment. *Ann. Rheum. Dis.* **54**, 613–614.
56. Ayral, X., Gueguen, A., Listrat, V., et al (1994) simplified arthroscopy scoring system for chondropathy of the knee (revised SFA score). *Rev. Rheum.* **61**, 88–90.
57. Ayral, X., MacKillop, N., Genant, H. K., et al. (2003) Arthroscopic evaluation of potential structure modifying drug in osteoarthritis of the knee. A multi-centre,

randomized, double-blind comparison of tenidap sodium vs piroxicam. *Osteoarth. Cart.* **11**, 198–207.
58. Ayral, X., Mayoux-Benhamou, A., and Dougados, M. (1996) Proposed scoring system for assessing synovial membrane abnormalities at arthroscopy in knee synovitis. *Br. J. Rheumatol.* **35 (suppl 3)**, 14–17.
59. Chang, R. W. and Sharma, L. (1994) Why a rheumatologist should be interested in arthroscopy. *Arthritis Rheum.* **37**, 1573–1576.
60. Ayral, X. and Dougados, M. (1998) Rheumatological arthroscopy or research arthroscopy in rheumatology. *Br. J. Rheumatol.* **37**, 1039–1041.
61. Weber, S., Abrams, J. S., and Nottage, W. M. (2002) Complications associated with arthroscopic shoulder surgery. *Arthroscopy* **18(Suppl 1)**, 88–95.
62. Wakefield, R. J., Gibbon, W. W., Conaghan, P. G., et al. (2000) The value of sonography in the detection of bone erosions in patients with rheumatoid arthritis: a comparison with conventional radiography. *Arthritis Rheum.* **43**, 2762–2770.
63. Karim, Z., Wakefield, R. J., Quinn, M., et al. (2004) Validation and reproducibility of ultrasonography in the detection of synovitis in the knee: a comparison with arthroscopy and clinical examination. *Arthritis Rheum.* **50**, 387–394.
64. Taylor, P. C. (2002) VEGF and imaging of vessels in rheumatoid arthritis. *Arthritis Res.* **4(Suppl 3)**, S99–S107.
65. Veale, D. J., Reece, R. J., and Parsons, W., (1999) Intra-articular primatised anti-CD4:efficacy in resistant rheumatoid knees. A study of combined arthroscopy, magnetic resonance imaging, and histology. *Ann. Rheum. Dis.* **58**, 342–349.
66. Reece, R. J., Kraan, M. C., Radjenovic, A., et al (2002) Comparative assessment of Leflunomide and methotrexate for the treatment of rheumatoid arthritis, by dynamic enhanced magnetic resonance imaging. *Arthritis Rheum.* **46**, 366–372.
67. Ostendorf, B., Peters, R., Dann, P., et al. (2001) Magnetic resonance imaging and miniarthroscopy of metacarpophalangeal joints: sensitive detection of morphologic changes in rheumatoid arthritis. *Arthritis Rheum.* **44**, 2492–2502.

3

Immunohistochemistry of the Inflamed Synovium

Martina Gogarty and Oliver FitzGerald

Summary

The development in the techniques for obtaining synovial tissue biopsy, especially through arthroscopy, have resulted in greater access to high-quality synovial tissue. The use of immunohistochemistry in arthritis research has greatly furthered our understanding of the varied immunological and biochemical pathways involved in inflammatory arthropathies such as rheumatoid and psoriatic arthritis. Immunohistochemistry provides a strikingly visual narrative of the essential elements involved in inflammatory arthritis, from the infiltrating inflammatory cells (e.g., T-cells, macrophages, B-cells, and neutrophils), their products (e.g., cytokines, metalloproteinases) and their varied receptor molecules. This chapter describes the standard three-stage immunoperoxidase technique used in our laboratory and widely in the literature. Some problems that may be encountered and how they may be overcome are commented on. Also described is a method for dual-labeled immunofluoresence staining.

Key Words: Immunohistochemistry; immunoperoxidase; antibody controls; CD3; dual-labeled immunofluorescence staining.

1. Introduction

Normal synovium is composed of two distinct layers, the lining and sublining (*see* **Fig. 1A**). The lining layer is in direct contact with the interarticular cavity, and is usually 1 to 2 cells deep. It secretes glycosaminoglycans (e.g., hyaluronic acid) into the articular cavity which gives synovial fluid it's characteristic viscosity and lubricant properties. The synovial lining layer is composed of two cell types, designated as type A and type B cells. Type A cells are characteristic of tissue macrophages, whereas type B cells have a fibroblast-like morphology and are of mesenchymal origin. The sublining layer is relatively acellular, with few blood vessels, fat cells, and fibroblasts.

Upon the initiation of the inflammatory process the synovial tissue mass increases dramatically (*see* **Fig. 1B,C**). Macroscopically the tissue has a folded

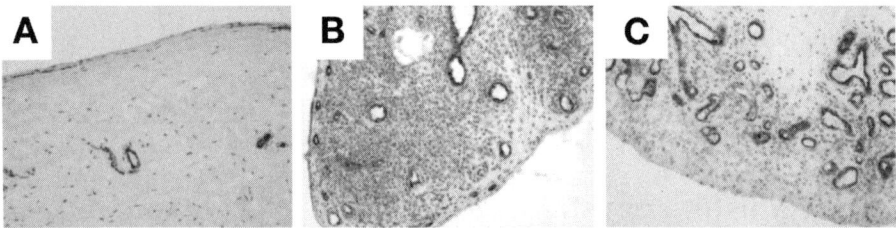

Fig. 1. Immunohistochemical staining of vessels with mouse antihuman Von Willebrand Factor (Factor VIII) in synovial tissue. The synovial tissue is representative of (**A**) normal synovium, (**B**) synovium obtained from a patient with psoriatic arthritis, and (**C**) synovium obtained from a patient with rheumatoid arthritis.

appearance and has a red hue resulting from the increase in the number of blood vessels. There is an initial increase in edema and a proliferation in the number of blood vessels leading to the migration of some lymphocytes and neutrophils. The cells in the lining layer proliferate, giving a characteristic hyperplastic appearance. As the inflammatory process progresses there is a dramatic increase in the number of macrophages, neutrophils, T-cells, and B-cells migrating from the increased number of blood vessels. Lining layer depth can increase up to 10 cells deep, especially in rheumatoid arthritis. Lymphocytes can often organize into distinct aggregates similar to lymph nodes, however this is not always the case and they can have a more dispersed pattern.

Synovial tissue may be obtained from a number of sources; blind needle biopsy, arthroscopy, and arthroplasty. Normal synovial tissue may be also obtained from amputated limbs. Arthroscopy is used widely for obtaining synovial tissue. The main advantage over blind needle biopsy is that the operator can take biopsies under direct vision from a number of different regions (e.g., suprapatellar pouch or adjacent to the cartilage-pannus junction) within the knee joint.

There are a vast number of antibodies against cell surface antigens and products commercially available. Immunohistochemistry provides us with a snap shot image of what is actually taking place within the joint. Immunohistochemical parameters have been shown to correlate with clinical indicators of disease activity *(1)*. A number of studies in our laboratory have shown a significant correlation between the level of synovial tissue macrophage infiltration and progressive joint damage *(1–3)*. Immunohistochemistry is a very valuable tool in elucidating the mode of action of pharmacological intervention, whether it is conventional nontargeted disease-modifying antirheumatic drugs (e.g., methotrexate *[4]* and leflunomide *[5]*), or targeted biological therapy (e.g.,

Immunohistochemistry of the Inflamed Synovium

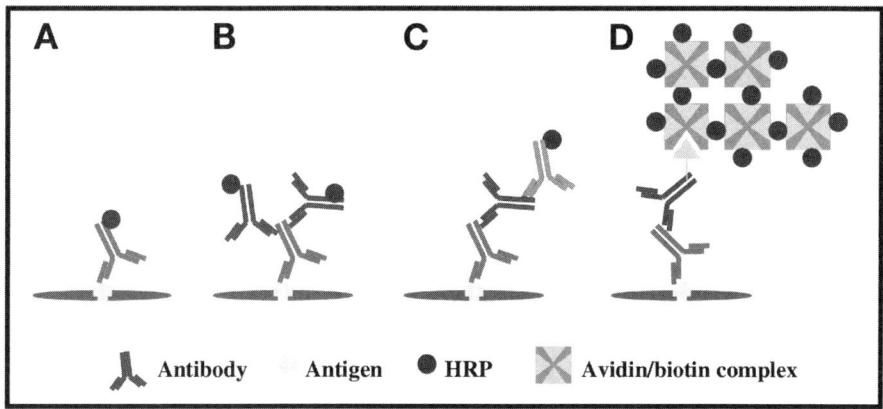

Fig. 2. Detection methods employed in immunohistochemical staining; (**A**) direct: HRP conjugated to a primary antibody, (**B**) two-step indirect: HRP conjugated to the secondary antibody, (**C**) three-step indirect: an antibody complex with the HRP enzyme conjugated to an antibody directed against the secondary antibody, and (**D**) avidin/biotin technology: the ABC(HRP) complex binds to the biotinylated secondary antibody.

interleukin [IL]-1 receptor antagonist *[6]*). Immunohistohemistry can provide information on the effects of the treatments on cell number, adhesion molecule, and cytokine expression. These measurements can then be compared with the clinical parameters of the patient before and following pharmacological intervention, thus providing valuable information on responders versus nonresponders.

Immunohistochemistry is used to detect antigens *in situ* and the approach used to detect any given antigen in any given tissue may vary widely. There are many methods to detect a specific antigen, from the direct method which uses an enzyme labeled primary antibody to the two and three step indirect methods with involve building a cascade of antibodies directed towards the primary antibody (*see* **Fig. 2A–C**). The build up of an antibody complex increases the detection sensitivity with the use of secondary antibodies that react with a number of different epitopes on the primary antibody.

The method used in our laboratory employs the use of Vector Laboratories, Peroxidase Vectrastain® Elite® ABC Kits. These kits are available directed towards goat, mouse, rabbit, rat, and sheep primary antibodies. This method of detection is based on the high affinity of avidin, a 68,000 molecular weight (MW) glycoprotein, for biotin. Avidin has four binding sites for biotin and most proteins (including antibodies and enzymes) can be conjugated with several molecules of biotin. This gives the potential for macromolecular complexes to be formed between avidin and a biotinylated enzyme (see **Fig. 2D**), thus achieving a strong signal.

Horseradish peroxidase (HRP) and alkaline phosphatase are the two main detection enzymes that are conjugated to biotin in the ABC complex. The use of HRP is described below. HRP has a number of substrates that form soluble and insoluble products. It is essential that the end product formed is insoluble and can be easily detected above the counterstain used. Here we use the substrate diaminobenzidine (DAB), which is converted to an insoluble brown precipitate, which gives good contrast against the blue Haematoxylin counterstain.

Whereas the use of serial sections aids in locating the cell of origin of an antigen of interest, dual-labeled immunofluorescence staining removes any uncertainty. Dual-labeled immunofluorescence staining incorporates the use of two specific antibodies directed against antigens of interest which can be incubated on the same section and detected separately by using two distinct fluorescence-labeled molecules under ultraviolet (UV) light. When performing dual staining it is imperative that the primary antibodies used together are of either a different immunoglobulin isotype (e.g., a mouse IgG with a mouse IgM antibody) or raised in different species (e.g., a mouse IgG with a rabbit IgG antibody).

Fluorescence of a molecule occurs when it is exposed to a specific wavelength of light (excitation wavelength) and the light it emits (emission wavelength) is always of a higher wavelength (e.g., Cy^3: excitation 550 nm and emission 680 nm). Specific filters are required to isolate the excitation and emission wavelengths of a fluorochrome (a fluorescent molecule). A dichroic beam splitter or partial mirror reflects the lower wavelengths and allows the higher wavelengths to pass. Only the emission light is required, this creates a dark background so that the fluorescence can be viewed easily.

The section is examined using a computerized fluorescent microscope with the relevant filters for the fluorescent tags employed in the assay; the fluorescein isothiocyanate (FITC) fluorochrome emits at 520 nm and Cy^3 at 680 nm. A photograph of the FITC staining is taken using a digital camera, then without moving the view under the objective, the filter is changed and Cy^3 fluorescence is viewed and imaged. The individual images are superimposed on each other and any yellow fluorescence obtained is due to colocalization of fluorescence signals derived from FITC and Cy^3, and therefore colocalization of the antigens of interest. A study from our department showed the novel colocalization of the corticotropin-releasing hormone receptor 1 (CRH-R1) to mast cells using an antibody against mast cell tryptase *(7)*. This method is described in detail in **Subheading 3.5.** and illustrated in **Fig. 3**.

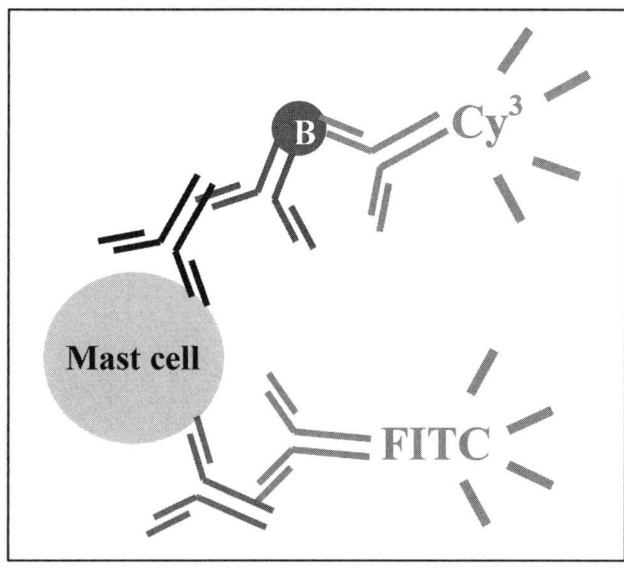

Fig. 3. A schematic diagram of the dual-labeled immunofluorescence detection of CRH-R1 (three step detection method) goat antihuman CRH-R1 antibody, biotinylated rabbit anti-goat IgG and finally a Cy^3 fluorochrome-conjugated antibiotin antibody that fluoresces red and mast cell tryptase (two step detection method) mouse antitryptase antibody and a fluorescein isothiocyanate (FITC)-conjugated antimouse IgG antibody that fluoresces green B-biotin.

2. Materials
2.1. Preparation of Tissue Slides for Immunohistochemistry
1. Industrial methylated spirits T100 (IMS): (Lennox Chemicals, Ireland).
2. 3-Aminopropyltriethoxy-silane (APES): (Sigma, USA; cat. no. A-3648).
3. Microscope slides (frosted), 1.2 to 1.5 mm thick: (Lames-Porte Objet, Germany; cat. no. 3492).
4. Tissue-Tek®, OCT™ compound (OCT): (Sakura, The Netherlands; cat. no. 4583).
5. Tissue-Tek, Cryomold® biopsy (10 × 10 × 5 mm): (Sakura; cat. no. 4565).

2.2. Immunohistochemistry: Three-Step Immunoperoxidase Staining of T-Cells
1. Mouse antihuman CD3, clone T3-4B5: (DAKO, UK; Cat. No. M0756).
2. Mouse IgG_1 negative control: (DAKO; cat. no. X0931).
3. Peroxidase VECTASTAIN® Elite ABC kit, mouse IgG: (Vector Laboratories, USA, cat. no. PK-6102).

4. DAKO pen: (DAKO; cat. no. S 2002).
5. Phosphate buffered saline (PBS): (Sigma, USA; cat. no. P-3813).
6. 3,3'-Diaminobenzidine (DAB): (Sigma; cat. no. D-9015).
7. Mayers haemalum: (BDH, England; cat. no. 350604T).
8. Hydrogen peroxide, 30% solution: (BDH; cat. no. 285194F).
9. Xylene: (BDH; cat. no. 102936H).
10. Acetone: (BDH; cat. no. 100034Q).
11. Industrial methylated spirits T100 (IMS): (Lennox Chemicals, Ireland).
12. DPX mountant for microscopy: (BDH; cat. no. 360294H) .

2.3. Tissue Controls (see Note 1)

1. Positive control tissue: tonsil.
2. Negative control tissue: normal un-inflamed synovial tissue.

2.4. Dual-Labeled Immunofluorescence Staining

1. Goat anti-human CRH-R1, C-20: (Santa Cruz, USA; cat. no. sc-1757).
2. Mouse antihuman mast cell tryptase (Accurate, USA; cat. no. YSRTMCA1438).
3. Mouse IgG_1 negative control: (DAKO; cat. no. X0931).
4. Normal goat IgG: (Santa Cruz Biotechnology; cat. no. sc2028).
5. FITC-conjugated goat antimouse IgG1: (Southern Biotechnology, USA; cat. no. 3050-02).
6. Cy^3 fluorochrome-conjugated mouse antibiotin antibody: (Sigma; cat. no. C5585).
7. Biotinylated rabbit anti-goat IgG: (Vector Laboratories, USA; cat. no. BA-5000).
8. Normal rabbit serum: (Vector Laboratories; cat. no. S-5000).
9. Normal goat serum: (Vector Laboratories, cat. no. S-1000).
10. DAKO pen: (DAKO; cat. no. S 2002).
11. PBS: (Sigma; cat. no. P-3813).
12. Fluorescent mounting media; (DAKO; cat. no. S 3023).
13. Normal human serum (donor in laboratory).
14. Paraformaldehyde: (BDH; cat. no. 294474L).

3. Methods

3.1. Preparation of Reagents for Immunohistochemistry

1. Prepare the DAB solution by adding 16 mL distilled water into the DAB isopac (100 mg). Store in 0.5 mL aliquots at −20°C, in a darkened container. Immediately prior to use thaw the 0.5 mL DAB aliquot and make up to 5 mL with PBS. Add 5 µL of 3% hydrogen peroxide just before use. DAB is a possible carcinogen.
2. Three percent hydrogen peroxide: 1/10 dilution of the 30% hydrogen peroxide solution prepared in distilled water.
3. Prepare the following reagents supplied in Vector kits as per manufacturers instructions:
   a. Normal horse serum (blocking serum): three drops (50 µL = one drop) in 10 mL.
   b. PBS.

c. Biotinylated secondary antibody (horse antimouse IgG): one drop in 10 mL PBS.
d. ABC solution: two drops of bottle A and B in 5 mL PBS. Prepare at least 30 min before use.
4. Preparation of primary (*see* **Note 2**) and negative control (*see* **Note 3**) antibodies.
    a. Anti-CD3 Ig concentration: 210 mg/L. The required dilution of primary antibody with PBS is 1/25, which results in a final Ig concentration of 8.4 mg/L.
    b. Control Ig concentration: 100 mg/L. A 1/11.9 dilution of the control antibody with PBS will give same Ig concentration as the primary antibody.

3.2 Preparation of Reagents for Dual-Labeled Immunofluorescence Staining

1. One percent paraformaldehyde prepared with distilled water.
2. Ten percent normal human serum prepared in PBS.
3. Preparation of normal serum; 150 µL of serum in 10 mL PBS.
4. Antibody preparation:
    a. Goat anti-CRH-R1; 1/10 dilution in the 10% normal human serum.
    b. Mouse antimast cell tryptase; 1/10 dilution in 10% normal human serum.
    c. Control mouse and goat immunoglobulins; diluted to the same Ig concentration as the relevant primary antibodies in 10% normal human serum.
    d. FITC-conjugated goat antimouse IgG1; 1/50 dilution in PBS.
    e. Cy^3-conjugated antibiotin antibody; 1/100 dilution in PBS.
    f. Biotinylated rabbit anti-goat IgG; 1/500 dilution in PBS.

3.3. Preparation of Tissue Slides for Immunohistochemistry

3.3.1. Preparation of 3-Aminopropyltriethoxy-Silane Coated Slides (see *Note 4*)

1. Place clean slides in a slide rack.
2. Immerse slides twice in a bath of IMS.
3. Then immerse five times in a bath of 2% APES prepared in IMS.
4. Immerse the slides twice in a bath of IMS.
5. Finally, immerse once in a bath of distilled water.
6. Tap the slides several times on tissue paper to remove excess liquid and dry slides overnight at 60°C. Discard any slides that have any white precipitate from the APES solution.

3.3.2. Preparation and Cryostat Sectioning of Synovial Biopsies

1. At arthroscopy, synovial biopsies are placed on a sterile saline moistened gauze. The biopsy is placed in a cryomould and OCT mounting media is added and immediately snap frozen in liquid nitrogen (*see* **Note 5**). The sample is removed from the cryomould, placed in a labeled cryovial and stored long term in liquid nitrogen.
2. When preparing to section the synovial biopsy remove it from liquid nitrogen and place in a refrigerated microtome –20°C for at least 20 min before use. The

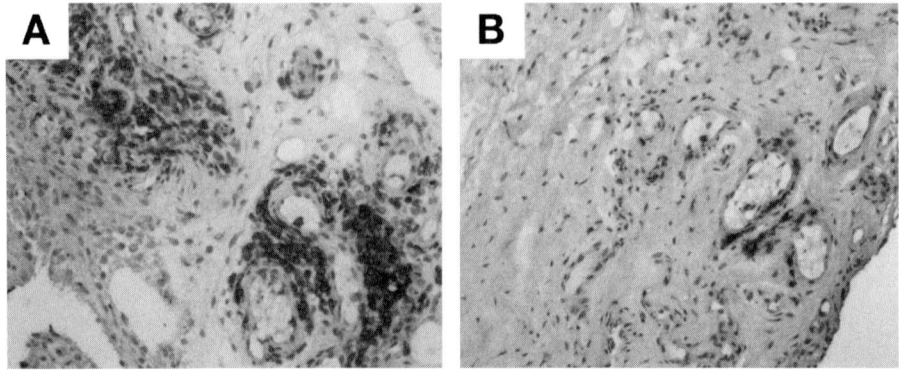

Fig. 4. Immunohistochemical staining of T cells (CD3$^+$) in synovial tissue from (**A**) a patient with active psoriatic arthritis and (**B**) the same patient following treatment with methotrexate, illustrating a dramatic decrease in T-cells after treatment.

biopsy temperature needs to increase because if you try to cut directly from liquid nitrogen the OCT may shatter, ruining the biopsy. Sections are cut at 7 µm and mounted on APES coated slides (*see* **Note 6**).

3. Performing a Haematoxylin and Eosin (H&E) stain at this stage and looking at the tissue microscopically will confirm that the tissue is representative of synovial tissue. For the analysis of synovial tissue by immunohistochemistry only tissue with an intact lining layer may be used for analysis. After cutting the tissue allow the slides to dry at room temperature overnight, then wrap individually in aluminum foil and store at –80°C.

3.4. Immunohistochemistry: Three-step Immunoperoxidase Staining of T Cells

Here we have described the staining of T cells (CD3$^+$) in synovial tissue (*see* **Fig. 4**).

1. Remove the slides from –80°C and allow to thaw at room temperature for 20 min before use.
2. Fix sections in acetone for 10 min and allow to air dry.
3. Encircle each section with a DAKO wax pen. If there is more than one section on the slide being stained by a panel of antibodies, encircling the tissue on a glass slide with a DAKO wax pen ensures that there will be no mixing of solutions.
4. Carry out the procedure in a sealed moistened container. The use of a sealed moistened container ensures that the small quantities of reagent placed on the tissue section will not evaporate during an incubation period.
5. Block sections with normal horse serum (*see* **Note 7**) for 15 min. Wash sections in PBS and drain off PBS (*see* **Note 8**).

6. Add the diluted anti-CD3 primary antibody and IgG$_1$ negative control to relevant sections and incubate for 1 h at room temperature. Wash in PBS and incubate with PBS for 5 min.
7. Drain off the PBS, add the biotinylated secondary antibody to sections, and incubate for 30 min. Wash sections with PBS and incubate with PBS for 5 min.
8. Drain off the PBS, and add the ABC solution for 30 min. Wash sections in PBS and incubate with PBS for 5 min.
9. Drain off the PBS, and treat sections with 3% hydrogen peroxide for 7 min. Wash in distilled water, incubate for 1 min with water and then incubate with PBS for 5 min (*see* **Note 9**).
10. Drain off the PBS, and add the DAB solution to the tissue sections for 12 min. Immerse the slides in tap water for 1 min to stop the chromogenic reaction.
11. Counterstain slides in Mayers Haemalum for 10 to 30 s and clear in running tap water for 1 to 3 min (*see* **Note 10**).
12. Dehydrate slides in two baths of IMS for 10 min each and clear in two baths of xylene for 10 min each. Mount slides in DPX (*see* **Note 11**).

3.5. Dual-Labeled Immunofluorescence Staining

1. The procedure is carried out at room temperature. When using fluorescence labeled antibodies all work should be carried out in darkened containers.
2. Synovial tissue sections are prepared using the same procedure as that used for immunohistochemistry.
3. Remove the slides from –80°C and allow to thaw for 20 min before use.
4. Fix sections in 1% paraformaldehyde for 20 min, wash sections with PBS and incubate with PBS for 15 min.
5. Block sections with normal rabbit serum for 1 h. Wash sections in PBS and incubate with PBS for 15 min.
6. Add the diluted anti-CRH-R1 antibody and normal goat IgG to relevant sections for 1 h.
7. Wash sections in a trough of PBS with agitation for 30 min.
8. Add the biotinylated rabbit antigoat antibody to the sections for 30 min. Wash sections in PBS and incubate with PBS for 5 min.
9. Block sections with normal goat serum for 1 h. Wash sections in PBS and incubate with PBS for 15 min.
10. Add the diluted antimast cell tryptase and normal mouse IgG to relevant sections for 1 h.
11. Wash sections in a trough of PBS with agitation for 30 min.
12. Add the diluted FITC-conjugated goat antimouse IgG1 to sections for 30 min.
13. Wash sections in a trough of PBS with agitation for 30 min.
14. Add the diluted Cy^3 fluorochrome conjugated mouse antibiotin antibody to sections for 30 min.
15. Wash sections in a trough of PBS with agitation for 30 min.
16. Drain slides by gently tapping on tissue paper and mount in fluorescent mounting medium. Store in the dark at 4°C.

3.6. Manual Quantification of Immunohistochemical Staining

The extent of synovial inflammation (i.e., synoviocyte hyperplasia, mononuclear cell accumulation, and increased vascularity) can be measured microscopically. The synovial tissue sublining is measured separately from the lining layer. Both quantitative and semiquantitative methods may be used for the analysis of immunohistochemical staining in synovial tissue. While one observer is required for quantitative analysis, the agreement of two observers is essential for semiquantitive anlysis.

3.6.1. Sublining Analysis

1. Quantitative analysis employs the use of an objective gracicule (approx 1 cm^2), which is calibrated against a stage micrometer. One gracicule area is equal to one high power field (hpf). The quantitative measurement involves counting the number of positive cells present in a hpf using the ×400 magnification on a binocular microscope. Greater than 90% of the hpf should be occupied by tissue. A study carried out in our laboratory showed that the 17 randomly picked hpf's from 3 biopsies/patient correlated significantly with the measured results obtained from the entire tissue section *(8)*.
2. Semiquantitative analysis of synovial tissue is based on a 5-point scale (0 = minimal infiltration and 4 = infiltration of numerous inflammatory cells). The evaluation of each cell type and cytokine requires a different sensitivity level *(9)*.

3.6.2. Lining Layer Analysis

1. Lining layer thickness is quantified at ×400 magnification. Lining layer thickness is measured as the number of cells in depth at three designated points (at the midpoint of each hpf, and 0.3 hpf to the left and to the right of the midpoint) in each hpf along the entire length of visible lining layer of the section. The lining layer depth is derived from a mean of all these measurements *(8)*.
2. Quantitative analysis of positive cells present in the lining layer is carried out by counting both the number of positive and negative stained cells in the lining layer. The result is presented as a percentage positive number.
3. Semiquantitative analysis of positive cells: estimate the percentage positive cells present or use a 5-point scale *(9)*.

Semiquantitative analysis is a more rapid means of analyzing specimens if large studies are being undertaken. Semiquantitative analysis has also been shown to be a sensitive and reproducible means of assessing differences between patient groups and pre- and post-therapeutic intervention *(5)*. However, quantitative analysis is more sensitive to subtle changes in synovium after treatment *(8)*. Ideally, the more biopsies that are analyzed per patient the more definitive the result. However, the analysis can be very time consuming and the time available for analysis must be taken into account. Taking biopsies from a number of different regions (e.g., suprapatellar pouch or adjacent to the

Immunohistochemistry of the Inflamed Synovium

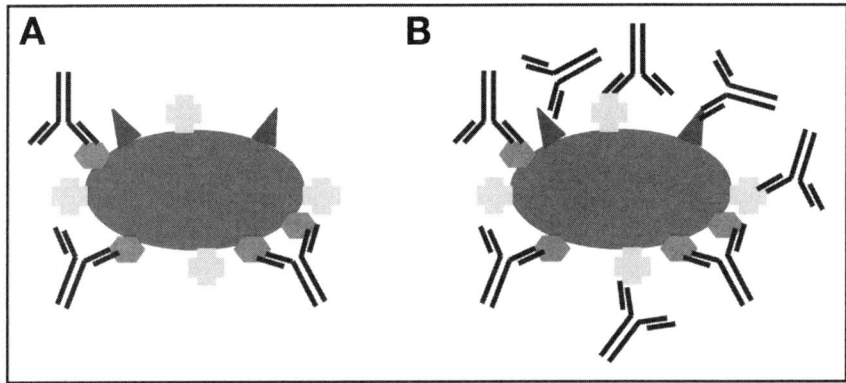

Fig. 5. Schematic diagram of (**A**) monoclonal antibodies reacting with a specific epitope on an antigen and (**B**) polyclonal antibodies reacting with various epitopes on an antigen.

cartilage-pannus junction) within the knee joint gives a more comprehensive impression of the inflammation occurring throughout the joint. With the advances in digital imaging systems (DIS) the above methods are gradually being eclipsed. DIS has been shown to correlate with both the quantitative and semiquantitative methods described above *(10)*. DIS requires less time to quantify large numbers of samples without the problems of inter- and intra-observer variability *(11)*. When setting up a DIS initially, it needs to be validated against a manual quantification method.

4. Notes

1. Tissues, which are either positive or negative for the antigen of interest, must be used as controls. All tissue used as controls must be processed and stained identically to the test tissue.
2. If one is interested in analyzing a particular protein by immunohistochemistry and are unsure about which antibody to choose an initial search of the literature and well know companies like DAKO, R&D Systems, and Santa Cruz should be undertaken. Mouse monoclonal IgG_1 antibodies would be the ideal antibody to choose because they tend to give extremely clean staining in synovial tissue. If a particular antigen expression is expected to be low then consider using a polyclonal antibody. A polyconal antibody, which is a mixture of antibodies raised against several epitopes on the antigen, may give more prominent detection when compared to a monoclonal (*see* **Fig. 5**). After deciding on the appropriate antibody follow method 3.4. with the appropriate Vectrastain Elite ABC Kit. The antibody datasheet will recommend an antibody working concentration. Work around the range provided (i.e., if a 1/200 dilution is recommended then try a 1/100, 1/200, 1/400 dilution range). If no dilution is recommended, use a range

of doubling dilutions that will give an immunoglobulin concentration between 1 and 10 μg/mL. Pick the dilution that gives the best specific staining without background for your antibody.
3. Antibody controls are necessary in immunohistochemistry to ensure that a protocol is followed correctly and to establish if there is any day-to-day or worker-to-worker variation. All controls and test samples should be treated in exactly the same way.

    It is necessary to test the primary antibody in immunohistochemistry so as to obtain the optimal antibody concentration on positive tissue and then against additional samples with known or unknown antigen. To establish the specificity of a primary antibody, the best way to obtain a valid negative control is by affinity absorption of the primary antibody with a highly purified antigen. Sometimes the purified antigen is either unavailable or it can be quite expensive. However some companies (e.g., Santa Cruz) produce blocking peptides that may be used to ensure the specificity of the staining of the primary antibody in a specific tissue. Preincubate the antibody with a 10X concentration of the blocking peptide before performing an immunohistochemical run. If there is any staining in the preabsorbed control, this implies that the antibody is binding nonspecifically to a component within the tissue. Generally nonimmune serum or its immunoglobulin fraction from the same species as the primary antibody is used as a negative control. The negative control should be used at the same Ig concentration as the Ig working concentration of the primary antibody. For monoclonal antibodies it is best to use a negative control that is the same Ig isotype as the primary control. An example of this is the CD3 antibody used in **Subheading 3.4.**, which is a mouse IgG_1 κ monoclonal antibody. A mouse IgG_1 κ antibody directed against *Aspergillus niger* glucose oxidase, which is not present nor inducible in mammalian tissues is used as a negative control.
4. To improve the adherence of the tissue to the microscope slides the use of an adhesive coating or subbing agent, such as poly-L-lysine, charging or silanization is recommended. Preteated slides may be purchased which have a positive charge that attracts the negatively charged tissue (e.g., poly-L-lysine or superfrost plus microscope slides [BDH]). The advantages of these slides are that they are nuclease free and therefore can be used for *in situ* hybridization. There is also a reduction in background staining which occurs with hematoxylin staining of tissue mounted on protein-coated slides. Whereas purchased slides may be more convenient but they can be expensive. Slides coated with APES give very good tissue adhesion with minimal background staining.
5. Biopsies may be placed singly into the cryomould but if a number of biopsies are to be analyzed per patient then they can be placed together (touching) in the cryomould. This helps to reduce the time required for cutting and staining the tissue. When adding the OCT mounting media to the cryomould ensure that there are no bubbles in the media, as this may cause difficulties when cutting the biopsy.
6. Cryostat sections may be cut at a range of 4 to 10 μm; paraffin sections can be cut as thin as 2 μm. A monolayer of cells is required for the quantitative analysis of

the tissue staining. Numbering the slides in sequence (serial sections) as the tissue is cut from the block may help identify which cell is expressing a particular antigen. This can then be further confirmed by dual-labeled immunofluorescence staining (*see* **Subheading 3.5.**).
7. The blocking serum used must be from the same species from which the secondary antibody is raised. This ensures that the secondary antibody will only bind to the primary antibody and not to any related antigen it may recognize in the test tissue. This serum may be washed off the tissue or drained by tapping the slide on the tissue paper.
8. Because cryostat sections are more delicate than paraffin sections, it is essential that great care be taken during the washing steps to ensure minimal tissue damage. Wash bottles may be used in manual procedures, but ensure that the spray is directed above the section (not directly at the section) and the wash reagent is allowed to flow gently over the tissue. Background and nonspecific binding of antibody may occur in damaged tissue. Tapping the edge of the slides on tissue paper removes any excess liquid in between steps, but at the same time make sure that the sections remain moist throughout. If sections dry out between steps, the staining obtained may have a brown background effect.
9. When using HRP detection method it is important that any endogenous peroxidase activity is blocked. Peroxidase activity results in the decomposition of hydrogen peroxide and is a common property of all hemoproteins such as haemoglobin, myoglobin, cytochrome and catalases. Quenching of endogenous peroxidase activity is accomplished by adding a dilute concentration of hydrogen peroxide to the tissue. Where there is peroxidase activity in tissue, a bubbling effect will be seen on the tissue surface. The addition of water stops the reaction, and the tissue section is allowed to come back to a neutral pH by incubating with PBS. The timing of the addition of the hydrogen peroxide step may vary between laboratories, for example addition before the primary antibody. The caustic nature of hydrogen peroxide may cause damage to the antigen of interest therefore addition at a later stage would avoid this concern.
10. The most important point to note when counterstaining tissue is that the main aim is to define the overall structure but not to stain so strongly that the detection of positive immunohistochemical (brown) staining is masked. Haematoxylin binds strongly to the acid nuclei and gives the cytoplasm a pale blue colour. The nuclei should be defined without the tissue being overstained.
11. The alcohol immediately preceding the first xylene must have no traces of water; if sections are not dehydrated correctly you may obtain a very hazy undefined tissue structure. When mounting the section with DPX, ensure there are no bubbles—remove any by gently pressing on the coverslip. The xylene in the DPX will evaporate overtime, any bubbles present will become larger and invariably they will end up obscuring the tissue section thus hindering future analysis.

5. Additional Immunohistochemistry Troubleshooting Notes

Where inappropriate or unexpected staining is obtained. It must be determined whether this is due to the antibody, tissue, technique, environment etc. The experiment should be examined in sequential manner, changing one step at a time.

5.1. Nonspecific or Background Staining

5.1.1. Primary Antibody

Optimize the concentration of the primary antibody to be used that will give distinct, specific staining without background.

1. The primary antibody may cross react with other tissue epitopes or bind nonspecifically. The addition of normal serum, BSA, nonfat dry milk, gelatin, or detergent may be added to the primary antibody diluent.
2. Nonspecific staining may be caused by ionic interactions of the primary antibody and the tissue. Increasing the sodium chloride concentration (0.15–0.6 M) of the primary antibody diluent may overcome this.
3. If the section shows small, undefined, punctuated staining this may result from the denatured precipitated immunoglobulin. This may occur where antibodies are subjected to repeated freeze-thaw cycles. The antibody may be centrifuged and the supernatant used. This centrifuged antibody may be useful for identifying a particular antigen in tissue however it should not be used if the quantification of staining is to be carried out. Where antibodies are to be stored frozen, aliquot them and do not subject them to freeze-thaw cycles.

5.1.2. Washing

Increasing the wash time after the primary antibody is removed from 5 min to 20 to 30 min with/without agitation can reduce background staining.

5.1.3. Endogenous Enzyme

1. The addition of the substrate to the tissue alone and the development of colour will indicate if the tissue contains endogenous enzyme. Use the same times and conditions as used in the procedure that is appropriate for the antibody staining. Colour that develops after this may not be seen within the development time for the specific staining.
2. Peroxidase: The method for overcoming this is described earlier. The use of a hydrogen peroxide/methanol solution is more suitable for paraffin embedded sections.

5.1.4. The ABC Reagent May Bind to Tissues for Three Main Reasons

1. Endogenous protein-bound biotin: add an avidin/biotin block to tissue.
2. Endogenous lectins: addition 0.2 M α-methyl mannoside to the ABC diluent.

3. Ionic interactions: prepare the ABC reagent in buffer containing 0.5 M sodium chloride.

All the above causes of ABC reagent binding to tissue can be overcome by using the avidin/biotin block (Vector Laboratories) after the normal serum blocking step. This has not been a problem for synovial tissue, however some control tissues eg. liver, may express endogenous biotin.

5.1.5. Biotinylated Secondary Antibody

Crossreactivity between the secondary antibody and endogenous immunoglobulins and other tissue proteins: add 2% or more normal serum from tissue species to the biotinylated secondary antibody diluent, and/or reduce the concentration of the biotinylated secondary antibody.

1. Nonspecific binding: add bovine serum albumin, nonfat dry milk, gelatin or 0.1% detergent (e.g., Tween-20).
2. Wrong species of blocking serum used: use serum from the species in which the biotinylated antibody was produced.

5.1.6. Problems Associated With Chromogen

1. Incubate the chromogen for a shorter time as DAB can develop very quickly.
2. If the chromogen has undissolved particles that have become associated with the tissue, filter the DAB before use.
   Background may occur where the tissue was allowed to dry during the procedure, ensure tissue is moist through out.

5.1.7. Fixative

Using a different fixative may improve staining with a particular antibody. For example fixing the tissue in 1% paraformaldehye solution for 20 min instead of acetone. After incubation slides should be washed with PBS and incubated for 10 min with PBS. Sections must not be allowed to dry.

5.2. Weak or Absent Staining

5.2.1. Procedure Check

1. Ensure that the procedure was followed correctly (e.g., no steps are omitted).
2. Check that correct reagents were used.
3. Check all dilutions.
4. Where a primary antibody is suddenly giving no staining in known positive tissue a good way to check the procedure is to stain with a similar antibody in the same experiment (e.g., mouse anti-CD3 not working then run an optimized mouse anti-CD4). This will inform you as to whether there is something wrong with the procedure (no staining with either antibody) or whether there is something wrong with your primary antibody (staining just occurring with the mouse anti-CD4).
5. Check the expiration dates and ensure correct storage of all reagents/antibodies

5.2.2. Primary Antibody

1. Check the concentration and dilution of primary antibody.
2. Check the pH of the diluent (PBS pH 7.4).
3. Where the primary antibody is being used for the first time, and there is no staining present in the positive control tissue it may be necessary to change the length of time or temperature the primary antibody is incubated on the tissue. For example, in our laboratory the staining with mouse anti-human tissue factor (American Diagnostics) required an overnight incubation at 4°C.
4. Ensure that the diluent does not contain the antigen of interest. If the primary antibody recognizes an antigen that is present in biological fluids, it may bind the antigen in solution rather than the tissue section. This may occur where normal serum, fetal bovine serum or nonfat dry milk is used in the diluent.

5.2.3. Secondary Antibody

1. Ensure the correct biotinylated secondary antibody was used.
2. Check the concentration and dilution of the secondary antibody.
3. Source of neutralizing antibodies (e.g., biotinylated antimouse IgG should not be diluted in mouse serum). The immunoglobulins in mouse serum will bind the biotinylated secondary antibody and prevent it binding to the primary antibody.
4. To ensure that the enzyme/substrate reaction is working add 1 to 2 drops of the ABC reagent to one ml of the DAB/hydrogen peroxide solution. The color of the mixture should change to brown in a few seconds.
5. Check specimen storage, compare the staining of an unknown sample with that of a known positive tissue in the same run.
6. The animal from which the blocking serum was obtained may have developed antibodies to the antigen of interest. If so, the antibodies may bind to the antigen and prevent the primary antibody binding.

References

1. Mulherin, D., FitzGerald, O., and Bresnihan, B. (1996) Synovial tissue macrophage populations and arthicular damage in rheumatoid arthritis. *Arth. Rheum.* **39(1),** 115–124.
2. Yanni, G., Whelan, A., Feighery, C., and Bresnihan, B. (1994) Synovial tissue macrophages and joint erosions in rheumatoid arthritis. *Ann. Rheum. Dis.* **53,** 39–44.
3. Cunnane, G., FitzGerald, O., Hummel, K. M., Youssef, P. P., Gay, R. E., Gay, S., and Bresnihan, B. (2001) Synovial tissue protease gene expression and joint erosions in early rheumatoid arthritis. *Arth. Rheum.* **44(8),** 1744–1753.
4. Dolhain, R. J. E. M., Tak, P. P., Dijkmans, B. A. C., De Kuiper, P., Breedveld, F. C., and Miltenburg, A. M. M. (1998) Methotrexate reduces inflammatory cell numbers, expression of monokines and of adhesion molecules in synovial tissue of patients with rheumatoid arthritis. *Br. J. Rheumatol.* **37,** 502–508.
5. Kraan, M. C., Reece, R. J., Barg, E. C., et al. (2000) Modulation of inflammation and metalloproteinase expression in synovial tissue by leflunomide and methotrexate in patients with active rheumatoid arthritis. *Arth. Rheum.* **43(8),** 1820–1830.

6. Cunnane, G., Madigan, A., Murphy, E., FitzGerald, O., and Bresnihan, B. (2001) The effects of treatment with interleukin-1 receptor antagonist on the inflamed synovial membrane in rheumatoid arthritis. *Rheumatology* **40,** 62–69.
7. McEvoy, A. N., Bresnihan, B., FitzGerald, O., and Murphy, E. P. (2001) Corticotropin-releasing hormone signaling in synovial tissue from patients with early inflammatory arthritis is mediated by the type 1α corticotropin-releasing hormone receptor. Arth. Rheum. **44(8),** 1761–1767.
8. Bresnihan, B., Cunnane, G., Youssef, P., Yanni, G., FitzGerald, O., and Mulherin, D. (1998) Microscopic measurement of synovial membrane inflammation in rheumatoid arthritis: proposals for the evaluation of tissue samples by quantitative analysis. *Br. J. Rheumatol.* **37,** 636–642.
9. Tak, P. T., Smeets, T. J. M., Daha, M. R., et al. (1997) Analysis of the synovial cell infiltrate in early rheumatoid synovial tissue in relation to local disease activity. Arth. Rheum. **40(2),** 217–225.
10. Kraan, M. C., Haringman, J. J., Ahern, M. J., Breedveld, F. C., Smith, M. D., and Tak, P. P. (2000) Quantification of the cell infiltrate in synovial tissue by digital image analysis. *Rheumatology* **39,** 43–49.
11. Youssef, P. P., Triantafillou, S., Parker, A., et al. (1997) Variability in cytokine and cell adhesion molecule staining in arthroscopic synovial biopsies: quantification using color video image analysis. *J. Rheumatol.* **24(12),** 2291–2298.

4

In Situ Hybridization of Synovial Tissue

Stefan Kuchen, Christian A. Seemayer, Michel Neidhart, Renate E. Gay, and Steffen Gay

Summary

The chapter focuses on the detection of specific mRNA by *in situ* hybridization (ISH) in synovial tissue specimens. This technique is widely applied, reliable, specific, and sensitive, because even small quantities of mRNA can be detected. Presented here contemporary protocols for ISH using a combined nonradioactive immunohistochemical detection system.

In overview, the following steps have to be covered to perform ISH. (1) mRNA probes (sense and antisense) are generated by in vitro transcription of cDNA utilizing digoxigenin-labeled UTP nucleotides, (2) fixed tissue sections are digested with trypsin and treated consecutively with prehybridization solutions, (3) hybridization with labeled riboprobes takes place at 50°C overnight in a humid chamber, (4) unbound riboprobe is removed by incubation with RNase A and additional washing with buffers, (5) stringent washing steps are performed with solutions of different sodium dodecyl sulfate, SSC, and formamide concentrations, (6) digoxigenin-labeled probes are detected immunohistochemically using antidigoxigenin antibodies linked with alkaline phosphatase and NBT/BCIP as detection system.

Key Words: Synovial tissue; rheumatoid arthritis; messenger RNA; *in situ* hybridization; probe.

1. Introduction

The technique of *in situ* hybridization (ISH) was developed in the 1970s. In the beginning it was used to localize specific DNA sequences on chromosomes *(1,2)*. Subsequently, this technique was modified for the detection of abundantly expressed cellular or viral RNA sequences (*see* **Fig. 1**). Within the last 15 yr the sensitivity of ISH has improved *(1–3)*. Accordingly, it is now possible to detect expression of specific messenger RNA (mRNA) in cells or tissues even in low copy numbers *(4)*. In contrast with Northern blotting (using extracted RNA for expression analysis), ISH of RNA allows localization of the

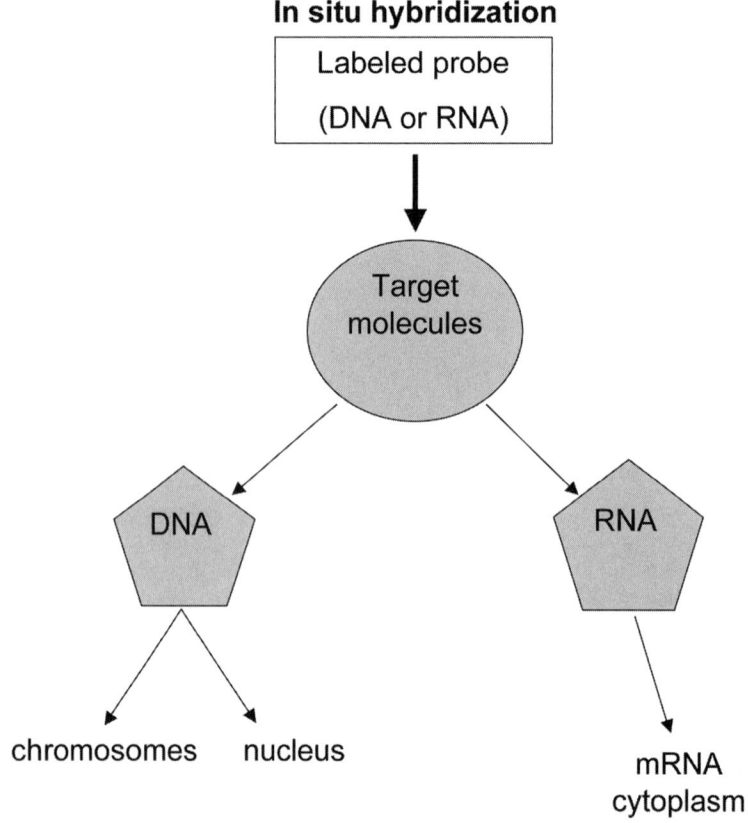

Fig. 1. Overview on the different applications of ISH in tissues. Depending on the method the probes can be either of DNA or RNA origin. Similarly, as target molecules can serve DNA, either on chromosomes or interphase DNA within the nucleus. Also RNA molecules can be addressed by ISH and specifically mRNA can be investigated that is expressed in the cytoplasm.

signals in the cytosol of a cell and thereby to characterize the expression pattern and the cell type *(5)*. Along this line, additional double labeling with cell specific markers using immunohistochemistry of the same section or of parallel sections helps to clarify unequivocally the expressing cell type *(7,8)*.

The broad range of application of ISH during the last twenty years resulted in a large number of modified protocols *(1–6)*. The present protocol focuses on mRNA ISH in synovial tissues. Therefore, the abbreviation ISH represents only *in situ* hybridization of RNA molecules. An overview of all necessary steps for mRNA ISH is given in **Fig. 2**.

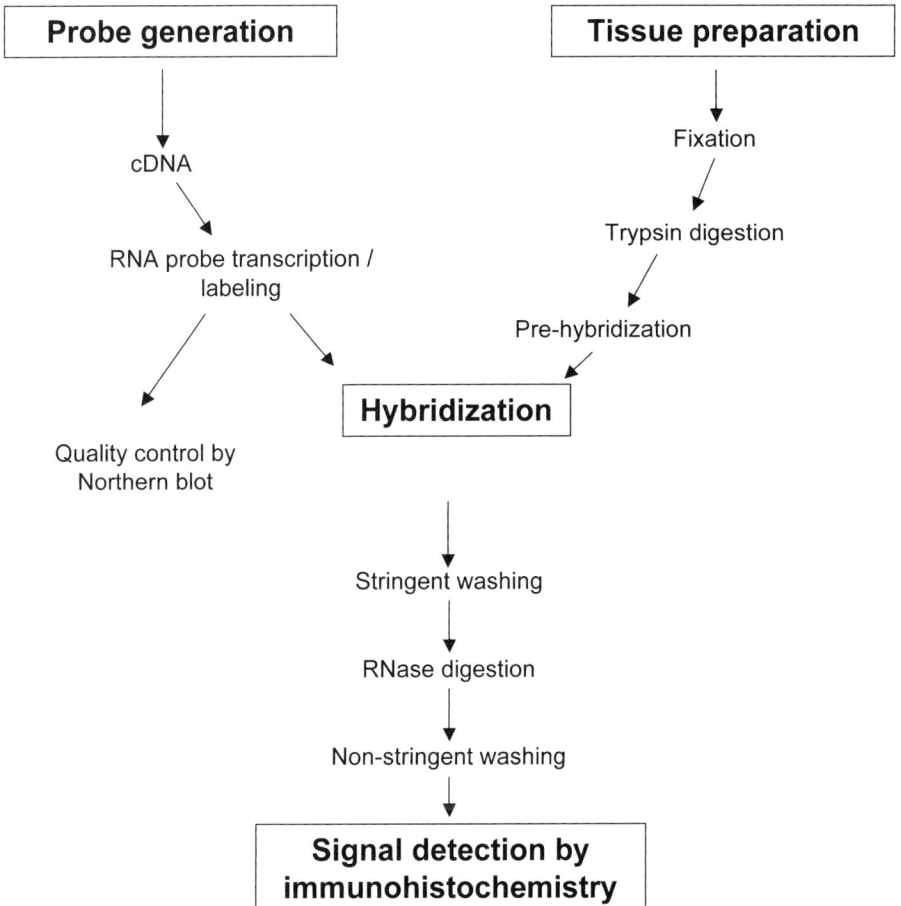

Fig. 2. Out-line of the working procedures to perform mRNA ISH in tissues. The RNA probe is generated by transcription from cDNA with labeled nucleotides. Successful transcription should be checked by Northern blot. Tissue specimens are properly fixed, digested with trypsin and treated with pre-hybridization solutions. Finally, hybridization takes place overnight and the next day unbound probe is removed by RNase digestion and washing. Specifically bound probes within the cytoplasm are detected by immunohistochemistry.

ISH can be performed on either paraffin embedded tissues or frozen tissue sections. Because the sensitivity of ISH depends greatly on the degree of RNA preservation, optimal procedures for tissue extraction and fixation are required. To work under RNase free working conditions until hybridization has been completed is a condition *sine qua non* to succeed. To increase the accessibility of mRNA transcripts for hybridization, the tissue is pretreated carefully to remove RNA binding proteins. Hybridization is performed with RNA probes of a length ranging from 30 to 300 nucleotides which encode the complementary sequence of the targeted gene transcript. The probe can be either labeled with ^{35}S or ^{32}P marked nucleotides for radioactive ISH, or alternatively nucleotides are coupled with digoxigenin or biotin for nonradioactive ISH. After hybridization has taken place, it is important to remove unbound and nonspecifically bound probes by digestion with RNase and stringent washing procedures. The signal detection of hybridized probes depends on the type of labeling of the probe. In contrast with the direct detection of radioactive labeled probes by film autoradiography or emulsion autoradiography, non-radioactive probes can only be visualized indirectly by fluorescence or immunohistochemical detection systems.

The following detailed protocol concentrates on nonradioactive ISH being optimized for detection of mRNA in paraffin embedded or frozen human synovial tissues (see **Fig. 2**). Digoxigenin-labeled RNA probes are detected immunologically by anti-digoxigenin-antibodies and visualized by a color reaction mediated by antibody-coupled alkaline phosphatase. In our hands, mRNA ISH of synovial tissues can be completed in two consecutive d including hybridization overnight. On day 1 (prehybridization) all steps are performed under RNase free conditions. On day 2 (post-hybridization) RNase is used to digest unspecific bound probes. To avoid contamination with RNase, the working place of the first and second day should be separated. All reagents and materials of day 1 including pipettes and glassware have to be RNase free.

2. Materials
2.1. Equipment
1. Slides superfrost.
2. Staining dishes/Coplin jars.
3. Slide racks.
4. Airtight slide box.
5. Water bath.
6. Incubator.
7. Heating block (up to 100°C).
8. Additional reagents and equipment for fixation and sectioning of tissues, probe preparation, staining and mounting.

2.2. Chemicals and Reagents
1. Anti-Digoxigenin-AP-Fab fragments: liquid, store at 2 to 8°C (Roche Biochemicals; 200 µL, 150 U).

2. BCIP (X-phosphate/5-Bromo-4-chloro-3indolyl-phosphate): liquid, store at −20°C in the dark (light protection); (Roche Biochemicals; 3 mL, 150 mg), harmful (*see* product data sheet for details).
3. Herring sperm DNA (10 mg/mL): liquid (viscous), store at −20°C (GibcoBRL; 1 mL).
4. Levamisol (inhibitor of endogenous alkaline phosphatase): liquid, store at 2 to 8°C (DAKO; 15 mL).
5. NBT (4-Nitro blue tetrazolium chloride): liquid, store in the dark at −20°C; (Roche Biochemicals; 3 mL, 300 mg), harmful (*see* product data sheet for details).
6. RNase A (from bovine pancreas): powder, store at 2 to 8°C (Roche Biochemicals; 25 mg).
7. Trypsin 1:125 (from porcine pancreas): powder, store at 2 to 8°C (Sigma; 25g).
8. Yeast-t-RNA (Ribonucleic acid transfer): liquid (viscous), store at −20°C (Sigma, 1 mL, 10.2 mg/mL).

2.3. Solutions

1. Acetic anhydride 0.25% in triethanolamin HCl solution: dilute 30 µL acetic anhydrid in 30 mL triethanolamine-HCl immediately before use. Vortex well. Cave: half-life of acetic anhydride in aqueous solutions is very short.
2. Anti-digoxigenin-alkaline phosphatase-antibody-solution: 1 mL Buffer 1, 10 µL horse serum, 2 µL anti-digoxigenin-AP-Fab fragments.
3. Buffer 1 (1000 mL): 100 mM TrisHCl pH 7.5, (100 mL 1 M Tris-HCl, pH 7.5). 150 mM NaCl (30.3 mL of 5M NaCl stock solution). Make up to 1 L ddH_2O, adjust pH to 7.6.
4. Buffer 2 (1000 mL): 100 mM Tris-HCl pH 7.5 (100 mL 1 M Tris-HCl, pH 7.5). 100 mM NaCl (20 mL of 5M NaCl stock solution). 5 mM $MgCl_2$ (5 mL 1 M $MgCl_2$). Fill up to 900 mL with ddH_2O. Adjust pH to 9.5 and make up to 1000 mL ddH_2O.
5. Diethylpyrocarbonate (DEPC) treated water: add 1 mL DEPC to 1000 mL ddH_2O and stir for at least 4 h in a hood. Incubate overnight at 37°C in open bottles (do not close the bottles!). Autoclave at 120°C, 1.21 bar, for 30 min).
6. Denhardt solution 50X: 5 g Ficoll 400, 5 g polyvinylpyrrolidone, 5 g bovine serum albumin (BSA), H_2O to 500 mL. Filter sterilize and store at −20°C in 25-mL aliquots.
7. Dextran sulfate 50% in 20X SSC (50 mL): Prepare 25 g dextran sulfate in a 50-mL tube and fill up to 50 mL with 20X SSC (need hours to days to dissolve!).
8. 10% Ehtylene diamine tetraacetic acid (EDTA) solution, pH 7.2 (only when decalcification is needed): 20 g EDTA in 40 mL distilled water. Add slowly 10 mL NaOH (40%) and mix. Add an additional 160 mL distilled water and adjust the pH to 7.2 (using HCl). Complete the volume to 200 mL and sterilize.
9. 0.2 M HCl (50 mL): 1 mL of hydrochloric acid fuming 37% in 50 mL DEPC water.
10. NBT/BCIP solution: 5 mL 10% polyvinyl alcohol (PVA) solution, 25 µL 1 N $MgCl_2$, 18 µL BCIP stock (50 mg/mL), 25 µL NBT stock (100 mg/mL), 5 drops Levamisol.

11. Normal serum 2% (1 mL): Dilute 20 µL horse serum in 1 mL buffer 1.
12. Paraformaldehyde solution 3% (100 mL): Preheat RNase free PBS to 60°C. Dissolve 3 g paraformaldehyde in 80 mL preheated PBS by adding 1 to 2 drops 10 N NaOH (until solution get clear). Adjust pH to between 7.2 and 7.4 with diluted HCl or NaOH. Make up to 100 mL and cool down. Filter through a 0.22 µm filter. Prepare always fresh or store at –20°C. Use only one time.
13. Prehybridization buffer (1 mL; 150–200 µL per slide, prepare fresh on ice): 500 µL formamide, 20 µL 50X Denhardt's solution, 400 µL 50% dextran sulfate in 20X SSC, 50 µL Herring-sperm DNA (10 mg/mL) denatured at 65°C for 10 min, 25 µL yeast-t-RNA (10 mg/mL).
14. PVA solution 10%: 10 g PVA. Fill up to 80 mL with 0.1 M Tris-HCl, 0.1 M NaCl pH 9.0. Heat the solution up to 90°C and stir well until the sticky solution clears. Cool down slowly while stirring.
15. RNase A stock solution (from bovine pancreas) (10 mg/mL): Dissolve the 25 mg in 2.5 mL STE. Store at 2 to 8°C. Dilute stock solution (10 mg/mL) in STE buffer to a final concentration of 40 µg/mL (1:250).
16. Sodium dodecyl sulfate (SDS) solution 20%: 20 g SDS, add ddH_2O to 100 mL.
17. STE buffer (1000 mL): 250 mM NaCl (100 mL 5 M stock solution), 10 mM Tris-HCl, pH 7.5 (20 mL 1 M stock solution), 0.5 mM EDTA (2 mL 0.5 M stock solution, pH 8.0), fill up to 1 L with ddH_2O.
18. Triethanolamine solution (50 mL): Dissolve 0.93 g triethanolamine in 50 mL DEPC water. Add 6 drops of 10N NaOH (adjust to pH 8.0, adjustment of the pH by pH meter is not recommended because of potential RNase contamination).
19. Trypsin solution: dissolve 1 mg/mL trypsin in DEPC treated water and preheat to 37°C before use.
20. Washing solution B1 (150 mL): 1.5 mL 20% SDS solution in 150 mL 1X SSC.
21. Washing solution B2 (150 mL): 1.5 m: 20% SDS in 150 mL 0.5X SSC.
22. Washing solution B3 (150 mL): 1.5 mL 20% SDS in 150 mL 0.1X SSC.

3. Methods

Adhere strictly to the principles of RNase free working conditions (*see* **Note 1**).

3.1. Dewaxing and Rehydration of Paraffin Embedded Sections or Fixation of Frozen Sections

Dewaxing and Rehydration of Paraffin Embedded Sections

1. Xylol (37°C), 15 min.
2. Xylol (RT), 2 × 10 min.
3. 100% Ethanol, 2 × 10 min.
4. 80% Ethanol (diluted with DEPC water), 3 to 5 min.
5. 50% Ethanol (diluted with DEPC water), 3 to 5 min.
6. DEPC water, 3 to 5 min.
7. Carry out dewaxing and rehydration in glass staining dishes.
8. For optimal dewaxing and rehydration, solutions should be gently shaken continuously.

9. Xylol and ethanol solutions from **steps 1–4** can be reused several times. The 50% ethanol solution with DEPC water should be used only once.

3.1.2. Fixation of Frozen Sections (see **Notes 2** and **3** for Fixation of Other Preparations)

1. Air dry the sections for 1 min on the slide.
2. Fix in freshly prepared 3% paraformaldehyde-PBS for 1 h, room temperature (RT).
3. Carry out fixation in a glass staining dish.
4. After fixation, wash with and store in PBS in a glass staining dish until trypsin digestion.

3.2. Pretreatment

1. Digest in trypsin solution (1 mg/mL), 20 to 40 min, 37°C. Carry out trypsin digestion in a closed (covered) glass staining dish (due to possible evaporation at 37°C).
2. Optional: fix in 3% paraformaldehyde-PBS for 10 min at RT. Carry out fixation in a glass staining dish. After fixation, wash with PBS. Place slides at level/flat slide rack in a humid, RNase free box for all further steps until hybridization.
3. Incubate with 2X SSC for 5 min at RT (1–2 mL/slide).
4. Incubate with 0.2M HCl solution for 8 to 10 min at RT (1–2 mL/slide).
5. Wash 2 times with triethanolamine solution (5–10 mL/slide and washing step).
6. Incubate with 0.25% acetic anhydrid/triethanolamine solution for 15 to 30 min at RT (1–2 mL/slide).
7. Prepare prehybridization buffer.
8. Wash 2 to 3 times with triethanolamine solution (5–10 mL/slide and washing step).

3.3. Prehybridization

1. Remove the superfluous triethanolamine solution carefully with clean paper towels; do not touch or let the tissue dry out!
2. Add 100 to 200 µL prehybridization buffer per slide and incubate for 1 h at RT.
3. After 30 min of incubation prepare hybridization solution.
4. Dilute riboprobe 1:10 in prehybridization buffer, 10 to 15 µL (total volume) per slide.
5. Denaturation of diluted riboprobes is achieved by incubating for 10 min at 80°C.
6. Put denatured probes on ice (*see* **Notes 4–6**).

3.4. Hybridization

1. Wash slides with 2X SSC (5–10 mL/slide and washing step). Let run and wipe off carefully the superfluous SSC with clean paper towels.
2. Add 10 to 15 µL of hybridization solution.
3. Cover with cover slips (corresponding to the size of the tissue specimen). Sealing of the cover slips with rubber cement is recommended when the box can not be closed airtight.

4. Humidified box for hybridization: paper towel soaked with 50% formamide-DEPC solution at the bottom of the box (100–200 mL 50% formamide-DEPC solution). Place slides on a rack in the box and close airtight.
5. Incubate the closed box at 50°C overnight.
6. Prepare washing solution for the next day in Coplin jars (150–200 mL each): 5X SSC and 50% formamide-2X SSC solution and preheat to 50°C (these solutions do not need to be RNase free).

3.5. Stringent Washing and RNase Treatment

1. Incubate in 5X SSC solution (150 mL in a coplin jar), 50°C, 20 min on a shaker.
2. Incubate in 50% formamide-2X SSC solution (150 mL in a coplin jar), 50°C, 30 min on a shaker.
3. Transfer slides on a flat rack in a new, not RNase-free chamber/box.
4. Wash 2 times with STE (5–10 mL/slide and washing step) at RT. Let run and wipe off carefully the superfluous STE solution with clean paper towels; do not touch and do not the tissue dry out!
5. Incubate with RNase A solution (100–200 µL/slide) for 1 h at 37°C.
6. Prepare washing solutions B1, B2, and B3.
7. Wash 3 times with STE (5–10 mL/slide and washing step) at RT.
8. Wash 3 times with 2X SSC (5–10 mL/slide and washing step).
9. Incubate with washing solution B1 (150 mL in a coplin jar) for 15 min at 50°C on a shaker.
10. Incubate with washing solution B2 (150 mL in a coplin jar) for 15 min at 50°C on a shaker.
11. Incubate with washing solution B3 (150 mL in a coplin jar) for 15 min at 50°C on a shaker.

3.6. Immunological Detection

1. Put slides on a flat racket and wash 3 times with buffer 1 (5–10 mL/slide and washing step).
2. Let run and wipe off carefully the superfluous solution with clean paper towels and incubate with 2% horse serum in buffer 1 (100–200 µL/slide) for 30 min at RT.
3. Wash with buffer 1 (5–10 mL/slide).
4. Let run and wipe off carefully the superfluous solution with clean paper towels and incubate with anti-digoxigenin-AP-antibody solution (100–200 µL/slide) for 1 h at RT (*see* **Note 7**).
5. Wash 3 times with buffer 1 (5–10 mL/slide and washing step).
6. Equilibrate with buffer 2 (150 mL in a coplin jar) for 10 min at RT on the shaker.
7. Led run and wipe off carefully the superfluous solution with clean paper towels and ad NBT/BCIP solution (300–500 µL/slide).
8. Incubate in the dark (light protected) at RT. Control staining development every 5 min in the first hour, then every 15 to 30 min after that (best control of staining

development is done by positive control tissue) (*see* **Note 8**). Color development can take hours. When overnight incubation is needed incubate at 4°C.
9. Stop color reaction by washing with buffer 1 (*see* **Note 9**).

4. Notes

1. Principles of RNase free working conditions *(1–5)*. Design RNase free zone, separate form the other lab and mark clearly. Clean bench surface regularly with RNase destroying solutions (e.g., RNase ZAP from Ambion). Wear powder free gloves (skin surface is contaminated with RNase). Use separated set of pipetes and filter tips. Store all reagents and materials for RNase free work in a separate RNase-free zone. Label them as RNase free. Use only RNase free declared commercial solutions. In the case the provider has not given such a declaration contact the company. Use only DEPC treated water for RNase free solutions. In addition do not let the tissues dry out, because drying causes high background. Slides need to be kept level during all incubation steps for homogenous treatment and have to be covered completely when incubated in glass staining dishes. Do not touch the tissue with pipet tip.
   During washing steps rinse the slides carefully by avoiding direct jetting on the tissue. For sufficient washing 5 to 10 mL per slide is needed.
2. Fixation of synovial tissue and synovial fibroblasts. Fix in 4% paraformaldehyde for 4 to 8 h. Resuspend the samples in 50% ethanol (diluted in distilled water). In the case of articular joint tissue, decalcify bone with 10% EDTA Solution (pH 7.2) for variable time periods. The decalcification process has to be controlled with a pin and/or a scalpel.
3. Advantage and disadvantage of paraffin embedded and frozen tissue *(4,5)*. The advantage of formaldehyde and paraffin embedded tissues is the high quality of visible cellular structures even in a simple haematoxylin and eosin (H&E) staining. The disadvantage is that the fixation with formaldehyde and paraffin embedment can hinder the access of the probe to the target RNA. Therefore, trypsin digestion as pretreatment is recommended. However, still the detection of low copy number target mRNA might be difficult. In general, the fixation of the tissue should be standardized and performed as early as possible to avoid RNA degradation. Frozen sections have the advantage that the targeting of the mRNA by the probe should be easier. However, repetitive freezing and thawing of the tissue reduces also the concentration of mRNA. The structural quality of frozen sections is lower when compared with paraffin embedded tissue sections.
4. Advantage of RNA probes. RNA probes reveal a higher specificity and lower background than DNA probes because non-specifically bound RNA probes are removed by RNase treatment.
5. Advantage and disadvantage of radioactive and non-radioactive probe labelling *(1,3)* are summarized in **Table 1**. Both, radioactive and non-radioactive ISH techniques are reliable, specific, and sensitive. However, for safety reasons we prefer the nonradioactive ISH.

Table 1
Advantages and Disadvantages of Radioactive and Nonradioactive Probe Labeling

	Advantage	Disadvantage
Nonradioactive	Easy production of high quantity by in vitro transcription Constant quality of labeling Stability/storage* Safety	Difficulties to assess quantities
Radioactive	Easy to count	Instability Difficulties to reproduce labeling Safety

*Digoxigenin labeled probes are stable for at least 2 yr at –20°.

6. Probe size. A probe size of 200 to 300 bases is recommended for optimal specificity and sensitivity.
7. Advantage and disadvantage of enzymatic and fluorescence based visualization methods (3,4). The advantage of enzymatic signal amplification is that the signal is stable and stained slides can be stored over years. However, this type of visualization is more time consuming and in certain situations might reveal a tendency of signal smearing. In such a case immunofluorescence is a good and fast alternative. However, immunofluorescence signals are not very stable and analysis of the signals should be performed as fast as possible after the staining is completed. Immunofluorescence slides should be stored at dark and optimally at 4°C.
8. Control experiments. Each experiment should include complimentary slides exposed either with sense or with antisense probes. The sense probe has to be blank (internal negative control) and the antisense probe should reveal a specific cytoplasmic staining. If available, also positive control tissues for the target molecule should be included. By detecting novel sequences one should be aware of the fact that same sequences might be present in certain cells in a reverse orientation.
9. Number of slides. The number of slides investigated in one experiment should be limited. The incubation times should not differ too much from slide to slide and all time consuming procedures should be taken into account. Therefore, we recommend, at least in the beginning, limiting experiments to 10 to 12 slides per experiment.

References

1. Schwarzacher, T. and Heslop-Harrison, P. (2000) *Practical in Situ Hybridization*, Garland Science (Bios Scientific Publisher).
2. Beesley, J. E. (2000) *Immunocytochemistry and in Situ Hybridization in the Biomedical Sciences*, first ed. Birkhauser, Boston Basel Berlin.

3. Darby, Ian, A. (ed.) (2000) *In Situ Hybridization Protocols (Series: Methods in molecular biology)*, second ed., Humana Press, Totowa, New Jersey.
4. Wilkinson, D. G. (ed.) (1999) *In Situ Hybridization: A Practical Approach,* second ed., Oxford University Press, New York.
5. Morel, G. and Cavalier, A. (2000) *In Situ Hybridization in Light Microscopy (Series: Methods of Visualization)*, CRC Press, Boca Raton Florida.
6. Online: *In situ hybridization and immunohistochemistry (Chapter 14 of Current protocols in molecular biology)*, Current protocols online, 2003 Wiley and Sons.
7. Kriegsmann, J, Keyser, G., Geiler, T., Gay, R. E., and Gay, S. (1994) A new double labeling technique for combined *in situ* hybridization and immunohistochemical analysis. *Lab Invest.* **71,** 911–917.

5

Subtractive Hybridization

Jörg H. W. Distler, Oliver Distler, Michel Neidhart, and Steffen Gay

Summary

Subtractive hybridization, like serial analysis of gene expression (SAGE), RNA arbitrarily primed polymerase chain reaction (RAP-PCR) and microarrays, is a screening method for differentially expressed genes. At first, poly-A$^+$-RNA is isolated and reverse transcribed into cDNA. With the SMART technology, total RNA can be used. For subtractive hybridization, two adaptors are ligated to the tester cDNA. The tester cDNA is then mixed twice with cDNA of the reference sample. Sequences that are present in equal levels in the tester and in the reference cDNA hybridize to each other and are then removed. In contrast, differentially expressed genes are highly enriched and then amplified by PCR. Because subtractive hybridization is a complex multistep procedure, the results should be controlled at each level and positive controls are recommended. Because all screening methods produce a significant number of false positives, the differential expression has to be confirmed by independent methods.

Key Words: Suppressive subtractive hybridization; SMART; mRNA; cDNA; reverse transcription; functional genomics.

1. Introduction

The term differential screening summerizes techniques which can be used to analyze a large number of genes for up- or downregulation under certain conditions. For differential screening, a number of newly developed methods can be used (for discussion of advantages and disadvantages of the different methods *see* **Note 1**):

- Microarray technology
- RNA arbitrarily primed polymerase chain reaction (RAP-PCR).
- Serial analyses of gene expression (SAGE).
- Subtractive hybridization (for commercially available kits and services (*see* **Note 2**).

This article will focus on the molecular basis, advantages and disadvantages and in particular material and methods for suppressive subtractive hybridization.

Suppressive subtractive hybridization (SSH) is a PCR-based technique for the identification of differentially expressed genes. In contrast to conventional subtractive hybridization, SSH requires fewer steps and enriches for mRNAs, which are expressed at a lower level.

1.1. RNA-Isolation and Quality Control

The first steps for SSH are RNA-isolation (*see* **Note 3**) and reverse transcription. Normally, poly-A^+-RNA is used for SSH. Alternatively, total RNA can be used, which then requires additional steps during reverse transcription. Good quality and high purity of the isolated RNA are essential for SSH and should be confirmed before SSH is started. This can be done in two ways. The conventional way is the analysis on a denaturating agarose gel containing ethidium bromide (EB). Total RNA from mammalians with good quality typically shows two bands at 4.5 kb and 1.9 kb corresponding to 28S- and 18S-RNA. The ratio of the intensity of the bands should be 1.5 to 2.5. A smear between the two bands indicates degraded RNA, which then should not be used for SSH. The second method uses RNA nanochips and analyzes the quality with a bioanalyzer (e.g., Agilent Technologies, Palo Alto, CA). This technique is much more sensitive than an agarose/EB gel. In addition, only 20 nanograms of RNA are sufficient for the analysis with a RNA chip compared to 1–2 µg for an agarose gel.

1.2. Reverse Transcription

If total RNA is isolated, there will be a reverse transcription of ribosomal RNA even if oligo-dT primers are used resulting in an insufficient subtractive hybridization. Therefore, if total RNA is used, a special technique must be chosen that enhances the transcription of poly-A^+-RNA. The most popular one of these techniques is the SMART technology (*see* **Fig. 1**). Modified anti-poly-A^+-primers (MOPA primers) bind to poly-A-regions at the 3'-end of mRNAs molecules. When the reverse transcriptase reaches the 5'-end of the mRNA, it adds additional deoxycytidine nucleotides. These additional nucleotides anneal a SMART oligonucleotide with an oligo-G-sequence at its 3'-end. The reverse transcriptase uses the SMART oligonucleotide as a template and completes reverse transcription of the single stranded cDNA *(1)*. These cDNAs are amplified by PCR with primers against the initial MOPA primer and the SMART oligonucleotide. The localization of the SMART primer at the 5'-end and the MOPA primer at the 3'-end guaranties the amplification of full-lenghts poly-A^+-RNA molecules.

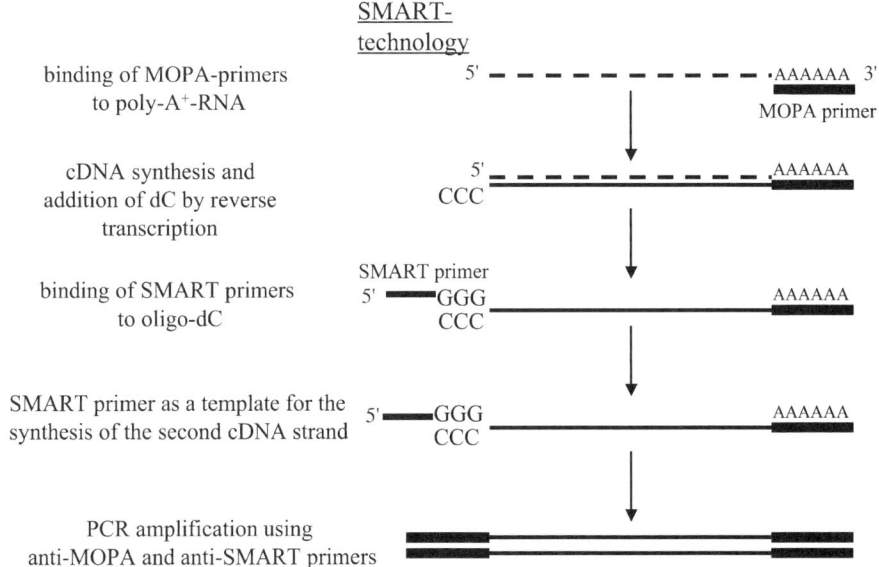

Fig. 1. With the SMART technology, poly-A$^+$-RNA is reverse transcribed and amplified by PCR. MOPA primer and SMART primer enable the selective amplification of poly-A$^+$-RNA out of a pool of total RNA.

1.3. Suppressive Subtractive Hybridization

The cDNA population, in which differentially expressed genes should be detected (e.g., synovial fibroblasts stimulated with TNF-α) is called tester cDNA. The reference cDNA (e.g., unstimulated synovial fibroblasts) is named driver cDNA. The general principle of SSH is a hybridization of sequences that are present in both cDNA populations. The hybridized sequences are removed and only cDNAs from differentially expressed genes are left (*see* **Fig. 2**) *(2)*. In a first step, the tester cDNA is divided into two portions, which are ligated to two different adaptors. Then, an excess of driver cDNA is added to both portions of the tester cDNA. During the first hybridization step, different molecules form, which are shown in **Fig. 2**.

Type A molecules are single stranded cDNAs of the tester cDNA, which are ligated to the specific adaptor. Because of hybridization kinetics of the second order, molecules with higher concentration hybridize faster than those with lower concentration, resulting in an enrichment of cDNA molecules with low concentrations among the type A molecules. Those cDNAs, which are only present in the tester, but not in the driver cDNA, are also enriched among the type A molecules, because genes with equal concentrations in tester- and

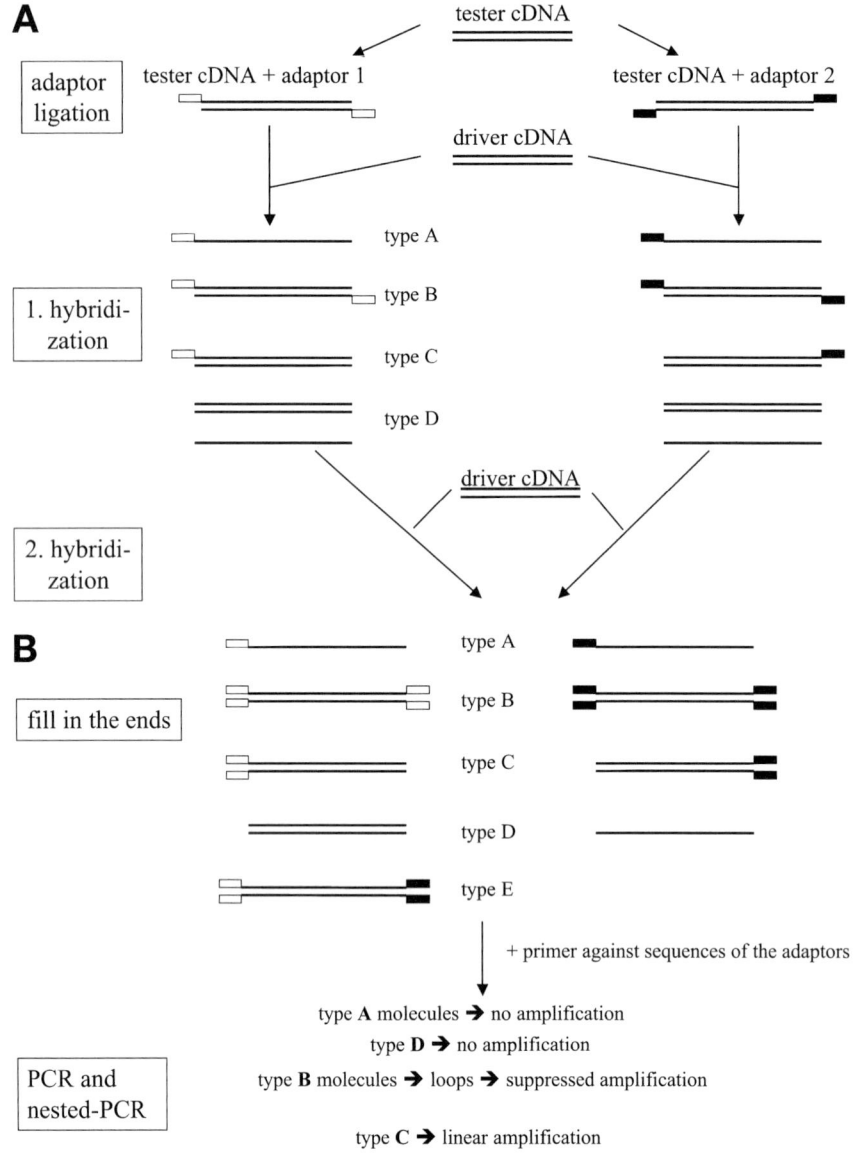

Fig. 2. For SSH, two hybridization steps with tester and driver cDNA are performed. During the following two PCRs, differentially expressed genes with different adaptors at both ends are amplified exponentially, whereas the amplification of all other sequences is suppressed.

driver-cDNA form type C molecules. During the second hybridization step, the two portions of the first hybridization are mixed. The addition of freshly denatured driver cDNA enriches further for differentially expressed type A molecules. Type A molecules of both portions can now hybridize with each other and form type B, C, and E molecules. Type E molecules are hybrids with adaptor 1 at one and adaptor 2 at the other end. Afterwards, the ends of single stranded adaptors are filled in to create binding sites for complementary PCR primers (*see* **Fig. 2**). Type A and type D molecules do not have binding sites for PCR primers and are therefore not amplified during PCR with adaptor primers. Type B molecules with the same adaptors at both ends form ring structures, which complicates binding of the primers and suppresses their amplification (*suppressive* subtractive hybridization). Type C molecules have only one binding site and are amplified linearly. Type E molecules have two binding sites for primers at the ends of their adaptors and are amplified exponentially resulting in a strong enrichment of differentially expressed genes. The resulting cDNA library can be cloned and sequenced using standard protocols.

2. Materials

For best results, 2 µg of poly-A$^+$-RNA of high purity and quality should be used. RNA should be stored at –70°C. Hybridization buffer is stored at room temperature, all other reagents are stored at –20°C.

2.1. Buffers and Solutions

1. 5X first strand buffer: 250 n*M* Tris-HCl (pH 8.5), 40 n*M* MgCl$_2$, 150 nM KCl, and 5 m*M* dithiothreitol (DTT).
2. 5X second strand buffer: 500 m*M* KCl, 50 m*M* ammonium sulfate, 25 m*M* MgCl$_2$, 0.75 m*M* nicotinamide adenine dinucleotide (β-NAD), 100 m*M* Tris-HCl (pH 7.5), and 0.25 mg/mL bovine serum albumin (BSA).
3. 10X Rsa I restriction buffer : 100m*M* bis tris propane-HCl (pH 7.0), 100 m*M* MgCl$_2$, and 1 m*M* DTT.
4. 5X DNA ligation buffer : 250 m*M* Tris-HCl (pH 7.8), 50 m*M* MgCl$_2$, 10 m*M* DTT, and 0.25 mg/mL BSA.
5. 4X hybridization buffer: 50 m*M* HEPES (pH 8.3), 0.5 *M* NaCl, 0.02 m*M* ethylene diamine tetraacetic acid (EDTA) (pH 8.0), and 10% polyethylene glycol (PEG 8000).
6. Dilution buffer: 20 mm HEPES-HCl (pH 8.3), 50 m*M* NaCl, and 0.2 m*M* EDTA (pH 8.0).
7. dNTP mix (10 m*M* of dATP, dCTP, dGTP, dTTP each).
8. 20X EDTA/glycogen mix: 0.2 *M* EDTA, 1 mg/mL glycogen.
9. 4 *M* Ammonium acetate (NH$_4$Oac).

10. Phenol:chloroform:isoamyl alcohol (25:24:1) and chloroform:isoamyl alcohol (24:1).
11. 50X TAE electrophoreses buffer: 242 g tris base, 57.1 mL glacial acetic acid, 37.2 g $Na_2EDTA \cdot H_2O$, in 1 L H_2O.
12. DEPC ddH_2O and ddH_2O.

2.2. Primers

1. cDNA synthesis primer: 5'-TTTTGTACAAGCT$_{31}$N$_1$N-3'
2. adaptor 1 (10 µM): 5'-CTAATACGACTCACTATAGGGCTCGAGCGGCCGCCCGGGGAGGT -3' and 3'-GGCCCGTCCA -5'
3. Adaptor 2 (10 µM): 5'-CTAATACGACTCACTATAGGGCAGCGTGGTCGCGGCCGAGGT -3' and 3'-GCCGGCTCCA -5'
4. PCR primer 1 (10 µM): 5'-CTAATACGACTCACTATAGGGC-3'
5. nested PCR primer 1 (10 µM): 5'-TCGAGCGGCCGCCCGGGCAGGT-3'
6. nested PCR primer 2 (10 µM): 5'-AGCGTGGTCGCGGCCGAGGT-3'

2.3. Enzymes

1. AMV reverse transcriptase.
2. 20X second strand enzyme cocktail:
    DNA polymerase I (6 U/µL).
    RNAse H (0.25 U/µL).
    Escherichia coli DNA ligase (1.2 U/µL).
3. T4 DNA polymerase (3 U/µL).
4. Rsa I (10 U/µL).
5. T4 DNA ligase (400 U/µL, containing 3 mM ATP).

In addition to the reagents listed above, two thermal cyclers, a water bath, and a heating block or a hybridization-oven are needed.

3. Methods

We follow the protocol provided by BD Bioscience/Clontech (Palo Alto, CA) with some modifications.

3.1. First Strand cDNA Synthesis

The following procedure has to be performed for each tester and driver cDNA. In addition, most commercially available kits provide poly-A$^+$-RNA as a positive control.

For this section, a thermal cycler must be preheated to 70°C, and an incubator to 42°C. Some steps are performed on ice. The time need is approx 2 h.

1. Add poly-A$^+$-RNA (total 2 µg), 2 to 4 µL cDNA synthesis primer, 1.0 µL, and DEPC water with a total volume of 5 µL to a microcentrifugation tube (do *not* use polystyrene tubes), mix thoroughly and spin down briefly.

2. Incubate for 2 min at 70°C in a preheated thermal cycler
3. Cool on ice for additional 2 min.
4. Add 2.0 µL 5X first strand buffer, 1.0 µL dNTP mix, 1.0 µL DEPC water, and 1.0 µL AMV reverse transcriptase. Mix thoroughly and spin down briefly.
5. Incubate the tubes in an air incubator at 42°C for 90 min: water baths of thermal cyclers should not be used because evaporation might reduce the volume of the mixture resulting in an inefficient reaction.
6. Place the tubes on ice and proceed *immediately*.

3.2. Second Strand cDNA Synthesis

The second strand synthesis is performed with each first strand tester, driver and control cDNA. For this section, a thermal cycler must be preheated to 16°C. The time need is 3.5 to 4 h.

1. Combine the following reagents to the tubes with the products from the first strand synthesis (containing a total volume of 10 µL): 48.4 µL dd H_2O, 16.0 µL 5X second strand buffer, 1.6 µL dNTP mix, 4.0 µL 20X second strand enzyme cocktail. Mix and spin down briefly.
2. Incubate tubes at 16°C in a thermal cycler for 2 h.
3. Add 2 µL of T4 DNA polymerase and mix well.
4. Incubate the tubes at 16°C for another 30 min in a thermal cycler.
5. Add 4 µL of 20X EDTA/ glycogen mix to terminate the reaction.
6. Add 100 µL of phenol:chloroform:isoamyl alcohol (25:24:1), vortex thoroughly and spin the tubes at >10,000*g* for 10 min at room temperature (RT).
7. You will see three phases: an aqueous layer on the top, an interphase, and a lower phase. Remove the aqueous layer and place it in a clean microcentrifugation tube. Avoid contaminations with the other two phases. The interphase and the lower phase can be discarded.
8. Add 100 µL of chloroform:isoamyl alcohol (24:1) to the new tube with the aqueous layer, spin the tubes at >10,000*g* for 10 min at room temperature and separate again the aqueous layer from the interphase and the lower layer.
9. Add 40 µL of 4 M NH_4OAc and 300 µL of 95 % ethanol, vortex thoroughly, and spin the tubes at >10000 g for 20 min at RT.
10. Remove and discard the supernatant.
11. Overlay the pellet with 500 µL of 80% ethanol, centrifuge at >10,000*g* for 10 min at RT.
12. Remove and discard the supernatant and air-dry the pellet for about 10 min until all ethanol is evaporated.
13. Redissolve the pellet in 50 µL of ddH_2O.
14. For controls:
    a. Measure the concentration of the cDNA in a spectrophotometer
    b. Remove an aliquot of 6 µL, digest it with Rsa I and analyze it on an agarose gel (*see* **Note 4**).

3.3. RsaI Digestion

This step, which is again performed with all tester, driver, and control cDNAs generates blunt-ended cDNA fragments. For this section, a thermal cycler must be preheated to 37°C. The time need is approx 3 to 3.5 h.

1. Mix 43.5 µL dscDNA, 5.0 µL 10X Rsa I restriction buffer, and 5.0 µL.
2. Incubate at 37°C for 90 min.
3. Remove 5 µL of the reaction mix to analyze the efficiency of the Rsa I digestion on an agarose/EB gel (*see* **Note 5**).
4. Add 2.5 µL of 20X EDTA/glycogen mix to stop the reaction.
5. Add 50 µL of phenol:chloroform:isoamyl alcohol (25:24:1), vortex thoroughly, and spin the tubes at >10,000g for 10 min at RT. Remove the top layer and place it into a new tube.
6. Add 50 µL of chloroform:isoamyl alcohol (24:1), vortex thoroughly, centrifuge the tubes at >10,000g for 10 min at RT and place the aqueous layer into a new tube.
7. Add 25 µL of NH_4OAc and 187.5 µL of 95 % ethanol, vortex thoroughly, and spin the tubes at >10,000g for 20 min at RT.
8. Remove and discard the supernatant, overlay the pellet with 80% ethanol and spin at >10,000g for 5 min.
9. Remove the supernatant carefully and air dry the pellet for 5 to 10 min.
10. Dissolve the pellet in 5.5 µL ddH_2O and store at –20°C until further use.

3.4. Adaptor Ligation

If genes that are upregulated in the tester cDNA (e.g., synovial fibroblasts stimulated with TNF-α) are of interest, it is sufficient to subtract only the driver cDNA from the tester cDNA (forward subtraction). For the identification of genes that are downregulated in the tester cDNA (e.g., genes that are suppressed after stimulation with TNF-α), the subtraction has to be performed additionally in the other direction (reverse subtraction). Each sample of cDNA, prepared so far, is now divided into two portions. One of those portions will be labeled with adaptor 1, the second one with adaptor 2. No adaptors are ligated to cDNAs used as driver. Unsubtracted tester cDNA will serve as a positive control for ligation and a negative control for subtraction. For this section, a thermal cycler with 16°C and 72°C must be available. The reaction incubates over night.

1. 1 µL of Rsa I digested cDNA from **Subheading 3.3., step 10** is diluted with 5 µL H_2O.
2. Prepare a mastermix (per reaction): 3 µL ddH_2O, 2 µL 5X ligation buffer, 1 µL T4 DNA ligase.
3. Combine the following reagents *in the order shown* for each tester cDNA. Be careful with the nomenclature. Label the first portion of tester 1 for example as 1a, the second portion as 1b, the first portion of tester 2 as 2a, and so on.

For the first portion of tester cDNA (1a): 2 µL diluted cDNA, 2 µL adaptor 1, and 6 µL master mix (VI/2).
For the second portion of tester cDNA (1b): 2 µL diluted cDNA, 2 µL adaptor 2, and 6 µL master mix (VI/2).
4. Combine 2 µL of both portions (1a and 1b) in a fresh tube and mix well. This will be used as an unsubtracted tester control later on.
5. Incubate over night at 16°C.
6. Stop the reaction by adding 1 µL of EDTA/ glycogen and heating to 72°C for 5 min.
7. Remove 1 µL from each unsubtracted tester control, dilute it in 1 mL ddH$_2$O and label it (e.g., as 1c, 2c). Before proceeding to the first hybridization, the ligation efficiency should be analysed on an agarose/EB gel (*see* **Note 6**).

3.5. First Hybridization

Before the hybridization is started, the 4X hybridization buffer must be warmed to room temperature. For this section, a thermal cycler with 98°C and 68°C must be available. The reaction incubates for 6 to 12 h.

1. Prepare the following master mix for each subtraction: Hybridization sample A: 1.5 mL *Rsa*I digested driver cDNA (III/10), 1.5 mL tester cDNA 1a (IV/6); 1.0 µL, 4X hybridization buffer Hybridization sample B: 1.5 µL *Rsa*I digested driver cDNA (III/10), 1.5 µL tester cDNA 1b (IV/6), 1.0 µL 4X hybridization buffer.
2. Heat the sample at 98°C for 90 s.
3. Incubate at 68°C for 6 to12 h.
Afterwards proceed *immediately* to the second hybridization (VI).

3.6. Second Hybridization

The two hybridization samples from the first hybridization (*see* **Subheading 3.5.**) are mixed together during the second hybridization and an excess of denatured driver cDNA is added. The following procedure has to be performed for each tester cDNA.

For this section, a thermal at 68°C is needed. The reaction incubates over night.

1. Combine the following reagents to freshly denature driver cDNA: 10. µL driver cDNA (III/10), 1.0 µL 4X hybridization buffer, and 2.0 µL ddH$_2$O.
2. Incubate at 98°C for 90 s.
3. To ensure that the two hybridization samples (A and B) are mixed only in the presence of an excess of freshly denatured driver cDNA, the following procedure is performed:
   a. Set a micropipettor to 15 µL and the entire hybridization sample B is drawn into the pipet tip.
   b. Create a separation space by drawing a small amount of air is into the tip.
   c. Draw freshly denatured driver into the tip.
   d. Add the entired mixture in the tip to hybridization sample A and mix thoroughly.

4. Incubate the reaction over night at 68°C.
5. Add 200 µL of dilution buffer.
6. Heat for another 7 min at 68°C.
7. Store the samples at –20°C.

3.7. PCR Amplification

As discussed in the introduction, only sequences with different adaptors at both ends are amplified exponentially during the two PCR amplification steps. Usually, a minimum of four different PCR reactions is performed: the forward subtracted experimental cDNA (e.g., to detect genes that are induced by TNF-α in synovial fibroblasts), the unsubtracted tester cDNA for the forward subtraction, the reverse substracted experimental cDNA (e.g., to identify genes that are downregulated in synovial fibroblasts after stimulation with TNF-α) and the unsubtracted tester cDNA for the reverse subtraction. Most commercially available kits also provide positive controls, which should be performed in addition.

For this section, a thermal cycler at 68°C is needed. The reaction incubates over night.

1. Aliquot 1.0 µL of each cDNA sample and add 24.0 µL of the following master mix (amount per reaction): 19.5 µL ddH$_2$O, 2.5 µL 10X PCR reaction buffer, 0.5 µL dNTP mix, 1.0 µL PCR primer 1, and 0.5 µL 50X advantage cDNA polymerase mix.
2. Incubate at 75°C for 5 min.
3. Start *immediately* afterward with thermal cycling: a hold at 94°C for 25 s, followed by 26 cycles of 94°C for 10 s, 66°C for 30 s, and 72°C for 90 s.
4. Analyze 8 µL of each reaction on an agarose/EB gel.
5. Dilute 3 µL of each primary PCR product in 27 µL of ddH$_2$O.
6. Aliquot 1 µL of each dilute product from the previous step and add 24 µL of the this master mix (for one reaction): 18.5 µL ddH$_2$O, 2.5 µL 10X PCR reaction buffer, 1.0 µL nested PCR primer 1, 1.0 µL nested PCR primer 2, 0.5 µL dNTP mix, and 0.5 µL 50X advantage cDNA polymerase mix,
7. Perform thermal cycling: 11 cycles of 94°C for 10 s, 68°C for 30 s, and 72°C for 90 s.
8. Analyze 8 µL of each reaction of this nested PCR on an agarose/EB gel.

The results of the PCR amplification should be controlled according to **Note 7**. Finally, the efficiency of the whole subtraction procedure can be controlled according to **Note 8**.

Differentially expressed genes are now highly enriched, whereas cDNAs corresponding to genes with equal expression levels have been subtracted. The cDNAs should now be cloned and transformed into *E. coli* for amplification. These sequences can then be identified by sequencing and BLAST search. As

Subtractive Hybridization 87

discussed in the introduction, SSH like other differential screening techniques produces a significant number of false positives. Therefore the results have to be confirmed by an independent method such as real-time PCR (*see* **Note 9**).

4. Notes

1. Subtractive hybridization offers the chance to identify unknown sequences (e.g., viral sequences that might be involved in the pathogenesis of certain diseases) and genes with yet unknown functions. In previous studies, 10 to 30% of the identified differentially expressed sequences were genes with unknown functions *(2)*. On the other hand, it is very cost- and time-intensive to generate a complete expression profile. Between 300 and 500 clones need to be analyzed from the cDNA library to find all differentially expressed genes including those with a low expression rate *(6)*. Among those 300 to 500 clones, many are picked several times, making the procedure ineffective for the analysis of a complete expression profile. Therefore, if the aim of the planed study is to generate a complete expression profile rather than searching for yet unknown sequences, other techniques such as microarrays are more appropriate. Because only one sample is used for subtractive hybridization, the results must be confirmed with additional patients/cultures to obtain statstically significant results and to draw reliable conclusions.
2. SSH can be performed using commercially available kits. The Clontech PCR-Select cDNA subtraction kit (BD Bioscience/ Clontech, Palo Alto, CA) is widely used by many researchers including ourselves. Besides the higher costs, these kits have a number of advantages; all reagents and primers are ready to use, the protocols are optimized, and poitive and negative controls are included. Because SSH is a rather complicated multi-step procedure, commercial cDNA subtraction services are offered (e.g., by Evrogen *http://www.evrogen.com*]and BD Bioscience [*http://www.bdbiosciences.com*].
3. Homogeneous samples are an essential requirement for all differential screening methods. SSH will be inefficient, if heterogeneous tissue with various populations of cells are investigated. In heterogeneous tissues, differences in the expression levels might simply due to different ratios between the cell populations.
   The synovium is a very heterogeneous tissue consisting of fibroblasts, endothelial cells macrophages, and other inflammatory cells. To avoid unreliable results with SSH, more homogeneous cell populations should be used. The gold-standard for selective isolation of small regions of a tissue or even of single cells is laser-captured microdissection (*see* Chapter 6). Prior to laser captured microdissection, cells of interest can be labeled by immunohistochemistry, for instance CD68 for the identification of macrophages in the synovial membrane. The labeled cells can be cut out, collected and used for the isolation of RNA.
4. Controls for the analysis of the cDNA synthesis: if the experimental cDNA appears as a smear of less than 2 kb on the agarose/EB gel, the RNA was probably degraded and RNA isolation and cDNA synthesis should be repeated. If the size of the experimental cDNA is similar to the cDNA produced from control RNA,

but the signal is weak, the cDNA is still usable, but some rarely expressed sequences might be lost.

5. Controls for the analysis of *RsaI* digestion: Rsa I digestion reduces the average size of the cDNA from 0.5 to 10 kb to 0.1 to 2 kb. If the size is not reduced, phenol/chloroform extraction, ethanol precipitation and Rsa I digestion should be repeated.

6. Analysis of ligation efficacy: A PCR is performed, which amplifies only fragments that span the adaptor/ cDNA junction of the tester cDNAs with gene specific primers:
   a. Dilute 1 µL of each ligated cDNA (VI/7) into 200 µL of dd H$_2$O.
   b. Set up four reactions:

i.	Tester cDNA 1a	1 µL
	Gene specific 3' primer	1 µL
	PCR primer 1	1 µL
ii.	Tester cDNA 1a	1 µL
	Gene specific 3' primer	1 µL
	Gene specific 5' primer	1 µL
iii.	Tester cDNA 1b	1 µL
	Gene specific 3' primer	1 µL
	PCR primer 1	1 µL
iv.	Tester cDNA 1b	1 µL
	Gene specific 3' primer	1 µL
	Gene specific 5' primer	1 µL

   c. Prepare the following master mix in the order shown (amount for one reaction): 18.5 µL dd H$_2$O, 2.5 µL 10X PCR reaction buffer, 0.5 µL dNTP mix, 0.5 µL 50X advantage cDNA polymerase mix, and add 22 µL of this master mix to each of the four reactions.
   d. Heat to 75°C for 5 min and perform thermal cycling *immediately*: a hold of 94°C for 30 s and 20 cycles of 94°C for 10 s, 65°C for 30 s, and 68°C for 2.5 min. Analyze 5 µL of each reaction on an agarose/EB gel: The products of the reactions using one gene specific primer and the PCR primer 1 (primers i and iii) should be about the same intensity as the PCR products of the reactions using two gene specific primers (primers ii and iv). If the intensity differs significantly, the ligation was ineffective and should be repeated. Ineffective ligation is often due to contamination with salts during the precipitation steps.

7. Controls for PCR amplification step: The experimental primary cDNA appears as a smear from 0.2 to 2 kb on an agarose/EB gel. If no product can be detected after 27 cycles, 3 additional cycles can be used. If there is still no signal after 30 cycles neither in the subtracted nor in the unsubtracted samples nor in controls, the polymerase might not work. If the polymerase is satisfactory, annealing and extension temperature might be decreased in small steps. However, this increases the background. If there is a signal after 30 cycles in the unsubtracted cDNA, but not in the subtracted sample, the cycles of the secondary PCR can be increased. Just like the primary cDNA, the secondary cDNA produces a smear. If no product is detected after 12 cycles, the number of cycles can be increased, but more

cycles also increase the background. If no smear is observed in the PCR control reaction, provided by most commercially available kits, the PCR conditions have to be optimized. If a signal is detected in the PCR control, but not in the experimental samples, there is a problem with the subtraction, most likely with the ligation.
8. Controls for analysis of subtraction efficiency: PCR with primers against a housekeeping gene that is not differentially expressed can be performed to estimate the efficiency of SSH. Avoid primers that amplify a cDNA fragment with a Rsa*I restriction site*. 5 cycles correspond to a 20-fold enrichment. If the difference is less than 5 cycles, we recommend checking another housekeeping gene, because some housekeeping genes can be present in different levels in tester and driver cDNAs and the low difference therefore does not necessarily indicate an insufficient subtraction. In addition, the levels of a gene, which is known to be differentially expressed under the experimental conditions showed be analysed. For analysis, a certain portion should be removed from the PCR reaction after 15, 20, 25, and 30 cycles and examined on an agarose/EBgel.

The control procedures above just give hints about the quality of the experiment. Even if all controls appear to be optimal, the SSH might still be insufficient. Therefore, we suggest to pick randomly 10 to 20 clones, identify the sequences and confirm the differential expression by a second method (e.g., SYBR Green Real-time PCR). If less than 25 % of these genes are confirmed, we recommend to repeat the SSH.
9. All differential screening techniques produce a significant number of false positives (for subtractive hybridization 5–95 % depending on the individual experiment). Therefore, the differential expression has to be confirmed by a second independent method, preferably with separate samples rather than retesting the original RNA. Methods commonly used for confirmation are:
   - Real-time PCR: SYBR Green or TaqMan,
   - RT-PCR,
   - Northern blot analysis.

Real-time PCR is thought to be the gold-standard. The expression of a large number of genes can be quantified in a rather short time and the sensitivity of Real-time PCR is much higher than that of conventional RT-PCR or Northern blot.

References

1. Zhu, Y. Y., Machleder, E. M., Chenchik, A., Li, R., and Siebert, P. D. (2001) Reverse transcriptase template switching: a SMART approach for full-length cDNA library construction. *Biotechniques* **30,** 892–897.
2. Desaj, S., Hill, J., Trelogan, S., Diatchenko, L., Siebert, and P. D. (2000) Identification of differentially expressed genes by suppression subtractive hybridization, in Hunt, S.P., Livesey, F. J. (eds.) *Functional genomics*. Oxford University Press, Oxford New York, pp. 81–112.
3. Diatchenko, L., Lau, Y. F., Campbell, A. P., et al. (1996) Suppression subtractive hybridization: a method for generating differentially regulated or tissue-specific cDNA probes and libraries. *Proc Natl Acad Sci USA* **93,** 6025–6030.

4. Diatchenko, L., Lukyanov, S., Lau, Y. F., and Siebert, P. D. (1999) Suppression subtractive hybridization: a versatile method for identifying differentially expressed genes. *Methods Enzymol* **303,** 349–380.
5. von Stein, O. D., Thies, W. G., and Hofmann, M. (1997). A high throughput screening for rarely transcribed differentially expressed genes. *Nucleic Acids Res* **25,** 2598–2602.

6

Laser Capture as a Tool for Analysis of Gene Expression in Inflamed Synovium

Ulf Müller-Ladner, Martin Judex, Elena Neumann, and Steffen Gay

Summary

Most current approaches used to analyze gene expression in tissue samples are based on RNA isolated either from cultured synovial cells or from synovial biopsies. However, this strategy does not distinguish between specific gene expression profiles of cells originating from discrete tissue areas. Therefore, we established the combination of laser-mediated microdissection and RNA arbitrarily primed polymerase chain reaction (RAP-PCR) for differential display to analyze profiles of gene expression in histologically defined areas of arthritic tissue.

Cryosections derived from synovial tissue were used to obtain cell samples from different tissue areas of both rheumatoid arthritis (RA) and osteoarthritis (OA) patients using a microbeam laser microscope. RNA was isolated and analyzed using nested RNA arbitrarily primed PCR to generate a fingerprint of the expressed gene sequences. Differentially expressed bands were isolated, cloned, and sequenced. Differential expression of identified sequences was confirmed by *in situ* hybridization and immunohistochemistry.

Key Words: Laser mediated microdissection; fingerprint; gene expression analysis, rheumatoid arthritis, osteoarthritis

1. Introduction

The sequencing of the human genome and the development of powerful and sensitive analytical tools have provided a framework for studying the intracellular mechanisms of cell activation at the molecular level. To achieve a better understanding of the pathogenesis and the course of arthritic joint diseases such as rheumatoid arthritis (RA) and osteoarthritis (OA), and to facilitate long-term monitoring of the effects of anti-inflammatory and disease-modifying drugs, it is desirable that this molecular analysis can be performed on specific cell populations that are crucial for the disease and that can be obtained from their natural or diseased environment.

From: *Methods in Molecular Medicine, Vol. 135: Arthritis Research, Volume 1*
Edited by: A. P. Cope © Humana Press Inc., Totowa, NJ

Current methods to identify genes that are differentially expressed between rheumatoid arthritis and related arthritic diseases such as osteoarthritis, and vice versa, are usually based on RNA isolated from either cultured cells or from synovial biopsies. In both cases, the examined cells consist of a mixed population of all cell types from the different areas and compartments of the synovium. In general, this problem can result in gene expression profiles that most likely superimpose each other.

To address this issue and to obtain detailed insights into the molecular processes occuring in exactly defined histological areas of the synovium, we adapted laser mediated microdissection (LMM) *(1)* for the analysis of RA synovium *(2)*. This method has recently been developed for molecular oncology *(3)* and is also being used in our laboratory for examination of gene expression profiles of colonic crypts, adenomas and (pre)-malignant stages of colon carcinoma *(4,5)*. Based on these experiences, it can be concluded that LMM can easily be adapted for the analysis of other types of tissue samples including synovia derived from OA joints, from other arthritides, and, even for the analysis of relatively acellular tissues such as articular cartilage.

Following the protocol described below, LMM allows isolation of histologically clearly defined regions of a cryosection *(1)*. After RNA isolation, subsequent nested RNA arbitrarily primed polymerase chain reaction (RAP-PCR) *(6,7)* amplifies a subpopulation of the expressed RNA sequences isolated from the dissected tissue. The "fingerprints" generated by this strategy can then be compared to identify genes that are differentially expressed between two distinct and histologically defined areas of the same synovial tissue or between the same area of different tissue samples. It is important that differences documented using this technique are confirmed at both the RNA and the protein level using in situ hybridization and immunohistochemistry.

2. Materials

2.1. Preparation of Tissue Sections

1. Synovial tissue samples.
2. Embedding medium (Tissue Tek OCT medium, VWR Scientific Products Corporation, San Diego, CA).
3. Custom-made aluminum mold (aluminum foil modeled into a mold by wrapping it around the end of a text marker; about 3 cm deep: autoclaved).
4. Cryostat (Reichert-Jung 2800 Frigocut, IMEBINC, San Marcos, CA or equivalent).
5. PEN membrane (PALM, Wolfratshausen, Germany).
6. RNase Zap (Ambion, TX).
7. Poly-L-lysine: 0.01% poly-L-lysine in sterile, DEPC treated H_2O.

8. Ultraviolet (UV) crosslinker (UV Stratalinker 2400, Stratagene, San Diego, CA or equivalent).
9. Hematoxylin solution: dissolve 1 g hematoxylin, 0.1 g sodium iodate, 50 g potassium aluminum sulfate in 1 L DEPC-treated H_2O. Add 50 g chloralhydrate and 1 g crystalized citric acid, filter sterilize, and store in the dark.

2.2. Laser Microdissection

1. Robot Microbeam laser microscope (PALM,) or equivalent.
2. Autoclaved LPC microcentrifuge tubes (clear, 500 µLvolume; PALM).

2.3. RNA extraction and RAP-PCR

1. RNeasy spin column purification kit (Qiagen, Hilden, Germany) or other RNA isolation kit.
2. Arbitrary oligonucleotide primers (10–12mer) (for examples *see* below and *[6]*).
3. 10X reverse transcription (RT) buffer (500 mM Tris-HCl (pH 8.3), 750 mM).
4. KCl, 30 mM $MgCl_2$, 200 mM dithiothreitol [DTT].
5. 100 mM dNTP mix.
6. MMuLV reverse transcriptase.
7. 10X RAP-PCR buffer: 100 mM Tris-HCl (pH 8.3), 100 mM KCl, and 40 mM $MgCl_2$.
8. *AmpliTaq*® DNA polymerase Stoffel fragment (Perkin Elmer, Norwalk, CT) [α-^{32}P] dCTP (3000 Ci/mmol).
9. Thermocycler (Model 9700, Applied Biosystems, Foster City, CA or equivalent).

2.4. Gel Electrophoresis and Cloning

1. Agarose gel and sequencing equipment.
2. Gel dryer (Model 583, BioRad, Hercules, CA or equivalent).
3. Gel loading buffer: 0.25% bromophenol blue, 0.25% xylene cyanol FF, 30% glycerol.
4. Tris-borate-ehylene diaimine tetraacetic acid (EDTA) buffer (TBE): 90 mM Tris-borate, 2 mM EDTA.
5. 8M urea, 5% polyacrylamide sequencing gel, prepared with TBE buffer (running buffer: 1X TBE).
6. SSCP gel: 25 mL 2X MDE gel solution (FMC, Rockland, MD), 6 mL 10X TBE, 10 mL 98% glycerol, H_2O to 100 mL, polymerized with 400 µL APS, and 40 µLTEMED.
7. SSCP running buffer: 0.6X TBE.
8. SSCP loading buffer: 0.25% bromophenol blue, 0.25% xylene cyanol FF, 30% glycerol, and 50% formamide.
9. BioMax X-Ray film (Kodak, Stuttgart, Germany) or equivalent.
10. TOPO-TA-Cloning® Kit DUALPromoter (Invitrogen, De Schelp, Netherlands) or other TA cloning system.

3. Methods
3.1. Tissue Preparation and Embedding

1. Rinse fresh rheumatoid synovial tissue samples in cold buffered saline to remove excess fibrin and clotted blood.
2. Remove adipose tissue as much as possible.
3. Dissect the tissue specimens into individual segments no larger than 5 mm using a sterile scalpel.
4. Fill a custom-made aluminum mold with TissueTek embedding medium, and place tissue segments in the mold so that they are completely submerged in embedding medium; the target surface of the tissue section should lie flat and face down to facilitate easier identification of the target tissue compartment for cryocutting.
5. Lower this mold slowly into liquid nitrogen to prevent uneven freezing which would generate ice crystals in the sample; this ultimately leads to cracks within the frozen embedding medium and reduced quality of tissue sections.
6. Label all cryoblocks with a cryomarker, wrap them tightly in aluminum foil and store at –80°C until sectioning. Ideally, tissue harvesting and processing should be performed as rapidly as possible to prevent excessive RNA degradation (*see* **Note 1**).

3.2. Preparation of the Slides

1. Clean glass slides with ethanol and RNAse Zap® to remove any residual RNAses.
2. Carefully mount a piece of PEN membrane of about the same dimensions as the slide onto the slide with tweezers (*see* **Note 2**).
3. Flatten the membrane using sterile cotton swabs to prevent the occurrence of creases.
4. Fix the membrane by using a small piece of autoclave tape on the barcode label side of the slide.
5. Dip the slide in 100% ethanol and allow it to air dry in a vertical position.
6. Smooth out creases and tape the membrane on the slide area opposite to the barcode. Slides prepared this way can be stored in a clean, closed container for several weeks at room temperature.

3.3. Preparing the Tissue Sections

1. Immediately before use, treat the slides with a UV crosslinker ($1200\ \mu J/cm^2$) and coat them with poly-L-lysine under RNase-free conditions by dipping them into 0.01% poly-L-lysine dissolved in DEPC-treated H_2O. This treatment facilitates adherence of tissue sections to the membrane. After air drying, the coated slides can be stored in an RNase-free container at 4°C for up to 2 wk.
2. Place the tissue in the cryostat for about 10 min prior to cutting to allow adjustment to the cutting temperature.
3. Prepare cryosections (7–8 µm) under RNase-free conditions (*see* **Note 3**).

4. After a short drying phase, fix the slides in 95% ethanol/5% acetic acid at –20°C for 5 to 10 min.
5. Airdry slides thoroughly and incubate in hematoxylin solution for up to 5 min to facilitate later identification of tissue areas by light blue cellular staining (*see* **Note 4**).
6. Discolor slides in DEPC-treated tap water (not in deionized water) and dehydrate in an ascending ethanol sequence (75, 95, and 100% ethanol, dipping the slides in ethanol for a few seconds is sufficient).
7. Finally, the slides have to be dried again, because residual ethanol interferes with laser dissection. To minimize RNA degradation, the time span between staining and microdissection needs to be kept as short as possible.

3.4. Laser Microdissection

3.4.1. General Principles

The Robot Microbeam laser microscope (PALM) uses a pulsed nitrogen laser with a wavelength of 337 nm (*see* **Fig. 1**). The small beam focus of the UV laser allows very accurate photoablation: in this process the focused laser beam cracks molecules at the cutting line to the atomic or low molecular level. Because of its short wavelength, this type of nonthermic ablation avoids energy dispension into adjacent cells and therefore preserves RNA integrity in the target cells and tissues. LMM is a rapid, reliable and precise technique. In addition, physical contact of the investigator with the tissue is negligible, which minimizes contamination. Also, multiple similar tissue areas can be harvested from the same section because surrounding tissue is left intact.

Dissected areas are collected in the cap of a microcentrifuge tube via laser pressure catapulting (LPC). LPC catapults the object of interest vertically off the membrane-coated slide using a high energy, defocused, short-duration laser pulse (*see* **Fig. 2**). Alternatively, larger pieces of tissue (e.g., sublining) can be picked with a 27-gage needle under a stereomicroscope and transferred to a microcentrifuge tube. The cap with the dissected tissue is then placed immediately on the microcentrifuge tube containing lysis buffer (RNeasy spin column purification kit or equivalent) and the tissue is lysed by mixing. This solution should be kept on ice or, for long term storage, at –80ϓC.

A critical parameter for the success of the entire procedure is the minimum amount of cells needed to obtain enough RNA to generate a stable fingerprint (i.e., a highly reproducible pattern of amplification products that range from 100 to approx 600 bp in length). This amount varies considerably for tissues of different origin and needs to be determined individually. As a starting point, areas containing 150, 300, 450, 600, and 900 cells should be microdissected from different tissue regions and used for RNA isolation (*see* **Note 5**).

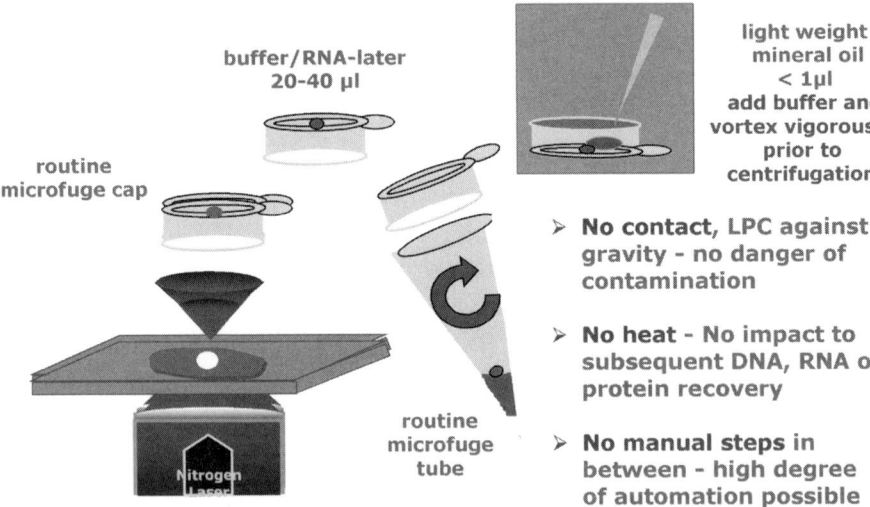

Fig. 1. Schematic diagram of the LMM procedure (generously provided by PALM).

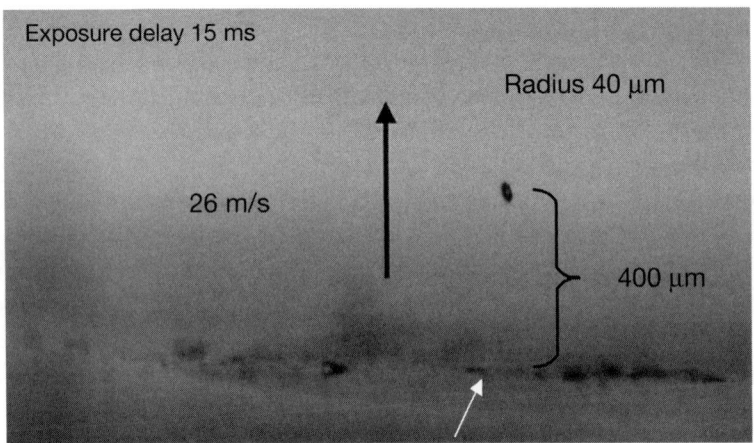

Fig. 2. Catapulting of a tissue fragment towards the cap of a microcentrifuge tube (Laser pressure catapulting, LPC; generous gift of Dr. A. Vogel, MLL, Lübeck, Germany). Black arrow: movement vector of catapulted fragment, red arrow: origin of catapulted fragment.

3.4.2. Microdissection

1. Count 150 cells of the region of interest.
2. Circle the respective area using the tissue-marking function of the PALM software (*see* **Figs. 3** and **4**).

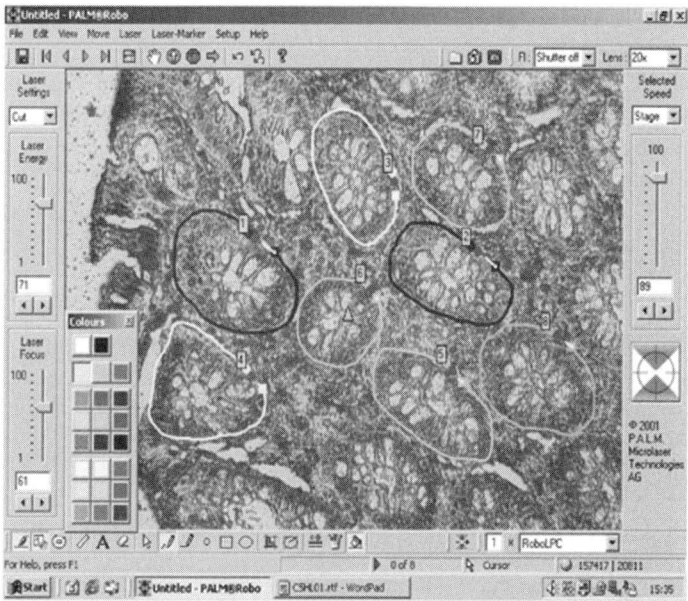

Fig. 3. Selection of the areas of interest (here: lining layer of rheumatoid synovium) using the tissue-marking feature of the PALM software (dark line). Marked areas can thereafter be automatically dissected and sampled by the PALM program. Note that the high magnification even allows to in- or exclude single cells. Triangle (in 6): starting point of laser beam.

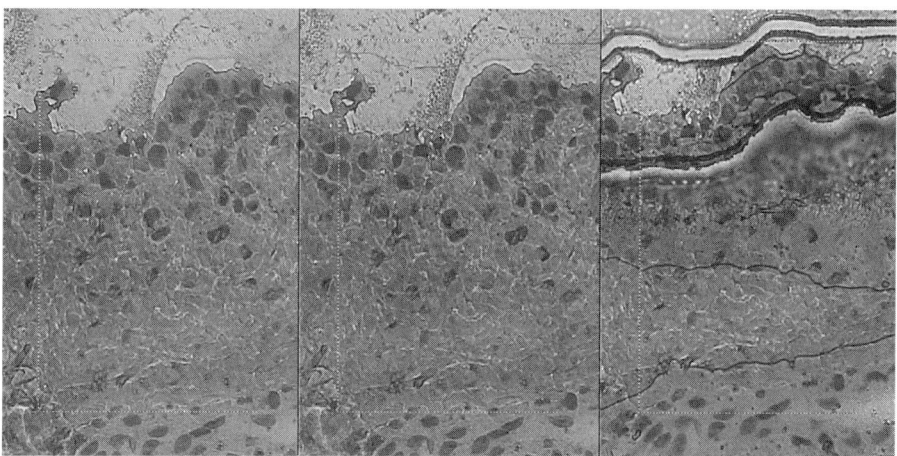

Fig. 4. Laser microdissection of the lining layer of osteoarthritis synovium. Note that owing to the thin lining layer a high magnification is essential for the microdissection of this compartment of the OA synovium. *Left*: native section; *middle*: the dark line is the tissue-marking feature of the PALM software; *right*: Status after cutting with the laser beam along the determined dissection line.

3. Sample 1, 2, 3, or 6 fragments of the same size and cell density sequentially by either laser pressure catapulting or picking them up with a sterile needle.
4. Transfer fragments to an LPC microcentrifuge tube containing 300 µ: lysis buffer immediately.
5. **Figure 5** illustrates the same section before and after removal of an inflamed region around a small terminal synovial vessel.

3.5. RNA extraction

1. Extract RNA by silica gel binding using the RNeasy spin column purification kit or any other method providing DNA-free high quality RNA.
2. Treat total RNA on the spin column with DNase I (Qiagen) at room temperature for 30 min to remove the remaining genomic DNA (*see* **Note 6**).
3. Perform the extraction according to the instructions of the manufacturer, and elute the RNA with 30 µL of RNase-free water (*see* **Note 7**).
4. Apply the eluate to the column a second time and spin through again to increase yield (*see* **Note 8**).

3.6. RAP-PCR

3.6.1. General Principles

RNA arbitrarily primed PCR (RAP-PCR) *(4,5,8)* results in amplification of an arbitrary subset of the total cellular RNA and has several advantages when compared with other commonly used amplification methods.

Unlike all methods using oligo dT priming, which tends to amplify preferentially the 3'-end of the RNA. Because the 3'-region is often untranslated, it is generally more variable and less informative than the translated part of the mRNA. In a typical RAP-PCR fingerprint, about 50 to 100 fragments can be visualized, including relatively rare messages that happen to match with the arbitrary primers. Therefore, rare messages that may not have been detected by other methods can also be detected *(6)*. Finally, RAP-PCR can be finished in a few hours and is a rather robust and easy-to-perfom method.

RAP-PCR consists of two major steps: (1) transcription of total RNA into cDNA using reverse transcriptase (RT reaction) and (2) a first arbitrary primer followed by amplification of a subset of the cDNA (arbitrarily primed PCR, or AP-PCR). This procedure uses short primers (10–12mers) of an arbitrarily selected sequence and low annealing temperatures of only 35°C. These conditions permit annealing of the primers to regions on the target nucleic acid when they match at least 5 or 6 bases at the 3'-end, a situation that on average takes place about every few hundred base pairs.

Nested RAP-PCR differs from the original protocol by the introduction of a second AP-PCR step of 35 amplification cycles using a nested primer. This primer is shifted with its complete base sequence one base to the 3'-end (and

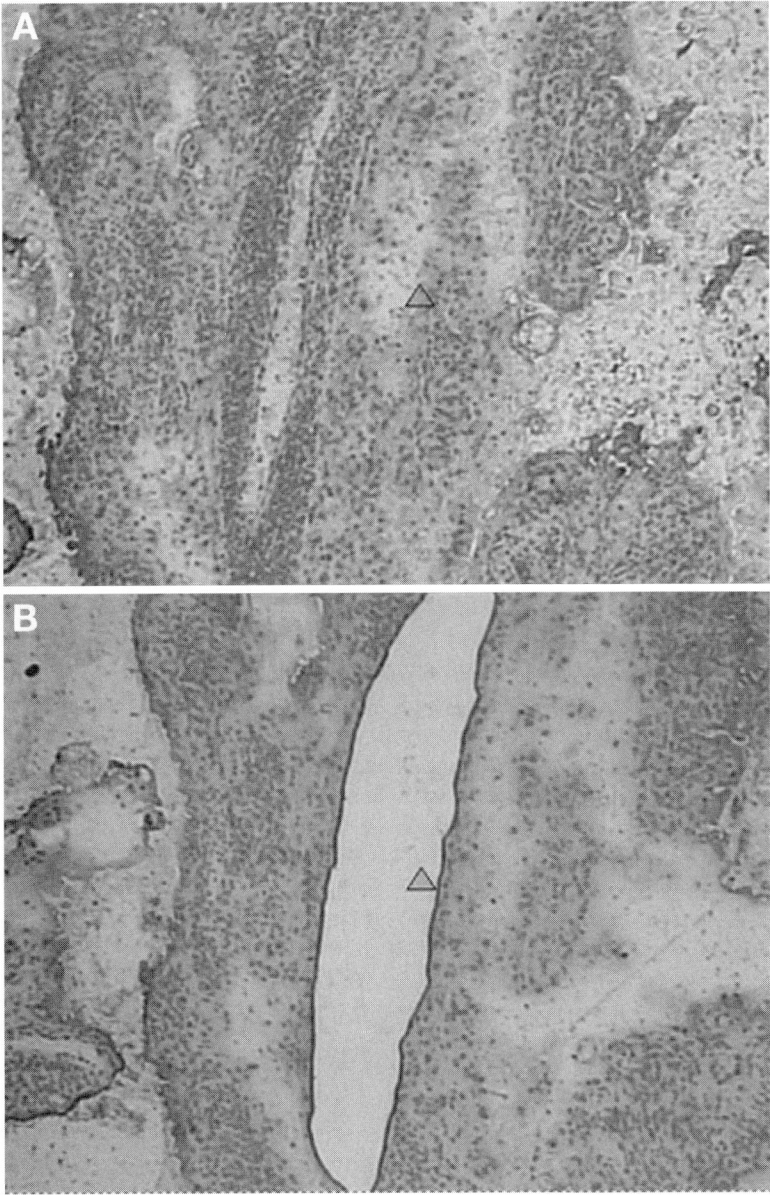

Fig. 5. Same tissue section before (**A**) and after (**B**) LMM of an inflamed region around a terminal vessel in rheumatoid synovium. *Open arrow*: lining layer; *bold arrow*: terminal vessel surrounded by inflammatory cells. *Triangle*: starting point of laser beam (experiment performed in collaboration with Dr. Ulfgren and Prof. Klareskog, Karolinska Insitute, Stockholm, Sweden).

therefore including this "nested" base into the amplification process) and a reduction of the first PCR step to 15 cycles. As a result of this altered primer sequence, the complexity of the fingerprint is reduced even further resulting in improved detection of rare messages.

3.6.2. RAP-PCR Fingerprinting Method

1. Mix RNA (approx 5 µg in 30 µL) with 20 µL toom temperature (RT) solution for a final volume of 50 µL containing 1X RT buffer, 0.2 mM dNTPs, 2 µM primer, and 100 U MMuLV reverse transcriptase (*see* **Note 9**).
2. To exclude DNA contamination, a reverse transcriptase-free reaction should be included in all RAP-PCR experiments.
3. Apply the following reaction conditions: 5 min ramp from 25°C to 37°C (primer annealing), 60 min at 37°C (reverse transcription), and 15 min at 68° (inactivation of enzyme).
4. Purify and concentrate the synthesized cDNA using the Microcon YM-100 kit (Millipore, Billerica, MA) according to the manufacturer's instructions. The final eluate contains the entire amount of cDNA in a volume of only 1 to 5 µL.
5. Perform AP-PCR for second strand synthesis using the nested PCR strategy. Preamplify for 15 cycles using the first arbitrary PCR primer (*see* **Note 9**) by mixing the cDNA (approx 5 µg in 1–5 µL, whole volume eluted from the Microcon column) with PCR solution for final concentrations of 1X PCR buffer, 0.2 mM dNTPs, 4 µM primer, and 2.5 U AmpliTaq® DNA polymerase Stoffel fragment in a volume of 20 µL.
6. Cycling conditions are as follows: 5 min at 94° (initial denaturation), 15 cycles of 94°C for 30 s, 35°C for 30 s, and 72°C for 60 s, and a final extension of 7 min at 72° (final extension).
7. Purify the PCR products using the Millipore Microcon YM-100 kit and bring to a total volume of 30 µL with nuclease-free water.
8. Use one third of the amplified products (10 µL) for the second amplification step with the nested 10mer arbitrary primer (*see* **Note 9**).
9. The reaction is performed in a volume of 20 µl under the same conditions as before with the addition of 2 µCi [α-32P] dCTP (3000 Ci/mmol).
10. Cycling conditions are the same as indicated above, except that this time 35 cycles are performed.
11. All reactions are carried out in duplicate to test the reproducibility and stability of the RNA fingerprint.

3.6.3. Gel Electrophoresis

1. Mix an aliquot of the PCR reaction 1:1 with loading buffer, denature it (94°C for 3 min, then put on ice immediately) and load onto an 8 M urea, 5% polyacrylamide sequencing gel prepared with TBE buffer.
2. Perform electrophoresis at 45°C and 100 W for about 90 min or until the bromophenol blue band reaches the lower third of the gel.

3. Peel the gel away from the glass support using a Whatman filter paper of about the same size as the gel and cover it with cling film.
4. Dry the gel under vacuum at 80°C for 1 to 2 h and expose it to a BioMax X-Ray film (Kodak, Stuttgart, Germany) overnight.

3.6.4. Band Excision and Elution of Fragments

1. Visually identify differentially expressed bands on the X-ray film.
2. Excise them from the gel by aligning the film with the dried gel, marking the bands of interest with a pin and cutting out the bands using a scalpel (*see* **Note 10**).
3. The corresponding region must also be isolated from the respective control lane, which should contain the product in significantly higher or lower levels.
4. Elute PCR products by soaking the gel fragments, with the Whatman filter paper still attached, in Tris buffer (10 m*M* Tris-HCl [pH 8.5]) at RT for 1 min, followed by a centrifugation step.
5. Transfer the supernatant containing the PCR fragments to a fresh tube and store it at –20°C or use it immediately for reamplification.

3.7. Purification of Differentially Expressed Bands by Gel Electrophoresis

3.7.1. SSCP Gel Electrophoresis—General Principles

As these eluates usually contain a mixture of several PCR products of more or less identical molecular weight, further purification by other techniques is required. Single strand conformation polymorphism (SSCP) gel electrophoresis can be used for this purpose. This method utilizes a characteristic phenomenon of single-stranded DNAs: they fold to form stable and metastable structures that affect their mobility in a nondenaturing acrylamide gel, thereby facilitating differentiation between PCR products of roughly the same molecular weight.

3.7.2. SSCP Gel Electrophoresis Method

1. Reamplify PCR products for 20 cycles using the same conditions and primers that were used in the second AP-PCR step.
2. Mix 5 μμL of this PCR reaction with 15 μL SSCP loading buffer and denature at 95°C, then put on ice immediately.
3. Load samples onto the SSCP gel.
4. Run the gel for 12 to 15 h at a constant power of 5 W.
5. Peel the gel away from the glass support using Whatman filter paper (*see* **Note 11**), dry it and expose it to BioMax X-Ray film.
6. Cut out differentially expressed bands and elute PCR products as before. Reamplify the products again, omitting the [α-^{32}P] dCTP.
7. Run the products on an agarose gel.
8. Cut out and elute fragments using the MinElute gel extraction kit (Qiagen).

3.8. Cloning and Sequencing of Differentially Expressed Genes

The eluate can then be used for cloning into the PCR®-II TOPO vector using the TOPO-TA-Cloning® Kit DUALPromoter according to the manufacturer's instructions and sequencing by standard methods *(4,5,8)*. At least five clones should be sequenced per cloning reaction to ascertain that only the desired sequence and not a mixture of different PCR products has been cloned.

3.9. Confirmation of Differential Expression

The differential expression of genes identified by the combination of LMM and RAP-PCR needs to be confirmed by other methods as there is always a chance that during the miscellaneous reamplification and gel electrophoresis steps an irrelevant gene sequence has been selected.

If an antibody to the respective protein is available, immunohistochemistry is the method of choice. Otherwise, expression of the corresponding mRNA may be confirmed by *in situ* hybridization. Because the TOPO-TA-Cloning® DUALPromoter vector used for cloning and sequencing already contains SP6 and T7 RNA polymerase promoters, RNA probes for the gene of interest can be easily generated.

4. Notes

1. Frozen tissue should be kept at very low temperatures and well-wrapped for as long as possible during the procedure (use dry ice for transportation. Due to slow but continuous RNA degradation even at $-20°C$, tissue specimens should not be used more than 3 times.
2. The electrostatically charged PEN membrane is very fragile and has to be handled with extra care.
3. Four to ten cryosections can be cut and placed on the same slide. However, the embedding medium from one section must not touch or overlap tissue from another section.
4. The hematoxylin solution should not be used more than three times. Dyeing takes less time using fresh hematoxylin solution compared with previously used solutions.
5. For rheumatoid arthritis synovial tissue, an area containing 600 cells (spanning between 200.000 and 2.000.000 μm^2, depending on cell density) was sufficient to obtain the amount of RNA necessary for stable fingerprinting. OA samples frequently require higher numbers because of the lower cellular activity of the synovial cells. Therefore, in all our subsequent experiments, tissue areas containing at least 600 cells were used. It is crucial to use only areas that represent a homogenous compartment of the tissue of choice (as opposed to "mixed" areas containing sublining as well as microvasculature, for example). Our experience suggests that laser microdissection can also be performed directly at the invasion zone of synovium into the adjacent cartilage.

6. Although RNA fingerprints are relatively insensitive to DNA contamination, precaution should always be taken. Therefore, a DNase digest should always be performed.
7. Addition of carrier RNA was tested but showed no improvement in yield and was thus discontinued.
8. To get a feeling for the amount of RNA that can be isolated, the RNA concentration of a few samples can be determined using the RiboGreen RNA Quantitation kit (Molecular Probes, Eugene, OR), which detects RNA in concentrations of down to 1 pg /μL.
9. Of the different arbitrary primers that were tested, for our purposes the combination of OPN23 (5'-CAG GGG CAC C-3') as RT primer and OPN21 (5'-ACCAGGGGCA-3')/OPN21 nested (5'-CCAGGGGCAC-3') for the AP-PCR steps resulted in the most stable and reproducible fingerprints. However, other arbitrarily chosen 10 to 12mer primers may work as well or even better for samples of different origin.
10. Cutting the bands from the gel has to be done no later than 2 wk after drying of the gel as the gel will shrink over time, making an exact alignment with the film difficult.
11. Due to the added glycerol, SSCP gels are very sticky. They are best peeled off from the glass support by bending the Whatman paper 180° and carefully pulling it parallel to the surface of the glass plate.

Acknowledgments

The authors wish to thank Wibke Ballhorn and Birgit Riepl for excellent technical assistance. The experiments were supported by grants awarded by the German Research Society (DFG Mu 1383/1-3 and Mu 1383/3-3).

References

1. Bonner, R. F., Emmert-Buck, M., Cole, K., Pohida, T., Chuaqui, R., Goldstein, S., et al. (1997) Laser capture microdissection: molecular analysis of tissue. *Science* **278,** 1481–1483.
2. Judex, M., Neumann, E., Lechner, S., et al. (2003) Laser-mediated microdissection facilitates analysis of area-specific gene expression in rheumatoid synovium. *Arthritis Rheum.* **48,** 97–102.
3. Wild, P., Knüchel, R., Dietmar, W., Hofstädter, F., and Hartmann, A. (2000) Laser microdissection and microsatellite analyses of breast cancer reveal a high degree of tumor heterogeneity. *Pathobiology* **68,** 180–190.
4. Lechner, S., Müller-Ladner, U., Neumann, E., Dietmar, W., Welsh, J., Schölmerich, J, et al. (2001) Use of simplified transcriptors for the analysis of gene expression profiles in laser-microdissected cell populations. *Lab. Invest.* **81,** 1233–1242.
5. Lechner, S., Müller-Ladner, U., Renke, B., Schölmerich, J., Rüschoff, J., and Kullmann, F. (2003) Gene expression pattern of laser microdissected colonic crypts of adenomas with low grade dysplasia. *Gut* **52,** 1148–1153.

6. Trenkle, T., Welsh, J., Jung, B., Mathieu-Daude, F., and McClelland, M. (1998) Non-stoichiometric reduced complexity probes for cDNA arrays. *Nucleic Acids Res.* **26,** 3883–3891.
7. Welsh, J., Chada, K., Dalal, S.S., Cheng, R., Ralph, D., and McClelland, M. (1992) Arbitrarily primed PCR fingerprinting of RNA. *Nucleic Acids Res.* **20,** 4965–4970.
8. Kullmann, F., Judex, M., Ballhorn, W., et al. (1999) Kinesin-like protein CENP-E is upregulated in rheumatoid synovial fibroblasts. *Arthritis Res.* **1,** 71–80.

7

Preparation of Mononuclear Cells from Synovial Tissue

Jonathan T. Beech and Fionula M. Brennan

Summary

In rheumatoid arthritis (RA), the synovium represents the predominant site of inflammation and joint destruction and is regarded as the key organ involved in disease pathogenesis. It has been studied in different ways over the last 30 yr, yielding information about the mechanisms involved in disease and remains the tool most proximal to understanding the pathogenesis of RA. This chapter outlines how both histological and in vitro studies of dissociated tissue played key roles in the development of biological anti-TNF-α therapy and provides detailed protocols used routinely in the laboratory to facilitate studies of RA synovium and its composite cell populations.

Key Words: Rheumatoid arthritis; synovium; cytokine; organ culture; synovial fluid; cell-lineage isolation; magnetic cell separation; flow cytometry.

1. Introduction

The healthy synovium is a simple structure composed predominantly of two different cell types forming a membrane only one or two cells deep. Historically these cells were defined histologically and termed type "A" cells which were phagocytic and secretory, exhibiting many characteristics typical of macrophages. The type "B" cells were fibroblast-like and secreted hyaluronate, important for maintaining the lubricant properties of synovial fluid. In patients with rheumatoid arthritis (RA) however, the synovial lining becomes hyperplastic and undergoes drastic morphological changes, becoming highly vascularised and infiltrated by cells from the blood including T-cells, B-cells, and macrophages.

It has long been observed that areas of destruction in the joints of RA patients lie predominantly in regions adjacent to masses of proliferating cells. Thus, a significant pathological role for the proliferative synovial tissue in RA has been implicated and this tissue has been studied in vitro since the early 1970s. However, obtaining large numbers of synovial tissue samples (from both whole joint and biopsy) can be fraught with difficulty and this has led to a significant proportion of RA research being performed using cells from the more accessible synovial fluids. Although this has undoubtedly yielded important information, it is clear that the cellular composition of fluids is quite different from that of the membrane. Whereas lymphocytes and macrophages predominate amongst the synovial membrane infiltrate, the neutrophils predominate in the fluid *(1)*. The synovial fluid is a viscous "soup" of enzymes, serum proteins (including cytokines and their inhibitors) and hyaluronic acid at very high concentration, all of which are likely to influence the properties of cells that pass through the membrane to accumulate within it. Furthermore, the retention of different cell populations in the membrane can be viewed as potentially indicative of their role in the destruction mediated by the synovium.

Early studies of synovial tissue used organ cultures of synovium to establish that tissue-destructive collagenases, as well as prostaglandins, were released into the culture media during this process *(2–7)*. However, concerns grew that the methods used for culturing synovial explants favoured the growth of particular cell populations (chiefly fibroblasts). Indeed, subsequent work has revealed the many drawbacks inherent in organ culture (primarily associated with nutrient exchange), not least of which are the formation of apoptotic bodies within and migration of particular cell-types away from the centre of the explant. Consequently, Dayer et al. *(8)* rapidly began using proteolytic enzymes to disaggregate and disperse the synovial cells. This enabled the production of collagenases and prostaglandins to be assigned largely to the adherent population. In the mid-1980s, we began isolating mononuclear cells from the synovium in a similar way in order to study the role of immune mediators in the pathogenesis of RA.

Initially, we focused on known mediators of prostaglandin E2 production, cartilage destruction, and bone resorption such as the proinflammatory cytokines Tumor necrosis factor (TNF)-α and Interleukin (IL)-1. Using a combination of specific cDNA probes for mRNA analysis and bioassays, we found that dissociated synovial membrane cells produced both mRNA and bioactive IL-1α, IL-1β, TNF-α, and TNF-β proteins in the absence of any exogenous stimulation *(9)*. Expression of IL-1α and to a lesser extent IL-1β mRNA was found to be prolonged in culture, suggesting that at least some of the mechanisms involved in the maintenance of chronicity of the joint disease are retained in culture *(10)*. Subsequent studies revealed that both granulocyte-macrophage colony-stimulating factor (GM-CSF) and IL-8 were similarly spontaneously produced in these cultures *(11,12)*. GM-CSF induces class II expression on monocytes and is considered to play an important role in macrophage activation and cytokine production *(13)*.

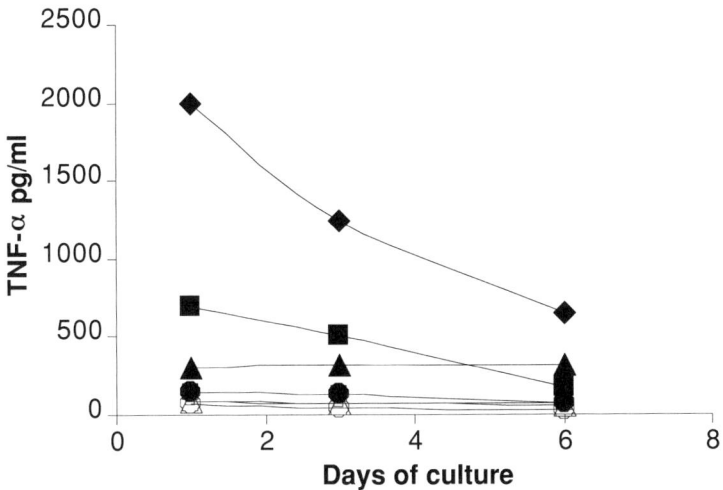

Fig. 1. TNF-α production in dissociated RA synovial cell cultures. Cells were isolated by enzymatic digestion of rheumatoid synovial tissue samples taken from seven different RA patients as described. Cells were subsequently placed in culture without exogenous stimulus. Supernatants were harvested after 1, 3, and 6 d in culture and assayed for TNF-α protein by ELISA. Results represent mean values of triplicate cultures.

With the ensuing availability of effective antibodies specific for these and other cytokines, we were not only able to measure their production directly in RA synovial cell cultures (using enzyme-linked immunosorbent assay [ELISA] and ELIspot assay) but also engage in inhibition studies to address their regulatory mechanisms. **Figure 1** illustrates typical kinetic profiles of TNF-α levels in dissociated RA synovial culture supernatants. Neutralizing antibodies to TNF-α, but not TNF-β, were shown to inhibit IL-1 production in RA synovial cultures *(14)*. Subsequent studies have revealed a key role for TNF-α in production of IL-6, IL-8 *(15)*, and GM-CSF (including the class II expression associated with it) *(11)*. Thus, a proinflammatory cytokine network for RA could be generated with TNF-α at its apex.

In 1994, Katsikis et al. demonstrated expression of the potent cytokine synthesis inhibitory factor and anti-inflammatory cytokine IL-10 in RA and OA joints *(16)*. Immunohistology revealed this protein to be localised to the synovial membrane lining layer and mononuclear cell aggregates. Not only was IL-10 spontaneously produced in RA and OA synovial membrane cultures, but subsequent neutralization of this endogenous protein by a monoclonal antibody resulted in a 2-3-fold increase in the protein levels of TNF-α and

IL-1β. The addition of exogenous recombinant IL-10 to the RA synovial cultures induced a corresponding decrease in levels of these proteins. Thus, IL-10 was revealed as an important immunoregulatory component in the cytokine network of RA. Subsequent studies utilising these cultures assigned similar anti-inflammatory roles to the soluble TNF receptors p75 and p55 *(17)* and IL-1 receptor anatagonist (IL1RA) *(15)*. Indeed, treatment of synovium with IL-10, or even more potently IL-4, has been shown to shift the balance between production of IL-1β and IL1RA in favour of the latter, as well as reducing IL-6 levels *(18,19)*.

These and many subsequent studies have now firmly established the key role played by TNF-α and other cytokines in regulating the many processes critical to the pathogenesis of RA. The importance of conducting these studies using the pathologically-relevant RA synovial tissue proved to be key in validating TNF-α as a therapeutic target. With the benefits of TNF-α-neutralisation in RA now well-established following the success of anti-TNF biologicals in the clinic, similar results are currently emerging in several other inflammatory diseases, including Crohn's disease, juvenile idiopathic arthritis, ankylosing spondylitis, psoriasis and psoriatic arthritis.

This chapter presents a comprehensive guide to the isolation and study of the mononuclear cell population of the RA joint. Our current protocol, developed over a period of many years allows the composite synovial cell populations to be studied in vitro both as a mixed population (i.e., within the context of eachother *[9,10,20]*) and in isolation *(21,22)*. The authors and others have isolated these populations from the synovial membrane for individual study *(18)*. **Figure 2** (from *[9]*) illustrates the cellular profile found within dissociated RA synovial cultures originating from seven different patients, as determined by flow cytometry. By far the predominant cell in most cultures is the $CD3^+$ lymphocyte, with monocyte macrophages ($CD14^+$) representing the second-most frequently seen cell-type. B-cells and natural killer (NK) cells are also present in variable numbers as more minor populations (data not shown). Interestingly, around half of the cells present were major histocompatibility complex (MHC) class II positive (DR^+), an indication of their highly activated state. Protocols for both the enzymatic digestion of synovial tissue and the subsequent purification of T-helper lymphocytes are outlined below. In principle these can be applied to purification of other cell populations (such as NK cells or monocytes) by simply altering the selective marker targeted during the isolation process.

2. Materials

It is perhaps an obvious, though critically important point that all procedures performed using cells/tissues from RA patients must be carried out under

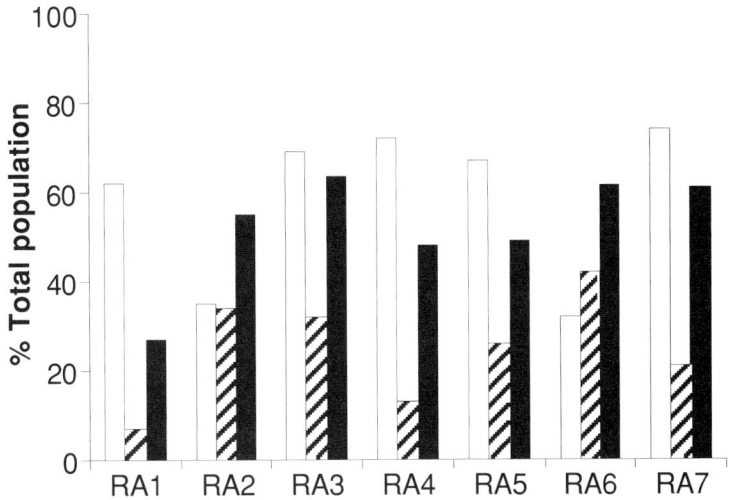

Fig. 2. Cellular composition of RA cultures. Cells were isolated by enzymatic digestion of rheumatoid synovial tissue samples taken from seven different RA patients as described. Cells were subsequently stained with fluorochrome-conjugated monoclonal antibodies specific for cell lineage markers and analysed by flow cytometry. □, %CD3$^+$; ▨, %CD14$^+$; ■, % DR$^+$.

sterile conditions. Equally important (though often overlooked) is the assurance that all reagents used are free of endotoxins which could potentially stimulate mononuclear cells and skew experimental results. All media, sera and reagents used in our laboratory are routinely tested for endotoxin using the Limulus Amoebocyte Lysate test (Biowhittaker Inc., Walkersville, MD) and rejected if the endotoxin concentration exceeds 0.1 U/mL.

All RA mononuclear cells we isolate are from synovial tissue specimens obtained either as a discarded tissue product of elective knee or hip joint replacement surgery, or as a result of wrist synovectomy. Refinement of the system over the years has enabled meaningful experiments to be performed using less tissue and consequently studies can now be carried out using tissue samples obtained as a result of diagnostic biopsies. All RA tissues/fluids are obtained from patients fulfilling the American College of Rheumatology (formerly the American Rheumatism Association) criteria for RA, with prior approval from the local research ethics committee.

2.1. Isolation of Mononuclear Cells From Synovial Fluids

1. Waste beaker containing anti-microbial agent, such as Virkon™ (Antec International, UK).
2. Sterile disposable pipettes (Costar®, Corning Inc., NY).

3. RPMI 1640 (BioWittaker, Cambrex BioScience, Belgium) supplemented with 5% (v/v) fetal calf serum (FCS) (heat inactivated; PAA laboratories, Austria) and 1% (v/v) penicillin/streptomycin solution (BioWittaker).
4. Hank's Balanced Salt Solution (HBSS) (BioWittaker).
5. Density gradient cell separation medium, such as Lympholyte®-H (Cedarlane, Ontario, Canada).
6. Sterile 50-mL conical tubes (Costar®, Corning Inc).
7. Haemocytometer and Trypan Blue solution (0.4%; Sigma-Aldrich, UK) for viable cell counting.

2.2. Isolation of Mononuclear Cells From Synovial Tissue

1. Waste beaker containing anti-microbial agent, such as Virkon™ (Antec International, UK).
2. Sterile scissors and forceps.
3. 10-mL Disposable syringe (Plastipak™ Becton Dickinson, S.A. Spain)
4. 0.2-µm filter.
5. Sterile disposable pipets (Costar®, Corning Inc.).
6. Sterile tissue culture (Petri-style) dish (Falcon® Becton-Dickinson, France).
7. 100 mg Collagenase (Roche Laboratories, Welwyn Garden City, UK).
8. 3 mg DNase (Sigma-Aldrich, UK).
9. RPMI 1640 (BioWittaker) supplemented with 10% (v/v) FCS (heat inactivated; PAA Laboratories) and 1% (v/v) penicillin/streptomycin solution (BioWittaker).
10. Sterile 50-mL conical tubes (Costar®, Corning Inc.).
11. Haemocytometer and Trypan Blue solution (0.4%; Sigma-Aldrich) for viable cell counting.
12. Sterile, autoclaved sieve beaker (*see* **Note 1**).
13. Sterile cell strainer (Nylon; 70 µm; BD Falcon™, BD Biosciences).

2.3. Isolation of Mononuclear Cell Lineages

2.3.1. Immunomagnetic Cell Isolation (Dynabeads)

1. Dynal magnetic particle separator® (or equivalent) (Dynal, Merseyside, UK).
2. Dynabeads® coated with antibodies to selective lineage marker of choice (in this case anti-CD4 mAb) (Dynal, Merseyside, UK).
3. HBBS (BioWittaker) supplemented with 2.5% (v/v) FCS (heat inactivated; PAA Laboratories).
4. Tube rotator (operated at 4°C).
5. Tissue culture plate (6 well; Falcon® Multiwell™ Becton Dickinson Labware).
6. RPMI 1640 (BioWittaker) supplemented with 5% (v/v) FCS (heat inactivated; PAA Laboratories).
7. Dissociated synovial tissue cells.

2.3.2. Fluorescence Activated Cell Sorting

1. Fluorescence-activated cell sorter (FACS) (we use FACSVantage™, BD Biosciences) with cooling facility (*see* **Note 2**).

2. Antibodies specific for selective lineage marker of choice. These antibodies must be conjugated (preferably directly) with a fluorochrome excited at the laser emission frequency for the sorter. In this case we use anti-CD4mAb directly conjugated to Phycoerythrin (PE) (BD Biosciences UK)
3. FACS staining buffer: phosphate-buffered saline containing 1% *(v/v)* FCS (heat inactivated; PAA Laboratories) and 1% *(v/v)* penicillin/streptomycin solution (BioWittaker) (*see* **Note 3**).
4. FACSFlow™ solution as sorting buffer (BD Biosciences).
5. Collection media: FCS containing 1% *(v/v)* penicillin/streptomycin solution (BioWittaker).
6. Dissociated synovial tissue cells.

2.4. Cell Culture of Synovial Mononuclear Cells

1. Complete culture media: RPMI 1640 (BioWittaker) supplemented with: 25 mM Hepes, 2 mM L-glutamine, 10% *(v/v)* FCS (heat inactivated) (PAA Laboratories), 2% *(v/v)* penicillin/streptomycin solution (BioWittaker).
2. Tissue culture plates: 24, 48, and 96 well flat-bottomed plates (Falcon® Multiwell™ Becton Dickinson Labware).
3. Incubator set at 37°C with an atmosphere of 5% CO_2.

3. Methods

3.1. Isolation of Mononuclear Cells From Synovial Fluids

As discussed above, synovial fluids are highly complex, containing many different cellular and molecular mediators of inflammation. Consequently, they can be notoriously difficult to work with; however the mononuclear cells present can be isolated using density gradient separation procedures. Using the right medium, it is possible to isolate viable lymphocytes and monocytes from the cellular "pool" that consists largely of neutrophils and dead cells in high numbers.

1. Transfer fluid to a sterile 50ml conical tube and centrifuge at 600g for 10 min at room temperature (22°C).
2. Remove supernatant (either discard into waste pot or aliquot and freeze if you have further use for it). Agitate tube gently to break up cell pellet and then resuspend in 25 mL supplemented RPMI 1640 (as outlined above).
3. Add 20 mL density gradient cell separation medium to a fresh conical tube and carefully layer the cell suspension over the top. Spin at 1000g for 25 min at room temperature (RT) (22°C) with the centrifuge brake off (to avoid disruption of the interface).
4. Carefully remove the interface layer containing cells into a new conical tube using a pipet. Discard the rest into waste pot.
5. Add HBSS to the tube up to 50 ml volume and centrifuge at 300g for 10 min at RT. Repeat once more to ensure all trace of the density gradient cell separation medium has been washed away.

6. Resuspend the cells in complete tissue culture media (*see* **Subheading 2.4.**) and count the cells in the presence of Trypan Blue (to differentiate between live and dead cells/debris).

3.2. Isolation of Mononuclear Cells From Synovial Tissue

To maximize cell viability, synovial tissue specimens taken from patients with RA undergoing joint surgery are placed immediately in RPMI supplemented with 10% (v/v) heat-inactivated foetal bovine serum (HIFBS) and 1% (v/v) penicillin/streptomycin prior to transportation. Briefly, the tissue is dissociated by cutting into small pieces and enzymatically digested with collagenase and DNase before subsequent culture or further purification processes.

1. Add 100 mg of collagenase and 3 mg of DNase to 20 mL of supplemented RPMI 1640 media. Mix by inverting several times until no crystals remain.
2. Filter the mixture using a 10-mL syringe and 0.2 µm filter into a new 50-mL conical tube. Set aside.
3. Carefully pour 2 to 3 mL of supplemented RPMI 1640 media into the tissue culture petri dish. This is to prevent the tissue drying-out during the next step.
4. Add the synovium, and using the sterile scissors and forceps, begin to cut up the tissue into small pieces of 4 to 5 mm^2. Discard any unnecessary tissue (i.e., fat, etc.) into the appropriate biohazardous waste bin. Continue until there are no large clumps.
5. Use the cell strainer (inserted into the neck of a conical tube) to isolate the diced tissue by quickly pouring the contents of the petri dish into it and allowing the media to drain through. *Note:* If the tissue is large, it may require several sieves. Be careful not to contaminate the tissue.
6. Using the forceps, scrape the tissue from the sieve into the conical tube containing the 20 mL of media, collagenase, and DNase. Mix tissue/enzyme thoroughly by gentle vortexing.
7. Incubate at 37°C for 1 h, periodically shaking vigorously. After 1 h, check the tissue. It should appear "gloopy" or "stringy." Shake mixture vigorously. If the tissue is still visibly clumpy, continue incubating for up to 30 additional min (*see* **Note 4**).
8. After the incubation, fill the conical tube to 50 mL volume with supplemented media to stop the reaction.
9. Filter contents of 50-mL conical tube through the sieve beaker (*see* **Note 1**) to remove the remaining pieces of tissue (they should be quite "gloopy" and/or "stringy"). Remove sieve material from beaker, being careful not to contaminate the spout. Dispose of the material as biohazard waste.
10. Spin down the sieved mixture at 200g (24°C) for 10 min. Remove any remaining fat floating on top of the supernatant using a pipette before carefully pouring off the remainder.
11. Resuspend the cells in complete tissue culture media (*see* **Subheading 2.4.**) and count them in the presence of Trypan Blue (to differentiate between live and dead cells/debris) (*see* **Note 5**) and adjust to 1 × 10^6 cells/mL.

3.3. Isolation of Mononuclear Cell Lineages

Using the digested synovial population, the individual composite populations can be isolated using a variety of techniques that utilise antibodies specific for markers of different cell lineages. The following protocol focuses on the purification of T-cells but can be adapted for almost any cell lineage. The protocols outline positive selection procedures, but can easily be adapted for negative selection by exchanging the selective antibodies for those specific for other "contaminating" lineages and using the remaining cell populations following fractionation. Such a negative selection procedure should be employed under conditions where antibody-mediated cell activation processes may be of concern. This possibility should be considered carefully when using CD3-mediated T-cell isolation, where we use negative selection wherever possible. However, both methods work well, yielding purified T-cell populations typically of >99% and >95% purity, respectively.

3.3.1. Immunomagnetic Cell Isolation (Dynabeads)

Briefly, anti-CD4-antibody-coated Dynabeads are incubated with the synovial cells before being passed through a magnetic field, which retains beads together with the cells bound to them. After thorough washing while in the magnetic field, the bead-bound cells are removed from the magnetic field, collected and cultured for a short period of time to allow the magnetic beads to detach from the cells before experimental use. Recently, Dynal have supplemented their protocol with the DETACHaBEAD reagent which will remove antibodies from the surface of cells by competitive binding techniques. We do not routinely perform this extra step, but find that both techniques perform equally well in our hands (*see* **Note 6**).

1. For selection of 1×10^7 CD4$^+$ cells, wash 100 µL of anti-CD4 coated beads 3 times in HBSS (containing 2.5% HIFCS) using the magnetic particle concentrator.
2. Add the magnetic bead solution to the washed mononuclear cells isolated from synovial tissue to a final volume of 2 mL. Incubate at 4°C on a rocking platform for 20 min.
3. Place the test tube in the magnetic particle concentrator (MPC) for 2 to 3 min to allow "rosetted" cells to attach to the wall of the test tube. Discard the supernatant and wash cells attached to beads 3 times in HBSS-HIFBS with the tube still in the magnetic particle concentrator.
4. Remove the test tube from the MPC, add supplemented RPMI media (approx 10 mL) and culture the cells for 6 h at 37°C in 3 to 4 wells of a 6-well plate to allow detachment of the cells from the magnetic beads.
5. Magnetic beads are then separated from detached cells by washing through the MPC.
6. After isolation, wash and count the cells before placing in culture. The purity and viability of the resulting isolated population should be confirmed using flow cytometry. This technique should typically yield CD4$^+$ enriched cells of high purity (>99% CD4$^+$) and viability (>95%).

3.3.2. Fluorescence Activated Cell Sorting

Briefly, fluorochrome-conjugated anti-CD4 antibodies are incubated with the synovial cells before being passed through a laser beam. Cells with bound antibody will fluoresce in the laser field and can be deflected differently enabling the populations to be separated and collected in separate tubes.

1. Centrifuge the mononuclear cells isolated from the RA membrane at 200g for 10 min at 4°C and resuspend the pellet in FACS staining buffer at 1×10^6 cells/mL.
2. Add the appropriate amount of fluorochrome-conjugated antibody(ies) (the BD anti-CD4-PE mAb is recommended for use at 20 µL per 1×10^6 cells, but we suggest titrating the reagent to find the optimal working dilution) and incubate on ice (or 4°C) for 20 to 30 min in the dark.
3. Spin down the cells at 200g for 10 min at 4°C. Resuspend the pellet in FACS staining buffer. Repeat this wash step a further two times and adjust to 5–10 × 10^6 cells/ml (see **Note 7**).
4. Set up the sorter (use FACSflow™ as the sorting media) and begin the sorting process. Collect the cells into tubes containing collection media, ensuring the tubes are maintained at 4°C throughout the sorting process by the cooling device (see **Note 2**).
5. After sorting, wash and count the cells before placing in culture. The purity and viability of the resulting isolated population should be confirmed using flow cytometry. This technique should typically yield CD4$^+$ enriched cells of high purity (>99% CD4$^+$) and viability (>95%).

3.4. Cell Culture of Synovial Mononuclear Cells

Once the synovial tissue is dissociated into a single-cell suspension, it can be cultured as a whole population, or in its individual cell components. If necessary, the cells may also be frozen and stored in liquid nitrogen for recovery and use at a later date (see **Note 8**). The total cell mixture is cultured at 37°C in complete medium for 2 to 5 d. The haemopoietic cells will die when synovial membrane cells are cultured for longer periods, resulting in the outgrowth of the fibroblastic population. The generation of fibroblastic cell lines through repeated passaging of RA cells is a widely used technique for studying the role of this population in RA pathology. However, the risk of substantial in vitro growth selection and phenotypic changes is high and consequently improved methods have been sought *(23,24)*. The propagation of synovial fibroblasts is covered in detail elsewhere in this book.

Cell cultures are seeded from a 1×10^6 cells/mL suspension (excluding red blood cells) in volumes of 1 mL, 0.5 mL, and 250 µL in 24-, 48-, or 96-well plates, respectively. **Figure 3** illustrates how these cultures look morphologically through the microscope. Here we can observe and identify many of the different composite populations, including macrophages, fibroblasts, T-cells, and red blood cells.

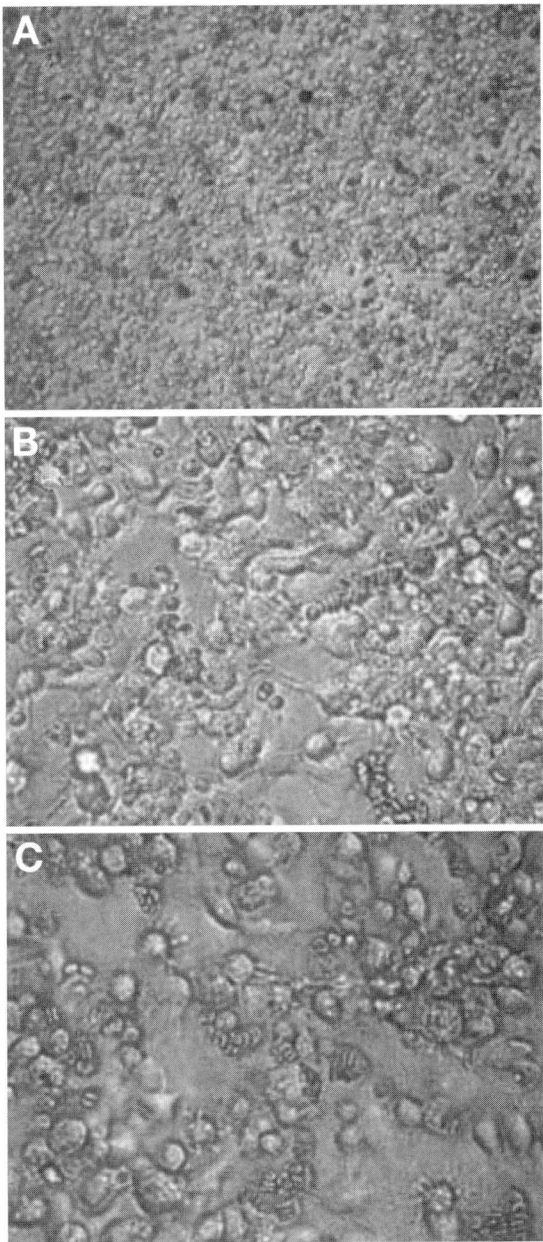

Fig. 3. Microscopic morphology of dissociated RA synovial membrane cells. Dissociated RA synovial cells were seeded in 48-well plates at 0.5×10^6 cells/well and cultured overnight at 37°C. Images were then taken at variable magnifications (increasing A→C). The full spectrum of constituent cell populations is visible, including lymphocytes, macrophages, fibroblasts and red blood cells.

4. Notes

1. *Preparation of sieve beakers:* A small square of approx 170 μm material (cheesecloth or sheer cloth) should be cut—enough to cover a 250-mL beaker. The material should be stretched over the beaker with enough slack to form a well in the middle. This will allow the digested synovium to sieve through to the beaker below. The material should be attached via autoclave tape. Cover the beaker with aluminium foil, tape, and autoclave. When subsequently drying the beaker, inverting it will reduce the condensation formed inside the beaker. Once dry, the beaker is now ready to be used.
2. *Maintaining sorted cells at low temperature:* Because of the relatively lengthy process of cell sorting (as compared with magnetic separation) techniques, we have found that the resulting cell viability can be quite severely compromised by allowing the sorted populations to sit at room temperature during the process. We do not therefore recommend using a cell sorter without a cooling facility. If this is unavoidable then high speed sorting (with the resulting compromised yield purity) should be the preferred mode of operation.
3. *FACS staining buffer :* Note that unlike conventional "FACS buffer" for cell scanning, this buffer does *not* contain sodium azide which, while preventing antibody "capping" in cells for scanning applications, is ultimately toxic for cells and consequently cannot be used where the cells are sorted for subsequent culture.
4. *Digestion:* It is critical not to digest for longer than 1.5 h because after this time, deeper tissue cell types will become dissociated.
5. *Red blood cells:* When performing the cell count, be sure to exclude red blood cells, which should appear smaller and darker than the other cells. If they are not easy to distinguish, adjusting the focus will often reveal their concave shape. If the number of red blood cells is large (i.e., above 20–30%), it is advisable to perform a red blood cell lysis step. We use Red Blood Cell Lysing Buffer (0.83% ammonium chloride in 0.01 M Tris buffer) following the manufacturers recommended protocol. Briefly, 1 mL of lysing solution is added to a loosened cell pellet (containing between 1 and 200×10^6 cells) and gently mixed for 1 min before diluting the buffer out with 15 to 20 mL medium and subsequent centrifugation. This protocol was developed for lysis of mouse splenic red blood cells but we find it works well for human red blood cells from synovial tissue. If the lysis is incomplete however, the process may be repeated.
6. *DETACHaBEAD reagent use:* As previously mentioned, we have not found it necessary to routinely utilise this additional step. The DETACHaBEAD reagent is a polyclonal anti-Fab antibody specific for the primary selection antibody attached to the Dynabeads. When the reagent is added to the bead-bound cells it competes with antibody/antigen binding at the cell surface and releases the antibody and bead from the cell. For in-depth protocol refer to manufacturers instructions. Briefly, the DETACHaBEAD reagent is incubated with the positively-selected cells for 45 to 60 min at RT, before the tube is placed in the magnetic particle concentrator for 2 min. Following removal of the supernatant (cells) and several washes the final cell population should be free of cell surface antibody.

7. *Cell sorting concentration:* The ideal cell concentration for sorting is proportional to the sort speed (i.e., high speed sorting can use cells at a higher concentration and has the obvious advantage of speed). However, low speed sorting is recommended if possible due to enhanced purity of the yields.
8. *Freezing of RA membrane cells:* We prefer to use the RA membrane cells fresh from isolation, without freezing wherever possible. Whereas use of freeze-thawed RA membrane cells is particularly suitable for studies using extracted cellular proteins, we find that such cells do exhibit slightly different characteristics in culture, most likely as a result of the cell damage/death and/or receptor shedding (particularly TNF-R) associated with the freeze-thaw process. Consequently, in such cases we usually "rest" the cells in culture for a few hours prior to using them in experiments. We find that freezing cells in a media consisting of 90% HIFBS and 10% dimethyl sulfoxide (DMSO) (as opposed to other media-HIFBS combinations) yields the best cell recovery on thawing.

Acknowledgments

Special thanks to Patricia Green, Parisa Amjadi, Sally Owen, Amy Peters, and Lauren Schewitz for valuable input. This work was funded by the arthritis research campaign (ARC), UK.

References

1. Zvaifler, N. J. (1973) The immunopathology of joint inflammation in rheumatoid arthritis. *Adv Immunol* **16,** 265–336.
2. Harris, E. D., Jr. and Krane, S. M. (1974) Collagenases (first of three parts). *N. Engl. J. Med.* **291,** 557–563.
3. Harris, E. D., Jr. and Krane, S. M. (1974) Collagenases (second of three parts). *N. Engl. J. Med.* **291,** 605–609.
4. Harris, E. D., Jr. and Krane, S. M. (1974) Collagenases (third of three parts). *N. Engl. J. Med.* **291,** 652–661.
5. Krane, S. M. (1975) Collagenase production by human synovial tissues. *Ann. N. Y. Acad. Sci.* **256,** 289–303.
6. Robinson, D. R., McGuire, M. B., and Levine, L. (1975) Prostaglandins in the rheumatic diseases. *Ann. N. Y. Acad. Sci.* **256,** 318–329.
7. Robinson, D. R., Tashjian, A. H., Jr., and Levine, L. (1975) Prostaglandin-stimulated bone resorption by rheumatoid synovia. A possible mechanism for bone destruction in rheumatoid arthritis. *J. Clin. Invest.* **56,** 1181–1188.
8. Dayer, J. M., Krane, S. M., Russell, R. G., and Robinson, D. R. (1976) Production of collagenase and prostaglandins by isolated adherent rheumatoid synovial cells. *Proc. Natl. Acad. Sci. USA* **73,** 945–949.
9. Brennan, F. M., Chantry, D., Jackson, A. M., Maini, R. N., and Feldmann, M. (1989) Cytokine production in culture by cells isolated from the synovial membrane. *J. Autoimmun.* **2 Suppl,** 177–186.
10. Buchan, G., Barrett, K., Turner, M., Chantry, D., Maini, R. N., and Feldmann, M. (1988) Interleukin-1 and tumour necrosis factor mRNA expression in rheumatoid arthritis: prolonged production of IL-1 alpha. *Clin. Exp. Immunol.* **73,** 449–455.

11. Haworth, C., Brennan, F. M., Chantry, D., Turner, M., Maini, R. N., and Feldmann, M. (1991) Expression of granulocyte-macrophage colony-stimulating factor in rheumatoid arthritis: regulation by tumor necrosis factor-alpha. *Eur. J. Immunol.* **21**, 2575–2579.
12. Brennan, F. M., Zachariae, C. O., Chantry, D., Larsen, C. G., Turner, M., Maini, R. N., Matsushima, K., and Feldmann, M. (1990) Detection of interleukin 8 biological activity in synovial fluids from patients with rheumatoid arthritis and production of interleukin 8 mRNA by isolated synovial cells. *Eur. J. Immunol.* **20**, 2141–2144.
13. Alvaro-Gracia, J. M., Zvaifler, N. J., and Firestein, G. S. (1989) Cytokines in chronic inflammatory arthritis. IV. Granulocyte/macrophage colony-stimulating factor-mediated induction of class II MHC antigen on human monocytes: a possible role in rheumatoid arthritis. *J. Exp. Med.* **170**, 865–875.
14. Brennan, F. M., Chantry, D., Jackson, A., Maini, R., and Feldmann, M. (1989) Inhibitory effect of TNF alpha antibodies on synovial cell interleukin-1 production in rheumatoid arthritis. *Lancet* **2**, 244–247.
15. Butler, D. M., Maini, R. N., Feldmann, M., and Brennan, F. M. (1995) Modulation of proinflammatory cytokine release in rheumatoid synovial membrane cell cultures. Comparison of monoclonal anti TNF-alpha antibody with the interleukin-1 receptor antagonist. *Eur. Cytokine. Netw.* **6**, 225–230.
16. Katsikis, P. D., Chu, C. Q., Brennan, F. M., Maini, R. N., and Feldmann, M. (1994) Immunoregulatory role of interleukin 10 in rheumatoid arthritis. *J. Exp. Med.* **179**, 1517–1527.
17. Brennan, F. M., Gibbons, D. L., Cope, A. P., Katsikis, P., Maini, R. N., and Feldmann, M. (1995) TNF inhibitors are produced spontaneously by rheumatoid and osteoarthritic synovial joint cell cultures: evidence of feedback control of TNF action. *Scand. J. Immunol.* **42**, 158–165.
18. Chomarat, P., Vannier, E., Dechanet, J., Rissoan, M. C., Banchereau, J., Dinarello, C. A., and Miossec, P. (1995) Balance of IL-1 receptor antagonist/IL-1 beta in rheumatoid synovium and its regulation by IL-4 and IL-10. *J. Immunol.* **154**, 1432–1439.
19. Chomarat, P., Banchereau, J., and Miossec, P. (1995) Differential effects of interleukins 10 and 4 on the production of interleukin-6 by blood and synovium monocytes in rheumatoid arthritis. *Arthritis Rheum.* **38**, 1046–1054.
20. Williams, L. M., Gibbons, D. L., Gearing, A., Maini, R. N., Feldmann, M., and Brennan, F. M. (1996) Paradoxical effects of a synthetic metalloproteinase inhibitor that blocks both p55 and p75 TNF receptor shedding and TNF alpha processing in RA synovial membrane cell cultures. *J. Clin. Invest.* **97**, 2833–2841.
21. Brennan, F. M., Hayes, A. L., Ciesielski, C. J., Green, P., Foxwell, B. M., and Feldmann, M. (2002) Evidence that rheumatoid arthritis synovial T cells are similar to cytokine-activated T cells: involvement of phosphatidylinositol 3-kinase and nuclear factor kappaB pathways in tumor necrosis factor alpha production in rheumatoid arthritis. *Arthritis Rheum.* **46**, 31–41.

22. Butler, D. M., Feldmann, M., Di Padova, F., and Brennan, F. M. (1994) p55 and p75 tumor necrosis factor receptors are expressed and mediate common functions in synovial fibroblasts and other fibroblasts. *Eur. Cytokine Netw.* **5,** 441–448.
23. Tsai, C., Diaz, L. A., Jr., Singer, N. G., Li, L. L., Kirsch, A. H., Mitra, R., Nickoloff, B. J., Crofford, L. J., and Fox, D. A. (1996) Responsiveness of human T lymphocytes to bacterial superantigens presented by cultured rheumatoid arthritis synoviocytes. *Arthritis Rheum.* **39,** 125–136.
24. Ermis, A., Henn, W., Remberger, K., Hopf, C., Hopf, T., and Zang, K. D. (1995) Proliferation enhancement by spontaneous multiplication of chromosome 7 in rheumatic synovial cells *in vitro*. *Hum. Genet.* **96,** 651–654.

8

Quantitative Image Analysis of Synovial Tissue

Pascal O. van der Hall, Maarten C. Kraan, and Paul Peter Tak

Summary

Quantitative image analysis is a form of imaging that includes microscopic histological quantification, video microscopy, image analysis, and image processing. Hallmarks are the generation of reliable, reproducible, and efficient measurements via strict calibration and step-by-step control of the acquisition, storage and evaluation of images with dedicated hardware and software.

Major advantages of quantitative image analysis over traditional techniques include sophisticated calibration systems, interaction, speed, and control of inter- and intraobserver variation. This results in a well controlled environment, which is essential for quality control and reproducibility, and helps to optimize sensitivity and specificity. To achieve this, an optimal quantitative image analysis system combines solid software engineering with easy interactivity with the operator. Moreover, the system also needs to be as transparent as possible in generating the data because a "black box design" will deliver uncontrollable results.

In addition to these more general aspects, specifically for the analysis of synovial tissue the necessity of interactivity is highlighted by the added value of identification and quantification of information as present in areas such as the intimal lining layer, blood vessels, and lymphocyte aggregates.

Speed is another important aspect of digital cytometry. Currently, rapidly increasing numbers of samples, together with accumulation of a variety of markers and detection techniques has made the use of traditional analysis techniques such as manual quantification and semi-quantitative analysis unpractical. It can be anticipated that the development of even more powerful computer systems with sophisticated software will further facilitate reliable analysis at high speed.

Key Words: Digital image analysis; histological quantification; image processing; histology; immunohistochemistry.

1. Introduction

This chapter provides a brief guide to quantitative image analysis of the inflamed synovium, a tool for the quantification of histology based on traditional microscopy and digital imaging techniques. Cytometry involves measurements in cells or tissues, describing planar aspects (e.g., area, numbers, length, profile, size) or photometric aspects (e.g., absorbance, fluorescence measurements, color intensity). Quantitative image analysis is a form of cytometry, which includes microscopic histological quantification, video microscopy, image analysis, and image processing.

For reliable, reproducible, and efficient measurements, quantitative image analysis relies on strict calibration and step-by-step control and documentation of the acquisition, storage, and evaluation of images by hardware as well as software. The hardware comprises the combination of a camera, (automated) microscope and a personal computer. With respect to software all steps are supported by general image analysis software programs such as VideoPro 32, ImagePro, Leica Qwin, or Zeis/Kontron. Dedicated algorithms may use general available image analysis software, or specifically designed programming.

The development of the solid-state cameras and relatively cheap computers in the second half of the 1980s was a major facilitator in the establishment and distribution of digital image cytometry as a scientific tool. In particular the development of the solid state sensors in the charged coupled device (CCD) type camera has been important in this perspective *(1)*. CCD sensors collect images from the object and transmits them as analog signal trough a computer that uses a frame grabber to digitize the images. Since the mid 1990s, completely "digital cameras" have become available. Although in most cameras the core is still a CCD sensor these cameras directly digitize the image increasing speed and reducing measurement error compared with cameras which output is analog and requires digitization in the PC.

The rapid increase of levels of performance by the central processor unit (CPU), random access memory (RAM) memory size, and hard disk storage capacity available in personal computers has facilitated and supported the use of quantitative image analysis to its current level.

Quantitative image analysis allows adequate calibration, user interaction, speed, and control of inter- and intraobserver variation. The controlled environment allows quality control and reproducibility, as well as optimization of sensitivity and specificity.

The generation of reliable measurements does not only require a rigid calibration of the system, but easy interaction with the end-user of the system is as important. Therefore, the system needs to combine solid software engineering with easy interactivity with the operator. The system needs to be as transparent as possible in generating the data. Obviously, a "black box design" will result in uncontrollable results. Variations need to be carefully controlled by the use of rigid protocols for processing (fixation, preservation, preparation of sections), handling, storage, and staining (immunohistochemistry, routine histological staining) *(2–5)*. Sufficient samples should be acquired and analyzed to reduce sample error in the light of variability *(6–8)*.

For the analysis of synovial tissue interactivity is required to identify areas such as the intimal lining layer, blood vessels, and lymphoid follicles. It has previously been demonstrated that the differentiation between distinct areas such as the intimal lining layer and synovial sublining may increase the sensitivity of measurements of macrophages and macrophage-derived cytokines such as tumor necrosis factor (TNF)-α and interleukin (IL)-1β *(9)*. Similar observations have been made for the evaluation of adhesion molecules *(10)*.

Speed is another essential aspect of digital image analysis. Analysis of large numbers of samples and immunohistological markers requires a system that is able to evaluate the sections in a time efficient fashion. To support this the system should be designed to perform measurements as simple as possible in an object that is as representative as possible *(11,12)*. This feature is especially important when serial synovial tissue sampling is employed in randomized controlled trials *(13–15)*.

Challenges in digital image analysis in the near future will mainly be found in the improvement of software algorithms and user interfaces, together with the application of relatively novel features such as confocal microscopy *(16)* and laser dissection microscopy *(17)*.

2. Hardware

A schematic representation of a typical system is given in **Fig. 1**. The primary piece of hardware besides the personal computer (PC) and microscope is the camera (**Table 1**). Modern cameras use in general a charged coupled device (CCD) chip as photoelectric element or sensor (**Fig. 2**). Knowledge of the principles of these sensors is essential to understand some of the advantages and pitfalls of digital image cytometry. CCD sensors were invented in

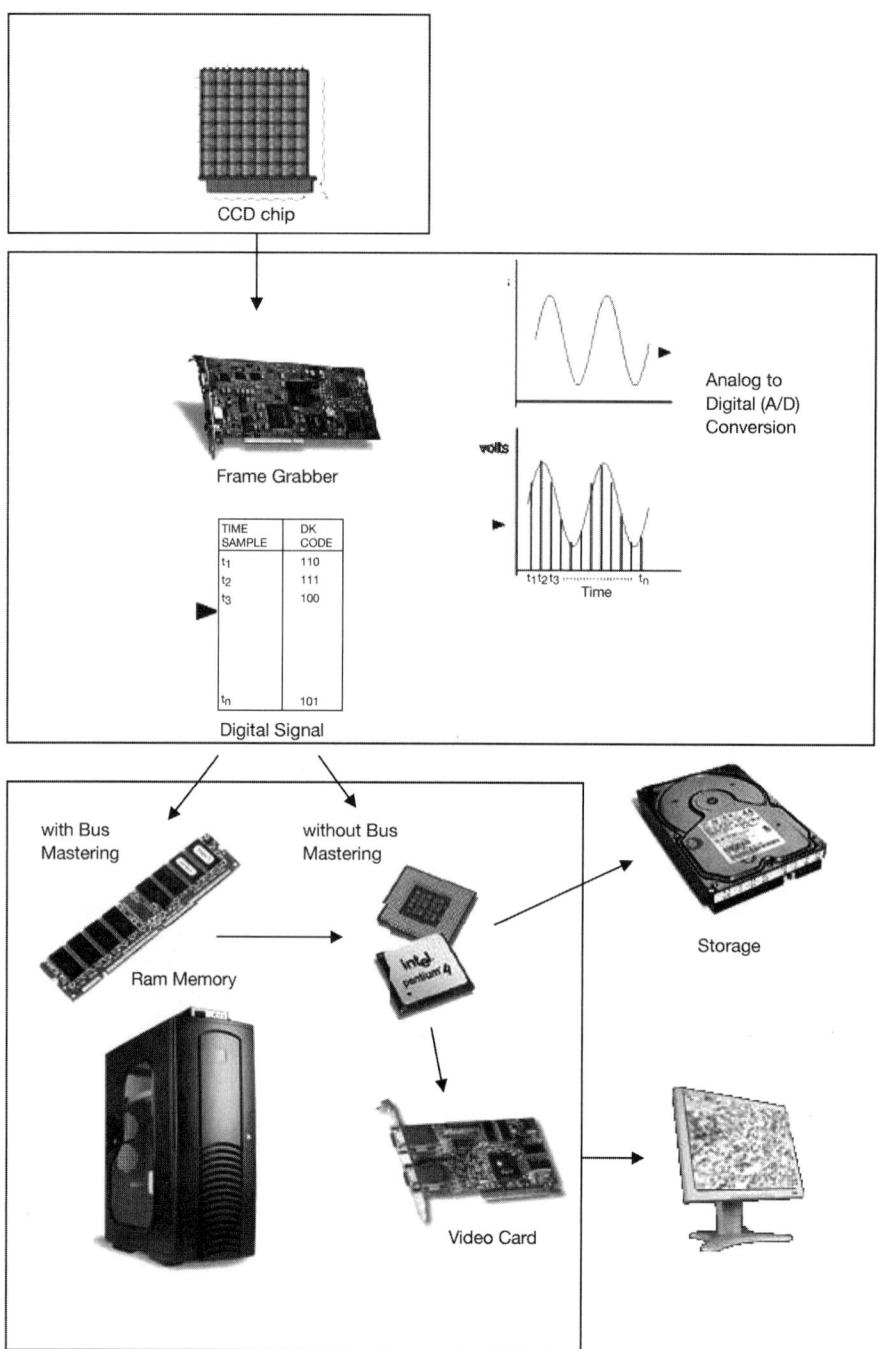

Fig. 1. Typical elements of an image analysis system.

Table 1
Components of a Typical Digital Image Analysis System

- A color or black/white digital camera.
- An (automated) microscope.
- A top of the bill personal computer.
- Storage media such as hard disks, ZIP disks, CD-ROM, or DVD-ROM.

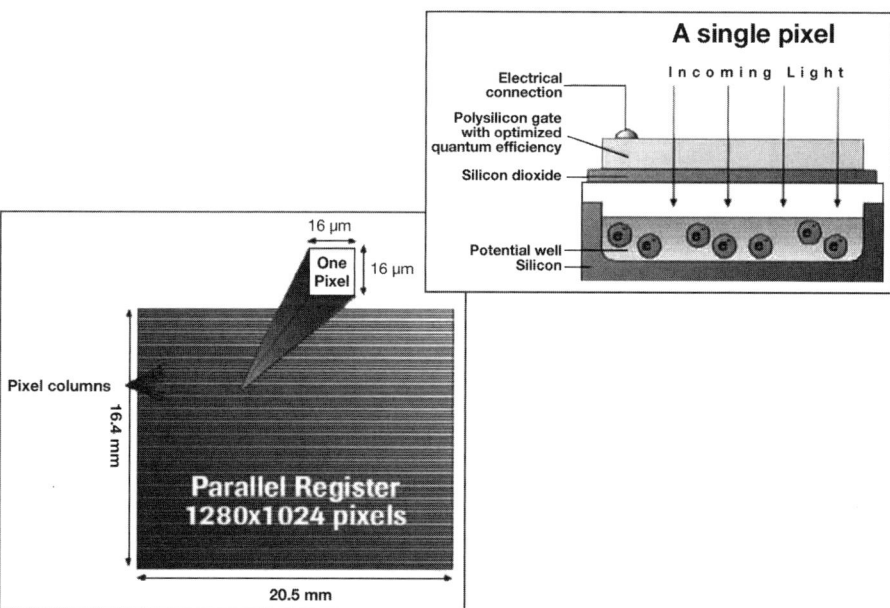

Fig. 2. Schematic presentation of a charged coupled device (CCD) with the typical elements of one pixel.

1969 at Bell laboratories and are silicon based chips composed of a series of closely arranged columns or channels, each subdivided along its length by tiny independent photo elements, parceling images of individual picture elements or pixels. Photoelements convert radiation (light) into electric energy by absorption by the crystalline silicon, where it breaks bonds, displacing a number of electrons proportional to the number of incident photons. Tiny polysilicon electrodes or gates create a buried electron depletion region or potential well that collects the charges displaced from the silicon substrate. After a preset time interval (microseconds to minutes), the packets of photoelectrons are transferred row by row to a register (usually a serial register) and then to an output amplifier where they are converted to a video signal representative of the optical image on the sensor (**Fig. 3**). It is important to realize that although the video signal originates from a solid-state silicon chip, the signal is analog and needs to be converted into a digital signal after the output amplifier. This can be done within the camera, as is done in a "digital" camera or can be performed by a digitizer card or frame grabber mounted in the personal computer housing (*see* **Note 1**).

The time intervals for readout of the potential wells are arbitrary, but one should appreciate that in the advent of constant illumination without readout, overflow in surrounding photoelements will occur, and so called "blooming" will be the result. Furthermore, photoelectrons in potential wells originating from signal are indistinguishable from thermal electrons originating from the silicon by heating (noise or black current). Thus, in any situation the readout will consist of signal and noise. For this reason cameras are often cooled and the operator is always advised to allow the camera to reach a constant operating temperature (*see* **Note 2**).

For actual video image construction the camera will mask the image sensitive area, this to enable the transfer of the charges row by row to the serial register. The simplest design would be an electronic shutter block but this will affect either price or performance and therefore usually the frame transfer CCD sensor is used where a second parallel register, permanently masked from light, is arranged in tandem with the parallel register for the image array. Simply put, two identical CCD sensors are used, one for imaging and one for storage. A third alternative is an interline transfer CCD where part of each photoelement is masked by electronic circuits. These circuits mask part of the photoelements and read from the image storage registry during read out. Because the areas of

Quantitative Image Analysis of Synovial Tissue

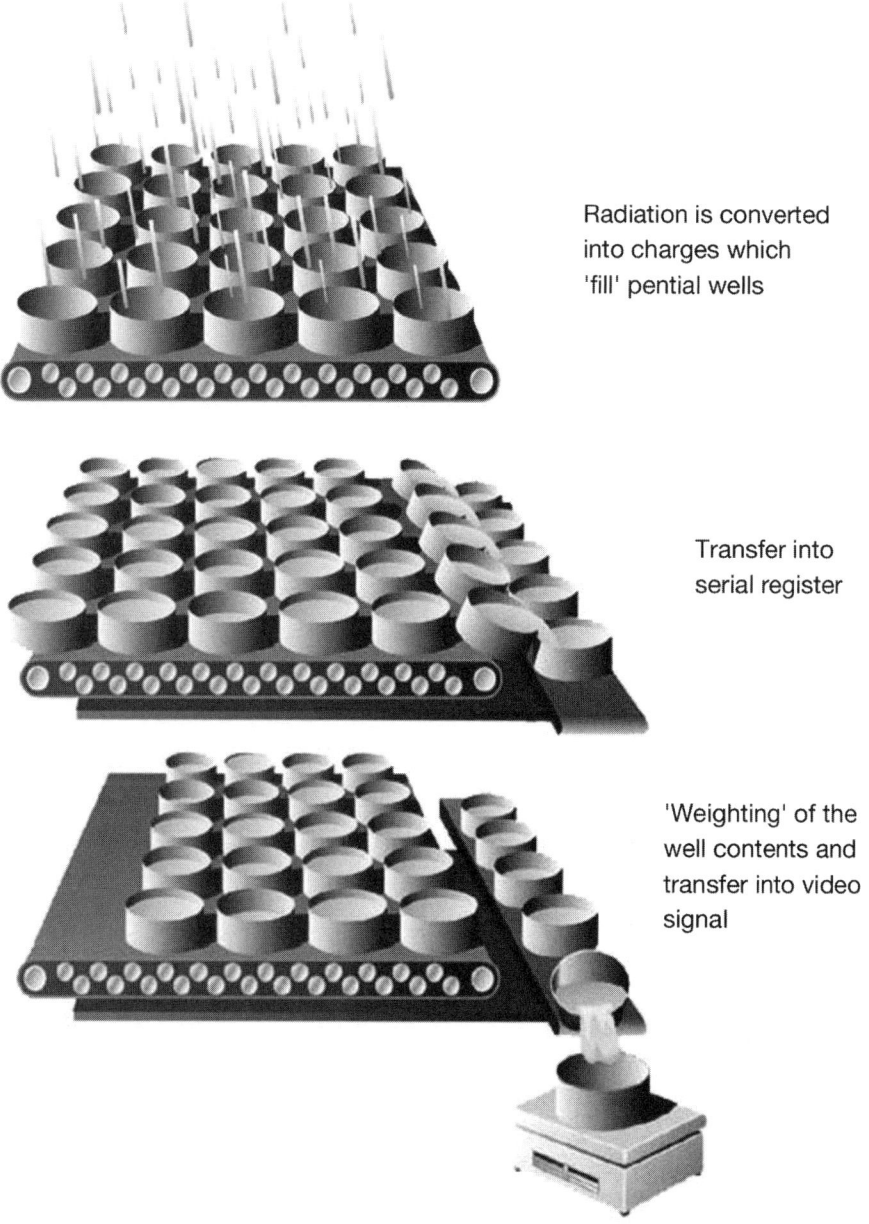

Fig. 3. Collection of photoelectrons, transfer row by row to a register and transfer to an output amplifier.

interline transfer CCD chips available for light collection is reduced to 50% or less, interline CCD chips are less sensitive to photons and are less suitable for low light conditions such as immunofluorescence imaging (*see* **Note 3**).

There is great variation in CCD sensor size from 3.5 mm to 35 mm². The individual rectangular or square photoelements vary between 6 and 30 µm. The size of the individual photoelements is important with respect to the dynamic range. The dynamic range of a single photoelement is the difference between the number of electrons collected at maximum saturation and the dark current. The dynamic range is relative to the size of the photoelements (the bigger the better) or the temperature of the chip (the cooler the better). The efficiency of CCD sensors to generate electrons from signal is in general very high in the visible light spectrum and the near infrared. The CCD sensor is placed into an environment or housing (camera) which allows the sensor to be placed behind a microscope objective. With regard to the choice of camera several issues will be addressed: (1) video or still camera; (2) analog or digital; and (3) monochrome or color (*see* **Note 4**).

As described above, the signal coming from a CCD sensor is always analog or, in other words, a linear sequence of electrical impulses. In analog cameras this electric signal is transported to the PC where it is digitized. In black and white cameras the signal only consists of information on the light intensity. In color cameras the video signal consists of information on light intensity and color. In a digital camera the analog signal is transformed into a binary format or digitized within the camera housing and this information is transported to the PC. For more detail on video information we refer to other sources.

The analog-to-digital converter (A/D converter or digitizer) is almost as important and vital as the CCD sensor. In general A/D converters use a pixel clock to dissect the signal of every image line. The A/D converter reads the luminance at each interval and transfers it into a luminance number or binary code. The lowest output value (0) represents no light and the highest output value 255 (for an 8 bit digitization) represents the maximal signal. For an RGB color signal this process is performed three times in parallel resulting in a signal between 0 (no signal) and 16,581,375 (maximal signal). A second A/D converter reads the light intensity, an 8-bit depth allows 256 gray levels. Very sensitive A/D converters allow a 12-bit depth with 4086 gray levels; this allows a high degree of accuracy in measurement of dark images. The next component is the framegrabber responsible for feeding the digital signal into the processing units of the PC. The framegrabber uses a frame transfer rate, the speed at which the captured image is transferred into the random access memory (RAM). This speed is determined by the chipset of the framegrabber, PC bus transfer rate, main CPU speed and RAM type. Currently most PCs use the peripheral component interconnect (PCI) parallel bus which allows a fast buffer and a direct access to either a direct memory access (DMA) mode or a unified memory architecture (UMA) mode. DMA transfers data direct to the normal RAM whereas UMA transfers data to a dedicated RAM memory space, both bypassing the main processor. Each transfer technique has its own advantages. The more time efficient UMA transfers are usually preferred (*see* **Note 5**).

Main techniques used in digital image analysis of synovial tissue are transmitted light or immunofluorescence; other possible future modalities are dark field, polarized light, phase contrast, and confocal microscopy. Most video cameras have a standard female C-mount attachment that allows the camera to be mounted to the male C-mount on the microscope. Additionally, the C-mount adaptor can harbor lenses to reduce the image if the image sensor requires so. Proper set-up of the microscope with proper alignment of optical elements, light source and object is essential. Applying transmitted light an external lamp housing with a 12 V 100 W tungsten Halogen lamp is advised with a digitally controlled DC power supply to allow constant calibration (with data storage possibilities) and thereby constant and controlled illumination (*see* **Note 6**).

In general, 3 beam splitters are used which allow direct access of all light to the camera, the eyepiece or 50/50 between the camera and eyepiece. The objectives used are in general a low power lens (×5) for the identification of areas of interest, mid power lenses (×10–40) for measurements, and high power immersion objectives (×100) (*see* **Note 7**).

The objectives generate the magnified image and act as the excitation light condenser in the epifluorescence microscope (**Table 2**). The objective influences the brightness, resolution, and general quality of any image. The greater an objective's correction of any kind, the greater the number of lens elements inside that objective, with concomitant light loss (about 3–4%) at every air-glass interface which by example can effect the strength of the signal derived with immunofluorescence techniques. Objectives are also classified by numeric aperture. Larger numeric apertures gather more light and produce brighter, more highly resolved images. Generally, numeric aperture increases with magnification (*see* **Note 8**).

The use of a motorized stage table allows the detailed control of the section analyzed together with the possibility to store the coordinates of the scanned

Table 2
Resolving power of Objectives in the Intermediate Image for ×0.63 and ×1.0 TV Coupling Adapters in Combination with a 2/3" CCD (8.5 mm × 6.4 mm)

Objective	Magnification	NA	Lp/mm (TV-Cpk ×1.0)	Necessary camera resolution	Lp/mm (TV-Cpk ×0.63)	Necessary camera resolution
Plan-Neofluar	1.25	0.04	96	1632 × 1229	152	2548 × 1946
Fluar	2.5	0.12	144	2448 × 1843	229	3893 × 2931
Plan-Neofluar	5	0.15	90	1530 × 1152	143	2431 × 1523
Achroplan	10	0.25	75	1275 × 960	119	2023 × 1523
Fluar	10	0.5	150	2550 × 1920	238	4046 × 3046
Plan-Neofluar	20	0.5	75	1275 × 960	119	2023 × 1523
Plan-Achromat	20	0.75	113	1921 × 1254	179	3040 × 2291
Plan-Neofluar M.-imm	25	0.80	96	1632 × 1229	152	2584 × 1984
Plan-Neofluar	40	0.75	56	952 × 717	89	1513 × 1139
Plan-Neofluar	40	1.3	98	1666 × 1254	155	2635 × 1984
Plan-Apochromat	63	1.4	67	1139 × 858	106	1802 × 1357
Epiplan-Neofluar	100	0.9	27	459 × 346	43	731 × 550
Plan-Apochromat	100	1.4	42	714 × 538	67	1139 × 858

Table 3
Characteristics of an Automated Microscope

- Motorized 8-position fluorescence cube changer.
- Fluorescence cubes with zero pixel shift.
- Electronic controlled shutter and motorized tube.
- Focus drive with 0.015 mm step size with readout.
- Digital light control with readout.
- Automated aperture, diaphragm and condenser settings with readout.

area. The rigid control of the position of the section allows the scanning of multiple images with minimal overlap, offering the opportunity to scan large areas with high power magnification. A second advantage of stage tables is the possibility to store the image coordinates. If the slides are mounted in a fixed frame on the stage table this storage enables exact repositioning of the exposed area on a later occasion. The major advantage of fully automated microscopes (**Table 3**) is the ability to standardize all processes and settings in a digital log file and it also allows storage of the settings used. These two features result in optimal control and a log of all processes involved.

3. Software

Digital image analysis programs can be divided into two groups: those that are commercially available such as ImagePro, Leica QWin and Kontron and those that are custom made such as VideoPro.

Despite the list of all features available in commercially available packages it is possible that a specific operation is not available. In that case custom creation of a program especially for that purpose is an option. This can be a stand-alone program or a plug-in for existing software, depending on the demands and the possibilities with existing software. The advantage of using existing software is that it is relatively cheap, tested and easy to implement. Specially designed software takes more time to implement, costs more but does correspond exactly to local needs and can be altered if needed.

Via a special client/server mechanism called Dynamic Data Exchange (DDE) two Microsoft based applications can communicate with each other, for instance image analysis software with Visual Basic, Excel, Access or Lotus (*see* **Note 9**).

4. Image Acquisition

Microscope calibration assists the operator in establishing a controlled and reproducible environment. As each different marker has its own outcome of color intensity and brightness and different types of cameras have different sensitivities the microscope needs to be calibrated before each set of tissue samples is acquired. This to ensure that there is as little diversity between the different images as possible and visibility is enhanced to a maximum, in order get the best result and be able to compare different sets of images (*see* **Note 10**).

The camera offset or black level is the voltage applied to the minimum value (V_{min}) of the video signal set to zero. Or, in other words, all data below V_{min} are digitized as black pixels. Consequently, if the black level is raised above the level of the darkest objects in the image, these objects will be clipped to black and will not be captured. This condition, known as "under saturation" should be avoided, as it cannot be corrected. Camera gain is the value used after the camera offset is applied. Main function is to multiply the amplitude of the signal, thereby increasing or decreasing the overall range of the signal (*see* **Note 11**).

White level is defined as the voltage above which all other voltages are digitized to white. In other words, if the white level is lowered below the level of the lightest objects in the image, these objects will be clipped to white and will not be captured. This condition, known as "over saturation" should be avoided, as it cannot be corrected, similar to the "under saturation" described earlier. During the calibration other factors are addressed as well, for instance determining the parafocality (a sharp picture both in the visual eyepiece and in the digital camera) to speed up the process during the actual acquisition. This has to be done for each objective used. In the past, storage capacity was a limiting factor. To control the costs, whole tissue samples were not acquired, but rather only representative regions. The last couple of years' storage costs have been significantly reduced, enabling the acquisition and storage of images covering whole tissue samples (*see* **Note 12**).

Any selection procedure still requires a well trained eye for histology and determination of relevant tissue areas (e.g., intimal lining layer, synovial sublining) and irrelevant areas (e.g., connective tissue, organized fibrin depositions, avascular regions, folds, non-specific staining). This selection will be used to either select representative regions or exclude nontissue areas. In the event of staged scanning the microscope will switch to the chosen magnification after the area to digitize has been selected and the stage table will position the first area to be digitized in front of the objective (*see* **Fig. 4**). The focus can

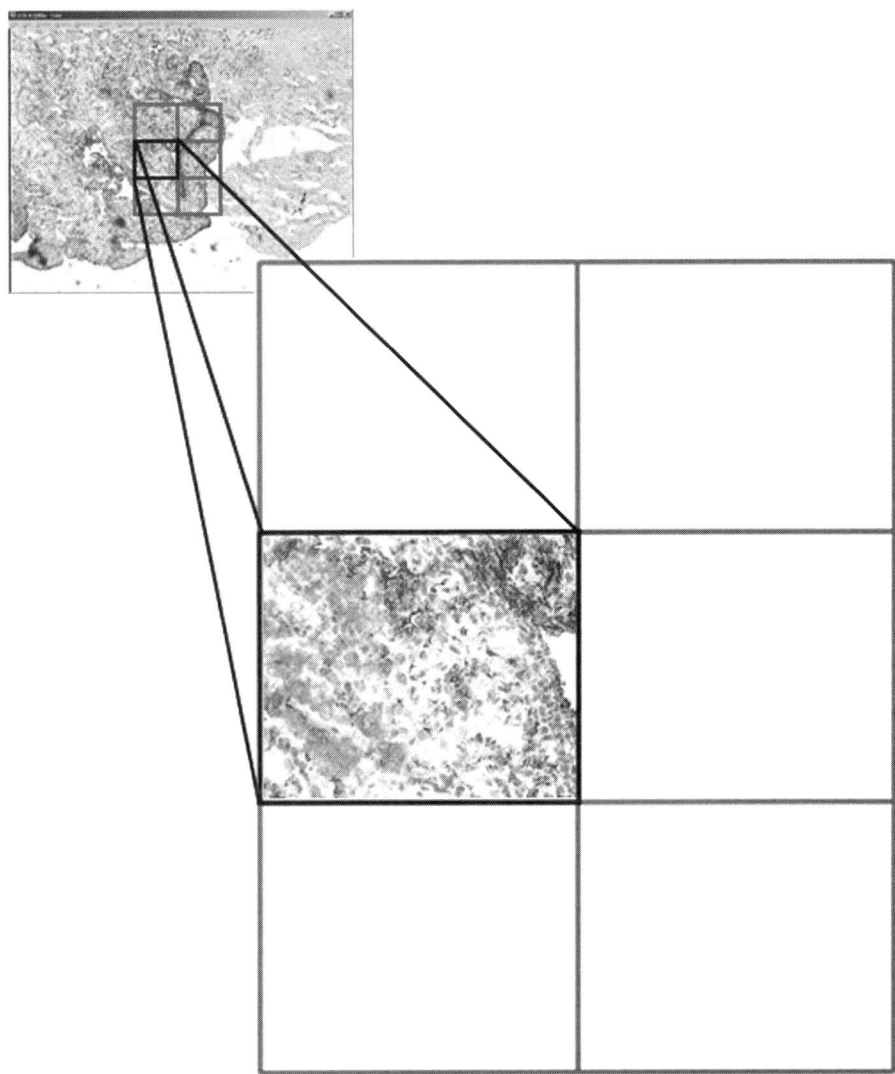

Fig. 4. The image acquisition process where a low power overview and raster identify the images which subsequently are acquired with an high power magnification using a stage table to transfer the tissue section.

be adjusted (manually or automatically) and subsequently all the areas of the grid are digitized automatically; the computer will move the stage table around, according to the grid used and will digitize each HPF. After completion either another Region of Interest can be selected on the same tissue section or a new tissue section may be started (*see* **Note 13**).

5. Image Analysis

Any applied individual step will have its implications for sensitivity and specificity of the image analysis. In addition, the order in which the various steps are scheduled is essential and this will vary based upon the marker analyzed and the procedures used for sectioning and staining of the tissue sample (*see* **Note 14**).

Optical noise is generated by the combination of the light source and optical parts of the microscope, before it generates an image on the CCD chip. Any contamination on the slides itself (e.g., dust particles, folds, unequal staining) will contribute to this noise as well. Besides this optical noise, there is hardware noise generated by the electronic parts, which is best defined as the visible effects of an electronic error (or interference) in the final image. The hardware noise in the image is a result of how well the sensor (CCD/CMOS) and digital signal processing systems inside the digital camera or PC can handle such errors (or interference). Most important sources of visible noise are temperature (the higher the worse) and ISO sensitivity (the lower the better) of the camera. Facing the challenge of reducing noise and producing a "cleaner" image, camera developers are improving the quality, allowing for higher sensitivity to be used without too much noise (*see* **Note 15**).

Noise in the originally acquired image is also prone to be amplified by the joint photographic experts group (JPEG) compression algorithm, which often introduces huge errors in already noisy images. For these reasons most image analysis software contains noise reduction algorithms to remove unwanted noise from a digital image (**Table 4**). There are two main categories, reduction or removal of noise from high ISO images and reduction or removal of noise from long exposure images (with "blooming"). Image compression is essential for transport and storage because a modern camera produces an uncompressed 48-bit/pixel image often in the multiple megabytes, certainly for a 5 megapixel camera in excess of 30 MB per image. The most commonly applied image format in digital cameras is JPEG, producing a relatively small file from a large amount of image data. JPEG compression discards certain information,

Table 4
General Specifications and Predefined Algorithms in Commercial Image Analysis Packages

Acquisition

- Capture images from cameras, other standard camera/video inputs and (TWAIN) scanners. Capture multiple images with user-specified interval time to image files, or to sequence file,
- Support for 8-, 12-, and 16-bit grayscale image capture, plus 24-bit RGB color image capture.
- Hardware supported gain and offset and fast frame averaging (on certain boards).

Calibration

- Work with intensity or spatial measurements. Create and display spatial calibration markers.
- Predefined spatial calibration units available.
- Save and recall calibrations.
- Apply intensity and spatial calibrations to measurements.
- Color channel processing.
- Extract RGB or hue, saturation and intensity channels for processing.
- Merge (mono) images for display and colocalization.
- Processing tools to extract multiple color channels simultaneously.
- Switch or copy channels within an image.

Enhancement

- Enhance color and contrast using equalization, gamma correction, contouring, display range and background subtract. Use low-pass, high-pass, Gauss, hi-Gauss, sharpen, flatten, median and local equalization filters.
- Image and data file format support.
- Read and write files in TIFF, BMP, GIF, PhotoCD, PCT, PCX, PNG, TGA, and FLAT (binary).
- Supports 1-, 8-, 12-, 16-, 24-, and 32-bit floating point images.
- JPEG, LZW, and RLE compression supported.
- Output data files to ASCII or XLS for spreadsheet input,
- Transfer images, data graphs/files via DDE or clipboard.

Measurement and Analysis

- Measure lengths, areas, perimeters and angles. Automatically outline objects.
- Quickly calculate max, min and average thickness between lines.
- Calculate line or area histograms.

- Calculate line (straight, circular and free-form), or thick (vertical or horizontal) profiles.
- Display data, line graphs, or bar histograms.
- Define/manage multiple areas of interest in an image.
- Calculate statistics.
- Analyze content of color image.
- Combine image with background correction for precise intensity or optical density measurements.
- Option to manually tag, count and classify objects.
- Output intensity map in ASCII format.
- Export measurements to statistical and spreadsheet packages (i.e., MS Excel) via DDE.
- Export measurements, histograms and profiles to file or via clipboard.

Operations
- Perform spatial (bilinear scale, decimation, rotate, transpose, reflect) and background correction operations. Image overlay and paste with blend/transparency options.

using a "lossy compression algorithm" deleting some not directly used image information (*see* **Note 16**).

A second alternative is the tagged image file ormat (TIFF). This is a complex but flexible image format; for more detail on TIFF we refer to other sources. In digital cameras, TIFF is used to provide a "ossless" image format, which does not delete information in the compression process. In photographic images, TIFF compression usually uses an 8- or 16-bit/color channel storage method.

A third alternative is using the uncompressed RAW format, which is the raw data as it comes directly off the CCD; no in-camera processing is performed. Typically this data is 8, 10, or 12 bits/pixel. In general, file sizes are considerably smaller (e.g., $2550 \times 1920 \times 12$ bits = 58,752,000 bits = 7,344,000 bytes). The image has not been processed or white balanced, allows correction, and it is a true representation of the original "digital negative." Main disadvantage is that these image files require an "acquire (TWAIN) module," which is generally not available in standard image analysis programs (**Table 5**).

Operative routines are applied in the algorithm to further improve the noise signal relation. Because the thresholds applied are always based on a consen-

Table 5
RAW Compression

Advantages of RAW format
- A true "digital negative," untouched by cameras processing algorithms.
- No sharpening applied.
- No gamma or level correction applied.
- No white balance applied.
- No color correction applied.
- Lossless yet considerably smaller than TIFF. Records data over a wider bit range (typically 10 or 12 bits) than JPEG or 8-bit TIFF

Disadvantages of RAW format
- Requires proprietary acquire module (typically TWAIN) or plugin to open images.
- Images can take 20 to 40 s to process on an average machine.
- No universally accepted RAW standard format, each manufacturer (even each camera) differs.

sus between images (comprising variation in color information) there is always noise present after the initial threshold process. Therefore, the algorithm applies calculations based upon individual pixels as identified after the initial threshold. The calculations used are erode, dilate, fill, and segment.

Masking is the function that can be used to separate the region of interest within an image from the background. Regions of interest or objects can be identified by: gray value, color, edges, and texture. Masking by color threshold has previously been developed for analysis of synovial tissue. The three main methods of segmentation are: threshold (*see* **Fig. 5**), edge finding, and region growing.

Segmentation by threshold/discrimination is the most simple and common method. At its simplest, all tones below a selected level are treated as of interest and all above as background. One may use more than one threshold, allowing several scales of grey tones or colors to be considered of interest. Threshold works well in situations where the illumination can be carefully controlled to have the same level across the entire scene, as in microscopy. Another adaptable technique is edge finding, which detects the regions of high rate of change of tone in an image, on the basis that these are likely to indicate edges of objects

Primary Staining Threshold

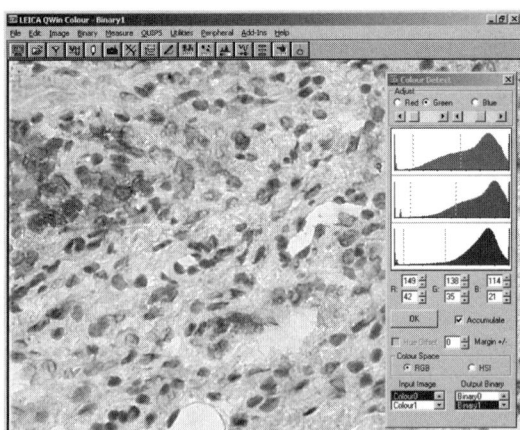

Nuclear Staining Threshold

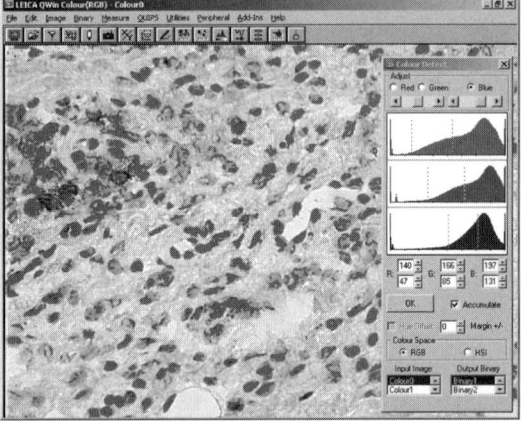

Background Threshold

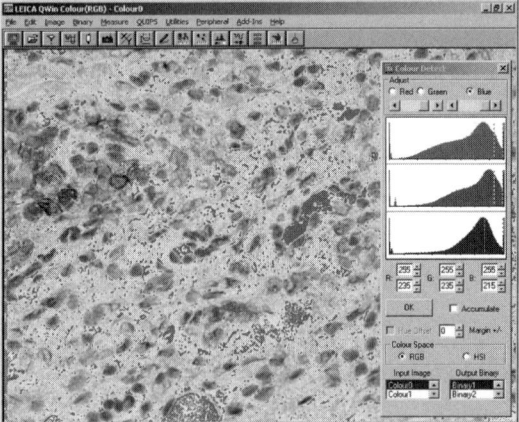

Fig. 5. Segmentation by color threshold and the three main methods of segmentation; threshold, edge finding, and region growing.

or regions. The third approach looks at neighboring pixels and groups them together if they are sufficiently similar. This process generates a binary image (each pixel has only one of two states: region of interest or background) that is used as a mask for further processing of the original image.

Segmentation uses a calculation that involves an ultimate erode command, meaning that the algorithm will apply the erode command until only one single pixel is left of every cluster. Next calculation is a dilate command until the average size of the object (identified by parameters) is met. If two objects are very close together the algorithm will leave one pixel row between the two objects, enabling counting of individual cells.

After identification of a region of interest, objective quantitative analysis can be performed. Such measurements may be used to quantify the signal defined in areas like the intimal lining layer, blood vessels, and cellular aggregates. Moreover, a number of new features may be added, such as volumetric estimations, and protein content in anatomically defined cells (for example in T-cell aggregates) to mention a few.

To measure the number of positive cells the image analysis system applies algorithms allowing the enumeration of positive cells. This combines the routines of signal-to-noise improvement together with segmentation steps resulting in the possibility to enumerate positive cells even within the context of dense structures such as lymphoid aggregates. To improve the quality of the segment algorithm and enable the quantification of positive cells an extra feature is incorporated: the nucleus control. In brief, this first implies the selection of positive staining and nuclear staining. Next step is signal/noise improvement as described above. The third step is a 1-cycle dilation of the positively stained region and the nuclear staining. The fourth step is the determination of overlap between the positive staining and the nuclear staining. The final step is the calculation of positive cells or joined staining and nucleus, nucleus of unstained cells, and noise or staining without a nucleus *(9)*.

For the measurement of expression of a positive signal region, as for example for the expression of cytokines and/or adhesion molecules, the measurement can be optimized with the use of the integrated optical density (IOD). The IOD is designed to improve the ratio between the identified positive signal regions *(10)*. In brief, this is a composite index of the number of pixels identified as positive and the color information contained within these pixels (optical density or OD). In the IOD calculation the color information is calculated as the mean of the RGBchannel. Each channel contains color information between 0 (maximal signal) and 256 (no signal), as a result the mean color information (or OD) will be a number between 0 and 256. Subsequently, the correction factor is calculated as 1-mean OD/255. The IOD is calculated as the number of positive pixels times the correction factor and will, therefore, be maximally the original number of pixels and minimally 0. For the use and interpretation of the IOD it should be taken into account that for transmitted light microscopy these measurements can only be performed within the limits of the optical properties of the system. For the evaluation of sections with immunofluorescence the use of the IOD is a more accurate representation of the actual expression of the protein analyzed.

6. Conclusions

Digital image analysis has improved the analysis of multiple histological sections in a reliable, reproducible and time efficient way allowing the analysis of high numbers of samples, as in randomized clinical trials including synovial tissue biomarkers. There is a significant investment involved in the initial setup and the nature of the system requires dedicated personnel and ongoing maintenance including the required infrastructure, which make it less interesting for incidental use or the quantification of small numbers of samples. It can be anticipated that the near future will see further improvement in both hardware and software, resulting in the development of even more sophisticated systems for quantification of stained tissue sections.

7. Notes

1. Digital and analog cameras both use a CCD chip for image acquisition but the analog signal is converted into a digital signal within the camera housing in a digital camera facilitating the interface with the PC.
2. The sensitivity of a CCD chip changes which its temperature and therefore reaching a stable operation temperature is crucial for reproducible data.
3. The type of CCD chip optimal for image acquisition is dependent on the procedures planned, immunofluorescence generally needs the most sensitive CCD setting whereas video imaging needs especially a fast CCD setting
4. Video cameras are designed to capture moving objects, or allow on line functions such as focusing whereas still cameras shoot still images. In general still cameras allow higher quality and detailed image because of longer exposure times.
5. The main benefit of direct digital output by the camera is that cabling cannot degrade the information and the A/D conversion clock is synchronized with the sensor clock (eliminating pixel jitter). Other advantages are found in optimized software control by universal serial bus (USB) connections.

6. For transmitted light microscopy it should be taken into account that a preferred light temperature of 5600 Kelvin should be established to make optimal use of the color spectrum. For fluorescence illumination, high pressure 50 to 100 W mercury or xenon lamps are used connected to a stabilized power supply.
7. The preferred magnification for the imaging of synovial tissue is depended on multiple factors such as CCD chip resolution and optical properties of the microscope. In general terms should the highest object of interest have a diameter of at least 15 pixels in the final image.
8. Recommended objectives are plano objectives optimized for fluorescence microscopy; high numeric aperture (approx 1.3) and objectives with a minimal number of lens elements; numeric aperture 75. Not recommended are objectives with many lens elements; numeric aperture 0.5.
9. The main advantage of a DDE link is that no data transformation to other formats or matched data structures is needed. However, the applications communicating through DDE must be running on the same computer; DDE does not support networks which limits the application in modern web based programs.
10. With calibration, the optimum white level of the background is determined by changing the light intensity, the gain, the offset, the brightness, and the exposure time.
11. Increasing camera gain produces a brighter image but also increases image noise. In most samples the gain is set such that the image is not saturated, but a few pixels reach peak brightness.
12. For synovial tissue it has been shown that with an average tissue area of 7.5 to 10 mm^2, digitization of 3 areas each ± 0.8 mm^2 (total of ± 2.4 mm^2) is sufficient to represent the entire tissue section *(10)*.
13. In general it is advised that only simple gray filters are applied to correct for illumination in transmitted light imaging.
14. Central theme in the image analysis process is signal-to-noise improvement.
15. It is important to realize that noise can affect certain color channels more than others, since the camera sensor (CCD/CMOS) is more sensitive to certain primary colors than to others. Often sensors are less sensitive to blue light. To compensate these differences in sensitivity, channels are differentially amplified.
16. Main disadvantage of JPEG compression is the exaggeration of noise and the relative imbalance of compression in the white area of the spectrum.

References

1. Goodman, A. H. (1986) CCD line-scan image sensor for the measurement of red cell velocity in microvessels. *J. Biomed. Eng.* **8,** 329–333.
2. Kraan, M. C., Reece, R. J., Smeets, T. J., Veale, D. J., Emery, P., and Tak, P. P. (2002) Comparison of synovial tissues from the knee joints and the small joints of

rheumatoid arthritis patients: Implications for pathogenesis and evaluation of treatment. *Arthritis Rheum.* **46,** 2034–2038.
3. Dolhain, R. J., Tak, P. P., Dijkmans, B. A., et al. (1998) Methotrexate reduces inflammatory cell numbers, expression of monokines and of adhesion molecules in synovial tissue of patients with rheumatoid arthritis. *Br. J. Rheumatol.* **37,** 502–508.
4. Tak, P. P., van der Lubbe, P. A., Cauli, A., et al. (1995) Reduction of synovial inflammation after anti-CD4 monoclonal antibody treatment in early rheumatoid arthritis. *Arthritis Rheum.* **38,** 1457–1465.
5. Kraan, M. C., Patel, D. D., Haringman, J. J., et al. (2001) The development of clinical signs of rheumatoid synovial inflammation is associated with increased synthesis of the chemokine CXCL8 (interleukin-8). *Arthritis Res.* **3,** 65–71.
6. Boyle, D. L., Rosengren, S., Bugbee, W., Kavanaugh, A., and Firestein, G. S. (2003) Quantitative biomarker analysis of synovial gene expression by real-time PCR. *Arthritis Res. Ther.* **5,** R352–R360.
7. Dolhain, R. J. E. M., Haar ter, N. T., Kuiper de, et al. (1998) Distribution of T cells and signs of T cell activation in the rheumatoid joint: implications for semiquantitative comparative histology. *Br. J. Rheumatol.* **37,** 324–330.
8. Rooney, M., Condell, D., Quinlan, W., et al. (1988) Analysis of the histologic variation of synovitis in rheumatoid arthritis. *Arthritis Rheum.* **31,** 956–963.
9. Kraan, M. C., Versendaal, H., Jonker, M., et al. (1998) Asymptomatic synovitis precedes clinical manifest arthritis. *Arthritis Rheum.* **41,** 1481–1488.
10. Kraan, M. C., Smith, M. D., Weedon, H., Ahern, M. J., Breedveld, F. C., and Tak, P. P. (2001) Measurement of cytokine and adhesion molecule expression in synovial tissue by digital image analysis. *Annals Rheum. Dis.* **60,** 296–298.
11. Kroustrup, J. P. and Gundersen, H. J. (1983) Sampling problems in a heterogeneous organ: quantitation of relative and total volume of pancreatic islets by light microscopy. *J. Microsc.* **132,** 43–55.
12. Gundersen, H. J. and Osterby, R. (1981) Optimizing sampling efficiency of stereological studies in biology: or 'do more less well!'. *J. Microsc.* **121,** 65–73.
13. Bresnihan, B., Tak, P. P., Emery, P., Klareskog, L., and Breedveld, F. (2000) Synovial biopsy in arthritis research: five years of concerted European collaboration. *Ann.Rheum.Dis.* **59,** 506–511.
14. Bresnihan, B. and Tak, P. P. (1999) Synovial tissue analysis in rheumatoid arthritis. *Best Pract. Res. Clin. Rheumatol.* **13,** 645–659.
15. Tak, P. P. and Bresnihan, B. (2000) The pathogenesis and prevention of joint damage in rheumatoid arthritis. Advances from synovial biopsy and tissue analysis. *Arthritis Rheum.* **43,** 2619–2633.
16. Sundarrajan, M., Boyle, D. L., Chabaud-Riou, M., Hammaker, D., and Firestein, G. S. (2003) Expression of the MAPK kinases MKK-4 and MKK-7 in rheumatoid arthritis and their role as key regulators of JNK. *Arthritis Rheum.* **48,** 2450–2460.

17. Wunder, A., Muller-Ladner, U., Stelzer, E. H., et al. (2003) Albumin-based drug delivery as novel therapeutic approach for rheumatoid arthritis. *J. Immunol.* **170,** 4793–4801.

II

CARTILAGE MATRIX AND BONE BIOLOGY

9

Cartilage Histomorphometry

Ernst B. Hunziker

Summary

In rheumatology and joint research, as in other fields, a purely descriptional appqoach to morphology cannot satisfy the exactions of modern clinical medicine. Investigators now appreciate the need to gauge pathological changes and their response to treatment by quantifying susceptible structural parameters. But the desired information respecting three-dimensional structures must be gleaned from either actual or virtual two-dimensional sections through the tissue. This information can be obtained only if the laws governing stereology are respected. In this chapter, the stereological principles that must be applied, and the practical methods that have been devised, to yield unbiased estimates of the most commonly determined structural parameters, namely, volume, surface area and number, are summarized.

Key Words: Stereology; morphometry; unbiased; reference volume; cartilage; joint; surface area; volume density; numerical density; counting; articular; systematic random sampling; statistics; animal experiments.

1. Introduction

Quantification is a most important and powerful tool, which is now being implemented more frequently in biological and pathological explorations. In the fields of joint and rheumatology research, the quantitative approach is classically applied to estimate molecular tissue components, such as the concentration of proinflammatory substances and the density of cell-surface receptors. But now the need to quantify also the structural components of joint tissues is being appreciated. However, in adopting a quantitative approach to morphology, many investigators tend to forget that the requisite information must be gleaned from two-dimensional (2D) microscopic images of an actually or a virtually sectioned structure which in itself may bear little or no relationship to the structure as a whole. Undoubtedly, the desired information can be obtained, but only after considering the spatial situation. Unfortunately, the expertise needed to solve these spatial problems is but too seldom sought, with the consequence that naïve and inappropriate methods of quantification are frequently adopted which yield biased and spurious information.

The science that describes the unbiased estimation of three-dimensional (3D) structures from 2D sections of these is known as stereology. The process of taking measurements from the 2D sections—which measurements are then fed into the appropriate equations defined by the laws of stereology—is called morphometry. This chapter describes a few of the basic principles that must be respected for an unbiased estimation of 3D biological structures. Emphasis will be placed on the practical methods that are currently employed to estimate the more commonly determined parameters, such as volume, surface area and number. The stereological approach has a statistical basis in that it is founded on calculations relating to geometrical probabilities. The practical methods that are now instigated to estimate such parameters, and which have been developed during the past two decades, are no longer based upon geometrical assumptions as was formerly the case, and they are thus unbiased.

Before embarking on a quantitative structural analysis, the parameter to be estimated must be clearly defined, which is not such a trivial task as it may appear to be. The quantified parameter must be biologically meaningful. For example, is it more meaningful to estimate the number of chondrocytes (1) per unit volume of articular cartilage tissue; (2) within the entire volume of the articular cartilage layer; or (3) per square millimeter of the articular cartilage surface? The answers to these questions will depend upon the particular experimental situation under consideration. But, as a general rule, absolute quantities are much more powerful than relative ones. Because the latter values do not take into account reference spaces, they may yield misleading information. Regrettably, it is precisely these relative quantities that are most frequently determined by investigators. This chapter will hopefully help workers in the fields of rheumatology and joint research to better appreciate the spatial issues confronting them and to base their quantitative structural analyses on the science of unbiased stereology.

2. Methods
2.1. Sampling Strategy

As aforementioned, a stereological approach to morphological analyses permits the investigator to obtain quantitative data relating to 3D objects from 2D images of these. Clearly, the methodology that is applied to yield this quantitative information must be more objective and rigorous than that which suffices for the descriptional approach yielding merely qualitative data. This rigour must be exercised from the very outset of the investigation, even during its planning, and should be brought to bear on the selection of well-defined parameters for quantification and the elaboration of an appropriate sampling strategy. The samples selected for evaluation in the microscope constitute but a minute portion of the tissue mass from which they are derived. These samples should thus be representative of the whole, but will be so only if the sampling strategy adopted is statistically sound.

A representative set of samples of the tissue under investigation can be obtained by instigating a uniform random sampling protocol, which requires that each structure of interest in the original specimen be sampled with an equal probability (viz., has the same chance of appearing in the final section upon which measurements will be made) *(1)*. If this simple principle is not rigorously respected at all times, a sampling bias will be introduced. To avoid this, a uniform random sampling protocol must be applied at each stage in the sampling hierarchy. Unbiased or design-based stereological methods rely heavily on the unbiasedness of the strategy instigated. But the advantage of these methods is that they are not based upon restrictive assumptions relating, for example, to the geometry of a structure.

Random sampling is the basis of all classical sampling theories. However, it is an inefficient process. It is well known that numbers chosen at random tend to cluster *(2)*. Furthermore, from a practical point of view, random sectioning of a tissue structure is virtually impossible. Systematic sampling is a much more efficient stereological approach. Although the very term "systematic" appears to contradict the popular notion of randomness, the sampling strategy is nonetheless mathematically sound and is, moreover, more readily applied in practice than is the repeated uniform random sampling protocol. Furthermore, estimates yielded according to the systematic random sampling strategy are frequently characterized by a lower variance *(3,4)*. Not surprisingly, a systematic random sample contains two components: a systematic element and a random one. The underlying principle of this strategy is that, with respect to an object's geometry, it generates a uniformly random location for the starting point of sampling within a defined initial segment of the said object (the *random* component). If this initial sample is "random," then the next one, taken at a defined distance from it, and all subsequent ones, separated by the same defined, viz., systematically repeated, distance (the *systematic* component), will be likewise random. The systematic random sampling process is best appreciated by example. Let it be supposed that a linear object is to be analyzed (*see* **Fig. 1**). The approximated length of the object is first mentally divided into a small number of equal units (more than one but usually fewer than 8), which will define the number of samples to be taken. In the present instance,

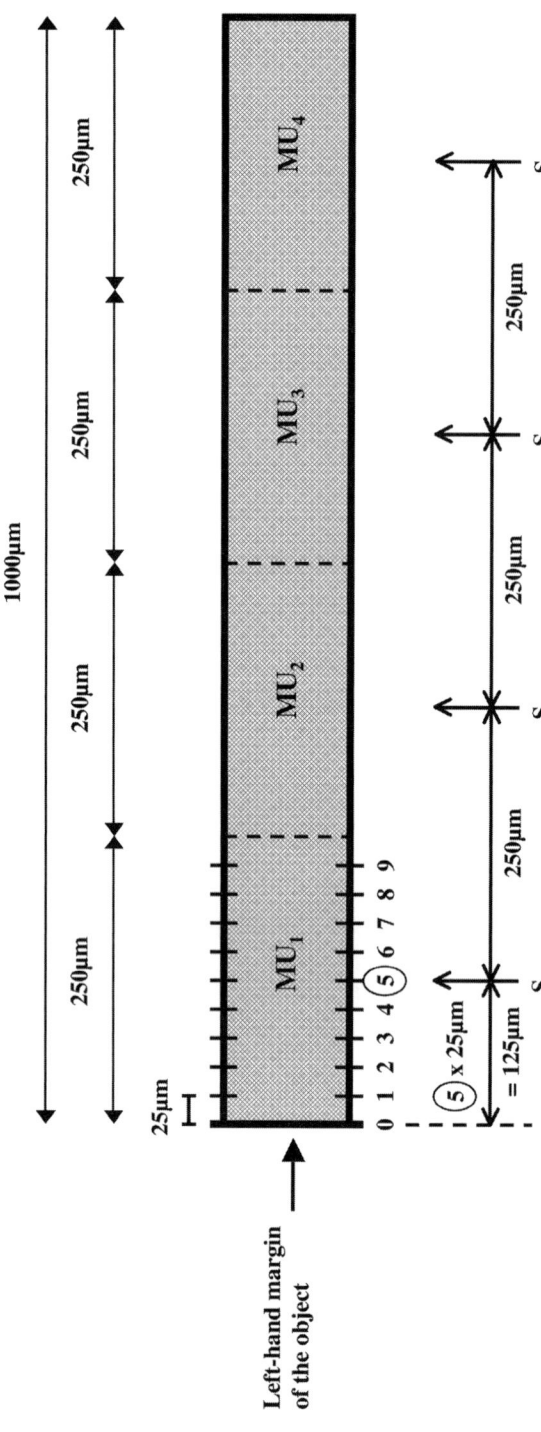

Fig. 1. The principle of systematic random sampling exemplified for a linear object. It is planned to take for this object 4 systematic random sections. In order to achieve this, the approximated total length of the object (1000 μm) is first mentally divided into 4 equal units (MU_{1-4}: each 250 μm in length). The initial unit (MU_1) is then mentally subdivided into 10 (this number will depend on the precision of the equipment used for subsampling) subunits (each 25 μm in length) and a number between "0" and "9" is chosen at random. This number ("5" in the example) defines the distance from the left-hand margin of the object (i.e., 5×25 μm = 125 μm) at which the first sample (S_1) is taken (using the available cutting equipment). The position of the second sample (S_2) is defined as the length of the object (1000 μm) divided by the number of mental units into which it has been divided (4) (i.e., 1000/4 μm = 250 μm). The third (S_3) and fourth (S_4) samples will be separated by the same (systematic) distance.

let the approximate length of the object be 1 mm and the number of units 4. The first unit (250 μm in length), within which the random starting point is to be defined, is first mentally subdivided into say 10 subunits (each 25 μm in length), and a number between 0 and 9 is then chosen at random. This number (N) defines the distance from the left-hand edge of the object (i.e., Nx25 μm) at which the first sample is cut. A second sample is then taken, whose distance from the first is defined as the approximate length of the object (1 mm) divided by the number of units into which it has been mentally divided (4) (i.e., 1 mm/ 4 = 250 μm). The third and fourth samples will be likewise separated by the same distance (250 μm). In the above example, the subdivision of the initial unit into 10 subunits was arbitrarily, although reasonably, chosen. Any range can in fact be selected, but the number of integers should take into account the length of the first unit (which will reflect that of the object itself) and the width of the blade of the cutting instrument (technical precision limitation). Similarly when analyzing a disc-like structure (*see* **Fig. 2**), the object's diameter is mentally divided and cut into a small number of equidistant units (more than one but usually fewer than 8) with a random start at the left-hand edge (within the initial unit). The systematic random sampling strategy can be readily applied either to actual sections or to virtual ones, such as those generated by magnetic resonance imaging (MRI), computed tomography (CT), or confocal light microscopy.

Data gleaned from sections yielded by the systematic random sampling strategy are generally analyzed using standard statistical approaches. During the statistical evaluation, however, a very small bias will be introduced, resulting from the systematic component of the sampling protocol. But this bias can usually be neglected. Its precise extent is a current topic of statistical research. Investigators should also be aware that each sampling step in an experiment will add to the statistical variance of the final estimate. A detailed appraisal of the variability that accumulates during the course of an investigation with a sampling hierarchy has revealed the relative contributions at each stage to be generally as follows *(5)*:

Variability between individuals (animals or human subjects):	70%
Variability between tissue blocks:	20%
Variability between sections:	5%
Variability between the microscopic fields analyzed:	3%
Variability between practical measurements:	2%
Total observed experimental variability:	100%

This breakdown discloses the variability between practical measurements to be very small. It is thus futile to attempt to improve measuring precision on the single-image level by installing expensive automatic image-analysis systems.

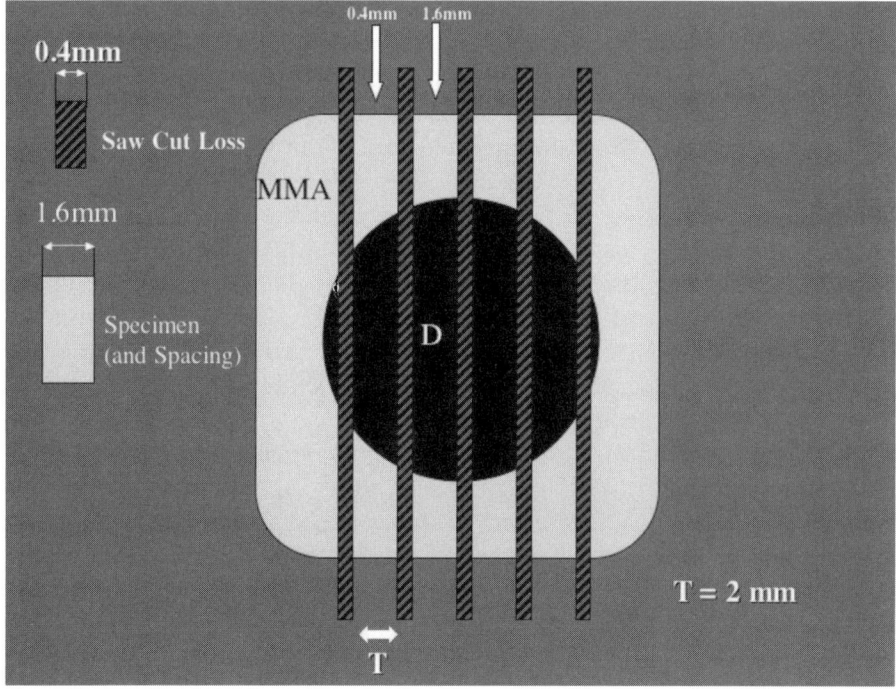

Fig. 2. Systematic tissue sampling strategy *(20)*, with a random start one-fifth of the way into the implanted material. Five consecutive sections, 0.6 mm in thickness and 1 mm apart, were prepared for the histological analysis. D = titanium-alloy disc; MMA = methylmethacrylate tissue-embedding material; T = the thickness of the tissue specimen (0.6 mm) plus the spacing (1 mm) plus the thickness of the tissue lost or damaged by the saw-blade incision (0.4 mm) *(25)*.

The increase in precision that can be thereby achieved will be maximally 2%. On the other hand, by including more animals or human subjects in the investigation (*see* **Note 1**), and by analyzing more tissue blocks per individual, the variability can be significantly reduced, because these two sampling steps contribute largely to the experimental precision. Hence, the greatest efforts should be invested in these directions. Investigators should also be aware that it suffices to measure maximally 200 points or line-intersections per experiment *(6,7)*. By remembering this rule of thumb, much fruitless effort can be avoided in a quantitative morphological analysis.

2.2. Volume Fraction of a Tissue Component

A morphometric or stereological parameter that is frequently estimated in biological studies is the volume fraction of a tissue component. This is defined by the equation:

$$V_v\,(Y, ref) = \frac{\text{Volume of component } Y \text{ in reference space}}{\text{Volume of reference space}} \qquad (1)$$

where V_V denotes the volume fraction of the component Y within the reference space (*ref.*).

The idea of measuring the volume fraction of a tissue component by a stereological method was conceived by the French mining engineer Delesse in 1847 *(8)*. He demonstrated that the volume fraction of a component within a 3D space was equal to its area fraction in a 2D section. About fifty years later, in 1898, Rosiwal established that volume density equates with linear density *(9)*. Thompson then showed that for a randomly positioned point grid, the number of points hitting the structure of interest divided by the total number of points covering the reference space gives an estimate of the volume fraction *(10)* (*see* **Fig. 3**).

Point counting is still the method most commonly used to estimate volume fraction. Indeed, for manual determinations, it is more efficient than either line or area measurements *(11,12)*. Volume fraction is a dimensionless ratio, which is often represented as a percentage. After having perused the previous section (sampling strategy), the reader should now be aware that stereological estimates of volume fraction will be unbiased only if the quantified object is represented uniformly randomly on the tissue sections, (viz., only if every portion of the object had an equal probability of being sampled). Strictly speaking, estimates of volume fraction are valid only if measurements have been made on infinitely thin 2D sections. Practically, of course, this situation cannot be realized. But if the sections are too thick, as may be the case in virtual CT images, overprojection and capping phenomena will interfere with the quantification. Overprojection leads to a systematic overestimation of volume fraction. To avoid these influences, the investigator should aim for a section thickness that is less than one-tenth of the height of the average-sized particle (usually a cell) in the tissue to be analyzed.

After having estimated the volume fraction of a tissue component within the reference space, its total volume therein can then be readily derived using Cavalieri's method (*see* **Subheading 3.5.**). For example, if the volume fraction of cartilage matrix within its referencc space has been estimated and the total volume of the articular cartilage layer is known, then the total volume of the

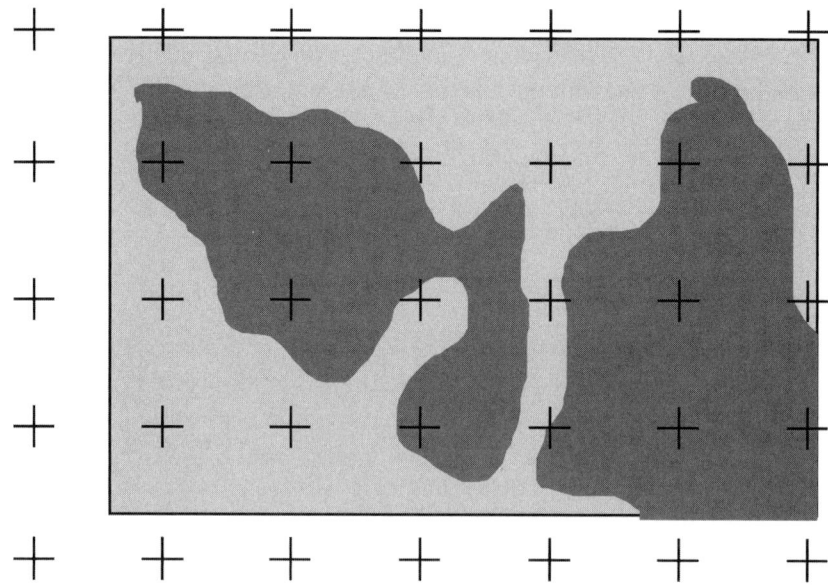

Fig. 3. An illustration of the basic idea of volume fraction estimation using a point grid. The number of points hitting the reference space, in this case, the dark-grey phase of interest Y plus the light-grey matrix, are counted (i.e., 18). The number of points that land just in Y (dark grey) are also counted (i.e., 9). The ratio of the number of points in Y divided by the number in the reference space (light-grey and dark-grey area together) is an estimate of the volume fraction (i.e., 9/18 = 0.5 or 50%). Note that points landing within Y also fall within the reference space! (Reprinted with permission from **ref. 24**.)

cartilage matrix can be derived by multiplying the volume fraction of the matrix by the total volume of the articular cartilage layer.

2.3. Surface Area and Surface Density of an Object

The surface area of an object can serve as a useful indicator of a surface-limited process's capacity to function. For example, an estimate of the surface area of a joint pannus can yield information respecting its resorptive activity and hence its destructive potential. The surface density of a set of interfaces $[S_V(Y, ref)]$ is defined as the surface area of interface per unit volume of the reference space (ref):

$$S_V(Y, ref) = \frac{\text{Area of interface of } Y \text{ in reference space}}{\text{Volume of reference space}} \qquad (2)$$

The dimensions of surface density, expressed in units of length (L), are L^2/L^3, which simplifies to L^{-1}. An estimate of the total surface area of the interface of interest $[\hat{S}(Y)]$ is derived by multiplying the surface density $[\hat{S}_V(Y, ref)]$ by the volume of the reference space $[\hat{V}_V(ref)]$:

$$\hat{S}(Y) = \hat{S}_V(Y, ref) \bullet \hat{V}(ref) \qquad (3)$$

A series of linear test probes is required to estimate an object's surface. Using such a system, a relationship has been found to exist between the number of intersections (I_L) and the surface area per unit volume $\hat{S}_V(Y)$:

$$\hat{S}_V(Y) = 2 \bullet I_L \qquad (4)$$

Surface density is thus equal to twice the number of intersections between the surface and the linear probe, per unit length of test line within the reference space *(13,14)*. The equation represents an unbiased estimator of surface density only if either the surface or the lines or both are isotropic.

For the estimation of volume fraction (*see* **Subheading 3.2.**), a point-counting test grid is applied. Using such a zero-dimensional point system, the orientation of the object of interest will not influence the number of points hitting it. But for the estimation of surface density, a linear test grid that projects across the reference space is required. The number of times that a particular object is intersected by the lines will be influenced not only by its surface area but also by its orientation. If a test line is orientated parallel to an object's surface, then it may intersect it more than once. Hence, the orientation of the lines relative to the object's surface must be taken into account. A cycloid-based (*see* **Fig. 4**) test grid (*see* **Fig. 5**) satisfies the prerequisite that all surface orientations within the reference space have an equal probability of being intersected. To ensure that this is indeed the case, a vertical axis within the object must be defined, which entails that tissue blocks and sections be prepared in strict geometrical relation to it. These needs being satisfied, surface density is estimated using the following equation:

$$\hat{S}_V(Y, ref) = \frac{2 \bullet \sum_{i=1}^{n} I_i}{1/p \bullet \sum_{i=1}^{n} P_i} \qquad (5)$$

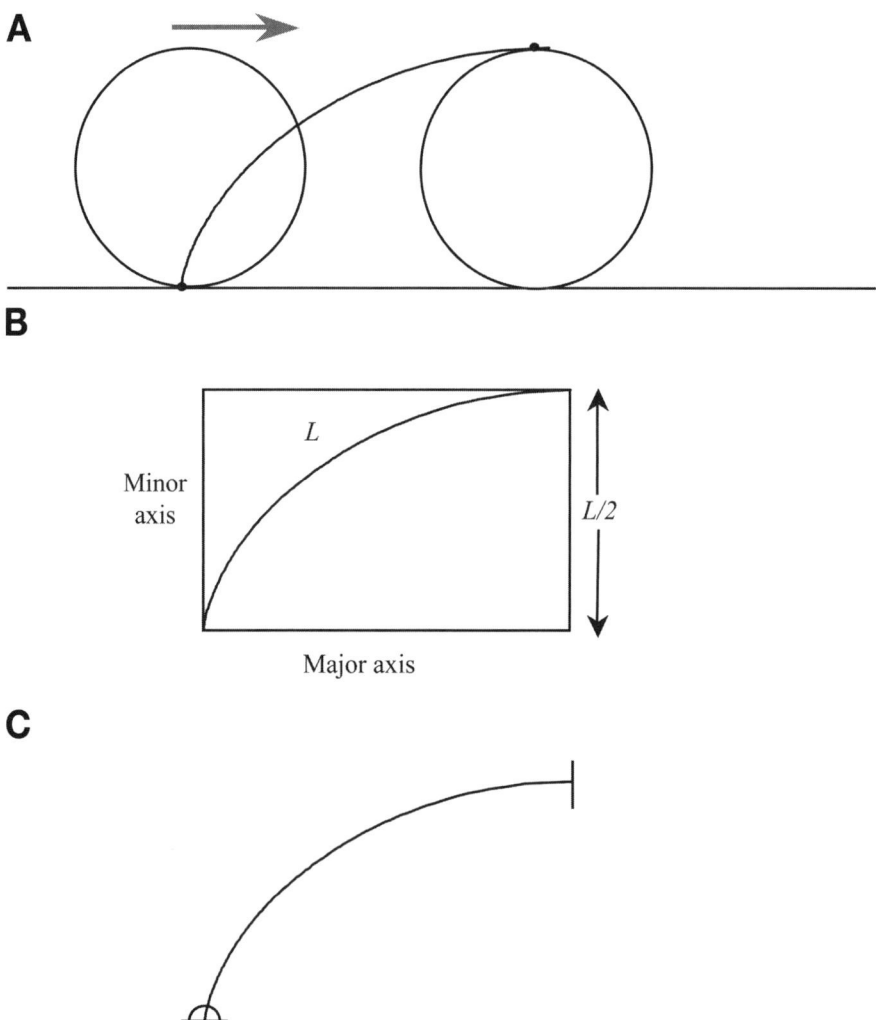

Fig. 4. (**A**) A cycloid is a curve described by the locus of a point on the periphery of a disc rolling along a straight edge. The length of line present in a particular segment of cycloid is proportional to the sine of its angle with respect to the normal to the straight edge. (**B**) If a cycloid arc is bound by a rectangle, the short and long sides of the rectangle describe the minor and major axes of the cycloid, respectively. The length of the cycloid is equal to twice the length of the minor axis. (**C**) In a practical estimation of surface density using vertical, uniformly random sections, cycloid arcs are arranged with their minor axes parallel to the defined vertical direction. Points associated with the cycloids permit an estimation of the total length of the cycloid used. (Reprinted with permission from **ref. 24**.)

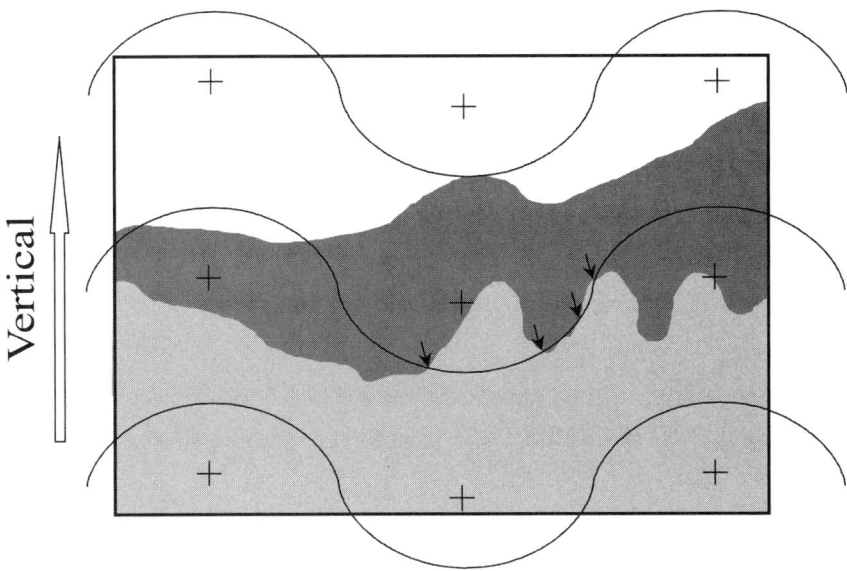

Fig. 5. A vertical section is shown with a cycloid test grid superposed. The minor axes of the cycloids are parallel to the known vertical direction, and the grid has been uniformly randomly translated with respect to both x and y. In this case, six points land within the reference space (light- and dark-grey phases). Each point is associated with two cycloid arcs. In this example, the top edge of the cycloid represents the "true" cycloid arc. Each time this top edge passes through the boundary of interest, in this case the interface between dark- and light-grey phases, an intersection is counted. The four intersections in this example are indicated by arrows. (Reprinted with permission from **ref. 24**.)

where I_i represents the number of intersections within the reference space (P_i) on each image and l/p is the length of the test line per grid point at the tissue level (i.e., corrected for the linear magnification).

It should be borne in mind that a vertical, uniformly random section is not an isotropic one in the 3D space, and that the bilaterally orientated test lines used in conjunction with such a section do not constitute an isotropic set. However, an unbiased estimate of surface density is nevertheless achieved because the length density of the set of lines used in this system is proportional to the surface area of the object represented in space *(15)*.

2.4. Counting Particles

The number of discrete objects of a particular kind (e.g., cells, nuclei) residing within a well-defined reference space can be estimated stereologically by first determining their numerical density (i.e., their number per unit volume of tissue) and then multiplying this value by the volume of the reference space.

In any attempt that is made to estimate the number of particles within a tissue, investigators should always endeavor to select unbiased stereological estimators. If they fail to do so, time and effort will be expended without yielding biologically meaningful or valid information. An example will perhaps help to emphasize this point. By way of illustration, pannus activity in a rheumatoid joint will be considered. Let it be supposed that the efficacy of a drug in reducing pannus activity and size is to be tested. For this purpose, the investigator should estimate the total volume of the pannus in the drug-treated and control groups as well as the numerical density of the inflammatory cells in each pannus. Using these values, the absolute number of inflammatory cells per joint can then be determined. An estimation of the number of inflammatory cells per unit volume of pannus will not yield the information that is required to truly assess the beneficial effect of the drug tested (*see* **Note 2**).

In order to estimate the number of particles within a 3D space using 2D sections, it is necessary to apply a 3D probe. A number of methods have been developed to render this task simple and efficient.

As is the case for the stereological estimators already described, namely, volume fraction (*see* **Subheading 3.2.**) and surface area (*see* **Subheading 3.3.**), each of the objects to be counted must have the same probability of being represented in the count, irrespective of its size and dimensions. But 2D sections through a tissue are more likely to hit large objects than small ones. And for this reason, counting the number of profiles of a particle within a section will never yield a meaningful estimate of the number of the particles themselves in space.

Although serial sectioning represents an unbiased approach to counting in 3D, it is a tedious process. A major breakthrough in stereological counting came with the publication of the disector principle by D.C. Sterio (a pseudonym and an anagram of "disector") in 1984 *(16)*. The disector consists of a pair of sections a known distance apart. The method relies on the principle that a particle's transect appears in one of the sections but not in the other. If this condition is fulfilled, the transect is counted. The disector is thus a directional counting method (*see* **Fig. 6**). It is efficient and yields an unbiased approximation of a particle's numerical density.

In order to apply the disector principle, the pair of sections must of course be parallel to each other *(17–19)*. The unbiased counting frame used for the purpose (*see* **Fig. 7**) is of known area and embodies an "acceptance" line (dashed) and an infinite "forbidden" line (unbroken). Any particle that is hit by the "forbidden" line is excluded from the count *(20)*. This rule can be implemented only if the counting frame is surrounded by a "guard area" (*see* **Fig. 7**),

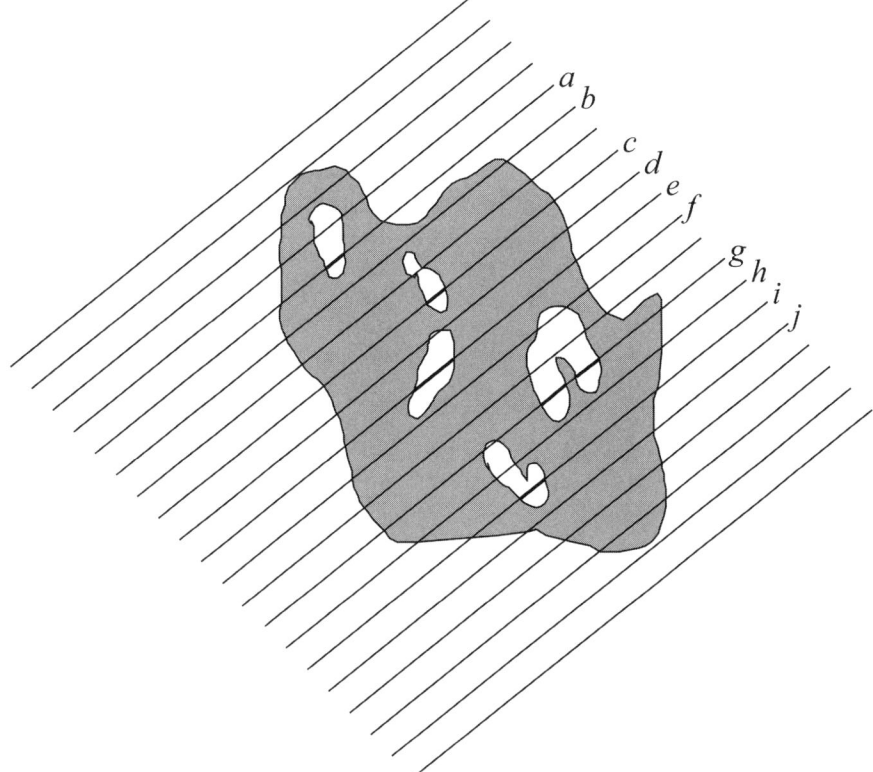

Fig. 6. The projected object has been exhaustively sectioned by a series of systematic uniformly random planes (shown as lines). The particles are counted by moving down the stack of sections and applying the rule that a particle is counted if it appears in one section but *not* in the next. For example, the uppermost particle is seen in section *a*, but not in section *b*. Note that for unbiased counting, it is necessary to know when a particle transect consists of more than one profile in a section (e.g., section *g*). (Modified with permission from **ref. 24**.)

and it is thus not applicable to a complete monitor image. The 2D transect-counting rule applied to the first or "reference" section is then extended to a 3D particle-counting one using the second or "look-up" section. For each of the transects counted in the "reference" section, a corresponding one is sought in the "look-up" one (*see* **Fig. 8**). If no corresponding transect is found in the latter, then the particle is counted. The count is effected within a volume of space that corresponds to the area of the counting frame multiplied by the distance between the two sections. This counting rule yields an unbiased estimation of the numerical density of the particle of interest.

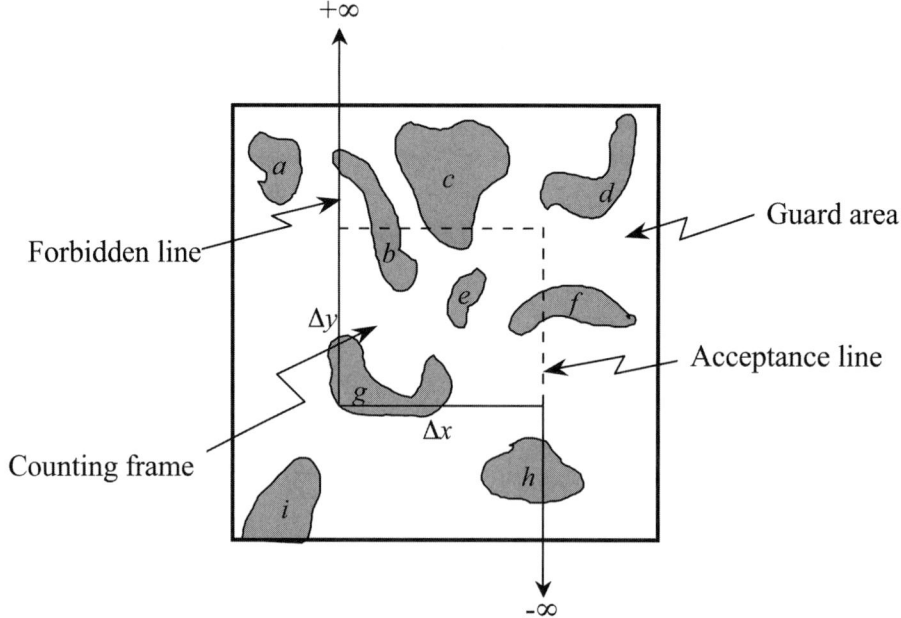

Fig. 7. A 2D field of view containing several 2D objects is shown together with an unbiased counting frame that is surrounded by a guard area. The counting frame consists of a solid forbidden line, which extends above and below the field of view to infinity, and a dashed acceptance line. The area of the counting frame is Δx multiplied by Δy units squared. Any 2D object that is cut by the forbidden line is not counted (i.e., *b*, *g* and *h*). 2D objects falling fully within the frame (i.e., *e*) or those that cut the acceptance line without also cutting the "forbidden" line (i.e., *c* and *f*) are counted. The application of this rule leads to an unbiased estimate of the number of 2D objects per unit area. (Reprinted with permission from **ref. 24**.)

The efficiency of the disector (i.e., a combination of the disector principle and the 2D counting frame) can be almost doubled by making a two-way use of it (i.e., by reversing the roles of the "reference" section and the "look-up" section). It is impossible for a particle counted in one direction to be also counted in the opposite direction (*see* **Fig. 9**). The numerical density of the particle counted ($\hat{N}_V$) is obtained using the following equation:

$$\hat{N}_v = \frac{1}{a/f \bullet h} \bullet \frac{\sum Q^-}{\sum P} \qquad (6)$$

where a/f is the area of the frame, h is the height of the disector (i.e., the distance between the two sections), ΣQ^- is the sum of the particles counted and ΣP is the sum of the frame-associated points hitting the reference space.

Arthritis Research: Methods and Protocols

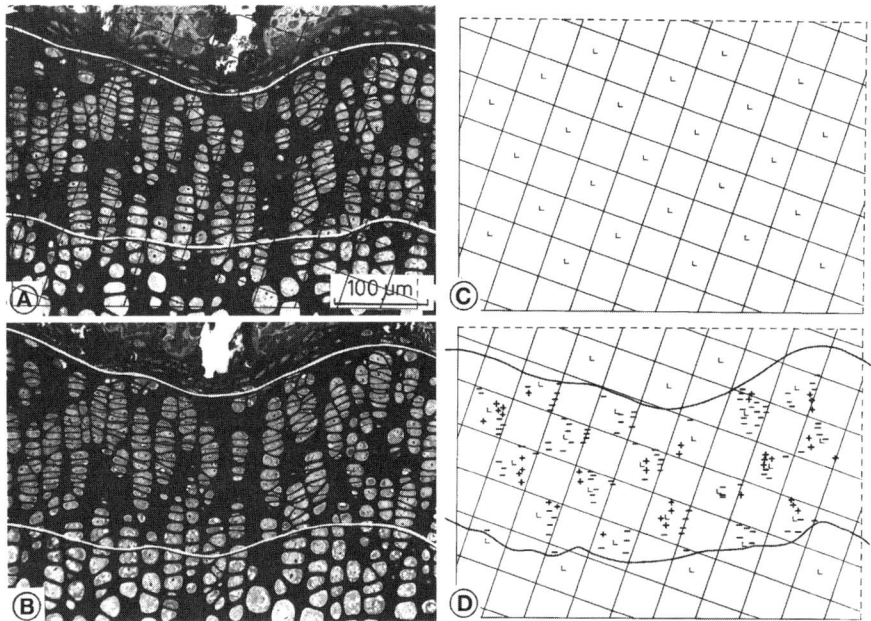

Fig. 8. Illustration of the *disector method* for estimating the numerical density of cells (N_v) within a stratum *(16)*. (**A**) Reference section (number 1, say, of a series) with the test system (**B**) superimposed upon it. (**B**) "Look-up" or lower section of the disector (number 5 of the same series). The distance between the upper faces of the first and fifth sections is thus $h = 4 \times 1.04 = 4.16$ μm, here represented approximately by the white strip between the pictures (**A**) and (**B**). Cell profiles in (**A**) that are no longer present in (**B**) are marked "+," in (**D**). The procedure was facilitated by looking at the intermediate sections (*2*, *3* and *4*). The total number N(+) divided by the product of *h* times the number of square centres hitting the reference space in (**A**) is, up to a constant factor, a practically unbiased estimator of N_v, irrespective of cell shape, section thickness and resolution conditions. (Reprinted with permission from **ref. *18*.**)

As a rule of thumb, the distance between the two sections used as a disector should be about 30% of the average projected height of the particle to be counted. For example, if the object to be counted is a cell with an average height of 15 μm, then the distance separating the two sections should be no greater than 5 μm. To ensure that the position of the section pair within the tissue specimen is uniformly random, a random number generator should be used for sampling.

A simpler alternative to working with actual tissue sections is to use optical disectors *(17,19)*. Indeed, this approach is statistically more efficient (lower variance), since the technology permits the rapid collection of many small optical disectors rather than fewer larger ones.

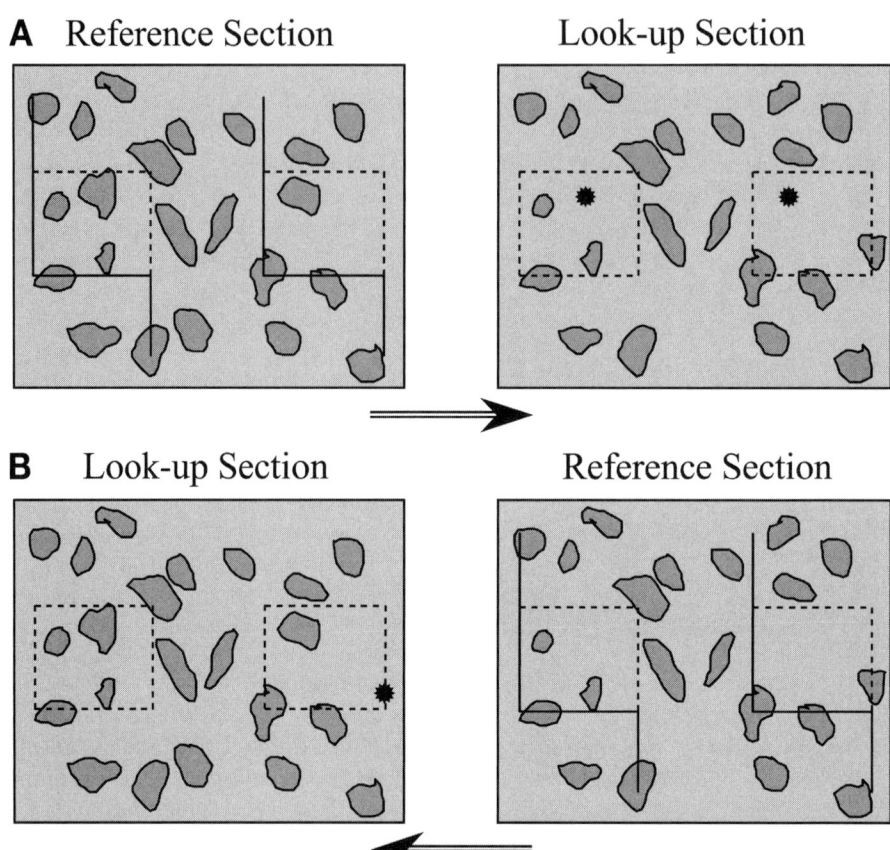

Fig. 9. An illustration of the two-way use of the physical disector. Both (**A**) and (**B**) show the same pair of perfectly registered sections, a known distance apart. In (**A**), the left-hand section, a pair of unbiased counting frames, which is superimposed, acts as the reference section. The count is made from this section to the right-hand section in the direction of the arrow, viz., transects properly sampled in the frames are sought in the look-up section. In this example, two particles are counted. In (**B**), the roles of the reference and the look-up sections have been reversed. The count now takes place from right to left, in the direction of the arrow. In this case, one particle is counted. Thus, in total, three particles have been counted. It is *impossible* for a particle counted in one direction to be also counted in the opposite direction. In (**A**), the volume of the disector used is twice the frame area times the distance between the sections. In (**B**), the volume of disector used is the same. Thus, in total, the volume of the disector used is the same. The volume of the disector used is four times the frame area times the distance between the sections. (Reprinted with permission from **ref. *24*.**)

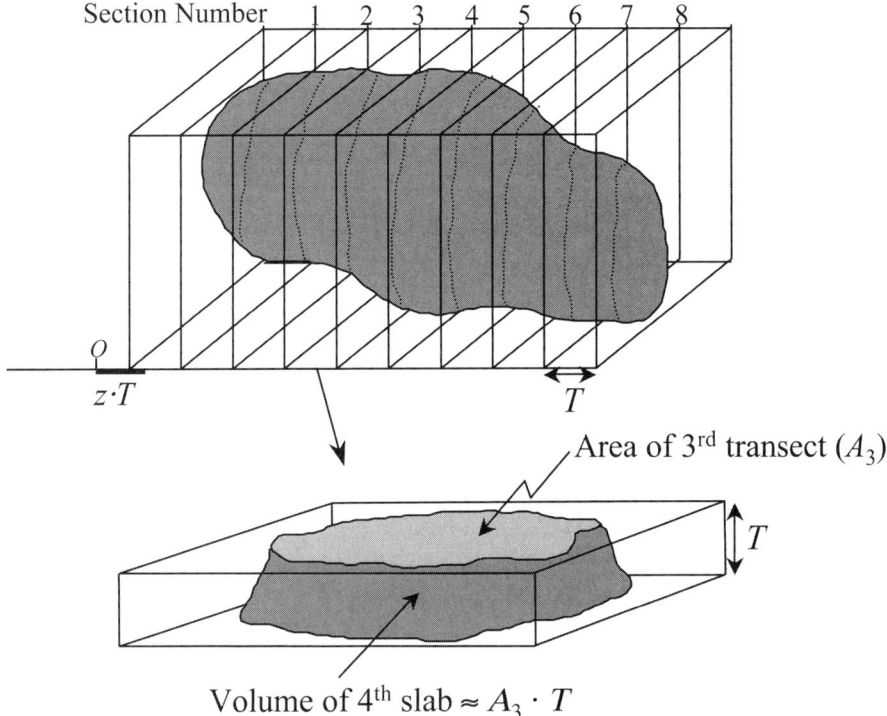

Fig. 10. Illustration of the Cavalieri method. An object of arbitrary shape is shown intersected by a series of parallel cutting planes (i.e., serial sections), a known and fixed distance apart of (T). The number of sections hitting the object in this example is eight, which gives rise to nine slabs of the object. The fourth slab has been extracted from the object. The thickness of the slab is T units, and the volume is given approximately by the cross-sectional area of the third transect times the distance T. (Modified with permission from **ref. 24**.)

Alternative methods to the disector include the fractionator technique *(21)* and the multistage fractionator *(22)*. It is also possible to use a single-section or "cheating" disector if the particles to be counted are very small, because, in this case, a model assumption can be made which greatly simplifies counting. However, being assumption-based, the approach is biased.

2.5. Reference Volume

Reference volume is a key estimator in most morphometric analyses. It is most readily determined using the method devised by Cavalieri *(4,23)*, which is illustrated in **Fig. 10**. It entails the preparation of an exhaustive series of parallel slices or slabs through the tissue or tissue component of interest at a fixed distance apart (T). The first slice should be uniformly random and should fall within a distance of 0-T from the left-hand border of the tissue or tissue component of interest. Each of the slices is laid flat, with the freshly-cut plane always facing downwards, and the cross-sectional area of the tissue or tissue component of interest is then measured or estimated. The volume of the tissue or tissue component of interest ($\hat{V}$) is estimated by summing the areas (A) and multiplying by the slice thickness (T):

$$\hat{V} = (T \bullet A_1) + (T \bullet A_2) + \ldots + (T \bullet A_m) = T \bullet \sum_{i=1}^{m} A_i \qquad (7)$$

where A_i is the cross-sectional area of the tissue or tissue component of interest on the ith slice.

The Cavalieri estimator of reference volume is guaranteed to be unbiased if the position of the first slice hitting the tissue or tissue component of interest is uniformly random. The cross-sectional areas do not need to be measured with great precision. Measurements can be made using a randomly translated point grid. Within such a grid, the area associated with each point is a known quantity. If this method is applied, the reference volume is estimated using the following equation:

$$\hat{V} = T \bullet \frac{a}{p} \bullet \sum_{i=1}^{m} P_i \qquad (8)$$

where a/p is the area associated with each point on the grid and P_i represents the number of points falling within the tissue or tissue component of interest on the ith slice.

As a general rule, maximally 200 points need to be counted per tissue or tissue component of interest (not per slice) in order to yield an estimate with a coefficient of error of 5 to 10%. The Cavalieri method can be applied to estimate the reference volume either of a large tissue expanse, such as the cartilage layer of a joint, or of a tissue component, such as a cell (which would be required to estimate the volume fraction of a subcellular organelle).

2.6. Other Estimators

Estimators other than those discussed in this chapter may be required. For example, it may be necessary to determine the length of a linear structure in space or to grade the size of particles. For most needs, an appropriate stereological method has been elaborated. For guidance in the choice of a suitable procedure, the reader is advised to consult the book written by Howard and Reed *(24)*, which is an excellent *vade mecum* for practical stereology.

3. Notes

1. It cannot be too often stressed that an unbiased stereological approach to the estimation of biological quantities within a 3D space is based upon the selection of an appropriate sampling strategy and the choice of a suitable experimental design. As was mentioned under **Subheading 2.1.**, biological variability between animals or human subjects represents the largest contribution (70%) to the total observed experimental variability. For this reason, investigators should endeavour to include a large number of subjects in their studies. Findings that are based upon the quantitative evaluation of three or four individuals will carry but little weight. For a meaningful study, the number of subjects included should be minimally six or seven. Generally speaking, standard statistical approaches can be used in conjunction with stereological measurements. When presenting their data, investigators are recommended to represent statistical variation as either the coefficient of variation or the coefficient of error, which are the most illustrative modes of expression for this quantity. For some stereological estimators, it will be necessary to apply special statistical tools, and for guidance in this matter the reader is referred to the chapter on "Statistics for Stereologists" in **ref. 24**.

2. Investigators are advised not to stint on the time invested in the planning of a stereological analysis. They are encouraged to pay attention to the design of their experiments and to the choice (and definition) of appropriate, biologically meaningful estimators. They are also recommended to seek the advice of a stereologist at this conception stage. Later, when the tissue has been sectioned, there will be no atonement for having unwittingly disregarded stereological, statistical and sampling principles. The experiments will have to be repeated. When biologically meaningful estimators are chosen and the study is carefully designed, stereology can be a very powerful tool in the quantification of biological processes under physiological, pathological and experimental conditions.

References

1. Stuart, A. (1984). *Basic Ideas of Sampling*, Griffin, London.
2. Diggle, P. J. (1983) *Statistical Analysis of Spatial Point Patterns*, Academic Press, London.
3. Cochran, W. G. (1977) *Sampling Techniques* (3rd Ed.), Wiley, New York.
4. Gundersen, H. J. and Jensen, E. B. (1987) The efficiency of systematic sampling in stereology and its prediction. *J. Microsc.* **147**, 229–263.
5. Gundersen, H. J. and Osterby, R. (1981) Optimizing sampling efficiency of stereological studies in biology: or 'do more less well! *J. Microsc.* **121**, 65–73.
6. Gundersen, H. J., Bagger, P., Bendtsen, T. F., et al. (1988a) The new stereological tools: disector, fractionator, nucleator and point sampled intercepts and their use in pathological research and diagnosis. *Apmis* **96**, 857–881.
7. Gundersen, H. J., Bendtsen, T. F., Korbo, L., et al. (1988b)Some new, simple and efficient stereological methods and their use in pathological research and diagnosis. *Apmis* **96**, 379–394.

8. Delesse, M. A. (1847).Procédé mécanique pour déterminer la composition des roches. *C. R. Acad. Sci. Paris* **25**, 544–545.
9. Rosiwal, A. (1898) Ueber geometrische Gesteinsanalysen. Verh. K. K. *Geol. Reichsanst. Wein* **143**.
10. Thompson, E. (1930) Quantitative microscopic analysis. *J. Geol.* **38**, 193.
11. Gundersen, H. J., Boysen, M., and Reith, A. (1981) A comparison of semiautomatic digitizer tablet and simple point counting performance in morphometry. *Virchows Arch. (Cell Pathol.)* **37**, 317–325.
12. Mathieu, O., Cruz-Orive, L. M., Hoppeler, H., and Weibel, E. R. (1981) Measuring error and sampling variation in stereology: comparison of the efficiency of various methods for planar image analysis. *J. Microsc.* **121**, 75–88.
13. Saltykov, S. A. (1946) *Zavodskaja Laboratorija* **12**, 816. Cited by Weibel, 1979.
14. Smith, C. S. and Guttman, L. (1953) Measurement of internal boundaries in three-dimensional structures by random sectioning. *Trans. AIME* **197**, 81.
15. Baddeley, A. J., Gundersen, H. J., and Cruz-Orive, L. M. (1986) Estimation of surface area from vertical sections. *J. Microsc.* **142**, 259–276.
16. Sterio, D. C. (1984) The unbiased estimation of number and sizes of arbitrary particles using the disector. *J. Microsc.* **134**, 127–136.
17. Wong, M., Wuethrich, P., Eggli, P., and Hunziker, E. (1996) Zone-specific cell biosynthetic activity in mature bovine articular cartilage: a new method using confocal microscopic stereology and quantitative autoradiography. *J. Orthop. Res.* **14**, 424–432.
18. Cruz-Orive, L. M. and Hunziker, E. B. (1986) Stereology for anisotropic cells: application to growth cartilage. *J. Microsc.* **143**, 47–80.
19. Hunziker, E. B., Quinn, T. M., and Hauselmann, H. J. (2002) Quantitative structural organization of normal adult human articular cartilage. *Osteoarthr. Cart.* **10**, 564–572.
20. Gundersen, H. J. (1977) Notes on the estimation of the numerical density of arbitrary profiles: the edge effect. *J. Microsc.* **111**, 219–223.
21. Gundersen, H. J. (1986) Stereology of arbitrary particles. A review of unbiased number and size estimators and the presentation of some new ones, in memory of William R. Thompson. *J. Microsc.* **143**, 3–45.
22. Ogbuihi, S. and Cruz-Orive, L. M. (1990) Estimating the total number of lymphatic valves in infant lungs with the fractionator. *J. Microsc.* **158**, 19–30.
23. Cavalieri, B. (1635) *Geometria Indivisibilibus Continuorum,* Typis Clemetis Feronij, Bononi. Reprinted (1966) as Geometria degli Indivisibili. Unione Tipografico-Editrice Torinese, Torino.
24. Howard, C. V. and Reed, M. G. (1998) *Unbiased Stereology. Three-Dimensional Measurement in Microscopy,* Bios Scientific Publishers, Oxford.
25. Liu, Y., de Groot, K., and Hunziker, E. B. (2005) BMP-2 liberated from biomimetic implant coatings induces and sustains direct ossification in an ectopic rat model. *Bone* **36(5)**, 745–757.

10

Image Analysis of Aggrecan Degradation in Articular Cartilage With Formalin-Fixed Samples

Barbara Osborn, Yun Bai, Anna H. K. Plaas, and John D. Sandy

Summary

Many studies in arthritis research require an evaluation of the cellular responses within the joint and the ensuing matrix degradation in articular cartilage. The early histochemical/histological scale of Mankin *(1)* has been widely used but recently challenged as insufficient *(2)*. Imaging techniques such as microscopic magnetic resonance imaging (MRI) *(3)*, polarized light microscopy *(3)*, atomic force microscopy *(4)*, and infrared spectral analysis *(5)* have opened new approaches to evaluating cartilage structure. Histological methods now include *in situ* hybridization for cell-specific gene expression and immunohistochemistry for the spatial organization of cartilage proteins and their processed forms.

This chapter details of a method for immunohistochemical analysis of aggrecan degradation in articular cartilage samples which have been prepared by standard methods of formalin fixation and paraffin embedding. The procedure focuses on the application of antibodies (e.g., anti-ADAMTS4, anti-MT4MMP) which detect some of the proteinases most likely involved, and anti-NITEGE which detects the terminal product of the aggrecanase-mediated cleavage of aggrecan at Glu392-Ala393 (bovine, human, dog, rat, pig, sheep, horse, mouse) or Glu393-Ala394 (chick).

Key Words: Aggrecanase; immunohistochemistry;cartilage; osteoarthritis; aggrecan; neoepitope; ADAMTS; chondroitin sulfate; antigen retrieval; proteolysis; extracellular matrix; degradation; joint disease.

1. Introduction
1.1. Aggrecanases and Arthritis

A major theme in arthritis research is the degradation of articular cartilage matrix by proteolytic enzymes *(6)*. Because aggrecan is the major structural element which confers reversible compressibility on the tissue *(7)*, and aggrecanolysis appears to be an early and important event in all forms of human arthritis *(8)*, the image analysis of aggrecan degradation in native cartilage tissue is central to an understanding of this process. Analysis of articular cartilage extracts has generated a detailed picture of the full-length and truncated aggrecan core protein species which are normally present in the mature cartilage matrix *(9)*. The proteinases responsible for the C-terminal truncations in normal cartilage appear to be largely calpain *(10)* and members of the MMP and aggrecanase families *(9,10)*. In contrast, destructive aggrecanolysis in arthritis is apparently does not result from calpain or MMP activity, but is rather the result of uncontrolled aggrecanase activity which caused by one or more of ADAMTS-1,4,5,8,9 and 15 *(11–13)*. These recently cloned glutamyl-C-endopeptidases cleave aggrecan at five Glu-X sites and the hyaluronan-associated terminal G1-domain product is G1-NITEGE392 (Glu392 is taken directly from the protein databases for aggrecan of all species analyzed. The more common usage of Glu373 for this site is derived by subtraction of the 19 residues of the bovine signal peptide established by N-terminal analysis of the native intact core protein. Cross-species references for residue numbers in aggrecan are more readily made without adjustment for the signal peptide). ADAMTS4 appears to be responsible for most, if not all of this activity in articular cartilages *(12,14–16)*, although ADAMTS-5-null mice, but not ADAMTS4-null mice, are highly resistant to experimentally-induced cartilage destruction in vivo and in vitro *(17)*. Processing of ADAMTS4 to its most destructive form appears to require cleavage by MT4MMP (MMP17) *(16,18)*. We therefore provide detailed methods for IHC of the product G1-NITEGE and both ADAMTS4 and MT4MMP. A model of the relationship between chondrocytes, MT4MMP, ADAMTS4 (p68 and p53 forms) and the tissue-associated product G1-NITEGE in cartilage matrix during aggrecanase-mediated aggrecanolysis is given in **Fig. 1**. This model is based on published data *(16,18–20)* and proposes that full-length ADAMTS4 (p100) is cleaved by furin to remove the pro-domain and then associates as ADAMTS-4 (p68) with MT4MMP on the cell surface. This form has the capacity to attack the aggrecan in the C-terminal CS-2 region only thereby treleasing the G3 domain. In addition, the p68 can be converted to a C-terminally truncated form of ADAMTS4 (p53) which can be found associated with cell surface syndecan and which now has acquired the capacity to destructively cleave the aggrecan within the interglobular domain to release the CS-1 region and generate the terminal G1-NITEGE product.

1.2. Immunohistochemistry for Aggrecan G1-NITEGE, ADAMTS4, and MT4MMP

The first descriptions of IHC with anti-NITEGE *(21–23)* were performed on 6 μM cryosections of human cartilage and mouse knee joints mounted on glass

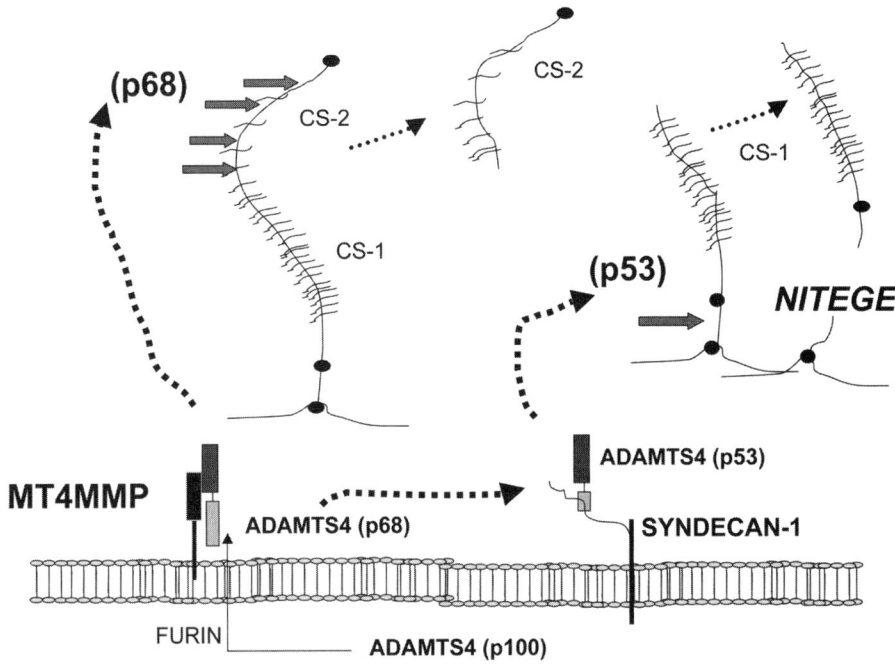

Fig. 1. Model based on published studies *(16,18)* depicting chondrocyte-mediated processing of ADAMTS-4 (aggrecanase-1) by MT4MMP and syndecan and the associated steps of aggrecan degradation. *See text* for description of structures. (Color illustration in insert following p. 268.)

slides and treated with chondroitinase ABC, fixed in periodate-lysine-paraformaldehyde fixative *(24)*, treated with 3% H_2O_2 in methanol and 0.1% Triton X-100. Bound IgG was detected via immunoperoxidase using the avidin-biotin complex method. In a different approach with human cartilage and mouse joint *(25–27)*, 8 μM cryosections were mounted on gelatin-coated glass slides and air-dried, treated with chondroitinase ABC as above, washed with phosphate buffered saline (PBS), blocked with normal serum and the bound IgG was detected as above. A protocol for detection of G1-NITEGE in rat growth plate samples *(28,29)* employed initial fixation in periodate-lysine-paraformaldehyde, decalcification for 12 d in 10% ehtylene diamine tetraacetic acid (EDTA) before infiltrating and embedding in a 2:1 volume of sucrose (20% *[w/v]* in PBS) and OCT compound. 8 μM sections were mounted on gelatin-coated slides, immersed in 4% formaldehyde in PBS to improve the adhesion of the section to the slide, exposed to 0.3% H_2O_2 and washed, followed by digestion with chondroitinase ABC. After treatment with normal serum and primary antibody, the IgG was visualized with an alkaline phosphate substrate and sections were counterstained with methyl green hematoxylin. Most recently *(15)*, bovine cartilages were fixed in formalin, embedded in paraffin and 10 µ*M* sections were mounted on glass slides, treated with chondroitinase ABC and blocked with 3% bovine serum albumin (BSA) before first antibody treatment and detection by fluorescein isothiocyanate (FITC)-labeled secondary. Immunohistochemistry for ADAMTS4 has employed routine formalin fixation of isolated chondrocytes *(15,30)* and cartilage sections *(15)* and the cell-surface localization of MT4MMP was confirmed by immunolocalization to the cell surface of formalin-fixed CHO cells *(31)*. Various histochemical methods for GAG and proteoglycan localization in cartilage have been used including toluidine blue, alcian blue 8GX, and Safranin O/Fast Green FCF/Hematoxylin for light microscopy and ruthenium red, cupromeronic blue, and cuprolinic blue for electron microscopy *(32)*. The protocol described here uses only Methyl Green as counterstain because this optimizes the visualization of the brown peroxidase product of DAB (3-3'-diaminobenzidine tetrahydrochloride) both associated with cells and in matrix. It also provides an indication of the abundance of negative charged glycosaminoglycans in the cartilage matrix.

2. Materials
2.1. Buffers, Reagents, and Solutions

1. 10% *(v/v)* neutral buffered formalin (37–40% formaldehyde) (Fisher Scientific, www.fishersci.com; cat. no. F75) diluted in 0.1 *M* sodium phosphate (pH 7.0).
2. Absolute ethanol (200 proof, reagent grade).
3. 95% Ethanol (190 proof, reagent grade).
4. Xylenes, histological grade (Histoprep; cat. no. HC700) .
5. Paraplast tissue embedding medium (Tyco Healthcare Group, www.kendallhq.com).
6. PBS (pH 7.4).
7. Vectastain *Elite* ABC Kit (Vector Laboratories, www.vectorlabs.com).
8. Peroxidase DAB Substrate Kit (Vector Laboratories).
9. Resinous mounting medium (Fisher Permount; cat. no. SP15-500).
10. Methyl green was from Vector Labs and used as supplied.

2.2. Equipment

1. Dermal biopsy punch (Miltex Instrument Company, Inc., Bethpage, NY).
2. Tissue processor, closed system (Hypercenter 2, Shandon Inc.) .
3. Slide warmer (Fisher; cat. no. 12-594).
4. Tissue-Tek uni-cassettes (Sakura Finetek, Inc., Torrance, CA).
5. Rotary microtome (Microm Inc. HM330).

6. Disposable microtome blades (Teflon coated low profile) (DuraEdge; cat. no. 7223).
7. Superfrost/plus slides (Fisher; cat. no. 12-550-15).
8. Embedder (Tissue Tek Embedding Center, Miles Scientific).
9. Metal embedding molds (Fisher Histoprep; cat. no. 15-182-505).

2.3. Antibodies

1. Anti-NITEGE (also called JSCNIT) is an anti-peptide neo-epitope antibody, prepared and affinity purified as described *(18)* and used at 20 µg IgG/mL in 1.5% goat serum/PBS (pH 7.4). It can be purchased from Affinity Bioreagents, Golden CO.
2. For ADAMTS-4, two separate antibodies (called JSCVMA and JSCYNH) were used at 20ug IgG/mL in 1.5% goat serum/PBS (pH 7.4). These antibodies were prepared and affinity purified as described *(18,19)*. They can be purchased from Affinity Bioreagents, Golden CO.
3. For MT4MMP, an affinity purified anti-peptide Ab to the N-terminal of human MT4MMP was obtained from Sigma Inc. (cat. no. M3684) and used at 20 µg IgG/mL in 1.5% goat serum/PBS (pH 7.4).
4. The specificity of the reactivity of these three antibodies was established by Western analysis of extracts of bovine articular cartilage for aggrecan G1-NITEGE *(33,34)* and for ADAMTS4 and MT4MMP *(16)*. These biochemical analyses confirmed that each antibody reacted only with the expected protein in bovine articular cartilages as follows: anti-NITEGE reacted only with the diffuse G1-NITEGE doublet at 64-68kDa; anti-ADAMTS4 (JSCVMA,JSCYNH) reacted only with the p68 and p53 forms of the enzyme and exhibited the expected species preference, JSCYNH for p68 and JSCVMA for p53; anti-MT4MMP reacted only with a tight protein band at approx 65 kDa.

3. Methods

3.1. Specimen Preparation

Routinely, the metacarpal-phalangeal joints from 6-mo-old bovines were used within 24 h of slaughter, however the same protocol can be used for cartilage obtained from any source. Where possible, full depth articular cartilage pieces (approx 5 mm × 5 mm in surface area) are dissected with a scalpel from a defined anatomic location, and disks (3 mm diameter and about 1 mm deep) punched out with a dermal punch, briefly washed in PBS and fixed in 10% neutral buffered formalin for at least 48 h.

3.2. Processing

The methods for histological processing are routine. Processing is in a closed system tissue processorusing vacuum in all stations and carried out at room temperature (RT) unless otherwise stated. Tissues are dehydrated through graded ethanols (70% for 60 min, 80% for 60 min, 3 changes of 95% for 60 min, changes of 100% for 60 min), cleared in xylenes (3 changes for 30 min, 45 min, and 45 min, respectively), infiltrated with paraffin, heated to 58°C (2 changes for 90 min) and embedded in Paraplast tissue embedding medium using Tissue-Tek Uni-Cassettes. Paraffin sections are cut at 4 µm on a rotary microtome and are floated on a distilled water bath at 48°C and then mounted on Superfrost/Plus slides. Slides are dried vertically at room temperature for 1 h and then placed flat on a slide warmer overnight at 37°C.

3.3. Pretreatments, Immunostaining, and Counterstaining

Sections are deparaffinized and hydrated through xylenes (2 × 5 min) as above, followed by 100% EtOH, 95% EtOH, and DI water and then rinsed in tap water for 5 min. Slides are placed in 10% neutral-buffered formalin for 30 min to improve tissue adhesion and rinsed for 5 min in tap water.

3.3.1. Pretreatments

We have obtained very useful data with cartilage sections which have not been pretreated with chondroitinase ABC or other "epitope retrieval" methods. When "native" fixed tissue is probed, possible treatment-induced variations in distribution and molecular association are avoided. However where epitope abundance is a problem and epitope-masking may be involved such treatments may be necessary. Chondroitinase ABC treatment of cartilage sections is done before the initial blocking step, by overlaying with 100 µL 1X Chase buffer containing 1.44 µU Chase ABC(PF) and incubation at 37°C for 1 h, followed by washing the slides (2 × 2min), with PBS.

3.3.2. Blocking and Antibody Incubation

Sections (native or Chase ABC-treated) are incubated for 20 min with 1.5% *(v/v)* normal goat serum diluted in PBS and, after tipping serum off slides, sections are incubated for 30 min at RT with primary antibody diluted in 1.5% *(v/v)* normal goat serum in PBS. Slides are washed in 2 changes of PBS for 2 min each and then incubated for 30 min at room temperature with biotinylated goat anti-rabbit IgG secondary antibody as per kit instructions.

3.3.3. Hydrogen Peroxide Treatment

When cartilage sections are pretreated with Chase ABC, we have found with a number of antibodies, including non-immune IgG, that the signal intensity generated with Vectastain Elite ABC Reagent increases markedly. This apparent nonspecific signal can generally be reduced by treatment with hydrogen peroxide (3% *[v/v]*) in tap water for 10 min. followed by a wash in PBS for 5 min.

3.3.4. Product Development

Slides are washed for 2 × 2 min in PBS and incubated for 30 min. at RT with Vectastain Elite ABC Reagent. After another 2 × 2 min. wash in kit buffer, sections are incubated in Vector DAB solution (diluted according to kit protocol in pH 7.5 buffer) until brown coloration of the sections is fully developed (approx 5–10 min.) Sections are then rinsed in tap water for 5 min.

3.3.5. Counterstaining

Slides are immersed in Methyl Green for 5 min at 60°C and then rinsed in DI water until water runs clear. The stained slides are coverslipped using Permount resinous mounting medium.

3.4. Visualization

Staining was observed via light microscopy (Olympus BH-2) and permanent photoimages taken with a Q-Color 5 digital camera with Q-Capture software (Olympus) and Adobe Photoshop Elements 2.0.

3.5. Application of Method

An extensive literature exists on the use of bovine cartilage explants to study different aspects of aggrecanolysis. It was this experimental system which led to the discovery of aggrecanase activity *(35)* and to the purification and cloning of ADAMTS4 (aggrecanase-1) *(36)* and ADAMTS5 (aggrecanase-2) *(37)*. Western analysis of the products of bovine explants treated with IL-1 *(16)* have supported a role for ADAMTS4 and MT4MMP in this process and here we provide novel immunohistochemical data on bovine explants treated with retinoic acid (RA) which are also consistent with this catabolic model (*see* **Fig. 1**). In the experiment described here explants were examined before explant culture (*see* **Figs. 2** and **3**), after 4 d of culture in serum-free medium alone (*see* **Fig. 4**) or supplemented with 3 µ*M* RA (*see* **Figs. 5** and **6**). The proportion of total aggrecan (as chondroitin sulfate) lost from the tissue over the 4 d of culture was about 20% in serum-fre medium alone and about 80% when supplemented with RA. It can be seen that the intensity of the Methyl Green counterstain correlates with the concentration of aggrecan in the tissue.

3.5.1. Effects of Chase ABC Digestion and Peroxide Treatment

The data shown in **Figs. 2** and **5** illustrate the effects of pretreatment with chondroitinase ABC (Chase) and also hydrogen peroxide treatment (PX). Only data with non-immune IgG and anti-NITEGE are shown (*see* **Fig. 2**) but similar results were obtained with the other immune IgGs tested. Unless otherwise indicated, all IgGs were used at 20 µg/mL.

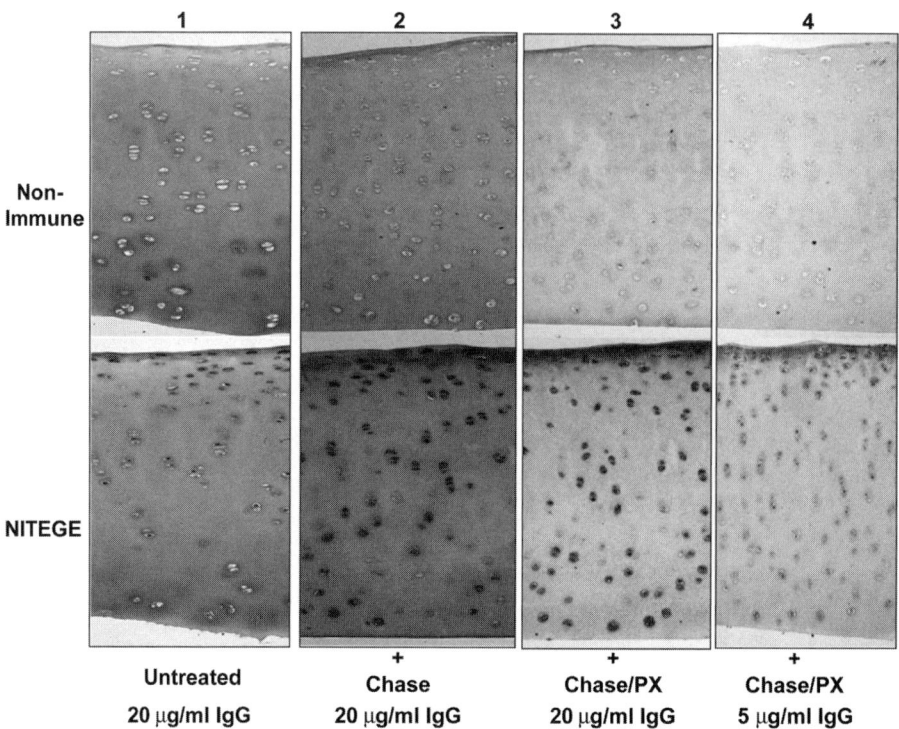

Fig. 2. Immunohistochemistry with non-immune IgG *(top)* and anti-NITEGE *(bottom)* of freshly excised bovine articular cartilage. The effects of pretreatment with Chase ABC and also incubation with hydrogen peroxide (PX) are shown. *See* text for detail and discussion. (Color illustration in insert following p. 268.)

In freshly isolated and untreated tissue (*see* **Fig. 2**, *panel1*) the non-immune IgG was non-reactive (top) whereas the anti-NITEGE (bottom) detected product in the superficial zone matrix and associated with cells at most depths. Following Chase digestion of sections (*see* **Fig. 2**, *panel 2*) the non-immune IgG (top) generated a signal in the superficial zone matrix and a diffuse signal throughout the full depth matrix , but did not show a signal associated with the cells. Chase ABC treatment enhanced NITEGE staining (bottom) throughout the cartilage matrix and in addition a strong cell-associated reactivity was now observed. This suggests unmasking by Chase of G1-NITEGE product which is both cell-associated and matrix-associated in freshly isolated tissue.

When sections were treated with peroxide in addition to Chase (*see* **Fig. 2**, *panel 3*) the signal was reduced for both non-immune and anti-NITEGE resulting in a clearer discrimination of specific anti-NITEGE signal. Indeed, this distinction between non-immune (upper) and anti-NITEGE (lower) was even

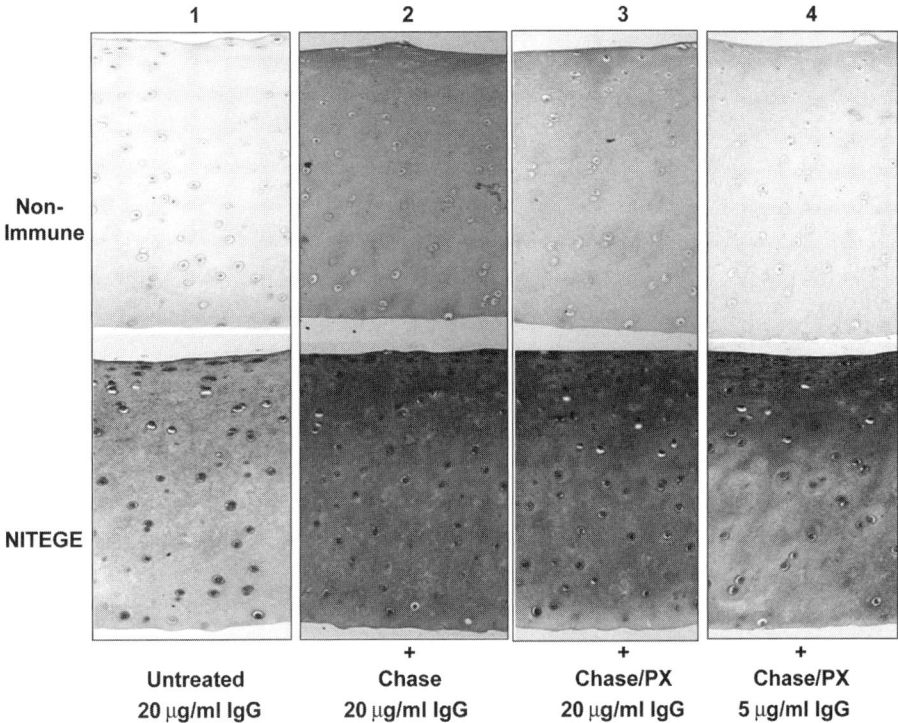

Fig. 3. Immunohistochemistry with non-immune IgG (top) and anti-NITEGE (bottom) of cartilage after explant culture in retinoic acid (RA). The effects of pretreatment with Chase ABC and also incubation with hydrogen peroxide (Px) are shown. *See* text for detail and discussion. (Color illustration in insert following p. 268.)

more evident when the IgGs were used at only 5 µg/mL (*see* **Fig. 2**, *panel 4*). It should also be noted that peroxide treatment greatly reduced the methyl green counterstain of the matrix which may be explained by the known degradative effects of peroxide on aggrecan structure *(38)*.

The same protocols were applied to tissue after 4 d of treatment with RA (*see* **Fig. 5**), and the data confirmed the usefulness of this approach in obtaining specific immunostaining data. The markedly higher signal intensity in the lower panels (anti-NITEGE) of **Fig. 5** relative to **Fig. 2** generally confirm the RA-mediated increase in tissue G1-NITEGE abundance, previously observed consistently by Western analysis. Chase digestion of these RA-treated samples appeared to specifically enhance matrix staining with no change in the cell-associated staining. This suggests that G1-NITEGE is largely cell-associated in native fresh tissue and that this product is generated by aggrecanolysis throughout the intercellular matrix during RA-treatment. Most importantly, a

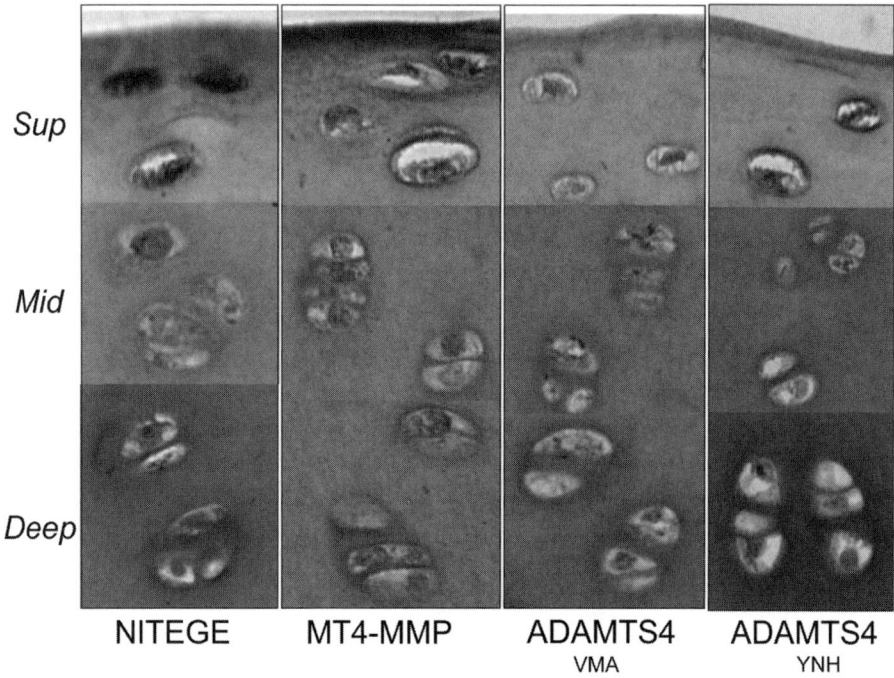

Fig. 4. Immunohistochemistry of freshly excised bovine articular cartilage. Antibodies used were anti-NITEGE, anti-MT4MMP, and anti-ADAMTS4 antibodies. Images of superficial (Sup), midzone (Mid), and deepzone (Deep) regions of the tissue are shown at ×40. *See* text for description. (Color illustration in insert following p. 268.)

comparison between the data in **Figs. 2–5** clearly indicate that the abundance of G1-NITEGE signal obtained by this approach is primarily a function of the abundance of the protein present in the tissue rather than being dependent on the extent of masking/unmasking of the epitope by matrix aggrecan. The immunohistochemical (IHC) data therefore fully support the parallel studies *(16,39)* on G1-NITEGE abundance done by tissue extraction and Western analysis.

In summary, it appears that the staining of native sections (without Chase digestion or peroxide treatment) provides sufficient sensitivity to generate useful images on both the location and abundance of the protein epitope under investigation. Whereas Chase digestion can clearly increase sensitivity it also appears to increase nonspecific IgG reactivity which renders strict interpretations on abundance and location more difficult. Interestingly, peroxide treatment can be used to reduce the nonspecific staining generated by Chase digestion. While peroxide treatment has been generally used to eliminate endo-

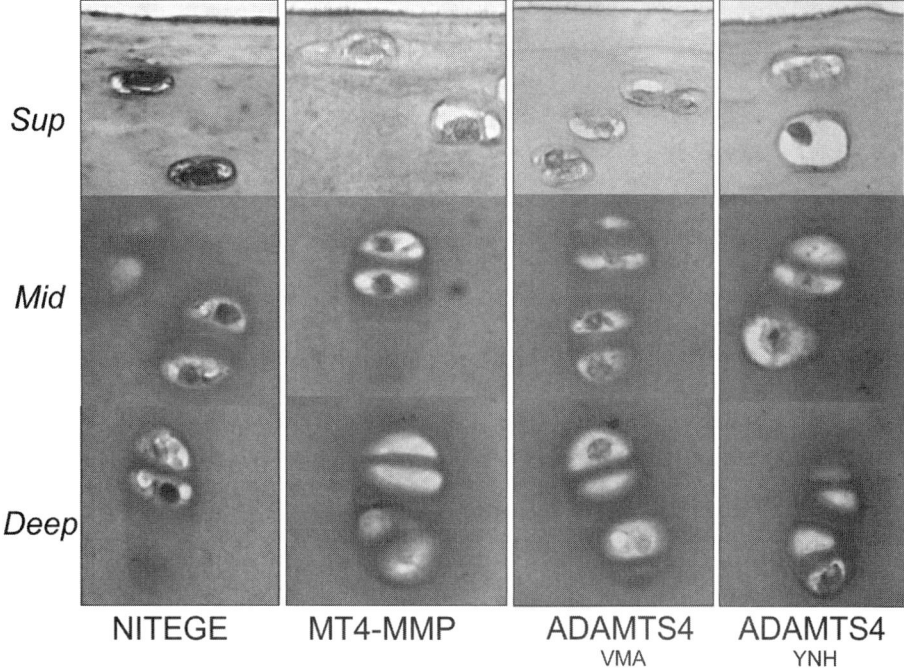

Fig. 5. Immunohistochemistry of bovine articular cartilage after 4 d of serum-free culture. Antibodies used were anti-NITEGE, anti-MT4MMP, and anti-ADAMTS4. Images of superficial (Sup), midzone (Mid) and deepzone (Deep) regions of the tissue are shown at ×40. *See* text for description. (Color illustration in insert following p. 268.)

genous peroxidase activity in the present context it also appears to eliminate much of the aggrecan from the cartilage matrix.

3.5.2. Localization of Proteinases and Aggrecan Cleavage Products in Cartilage Explants

In this study (*see* **Figs. 3–5**), sections were stained with anti-NITEGE, anti-MT4MMP, and anti-ADAMTS4 (JSCVMA and JSCYNH separately) without the need to use Chase ABC or hydrogen peroxide. These images were obtained at ×40 magnification and are arranged to illustrate details of the staining for cells and matrix in superficial, mid-, and deep-zone regions of the tissue. In fresh tissue (*see* **Fig. 3**) all protein epitopes were detected to some extent in the matrix of the superficial zone and localized within cells or in the pericellular space throughout the full depth of the tissue. After 4 d of explant culture in serum-free medium (*see* **Fig. 4**) all staining was reduced, suggesting a net loss of the subject proteins from the tissue during the culture period. On the other

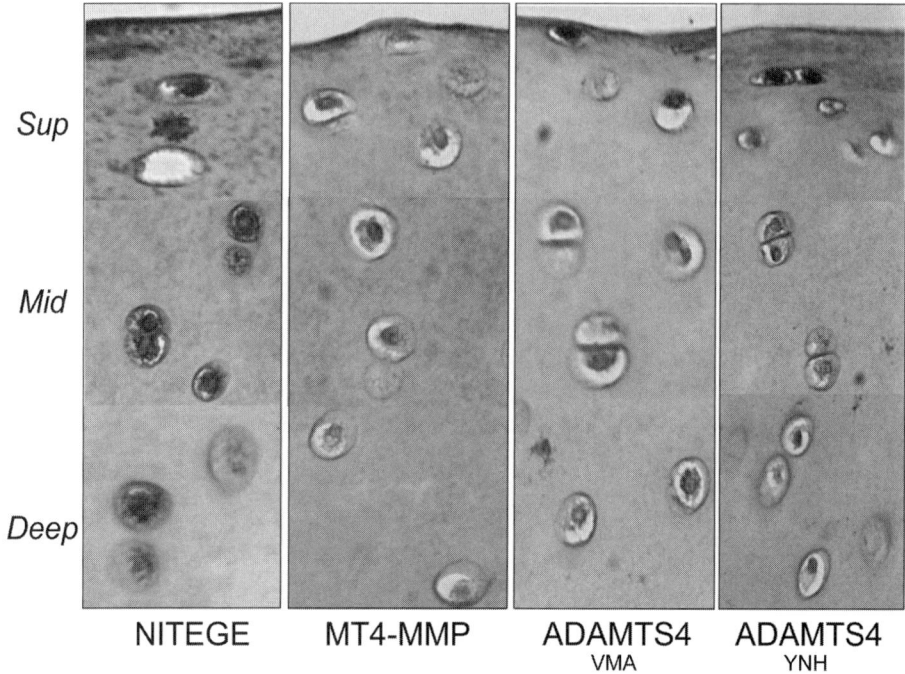

Fig. 6. IHC of bovine articular cartilage after 4 d of serum-free culture with 3 µM retinoic acid. Antibodies used were anti-NITEGE, anti-MT4MMP, and anti-ADAMTS4. Images of superficial (Sup), midzone (Mid) and deepzone (Deep) regions of the tissue are shown at ×40. *See* text for description. Color illustration in insert following p. 268.)

hand, tissue which was maintained for 4 d in RA (and which underwent extensive aggrecanase-mediated degradation) (*see* **Fig. 6**) showed marked changes in cell shape and in the abundance and distribution of the G1-NITEGE, MT4-MMP, and ADAMTS4 (with both antibodies). In addition, the cells appeared to have "rounded up" relative to controls, presumably due to loss of matrix attachments associated with the catabolic cascade in the pericellular space. The abundance of all components was markedly increased and each was now found associated with matrix in addition to cells at all depths. This observation is consistent with catabolic model (*see* **Fig. 1**) in which a cell-surface complex of ADAMTS4 and MT4-MMP is released into the matrix and this is accompanied by conversion of the p68 ADAMTS4 into the p53 form followed by cleavage of the NITEGE-ARGSV bond by p53 ADAMTS4.

4. Notes

1. *Specimen preparation*. The protocol is described for cartilage disks of approx 3 mm × 1 mm. Larger tissue samples may require longer fixation and processing

times. Smaller samples may be more difficult to maintain on slides. Adherence is improved by cutting thin sections (4–5 µm), using the drying technique described, and by mounting sections on positively charged slides such as Superfrost/Plus. We have also found that problems with adhesion of cartilage sections to slides can be largely overcome by treatment in 10% neutral-buffered formalin for 30 min and rinsing for 5 min in tap water. If working with cartilaginous tissues that require decalcification, fixation in 10% formalin containing 0.5% cetylpyridinium chloride (CPC) may be necessary to preserve proteoglycans so that they are not leached out of the tissue during decalcification. Sufficient decalcification can usually be accomplished in 5% *(w/v)* EDTA , PBS (pH 7.5) for 10 d or in 7.5% *(v/v)* formic acid until endpoint check shows completion.

2. *Immunostaining controls.* To ensure that positive immunostaining is specific for the IgG in use, the antibodies must be characterized by Western analysis of extracts of the tissue under study (*see* **Subheading 2.3., step 4**). In the absence of this biochemical test, the validity of the IHC data can be questioned. In addition, appropriate negative IHC controls are essential. These should include omission of primary antibody, replacement of primary antibody with non immune rabbit IgG at 20ug/ml or preabsorption of IgG from serum with a molar excess of immunizing peptide.

References

1. Mankin, H. J., Dorfman, H., Lippiello, L., and Zarins, A. (1971) Biochemical and metabolic abnormalities in articular cartilage from osteo-arthritic human hips. II. Correlation of morphology with biochemical and metabolic data. *J. Bone Joint Surg. Am.* **53,** 523–537.
2. Ostergaard, K., Petersen, J., Andersen, C. B., Bendtzen, K., and Salter, D. M. (1997) Histologic/histochemical grading system for osteoarthritic articular cartilage: reproducibility and validity. *Arthritis Rheum.* **40,** 1766–1771.
3. Alhadlaq, H. A., Xia, Y., Moody, J. B., and Matyas, J. R. (2004) Detecting structural changes in early experimental osteoarthritis of tibial cartilage by microscopic magnetic resonance imaging and polarised light microscopy. *Ann. Rheum. Dis.* **63,** 709–717.
4. Stolz, M., Raiteri, R., Daniels, A.U., VanLandingham, M. R., Baschong, W., and Aebi, U. (2004) Dynamic elastic modulus of porcine articular cartilage determined at two different levels of tissue organization by indentation-type atomic force microscopy. *Biophys. J.* **86,** 3269–3283.
5. West, P. A., Bostrom, M.P., Torzilli, P. A., and Camacho, N.P. (2004) Fourier transform infrared spectral analysis of degenerative cartilage: an infrared fiber optic probe and imaging study. *Appl. Spectrosc.* **58,** 376–381.
6. Koshy, P. J., Lundy, C. J., Rowan, A. D., et al. (2002) The modulation of matrix metalloproteinase and ADAM gene expression in human chondrocytes by interleukin-1 and oncostatin M: a time-course study using real-time quantitative reverse transcription-polymerase chain reaction. *Arthritis Rheum.* **46,** 961–967.
7. Ng, L., Grodzinsky, A. J., Patwari, P., Sandy, J., Plaas, A., and Ortiz, C. (2003) Individual cartilage aggrecan macromolecules and their constituent glycosaminoglycans visualized via atomic force microscopy. *J. Struct. Biol.* **143,** 242–257.

8. Lohmander, L. S., Neame, P. J., and Sandy, J. D. (1993) The structure of aggrecan fragments in human synovial fluid. Evidence that aggrecanase mediates cartilage degradation in inflammatory joint disease, joint injury, and osteoarthritis. *Arthritis Rheum.* **36,** 1214–1222.
9. Sandy, J. D. and Verscharen, C. (2001) Analysis of aggrecan in human knee cartilage and synovial fluid indicates that aggrecanase (ADAMTS) activity is responsible for the catabolic turnover and loss of whole aggrecan whereas other protease activity is required for C-terminal processing in vivo. *Biochem. J.* **358,** 615–626.
10. Oshita, H., Sandy, J. D., Suzuki, K., et al. (2004)Mature bovine articular cartilage contains abundant aggrecan that is C-terminally truncated at Ala719–Ala720, a site which is readily cleaved by m-calpain. *Biochem. J.* **382,** 253–259.
11. Sandy, J. D., Flannery, C. R., Neame, P. J., and Lohmander, L. S. (1992) The structure of aggrecan fragments in human synovial fluid. Evidence for the involvement in osteoarthritis of a novel proteinase which cleaves the Glu 373-Ala 374 bond of the interglobular domain. *J. Clin. Invest.* **89,** 1512–1516.
12. Malfait, A. M., Liu, R. Q., Ijiri, K., Komiya, S., and Tortorella, M. D. (2002) Inhibition of ADAM-TS4 and ADAM-TS5 prevents aggrecan degradation in osteoarthritic cartilage. *J. Biol. Chem.* **277,** 22,201–22,208.
13. Collins-Racie, L.A., Flannery, C.R., Zeng, W., et al. (2004) ADAMTS-8 exhibits aggrecanase activity and is expressed in human articular cartilage. *Matrix Biol.* **23,** 219–230.
14. Arner, E. C., Pratta, M. A., Trzaskos, J. M., Decicco, C. P., and Tortorella, M. D. (1999) Generation and characterization of aggrecanase. A soluble, cartilage-derived aggrecan-degrading activity. *J. Biol. Chem.* **274,** 6594–6601.
15. Pratta, M. A., Scherle, P. A., Yang, G., Liu, R. Q., and Newton, R. C. (2003) Induction of aggrecanase 1 (ADAM-TS4) by interleukin-1 occurs through activation of constitutively produced protein. *Arthritis Rheum.* **48,** 119–133.
16. Patwari, P., Gao, G., Lee, J. H., Grodzinsky, A. G., and Sandy, J. D. (2005) Analysis of ADAMTS4 and MT4-MMP indicates that both are involved in aggrecanolysis in interleukin-1-treated bovine cartilage. *Osteoarth. Cartilage* **13,** 269–277.
17. Stanton, H., East, C., Golub, S., et al. (2005) ADAMTS-5 is the major aggrecanase in cartilage: in vitryo studies with ADAMTS-4 and ADAMTS-5 deficient mice. Orthopedic Research Society, Washington, DC, 206.
18. Gao, G., Plaas, A., Thompson, V. P., Jin, S., Zuo, F., and Sandy, J. D. ADAMTS4 (aggrecanase-1) activation on the cell surface involves C-terminal cleavage by glycosylphosphatidyl inositol-anchored membrane type 4-matrix metalloproteinase and binding of the activated proteinase to chondroitin sulfate and heparan sulfate on syndecan-1. *J. Biol. Chem.* **279,** 10,042–10,051.
19. Gao, G., Westling, J., Thompson, V. P., Howell, T. D., Gottschall, P. E., and Sandy, J. D. Activation of the proteolytic activity of ADAMTS4 (aggrecanase-1) by C-terminal truncation. *J. Biol. Chem.* **277,** 11,034–11,041.

20. Kashiwagi, M., Enghild, J.J., Gendron, C et al. (2004) Altered proteolytic activities of ADAMTS-4 expressed by C-terminal processing. *J. Biol. Chem.* **279**, 10,109–10,119.
21. Singer, II, Scott, S., Kawka, D.W., et al. (1997) Aggrecanase and metalloproteinase-specific aggrecan neo-epitopes are induced in the articular cartilage of mice with collagen II-induced arthritis. *Osteoarthritis Cartilage* **5**, 407–418.
22. Lark, M.W., Bayne, E.K., Flanagan, J., et al. (1997) Aggrecan degradation in human cartilage. Evidence for both matrix metalloproteinase and aggrecanase-activity in normal, osteoarthritic, and rheumatoid joints. *J. Clin. Invest.* **100**, 93–106.
23. van Meurs, J. B., van Lent, P. L., Holthuysen, A. E., Singer, II, Bayne, E. K., and van den Berg, W. B. (1999) Kinetics of aggrecanase- and metalloproteinase-induced neoepitopes in various stages of cartilage destruction in murine arthritis. *Arthritis Rheum.* **42**, 1128–1139.
24. McLean, I. W. and Nakane, P. K. (1974) Periodate-lysine-paraformaldehyde fixative. A new fixation for immunoelectron microscopy. *J. Histochem. Cytochem.* **22**, 1077–1083.
25. Chambers, M. G., Cox, L., Chong, L., et al. (2001) Matrix metalloproteinases and aggrecanases cleave aggrecan in different zones of normal cartilage but colocalize in the development of osteoarthritic lesions in STR/ort mice. *Arthritis Rheum.* **44**, 1455–1465.
26. Bayliss, M. T., Hutton, S., Hayward, J., and Maciewicz, R. A. (2001) Distribution of aggrecanase (ADAMts 4/5) cleavage products in normal and osteoarthritic human articular cartilage: the influence of age, topography and zone of tissue. *Osteoarthritis Cartilage* **9**, 553–560.
27. Clements, K. M., Price, J. S., Chambers, M. G., Visco, D. M., Poole, A. R., and Mason, R. M. (2003) Gene deletion of either interleukin-1beta, interleukin-1beta-converting enzyme, inducible nitric oxide synthase, or stromelysin 1 accelerates the development of knee osteoarthritis in mice after surgical transection of the medial collateral ligament and partial medial meniscectomy. *Arthritis Rheum.* **48**, 3452–3463.
28. Lee, E. R., Lamplugh, L., Leblond, C. P., Mordier, S., Magny, M. C., and Mort, J. S. (1998) Immunolocalization of the cleavage of the aggrecan core protein at the Asn341-Phe342 bond, as an indicator of the location of the metalloproteinases active in the lysis of the rat growth plate. *Anat. Rec.* **252**, 117–132.
29. Lee, E. R., Lamplugh, L., Davoli, M. A., et al. (2001) Enzymes active in the areas undergoing cartilage resorption during the development of the secondary ossification center in the tibiae of rats ages 0-21 days: I. Two groups of proteinases cleave the core protein of aggrecan. *Dev. Dyn.* **222**, 52–70.
30. Wang, P., Tortorella, M., England, K., et al. (2004) JBCA-EJ. Proprotein convertase furin interacts with and cleaves pro-ADAMTS4 (Aggrecanase-1) in the trans-Golgi network. *J. Biol. Chem.* **279**, 15,434–15,440.
31. Itoh, Y., Kajita, M., Kinoh, H., Mori, H., Okada, A., and Seiki, M. (1999) Membrane type 4 matrix metalloproteinase (MT4-MMP, MMP-17) is a glycosylphosphatidylinositol-anchored proteinase. *J. Biol. Chem.* **274**, 34,260–34,266.

32. Hunziker, E. B., Michel, M., and Studer, D. (1997) Ultrastructure of adult human articular cartilage matrix after cryotechnical processing. *Microsc. Res. Tech.* **37,** 271–284.
33. Sandy, J. D., Gamett, D., Thompson, V., and Verscharen, C. (1998) Chondrocyte-mediated catabolism of aggrecan: aggrecanase-dependent cleavage induced by interleukin-1 or retinoic acid can be inhibited by glucosamine. *Biochem. J.* **335,** 59–66.
34. Patwari, P., Kurz, B., Sandy, J. D., and Grodzinsky, A.J. (2000) Mannosamine inhibits aggrecanase-mediated changes in the physical properties and biochemical composition of articular cartilage. *Arch. Biochem. Biophys.* **374,** 79–85.
35. Sandy, J. D., Neame, P. J., Boynton, R. E., and Flannery, C.R. (1991) Catabolism of aggrecan in cartilage explants. Identification of a major cleavage site within the interglobular domain. *J. Biol. Chem.* **266,** 8683–8685.
36. Tortorella, M. D., Burn, T. C., Pratta, M. A., and Abbaszade, I., (1999) Purification and cloning of aggrecanase-1: a member of the ADAMTS family of proteins. *Science* **284,** 1664–1666.
37. Abbaszade, I., Liu, R. Q., Yang, F., et al. (1999) Cloning and characterization of ADAMTS11, an aggrecanase from the ADAMTS family. *J. Biol. Chem.* **274,** 23,443–23,450.
38. Roberts, C. R., Roughley, P. J., and Mort, J. S. (1989) Degradation of human proteoglycan aggregate induced by hydrogen peroxide. Protein fragmentation, amino acid modification and hyaluronic acid cleavage. *Biochem. J.* **259,** 805–811.
39. Sandy, J. D., Thompson, V., Verscharen, C., and Gamett, D. (1999) Chondrocyte-mediated catabolism of aggrecan: evidence for a glycosylphosphatidylinositol-linked protein in the aggrecanase response to interleukin-1 or retinoic acid. *Arch. Biochem. Biophys.* **367,** 258–264.

11

In Situ Detection of Cell Death in Articular Cartilage

Samantha N. Redman, Ilyas M. Khan, Simon R. Tew,
and Charles W. Archer

Summary

Necrosis and apoptosis have been demonstrated in articular cartilage in response to trauma and disease. However, cell death in articular cartilage may also be thought of as a scale of cell death culminating in secondary necrosis with the failure to remove apoptotic cells from the tissue. The *in situ* detection of cell death is an important technique in studying articular cartilage as it most closely resembles the in vivo situation. The methods described here involve the use of light microscopy and electron microscopy in conjunction with fluorescent and biochemical methods to correctly ascertain the type of cell death that has occurred.

Key Words: Articular cartilage; apoptosis; necrosis; detection techniques; cell death.

1. Introduction
1.1. Cell Death

Cell death has traditionally been described as either necrotic or apoptotic. Necrosis usually occurs as the result of traumatic insult to the cell. The cell loses osmotic regulation and swells uncontrollably, resulting in the rupture of the cell membrane and the loss of cytosolic contents into the extracellular space. This loss results in a local inflammatory response as the cell debris is then removed from the tissue *(1)*. Apoptosis, which was first described by Kerr et al. *(2)*, is a process in which the cell actively initiates and executes its own destruction. This destruction is an important biological process in development, tissue homeostasis, and in the removal of damaged cells from the system before they pose a threat to the organism *(3)*. Apoptosis is characterized by key morphological and biochemical features. The cell shrinks and loses focal contact with the extracellular matrix and neighboring cells. The chromatin condenses and internucleosomal DNA fragments. Aside from membrane blebbing and the formation of apoptotic bodies, the plasma membrane remains intact which, in contrast to necrosis, enables the removal of the dying cell without initiating an inflammatory response *(3)*. The apoptotic cell presents cell surface markers such as phosphotidylserine, which mediate the recognition of the cell by phagocytes. If the apoptotic cells are not removed from the system they may undergo secondary necrosis, such that plasma membrane integrity is lost and the contents of the cell are expelled to the extracellular space *(4,5)*. Because of the avascular nature of articular cartilage, apoptotic chondrocytes are unlikely to be removed from the tissue. Therefore, this mechanism of cell death should also be taken into account when discussing cell death in articular cartilage.

The signal to undergo apoptosis may arise from a number of internal or external stimuli including DNA damage and ligand binding to cell surface "death receptors." Receptor mediated apoptosis arises from ligand binding to receptors such as Fas (apo-1 or CD95) and tumor necrosis factor receptor (TNFR-1) *(6)*. These receptors contain a cytosolic death domain (DD) and ligand binding to these receptors initiates an intracellular cascade resulting in cell death. The regulation of apoptosis was first determined in the nematode *Caenorhabditis elegans (7)* where two genes were identified that were essential for apoptosis; *ced-3* and *ced-4*. The product of a further gene *ced-9* was shown to have an inhibitory effect on the gene products of *ced-3* and *ced-4* and was, therefore, determined to be an antiapoptotic protein. The mammalian homologue of *ced-3* was found to be interleukin-1β converting enzyme (ICE) which belongs to a family of cysteine proteases, known as caspases, which are intimately involved in the initiation and execution of apoptosis *(8,9)*. Caspases are synthesised as proenzymes which are activated by proteolytic cleavage by other caspases. Activated caspases cleave and activate other effector caspases which then degrade cellular substrates to give rise to the morphological features of apoptosis. To date, 14 mammalian caspases have been identified *(10)*. The mammalian homolog of the antiapoptotic gene product of *ced-9* is the protein Bcl-2. A second member of the Bcl-2 family is the proapoptotic protein, bax. In essence, it is the balance between these two proteins that determine the cell's susceptibility to undergo apoptosis *(11)*.

It has been shown in many different cell systems that a loss of cell-matrix interactions may initiate apoptosis through a loss of survival signals to the cell *(12)*, and that chondrocyte-ECM interactions mediate chondrocyte survival *(13)*. In vitro and in vivo studies have demonstrated that integrin mediated interactions between the chondrocyte and the collagen architecture are vital for chondrocyte survival *(14,15)*.

1.2. Apoptosis in Articular Cartilage

Apoptosis occurs in articular cartilage in response to impact loading, injurious compression, experimental wounding, and osteoarthritis. Studies of impact damaged cartilage have demonstrated that in response to injurious compression, chondrocytes die by a combination of necrosis and apoptosis *(16)*. It has been shown that chondrocytes undergo apoptosis in response to impact loading and that the extent of apoptosis is directly related to the amount of load applied *(17,18)* It is unclear from these studies however, whether chondrocyte death drives subsequent matrix degradation or whether matrix degradation causes cell death, as apoptosis occurs at impact levels lower than those required to initiate matrix degradation *(19,20)*. Histological studies have previously demonstrated cell death adjacent to the wound margin in models of experimental wounding of articular cartilage both in vivo *(21,22)* and in vitro *(23)*. Cell death at the wound margin has classically been described as necrotic *(24)*. Recent studies have, however, demonstrated that cell death at the wound site may be a combination of both necrosis and apoptosis *(23,25)*.

1.3. Osteoarthritis and Cell Death

Osteoarthritis is a disabling disease that is experienced to some degree by over 70% of adults over 70 yr of age. Wear and tear as a consequence of aging is usually the reason given for the loss or reduction of functional activity of the synovial joint. Biochemical analysis of diseased tissue from human subjects undergoing surgical treatment and the use of well defined animal models of OA has demonstrated that articular cartilage degeneration proceeds in several distinct stages. Chondrocytes in diseased cartilage may undergo cell death, proliferation, or remain unchanged. There is also a phase of hyper-anabolism, phenotypic alteration of chondrocytes, extracellular matrix degradation, and the formation of osteophytes *(26)*. Many studies have cited chondrocyte cell death as of fundamental importance in disease initiation principally because it is hypothesized that local homeostatic maintenance of ECM is perturbed causing an imbalance in the ratio of anabolism to catabolism leading to net proteoglycan loss *(27,28)*. As chondrocytes constitute only 5% of the volume of articular cartilage, cell death (either slowly through aging or rapidly following acute injury) of even a small percentage of cells may initiate OA through the lack of maintenance of surrounding ECM. Biomechanical failure is then precipitated by an inability to respond to normal mechanical loading followed by further cell death (and cellular proliferation in the form of chondrocyte clusters) and another cycle of degeneration leading inexorably to joint destruction. Therefore chondrocyte viability is critical for maintaining normal tissue function.

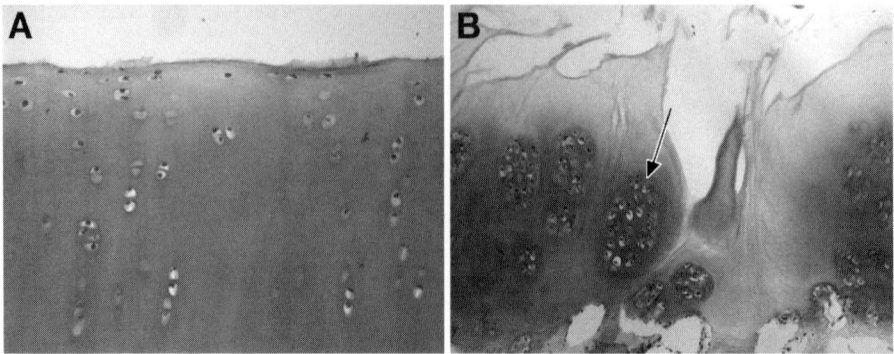

Fig. 1. Normal articular cartilage from the metacarpalphalangeal joint of 18-mo-old bovine steer has an intact surface and readily stains for safranin-O indicative of the presence of proteoglycans (**A**). Osteoarthritic cartilage from the same joint (**B**) displays several characteristic pathological signs such as cell death, chondrocyte clustering (*arrow*), fissuring of the tissue and hyperanabolism as determined by intense safranin-O staining of the clustered chondrocytes.

1.4. Detection of Chondrocyte Cell Death

There are numerous methods for the detection of apoptosis in articular cartilage. The most fundamental and reliable is histological analysis at both the light and electron microscope level. Evidence for cell death in OA was first uncovered by examination of diseased tissue under light microscope where it was observed that nuclear staining was absent and empty lacunae were often present (probably caused through cell shrinkage and detachment from the surrounding extracellular matrix following apoptosis). The use of high-magnification optical microscopy has made recognizing the morphological characteristics of cells undergoing apoptosis (such as intense nuclear staining, cell shrinkage, membrane blebbing, and budding of apoptotic bodies), a relatively straightforward process. OA affected tissue is characterized by mild to severe fibrillation of the articular surface and this feature is associated with a significant reduction in cellular density especially in later stages of disease. Cellular proliferation, a reactive response to injury, is also evident in early and late OA, in early OA the tissue can have a hypercellular appearance interspersed by hypocellular areas; in the later stages of OA, despite a marked reduction in cellularity in general, large multicellular clusters are present (*see* **Fig. 1**).

Early studies attempted to gain more understanding of cell death by studying the ultrastructural features of chondrocytes within fibrillated cartilage *(29)*. Features consistent with a necrotic phenotype, such as swollen mitochondria and abnormally dilated irregular endoplasmic reticulum, were observed in chondrocytes in deeply fibrillated cartilage. The introduction of in vivo animal models of disease, such as anterior cruciate ligament transaction models *(30)*, has made the dissection of disease mechanism at different stages more accessible. Using such models, chondrocytes with characteristic apoptotic features have been consistently identified using electron microscopy in early, mid, and late stages of disease.

Further mechanisms of detection exist; these consist predominately of the use of fluorescent dyes which take advantage of distinct biochemical and morphological features of apoptosis. The most common fluorescent dyes used to detect apoptosis are discussed in this chapter. Fluorescent dyes specific for detection of DNA have been shown to discriminate between living, apoptotic, and necrotic cells using fluorescent microscopy. The fluorescent dyes 4,6-diamidino-2-phenylindole-2-HCl (DAPI: excitation 358 nm, emission 461 nm) and propidium iodide (ex 536nm, em 615nm) detect apoptotic nuclei in sectioned cartilage by the condensed chromatin gathering at the periphery of the nuclear membrane or a total fragmented morphology of nuclear bodies; in practice, fluorescently labeled apoptotic nuclei appear much brighter because they are more compact than their normal counterparts. This technique should be used in conjunction with light microscopy

The plasma membrane of live cells prevents cationic fluorescent dyes such as propidium iodide and ethidium homodimer-1 (ex 528 nm, em 617 nm) interacting with nuclear DNA, the detected fluorescence signal is inversely proportional to the integrity of the membrane and also, therefore, to cellular viability. Fluorescent dye exclusion has been used successfully by our laboratory to label dead cells in OA and wounded cartilage. Using this technique we estimate that less than 1% of "intact" cells are labeled in mildly diseased tissue, however the dyes also label cellular apoptotic debris containing nuclear bodies called "micronuclei," that litter the diseased cartilage and revealing evidence of previous apoptotic events (unpublished observations).

In order to identify cells undergoing early apoptotic events, smaller dyes such as cyanine-based fluorescent stain YO-PRO-1 (ex 491nm, em 509nm) can be used *(31)*. The rationale here is that cells in early apoptosis are unable to pump out cell-permeant YO-PRO-1 but are impermeable to other dead cell discriminatory dyes. Later stages of apoptosis are accompanied by an increase in membrane permeability, which allows larger dyes such as ethidium homodimer-1 to enter cells. The utility of using smaller dyes to detect apoptotic nuclei in OA tissue has not as yet been demonstrated.

Tdt-mediated dUTP nick end labeling (TUNEL) is a technique that enables the specific labelling of 3'-hydroxyl ends generated by intranucleosomal digestion of DNA within apoptotic nuclei. Using TUNEL, the percentage of apoptotic cells reported in OA tissue has varied from <1 to as high as 22%

(27,32). Variations in the percentage of apoptotic cells may result from many factors, such as sampling of different stages of disease, topographic variation of tissue used, technical differences in removing tissue for analysis, and differences in TUNEL methodology. The reported wide variation in values has also generated debate into the use of appropriate controls for TUNEL labeling and also the use of data where TUNEL is the sole experimental evidence for apoptosis *(33)*. It is argued that for TUNEL labeling to be of any significance an appropriate positive control such as the use of fetal growth plate must be used in parallel. TUNEL labeling should only be observed in the lowest hypertrophic zone consistent with histomorphological data of apoptosis and not in the proliferative and prehypertrophic zones above *(33)*. As a consequence of articular cartilage being avascular, dead cells are unable to be removed from diseased and damaged tissue and DNA degradation of these nuclei may generate false positive labeling, leading to overestimated values for apoptotic cells. It has been suggested in numerous reports that reliance on TUNEL alone is unsatisfactory and that supplementary evidence using Annexin V labeling, DNA laddering, or electron microscopy is necessary. In the assessment of the type of cell death that has occurred, a combination of histological, dye inclusion vs exclusion and biochemical methods should be used for a definitive conclusion

2. Materials

All chemicals obtained from Sigma (Poole, UK) unless otherwise stated.

2.1. Light Microscopy

1. Xylene (Fisher, Leicester UK).
2. Industrial methylated spirit (IMS) (Fisher, Leicester UK).
3. Haematoxylin (R. A. Lamb, Eastbourne UK).
4. Eosin (R. A. Lamb).
5. Distroplasticiser xylene (DPX) (R. A. Lamb).

2.2. Electron Microscopy

1. Glutaraldehyde ruthenium (III) hexamine trichloride (RHT).
2. 1% Osmium tetroxide.
3. 0.05 M Sodium cacodylate (pH 6.75).
4. Propylene oxide.
5. Araldite.
6. Uranyl acetate.
7. Lead citrate.

2.3. Fluorescent Detection of Cell Death

1. 10% Neutral buffered formal saline (NBFS).

Table 1
Detection of Cell Viability/Cytotoxicity

	Live cells	Necrosis	Early apoptosis	Late apoptosis	Source
Dye inclusion					
CalceinAM	✓	✗	✗	✗	Molecular probes
PI	✗	✓	✗	✓	vector
Ethidium homodimer	✗	✓	✗	✓	Molecular probes
YO PRO-1	✗	✗	✓	✗	
Biochemical detection					
TUNEL	✗	✗	✓	✗	Roche Diagnostics

2. Phosphate buffered saline.
3. Vectashield containing DAPI (1.5 µg/mL)/propidium iodide (1.5 µg/mL) (Vector labs, Peterborough UK).
4. Live/Dead™ kit (Molecular Probes, Netherlands).

2.4. Biochemical Detection of Cell Death

1. Proteinase K: 10 µg/mL in 10 mM Tris-HCl (pH 7.8).
2. TUNEL kit (Roche Diagnostics, Lewes UK).
3. Paraformaldehyde .

Table 1 illustrates the use of dye inclusion and biochemical methods in detecting cell viablility/cytotoxicity

3. Methods

The following methods are described for explant culture of articular cartilage and histological tissue sections derived from explant culture or immediately excised tissue.

3.1. Histological Analysis

3.1.1. Light Microscopy

Light microscopy can be used to a certain extent to examine necrotic vs apoptotic cell death. In order to view the tissue under light microscopy the sections must be stained with a general histological stain. The most common staining method used is haematoxylin and eosin which results in blue staining of the nuclei and pink staining of the extracellular matrix (*see* **Fig. 2**).

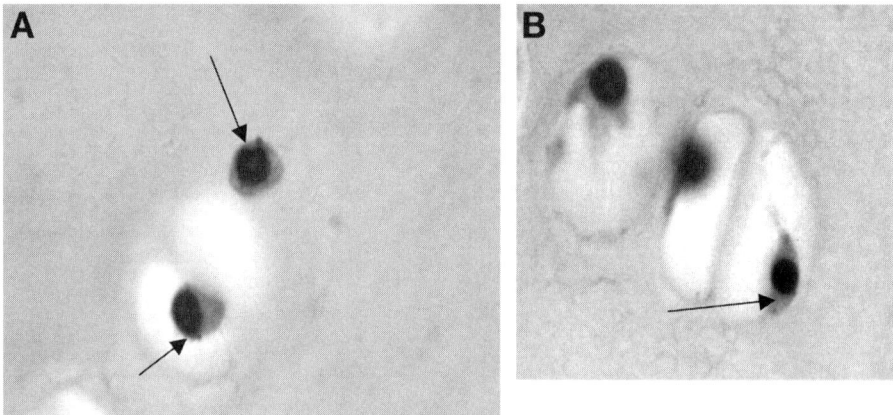

Fig. 2. Apoptotic nuclei from sections of cartilage from the metacarpalphalangeal joint of 18-mo-old bovine steer as viewed under light microscopy. The images show membrane blebbing (A) homogenous eosinophillic cytoplasm and chromatin condensation (**B**).

1. Dewax paraffin wax sections in 2 changes of xylene (2 min) (*see* **Note 1**).
2. Rehydrate sections through descending alcohols comprising.
    a. Two changes of 100% IMS (2 min).
    b. 95% IMS for 2 min.
    c. 70% IMS for 2 min.
    d. Running water for 2 min. (*See* **Note 2**.)
3. Stain in Mayers haematoxylin for 2 min.
4. Wash in running tap water for 5 min.
5. Stain in 1% aqueous eosin for 5 min.
6. Wash in running tap water for 20 s.
7. Dehydrate through ascending concentrations of alcohols:
    a. 70% IMS for 20 s.
    b. 95% IMS for 30 s.
    c. 100% IMS for 1 min.
    d. 100% IMS for 2 min.
8. Clear in 2 changes of xylene (2 min).
9. Mount under a coverslip with DPX.

3.1.2. Electron microscopy

Transmission electron microscopy (TEM) can be used to detect apoptotic vs necrotic cell death through high resolution ultrastructural analysis (*see* **Fig. 3**). The standard procedure used in our lab for the preparation of tissue for TEM is as follows.

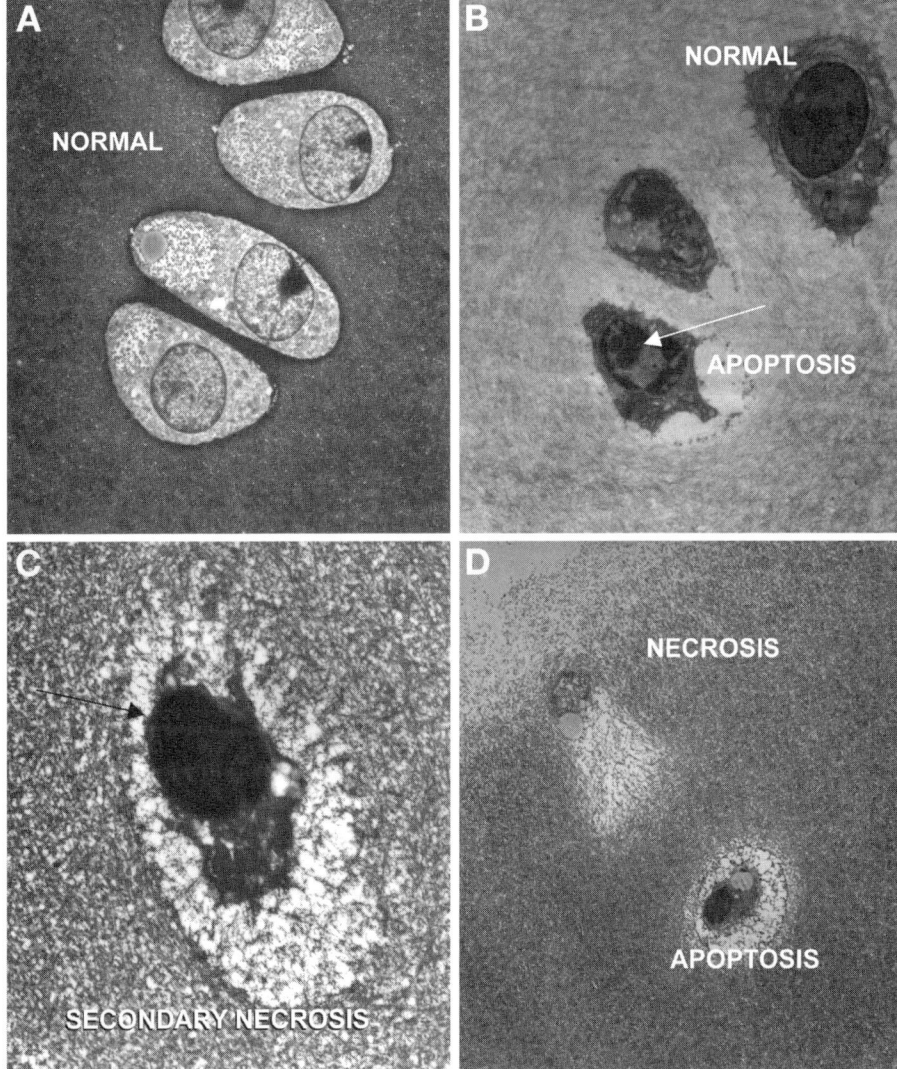

Fig. 3. Electron micrographs demonstrating normal chondrocytes (**A**), a normal chondrocyte and an adjacent apoptotic chondrocyte (**B**) showing cell shrinkage and chromatin condensation (*arrow*). A chondrocyte which has lost its cytosolic contents to the extracellular space indicating necrosis but also exhibits a morphology characteristic of apoptosis with chromatin condensation (*arrow*) demonstrating a scale of cell death progressing to secondary necrosis (**C**), and finally an apoptotic and necrotic cell adjacent to each other (**D**).

1. Fix explants in 2% glutaraldehyde and 0.7% ruthenium (III) hexamine trichloride (RHT) in 0.05 M sodium cacodylate (pH 6.75) for 2 to 3 h.
2. Wash in 0.05 M sodium cacodylate.
3. Post-fix in 1% osmium tetroxide and 0.7% RHT in 0.05 M sodium cacodylate (pH 6.75) for 1 h.
4. Wash in 0.05M sodium cacodylate.
5. Dehydrate in ascending concentrations of ethanol:
    a. 50% ethanol for 30 min.
    b. 70% ethanol for 30 min.
    c. 95% ethanol for30 min.
    d. Three changes of 100% ethanol for30 min each.
6. Place in propylene oxide
7. Place in a 50:50 propylene oxide Araldite mix.
8. Embed in Araldite.
9. Heat the araldite to 60°C to polymerize.
10. After cooling cut ultra-thin sections (50 nm) on an ultratome.
11. Mount sections on copper grids.
12. Stain sections with uranyl acetate for 10 min.
13. Wash with distilled water.
14. Stain with lead citrate for 10 min.
15. Wash in distilled water.
16. View under a transmission electron microscope.

Table 2 illustrates key morphological features of apoptosis visible histologically at both the light microscope and the electron microscope level.

3.2. Fluorescent Detection of Cell Death

3.2.1. DAPI and Propidium Iodide Analysis of Cell Death

There are two methods for cell death detection using these two dyes. The first of which utilizes morphological characteristics of apoptosis. The use of "Vectashield" containing DAPI or propidium iodide on sections of articular cartilage will stain all the nuclei for viewing under fluorescence microscopy. As described above, however, apoptotic nuclei will appear much brighter than normal nuclei (*see* **Fig. 4A**).

1. Dewax paraffin wax sections in two changes of xylene for 2 min each (*see* **Note 1**).
2. Rehydrate sections through descending alcohols comprising:
    a. Two changes of 100% IMS for 2 min each.
    b. 95% IMS for 2 min.
    c. 70% IMS for2 min.
    d. Running water for 2 min (*see* **Note 2**).
3. Mount under a coverslip with Vectashield containing DAPI/propridium iodide (*see* **Note 3**).
4. View under fluorescence microscopy.

Table 2
Morphological Features of Apoptosis

Light Microscopy (see **Fig. 2**)	Electron microscopy (see **Fig. 3**)
Retraction of cytoplasm from the pericellular matrix.	Margination of the chromatin, compaction to the nucleus periphery into a characteristic cup-shaped morphology.
Homogenously condensed chromatin appearing as solid round or oval masses resulting from nuclear condensation.	Membrane blebbing.
Deeply staining homogenous eosinophillic cytoplasm.	Mitochondrial morphology is preserved.
Membrane blebbing.	Formulation of membrane bound vesicles.
Cell fragmentaion into apoptotic bodies that also stain intensely with haematoxylin.	Loss of microvillus structure (membrane smoothing).
	Desmosome complexes become fragmented.
	Formation of electron-dense micronuclei in extracellular space.

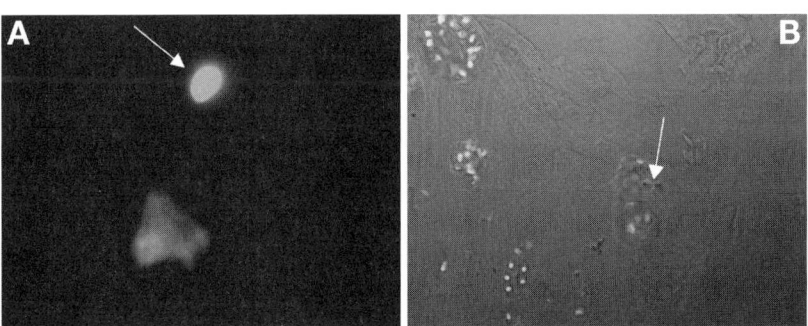

Fig. 4. Illustration of the use of DAPI to detect apoptic nuclei under fluorescence on fixed sections (**A**). Note that the apoptotic nucleus (*arrow*) labels more intensely than its normal counterpart. Propidium iodide (red) was used before fixation to detect dead cells and DAPI (blue) was used after fixation to label the remaining nuclei (**B**). Dual labeling with DAPI and propidium iodide demonstrates that some of the chondrocytes in the OA clusters are undergoing cell death (*arrow*). (Color illustration in insert following p. 268.)

The second method utilising these dyes involves incubation of the tissue prior to processing with propidium iodide (red) which will be incorporated into cells with compromised cell membranes (i.e., during necrosis and secondary apoptosis). Remaining nuclei can then be counterstained with DAPI (blue) once the tissue is processed and sectioned (*see* **Fig. 4B**).

1. Prior to fixation, incubate explants individually in 12-well plates with 1 mL phosphate buffered saline (PBS) containing 1 μM propidium iodide for 30 min at 37°C.
2. Wash the explants in 3 changes of PBS for 5 min each (*see* **Note 4**).
3. Fix explants in NBFS overnight at 4°C (*see* **Note 5**).
4. Process the explants through to wax.
5. Take 8 μm sections using a microtome.
6. Dewax paraffin wax sections in 2 changes of xylene for 2 min each.
7. Rehydrate sections through descending alcohols comprising:
   a. Two changes of 100% IMS for 2 min each.
   b. 95% IMS for 2 min.
   c. 70% IMS for 2 min.
   d. Running water for 2 min.
8. Mount under a coverslip with Vectashield containing DAPI (*see* **Note 3**).
9. View under fluorescence microscopy.

3.2.2. Ethidium Homodimer Analysis of Cell Death

Ethidium homodimer is the "dead" component of a commercially available kit for cell viability/cytotoxicity assays; Live/Dead™ (Molecular Probes, Poortgebouw, Netherlands). Ethidium homodimer enters cells with compromised cell membranes (i.e., necrosis and secondary apoptosis), and undergoes a 40-fold enhancement of the fluorescence after binding to nucleic acids producing a bright red fluorescence.

1. Prior to fixation, incubate explants individually in 12-well plates with 1 mL PBS containing 2 μm ethidium homodimer at 37°C for 1.5 h.
2. Wash the explants in 3 changes of PBS for 5 min each (*see* **Note 4**).
3. Fix explants in NBFS, overnight at 4°C (*see* **Note 5**).
4. Process the explants through to wax.
5. Take 8 μm sections using a microtome.
6. Dewax paraffin wax sections in 2 changes of xylene for 2 min each.
7. Rehydrate sections through descending alcohols comprising:
   a. Changes of 100% IMS for 2 min each.
   b. 95% IMS for 2 min.
   c. 70% IMS for 2 min.
   d. Running water for 2 min.
8. Mount under a coverslip with Vectashield (*see* **Note 3**).
9. View under fluorescence microscopy.

3.2.3. Live/Dead™ Analysis of Cell Death

The Live/Dead™ kit comprises the "dead" component ethidium homodimer, as described above, and the "live" component calcein-AM. The nonfluorescent

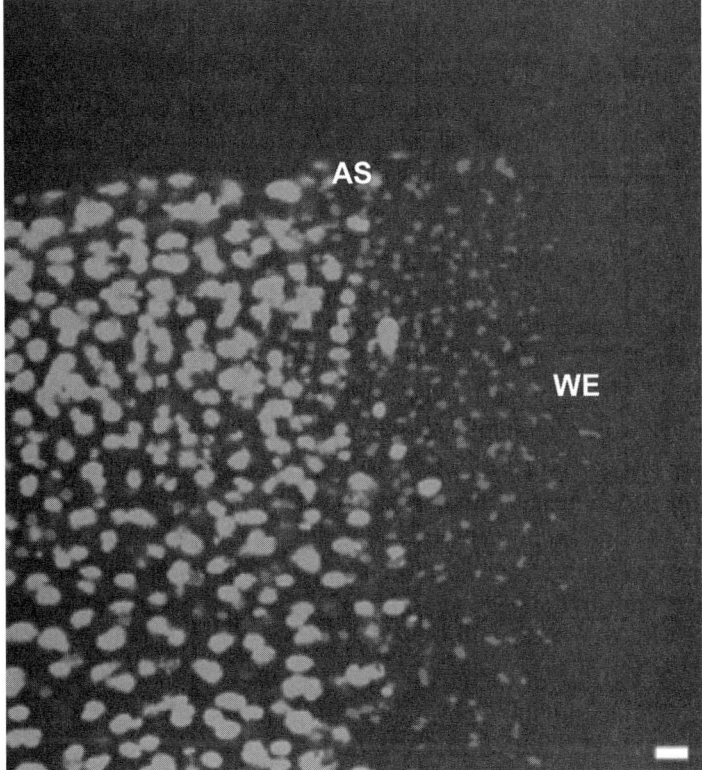

Fig. 5. The use of the Live/Dead™ kit to detect live cells (green) and dead cells (red) utilizing calcein am and ethidium homodimer respectively. The image shows the articular surface (AS) and the wound edge (WE) of 7-d bovine articular cartilage after wounding with a trephine. The labeling demonstrates cell death at the wound edge with viable cells behind the region of cell death. The labeling was visualized on unfixed tissue using confocal laser scanning microscopy. Scale bar = 50µm. (Color illustration in insert following p. 268.)

calcein-AM permeates the cells and is converted by intracellular esterase activity to the highly fluorescent calcein resulting in the observation of a uniform green fluorescence in live cells (*see* **Fig. 5**).

1. Incubate explants individually in 12-well plates with 1 mL PBS containing 2 µM ethidium homodimer and 10 µM calcein-AM at 37°C for 1.5 h.
2. Wash the explants in 3 changes of PBS for 5 min each (*see* **Note 4**).
3. Because of the nature of the "live" component, the tissue can not be processed and sectioned as discussed for ethidium homodimer analysis of cell death. Therefore, take slices of tissue of approximately 100 µm thickness using a dissecting microscope (*see* **Note 6**).

4. Mount the slices of tissue in PBS on a microscope slide and view using a confocal laser scanning microscope.

3.3. Biochemical Detection of Cell Death

3.3.1. Tdt-Mediated dUTP Nick-End Labelling Analysis of Cell Death

TUNEL detects the DNA fragmentation exhibited when cells undergo apoptosis. The assay conducted here uses a commercially available kit that label the 3'-OH ends of these DNA fragments with a fluorescein isothiocyanate (FITC)-conjugated probe (Roche Diagnostics, Lewes, UK) (*see* **Fig. 6**).

1. Dewax paraffin wax sections in 2 changes of xylene for 2 min each.
2. Rehydrate sections through descending alcohols comprising:
    a. Two changes of 100% IMS for 2 min each.
    b. 95% IMS for 2 min.
    c. 70% IMS for 2 min.
    d. Running water for 2 min (*see* **Note 2**),
3. Equilibrate sections in PBS for 2 min.
4. Outline sections with a water repellent wax pen.
5. Incubate with proteinase K (10 µg/mL) for 30 min at 37°C.
6. Wash in 2 changes of PBS for 5 min each.
7. Incubate with the TUNEL reaction mixture at room temperature for 30 min.
    a. TUNEL reaction mixture comprises terminal transferase enzyme solution and label solution in combination, prepared immediately prior to use.
    b. The negative control comprises only the label solution in the absence of the enzyme solution.
8. Wash in 2 changes of PBS for 5 min each.
9. Mount under a coverslip with Vectashield (*see* **Note 5**).
10. View under fluorescence microscopy.

4. Notes

1. Paraffin wax sections on APES or poly-L lysine coated slides have been used here as the sections are less likely to lift or fold.
2. Cryosections may be introduced at this point.
3. The use of Vectashield protects the fluorescence from bleaching. While waiting to view the slides under fluorescence it is also advisable to store the slides at 4°C wrapped in foil, again to prevent loss of fluorescence.
4. If possible wash the explants on a shaker to ensure all unbound label has been removed in order to reduce background staining.
5. If required the explants may be snap frozen at this point and cryosections taken, the sections may then be viewed directly once **Subheading 3.2.2., step 8** has been carried out.
6. Snap freezing and the taking of cryosections does not work in this instance, the calcein AM, is not retained in the cells during these procedures and will not be visible under fluorescent or confocal microscopy.

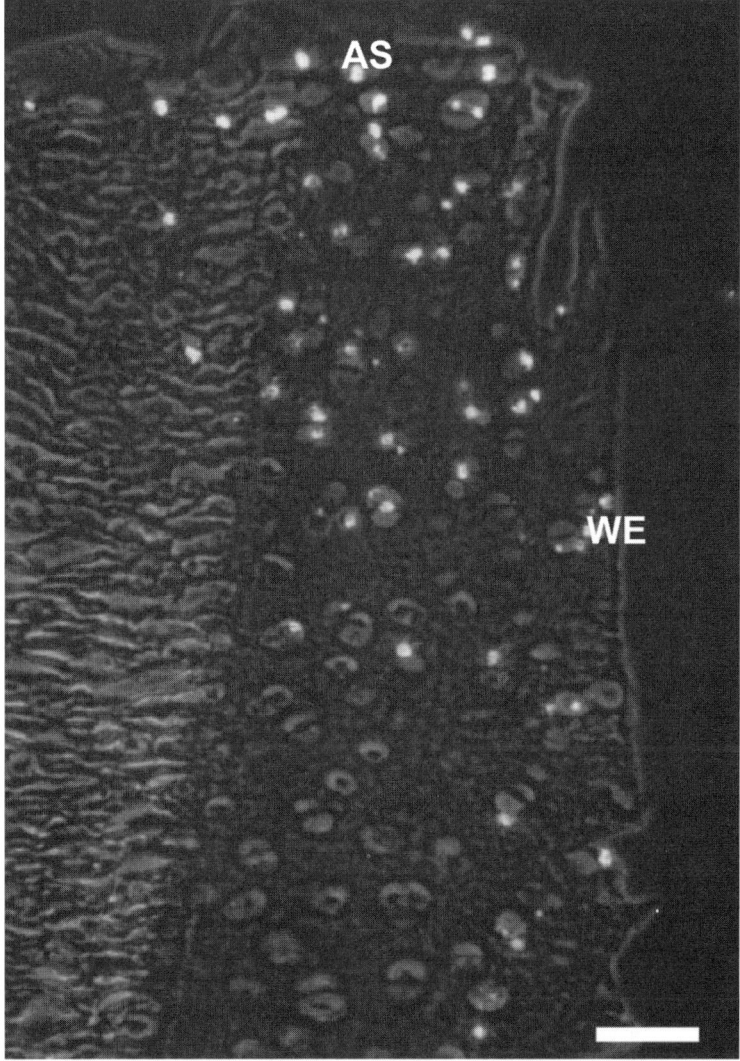

Fig. 6. Image shows phase/fluorescent overlay image of TUNEL labelling of 7-d bovine articular cartilage again wounded with a trephine. Image shows the articular surface (AS) and the wound edge (WE). Note the difference in labeling of the apoptotic cells compared with the ethidium homodimer labelling shown in **Fig. 4** which shows cells with compromised cell membranes. Scale bar = 20 µm.

References
1. Majno, G. and Joris, I. (1995) Apoptosis, oncosis, and necrosis. An overview of cell death. *Am. J. Pathol.* **146,** 3–15.

2. Kerr, J. F., Wyllie, A. H. and Currie, A. R. (1972) Apoptosis: a basic biological phenomenon with wide-ranging implications in tissue kinetics. *Br. J. Cancer* **26**, 239–257.
3. Bowen, I. (1999) Apoptosis and programmed cell death., in *Methods in aging research.* (Yu, B., ed.), CRC Press, pp. 453–473.
4. Kroemer, G. (1995) The pharmacology of T cell apoptosis. *Adv Immunol.* **58**, 211–296.
5. Thompson, C. B. (1995) Apoptosis in the pathogenesis and treatment of disease. *Science* **267**, 1456–1462.
6. Ashkenazi, A. and Dixit, V. M. (1998) Death receptors: signaling and modulation. *Science* **281**, 1305–1308.
7. Liu, Q. A. and Hengartner, M. O. (1999) The molecular mechanism of programmed cell death in C. elegans. *Ann. N. Y. Acad. Sci.* **887**, 92–104.
8. Thornberry, N. A. and Lazebnik, Y. (1998) Caspases: enemies within. *Science* **281**, 1312–1316.
9. Alnemri, E. S. (1997) Mammalian cell death proteases: a family of highly conserved aspartate specific cysteine proteases. *J. Cell Biochem.* **64**, 33–42.
10. Strasser, A., O'Connor, L., and Dixit, V. M. (2000) Apoptosis signaling. *Annu. Rev. Biochem.* **69**, 217–245.
11. Adams, J. M. and Cory, S. (1998) The Bcl-2 protein family: arbiters of cell survival. *Science* **281**, 1322–1326.
12. Meredith, J. E., Jr., Fazeli, B., and Schwartz, M. A. (1993) The extracellular matrix as a cell survival factor. *Mol. Biol. Cell* **4**, 953–961.
13. Svoboda, K. K. (1998) Chondrocyte-matrix attachment complexes mediate survival and differentiation. *Microsc Res. Tech.* **43**, 111–122.
14. Cao, L., Lee, V., Adams, M. E., Kiani, C., Zhang, Y., Hu, W. and Yang, B. B. (1999) beta-Integrin-collagen interaction reduces chondrocyte apoptosis. *Matrix Biol.* **18**, 343–355.
15. Hirsch, M. S., Lunsford, L. E., Trinkaus-Randall, V. and Svoboda, K. K. (1997) Chondrocyte survival and differentiation in situ are integrin mediated. *Dev. Dyn.* **210**, 249–263.
16. Chen, C. T., Burton-Wurster, N., Borden, C., Hueffer, K., Bloom, S. E. and Lust, G. (2001) Chondrocyte necrosis and apoptosis in impact damaged articular cartilage. *J. Orthop. Res.* **19**, 703–711.
17. Borrelli, J., Jr., Tinsley, K., Ricci, W. M., Burns, M., Karl, I. E. and Hotchkiss, R. (2003) Induction of chondrocyte apoptosis following impact load. *J. Orthop. Trauma* **17**, 635–641.
18. D'Lima, D. D., Hashimoto, S., Chen, P. C., Colwell, C. W., Jr., and Lotz, M. K. (2001) Human chondrocyte apoptosis in response to mechanical injury. *Osteoarthritis Cartilage* **9**, 712–719.
19. Duda, G. N., Eilers, M., Loh, L., Hoffman, J. E., Kaab, M. ,and Schaser, K. (2001) Chondrocyte death precedes structural damage in blunt impact trauma. *Clin. Orthop.* 302–309.

20. Loening, A. M., James, I. E., Levenston, M. E., Badger, A. M., Frank, E. H., Kurz, B., Nuttall, M. E., Hung, H. H., Blake, S. M., Grodzinsky, A. J., and Lark, M. W. (2000) Injurious mechanical compression of bovine articular cartilage induces chondrocyte apoptosis. *Arch. Biochem. Biophys.* **381,** 205–212.
21. Bennet, G. A., Bauer, M. D., and Maddock, S. J. (1932) A study of the repair of articular cartilage and the reaction of normal joints of adult dogs to surgically created defects of articular cartilage, "joint mice" and patellar displacement. *American Journal of Paathology.* **8,** 449–524.
22. Key, J. A. (1931) Experimental arthritis: the changes in joints produced by creating defects in the articular cartilage. *J. Bone Joint Surg.* **13,** 725–739.
23. Tew, S. R., Kwan, A. P., Hann, A., Thomson, B. M. and Archer, C. W. (2000) The reactions of articular cartilage to experimental wounding: role of apoptosis. *Arthritis Rheum.* **43,** 215–25.
24. Stockwell, R. A. (1978) Chondrocytes. *J. Clin. Pathol. Suppl (R. Coll. Pathol.).* **12,** 7–13.
25. Redman, S. N., Dowthwaite, G. P., Thomson, B. M., and Archer, C. W. (2004) The cellular responses of articular cartilage to sharp and blunt trauma. *Osteoarthritis Cartilage.* **12,** 106–116.
26. Sandell, L. J. and Aigner, T. (2001) Articular cartilage and changes in arthritis. An introduction: cell biology of osteoarthritis. *Arthritis Res.* **3,** 107–113.
27. Hashimoto, S., Ochs, R. L., Komiya, S., and Lotz, M. (1998) Linkage of chondrocyte apoptosis and cartilage degradation in human osteoarthritis. *Arthritis Rheum.* **41,** 1632–1638.
28. Sharif, M., Whitehouse, A., Sharman, P., Perry, M. and Adams, M. (2004) Increased apoptosis in human osteoarthritic cartilage corresponds to reduced cell density and expression of caspase-3. *Arthritis Rheum.* **50,** 507–515.
29. Roy, S. and Meachim, G. (1968) Chondrocyte ultrastructure in adult human articular cartilage. *Ann Rheum Dis.* **27,** 544.
30. Pond, M. J. and Nuki, G. (1973) Experimentally-induced osteoarthritis in the dog. *Ann. Rheum. Dis.* **32,** 387–388.
31. Briant, L., Robert-Hebmann, V., Sivan, V., Brunet, A., Pouyssegur, J., and Devaux, C. (1998) Involvement of extracellular signal-regulated kinase module in HIV-mediated CD4 signals controlling activation of nuclear factor-kappa B and AP-1 transcription factors. *J. Immunol.* **160,** 1875–1885.
32. Aigner, T., Hemmel, M., Neureiter, D., Gebhard, P. M., Zeiler, G., Kirchner, T., and McKenna, L. (2001) Apoptotic cell death is not a widespread phenomenon in normal aging and osteoarthritis human articular knee cartilage: a study of proliferation, programmed cell death (apoptosis), and viability of chondrocytes in normal and osteoarthritic human knee cartilage. *Arthritis Rheum.* **44,** 1304–1312.
33. Aigner, T. and Kim, H. A. (2002) Apoptosis and cellular vitality: issues in osteoarthritic cartilage degeneration. *Arthritis Rheum.* **46,** 1986–1996.

12

Measurement of Glycosaminoglycan Release from Cartilage Explants

John S. Mort and Peter J. Roughley

Summary

Quantitation of glycosaminoglycans (GAGs) in the form of aggrecan fragments released from cartilage in culture is a simple way to determine the efficacy of different cytokines alone or in combination in simulating cartilage catabolism. Two approaches for GAG assay are described, with special attention being paid to the advantages and limitations of each method

Key Words: Aggrecan; glycosaminoglycan; dimethylmethylene blue; carbazole; assay.

1. Introduction

Study of cartilage catabolism in culture has received considerable attention over the past three decades because of the pivotal role that this process plays in cartilage degeneration in arthritis. With the appreciation that such degeneration is driven by cytokines, cartilage organ culture became a standard technique for monitoring the effects of cytokines on chondrocyte metabolism. Under such conditions proteoglycan degradation was found to be an early event, with the glycosaminoglycan (GAG)-rich degradation products being readily released from the tissue into the culture medium. Means to analyze such released products were therefore developed in order to assess the rate and extent of cartilage degradation. Initially, colorimetric methods were used to detect chemically modified uronic acid derived from the released GAGs, but in the early 1980s these methods were superceded by direct GAG-dye binding assays. Here we describe both of these methodologies pointing out their strengths and weaknesses.

2. Materials
2.1. Cartilage Culture

1. Cartilage: tissue containing cartilage obtained fresh for processing in the laboratory.
2. Culture medium: Dulbecco's modified Eagle's medium (DMEM) (Gibco) buffered with 44 mM sodium bicarbonate and 25 mM Hepes, and containing 0.1 mg/mL bovine serum albumin (BSA) (Sigma) and penicillin G (100 units/mL) and streptomycin sulfate (100 µg/mL) (*1*). Filter sterilize.
3. Fungizone (GibcoBRL): stock solution of amphotericin B (250 µg/mL) and desoxycholate (205 µg/mL) in water.
4. Cytokines: recombinant cytokines such as human interleukin (IL)-1β (R&D Systems).

2.2. DMMB Assay

1. Chondroitin sulfate standard (Sigma C 9819). Dissolve in water (4.0 mg/mL) leaving overnight to ensure complete dissolution. Store at 4°C.
2. Dimethylmethylene blue reagent. 1,9-dimethylmethylene blue can be purchased from Aldrich (34,108-9) (*see* **Note 1**). The working solution is made by dissolving 16 mg of dye in 1 L of water containing 3.04 g glycine, 2.37 g NaCl and 95 mL 0.1 M HCl (*2,3*). The solution should be stored at room temperature in a brown bottle.
3. 96-Well plates. U-shaped bottom plates (Evergreen Scientific) are used.

2.3. Uronic Acid Assay

1. Disodiumtetraborate decahydrate (0.025M) dissolved in concentrated sulfuric acid (22 g added to bottle containing 4.25 kg sulfuric acid).
2. Carbazole in absolute ethanol (125 mg/100 mL), stored in brown bottle at 4°C (Sigma C 5132).
3. Glucuronolactone (1 mg/mL) in water as standard (Sigma/Aldrich 85,145-0).

3. Methods
3.1. Cartilage Culture

Cartilage from various species and various sites have been used. It should be pointed out that the choice of cartilage source has been driven by considerations of economy and availability. Thus some laboratories with ready access to porcine material have used pig articular cartilage. Others use bovine articular cartilage derived from the metacarpophalangeal joints of the feet, which are usually discarded by the slaughter house. The most commonly used tissue is cartilage derived from the bovine nasal septum which is available in large quantities and responds rapidly to various cytokines with essentially complete release of the tissue GAG content. Various tissue culture media have been used for cartilage organ culture, including DMEM, Ham's F12, RPMI-1640, and Iscove's. Currently, DMEM is probably the most widely used of these media. Thus bovine nasal cartilage in DMEM has become a standard system for the study of chondrocyte metabolism in cartilage because of its availability in large quantities and its rapid and extensive response to perturbation by cytokines (*see* **Note 2**).

3.1.1. Retrieval of Cartilage

1. Cartilage is retrieved from the joint surfaces (articular cartilage) or the nasal septum (nasal cartilage) by sharp dissection under sterile conditions, and placed into tissue culture medium supplemented with twice the normal concentration of antibiotics and 10 mL/L Fungizone stock solution.
2. The tissue is chopped into small cubes (approx 2 mm in each dimension) and aliquots equivalent to 100 mg tissue are placed in 24-well culture plates together with 2 mL of culture medium.

3.1.2. Cartilage Culture

1. Cartilage is allowed to equilibrate for 24 h prior to addition of modulators of chondrocyte metabolism (for example, IL-1 at 1–10 ng/mL).
2. Culture is maintained for the requisite number of days with medium change every 48 h. Tissue can be maintained for several weeks if required.
3. Media are collected and stored at –20°C for GAG analysis.

All culture is performed at 37°C in an atmosphere of 95% air 5% CO_2 (*see* **Note 3**).

3.2. Analysis of GAG Release by DMMB Assay

At the present time, the most widely used method for GAG quantitation is based on the shift in absorption observed when the DMMB dye (*see* **Fig. 1**) associates with repeating negative charges (sulfate) on the GAGs resulting in stacking of the dye molecules and a shift in the absorption maximum (metachromatic shift). This method was originally developed as a tube assay *(4)* and later modified for specificity by changing the buffer *(2)*, and subsequently adapted to the 96-well plate format *(5)*. This latter format is currently the method of choice because of its convenience and small scale.

3.2.1. Procedure

1. 20 µL Aliquots of samples, in duplicate, are pipetted into a microtiter plate.
2. 180 µL of the DMMB reagent is added to each well, preferably using a multichannel pipet.
3. The absorbance at 530 nm measured immediately using a microtiter plate reader.

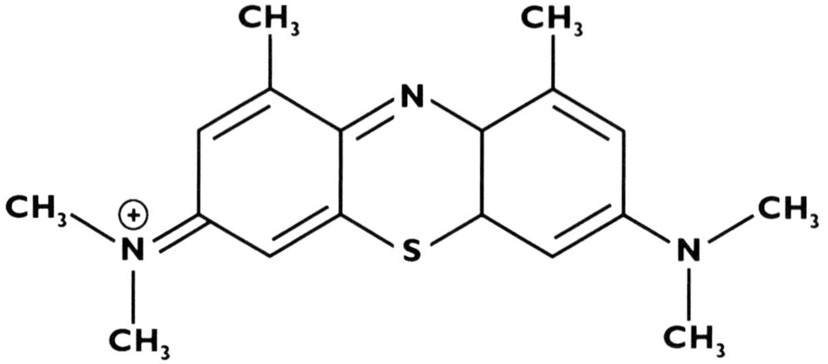

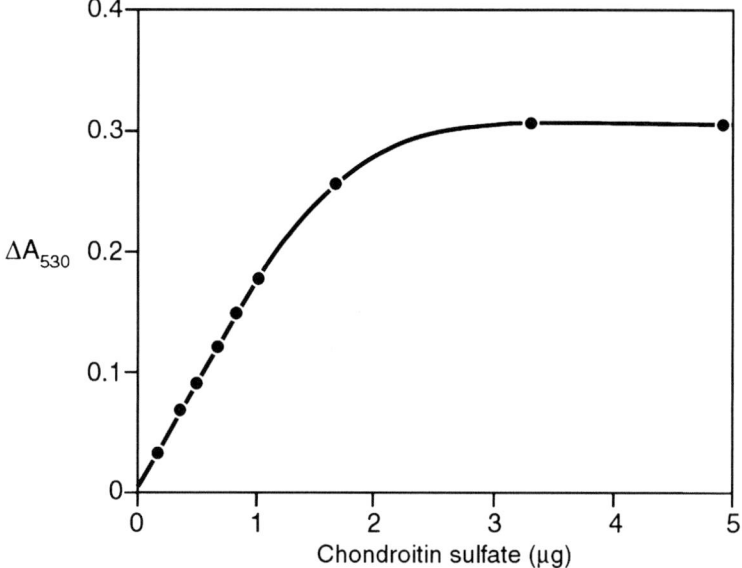

Fig. 1. Structure of 1,9 Dimethyl Methylene Blue (DMMB).

Fig. 2. Standard curve for DMMB analysis. The figure depicts change in absorbance at 530 nm relative to the appropriate blank (ΔA_{530}) plotted against increase in chondroitin sulfate level. Note the very limited linear range ($\Delta A_{530} < 0.2$) of the assay and the plateau effect (at ΔA_{530} approx 0.3) because of exhaustion of dye.

3.2.2. Standard Curve

1. For accurate quantitation it is essential that an appropriate standard curve be using as the assay is not linear over a large concentration range (*see* **Fig. 2**). Dilutions of chondroitin sulfate are prepared in water or when culture medium samples are to be analyzed in the same medium. The stock chondroitin sulfate solution

(4 mg/mL) is diluted 80-fold to prepare a working solution of 50 µg/mL. In a 96-well microtiter plate the following dilutions are made and 20 µL aliquots used for analysis. Blanks must always be subtracted.

µg CS in 20 µL	µL working solution	µL water
1.0	10	–
0.8	80	20
0.6	60	40
0.4	40	60
0.2	20	80
0.1	10	90
0	–	100

3.2.3. Limitations

1. The linear range of the assay is very limited and only samples and standards containing less than 1 µg GAG are suitable for quantitation. It is critical that the amount of sample be adjusted so that the resulting absorbance falls within the limited linear range of the assay. A good approach is to assay a series of dilutions of samples expected to contain the maximum amount of GAG and thus determine the most appropriate dilution for the rest of the series.
2. The solubility of the dye/CS complex is low at high CS concentrations leading to eventual precipitation. Whereas solubility increases as the CS concentration decreases, the complex can eventually come out solution. Hence the need to read absorbance as soon as possible.
3. Only sulfated GAGs can be monitored using this assay. It is not suitable for the quantitation of hyaluronic acid (or any other nonsulfated GAG).
4. Different sulfated GAGs produce different color yields with DMMB because of variation in their degree and position of sulfation. Because of this the assay does not result in absolute quantitation but provides values relative to the standard being used.
5. Whereas interference by other polyanions (for example nucleic acids) is minimal and there is no inference by components of the culture medium, the assay does not tolerate the high concentrations of guanidinium chloride generally used to extract proteoglycans from cartilage residues. However, the high sensitivity of the assay usually allows this limitation to be resolved by sample dilution.

3.2.4. Advantages

The DMMB assay has gained wide acceptance because it is quick, easy and cheap to perform, and can be carried directly on culture medium with no intervening sample preparation.

3.3. Uronic Acid Analysis for GAG Quantitation

Prior to development of the DMMB assay, the most widely accepted technique for GAG quantitation used uronic acid analysis via the carbazole

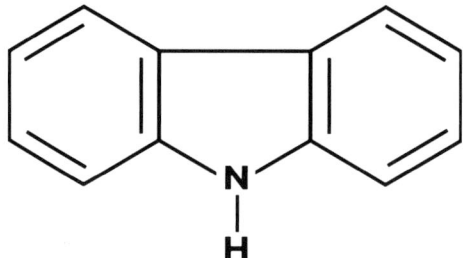

Fig. 3. Structure of carbazole.

reaction. This technique is based on the color reaction that results when uronic acids are treated with carbazole (*see* **Fig. 3**) in the presence of sulfuric acid *(6,7)*. Both monomeric and polymeric forms of uronic acids produce a colour reaction, though the color is unstable and the extinction coefficient varies with different uronic acids and GAGs. The use of borate in the color reaction helps increase sensitivity, stabilize the chromogen, and diminish the variation in color yield between different GAGs *(8–11)*. For several decades the carbazole reaction described by Bitter and Muir *(9)* was the accepted technique for GAG analysis, being reliable, cheap, and relatively easy to perform.

3.3.1. Procedure

1. Place a 0.5-mL sample (containing 1 to 20 µg uronic acid) in a glass tube. Cool in ice bath.
2. Add 3 mL borate/sulfuric acid reagent. Mix well.
3. Heat at 100°C for 10 min. Cool in ice bath.
4. Add 100 µL carbazole/ethanol reagent. Mix well.
5. Heat at 100°C for 15 min. Cool in ice bath to room temperature (*see* **Note 4**).
6. Read absorbance at 530 nm

3.3.2. Sensitivity

The assay will reliably detect 1 µg glucuronolactone and is linear up to 20 µg (*see* **Fig. 4**). Thus the carbazole assay is not as sensitive as the DMMB assay, but its linear range is much greater.

3.3.3. Limitations

1. Whereas pentoses and hexosamines do not interfere in the carbazole reaction, hexoses do generate an absorption at 530 nm, albeit at much higher concentrations. About 100 µg hexose is needed to give a similar colour yield to 10 µg uronic acid. This is of particular concern when assaying GAG present in tissue culture medium where glucose is always present. Typical media used for cartilage culture contain between 1 and 10 mg/mL glucose. If GAG release is to be quantitated in culture medium, it is first necessary to remove the glucose.

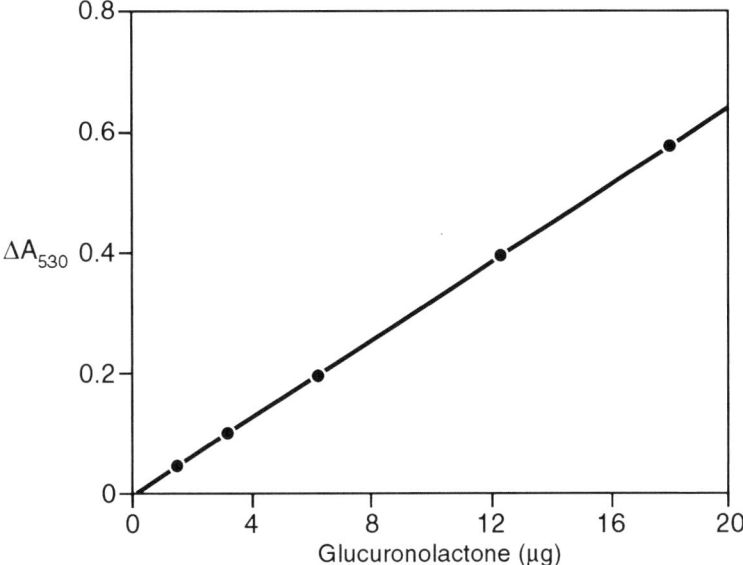

Fig. 4. Standard curve for carbazole analysis. The figure depicts change in absorbance at 530 nm relative to the appropriate blank (ΔA_{530}) plotted against increase in glucuronolactone level. Note the extensive linear range of the assay.

2. Chloride ion concentrations up to 0.4 M do not interfere with the analysis of CS, DS, or KS present on cartilage matrix proteoglycans, though it has been reported to interfere with the analysis of heparin (9). If proteoglycans are extracted from cartilage with 4 M guanidinium chloride following tissue culture, the extremely high chloride concentration in the extract will interfere with the uronic acid assay, unless diluted. When working with high chloride ion concentrations, care must be taken as HCl gas can be (violently) released upon adding concentrated sulfuric acid.
3. Glucuronic acid and iduronic acid behave differently in the carbazole reaction with iduronate having a lower colour yield than glucuronic acid. This difference can be minimized by increasing the borate concentration and by using anhydrous disodium tetraborate rather than the decahydrate (10,11). If such modifications are made then different GAGs behave in a similar manner, whereas if the routine assay conditions are used each GAG will behave differently and require unique standards for accurate quantitation.
4. This assay is not readily adaptable to a 96-well plate format for ease of use.

3.3.4. Advantages

The uronic acid assay does offer some advantages over DMMB analysis.

1. It is independent of GAG size and degree of sulfation, and hence can detect all uronic acid-containing products of GAG catabolism.

2. It is able to detect both hyaluronic acid and sulfated GAGs. Thus, although the uronic acid assay may be much more tedious to carry out than the DMMB assay, there may still be occasions when its unique features warrant its use.

4. Notes

1. There have been reports of variations in DMMB quality *(12)* because of the presence of contaminants originating from its synthesis *(13)*.
2. Different cartilages can respond differently to cytokines, with the extent of GAG release varying considerably. This relates to variation in the expression of cell surface cytokine receptors and downstream signaling events with cartilage species, age, and anatomical site.
3. The oxygen tension in the joint is lower than in air, and culture under hypoxic conditions (but physiologically normal to the joint) could influence cartilage metabolism.
4. When performing the carbazole assay care must be taken to ensure that the final solution reaches room temperature prior to absorbance determination otherwise light scattering will occur as a result of the generation of convection currents (Schlieren lines) when samples are poured into the cuvet.

Acknowledgments

This work was supported by the Shriners of North America, the Canadian Institutes of Health Research and the Arthritis Society of Canada.

References

1. Sztrolovics, R., White, R. J., Roughley, P. J. and Mort, J. S. (2002) The mechanism of aggrecan release from cartilage differs with tissue origin and the agent used to stimulate catabolism. *Biochem. J.* **362**, 465–472.
2. Farndale, R. W., Buttle, D. J., and Barrett, A. J. (1986) Improved quantitation and discrimination of sulphated glycosaminoglycans by use of dimethylmethylene blue. *Biochim. Biophys. Acta* **833**, 173–177.
3. Handley, C. J. and Buttle, D. J. (1995) Assay of proteoglycan degradation. *Methods Enzymol.* **248**, 47–58.
4. Farndale, R. W., Sayers, C. A. and Barrett, A. J. (1982) A direct spectrophotometric microassay for sulfated glycosaminoglycans in cartilage cultures. *Connect. Tissue Res.* **9**, 247–248.
5. Hollander, A. P., Atkins, R. M., Eastwood, D. M., Dieppe, P. A., and Elson, C. J. (1991) Human cartilage is degraded by rheumatoid arthritis synovial fluid but not by recombinant cytokines *in vitro*. *Clin. Exp. Immunol.* **83**, 52–57.
6. Dische, Z. (1947) A new specific color reaction of hexuronic acids. *J. Biol. Chem.* **167**, 189–198.
7. Dische, Z. (1955) New color reactions for determination of sugars in polysaccharides. *Methods Biochem. Anal.* **2**, 313–358.
8. Gregory, J. D. (1960) The effect of borate on the carbazole reaction. *Arch. Biochem. Biophys.* **89**, 157–159.

9. Bitter, T. and Muir, H. (1962) A modified uronic acid carbazole reaction. *Anal. Biochem.* **4,** 330–334.
10. Kosakai, M. and Yosizawa, Z. (1978) Study on the factors yielding high color in the carbazole reaction with hexuronic acid-containing substances. *J. Biochem. (Tokyo)* **84,** 779–785.
11. Kosakai, M. and Yosizawa, Z. (1979) A partial modification of the carbazole method of Bitter and Muir for quantitation of hexuronic acids. *Anal. Biochem.* **93,** 295–298.
12. Stone, J. E., Akhtar, N., Botchway, S., and Pennock, C. A. (1994) Interaction of 1,9-dimethylmethylene blue with glycosaminoglycans. *Ann. Clin. Biochem.* **31,** 147–152.
13. Taylor, K. B. and Jeffree, G. M. (1969) A new basic metachromatic dye, 1:9-Dimethyl Methylene Blue. *Histochem. J.* **1,** 199–204.

13

Assessment of Collagenase Activity in Cartilage

Tim E. Cawston and Tanya G. Morgan

Summary

Assay of collagenase activity involves the use of radiolabeled collagen. Stimulation of cartilage with proinflammatory cytokines results in the upregulation of collagenases and the subsequent release of degraded collagen fragments. These enzymes can be localized in both osteoarthritic and rheumatoid arthritis cartilage and synovial tissues.

Key Words: Collagen; MMP; TIMP-1; proteoglycan.

1. Introduction

Mammalian collagenases cleave all three polypeptide chains of the triple helical collagen molecule at a specific site to give characteristic one quarter and three quarter fragments. These denature at 37°C becoming susceptible to digestion by less specific proteinases.

The most widely used forms of collagenase assay depend on the measurement of fragments released from radiolabeled collagen *(1,2)*. Collagen is extracted from skin or tail tendons and labelled with [1-^{3}H] acetic anhydride. This labeled collagen is incubated at neutral pH and at 37°C to form collagen fibrils. Collagenases degrade this fibrillar substrate and at the end of the assay period, cleaved fragments are separated from undigested collagen by centrifugation. Other methods for the separation of uncleaved substrate from the products of digestion have included precipitation with dioxane *(3)* or the use of radiolabeled native collagen immobilized in microtitre plates *(4)*.

These assays for collagenases using radiolabeled substrate can be used for conditioned culture medium from chondrocytes or from cartilage and have the advantage of measuring enzyme activity. They do not distinguish between the different collagenases (MMP-1, MMP-8, and MMP-13) or indeed other enzymes that can also cleave triple helical collagen (MMP-2 and MT1-MMP). They also have the disadvantage that measurement of activity reflects the overall balance between enzymes that can degrade collagen at neutral pH and the level of the tissue inhibitor of metalloproteinases (TIMPs), and other inhibitors, that are present in the sample. Different procedures have been recommended in an attempt to overcome these problems *(5,6)*. Immunological based assays (e.g., enzyme linked immunosorbent assay [ELISA]) can accurately measure the amount of an individual MMP present but often do not distinguish between proenzyme, active enzyme or inhibitor-complexed enzyme. These assays have been previously described in this series *(7)*.

Stimulation of cartilage in organ culture by cytokines has been used as a model system to demonstrate the release of proteoglycan and collagen. Whereas proteoglycan is readily released from cartilage by a variety of stimuli it is difficult to reproducibly initiate the degradation of collagen. We report a method using a combination of interleukin 1 (IL-1) and oncostatin M that can reproducibly stimulate collagen release from bovine nasal cartilage by day 14 of culture. This release is preceeded by upregulation of procollagenases and their subsequent activation. There is a direct correlation between the level of active collagenase and collagen release and the release of collagen is blocked by the inclusion of TIMP-1 and TIMP-2 showing that collagen release is MMP driven in this model *(8)*.

Human cartilage from patients with rheumatoid and osteoarthritis can be sectioned and stained using immunohistochemical techniques to demonstrate the presence of MMP-1, MMP-8, MMP-13, and TIMP-1 suggesting that these collagenases are involved in the turnover of cartilage collagen found in diseased tissue.

2. Materials

All chemicals and biochemicals used were reagent or analytical grade and obtained from general laboratory suppliers such as VWR International and Sigma-Aldrich Company Ltd., unless otherwise indicated.

2.1. Collagenase Activity Assay

1. Aminophenylmercuric acetate (APMA).
2. Diisopropylphosphoflouridate (DFP).
3. Trypsin type II-S (porcine pancreas).
4. Soyabean trypsin inhibitor.
5. Brij35.
6. Acetic acid.
7. Hammmerstein grade casein (Merck Ltd, Poole, BH15 1TD).
8. [^{3}H] acetic anhydride (25 mCi; Amersham PLC, Little Chalfont, UK).
9. Cheese cloth or fine mesh-beer making bag (Boots Chemist).
10. Electric mincer (Lynx Asco meat grinder).
11. Dry dioxane.

12. 10 mM disodium tetraborate (pH 9.0), 0.2 M calcium chloride (adjust with NaOH, keep at 4°C).
13. 50 mM Tris-HCl (pH 7.6), 0.2 M sodium chloride, 5 mM calcium acetate and 0.03% (v/v) toluene at 4°C.
14. 50 mM Tris-HCl (pH 7.6), 0.2 M sodium chloride and 0.03% (v/v) toluene at 4°C.
15. Buffer A: 25 mM sodium cacodylate buffer (pH 7.6), 0.05% (v/v) Brij35, at 4°C.
16. Buffer B: 100 mM Tris/HCl (pH 7.6), 15 mM calcium chloride at 4°C.
17. 20 mM Tris/HCl (pH 8.0–9.5).
18. Ultima Gold scintillation fluid (Packard BioScience, Groningen, The Netherlands).
19. Supermix scintillation fluid (Perkin Elmer LAS (UK) Ltd., Cambridge, UK).

2.2. Bovine Nasal Cartilage Assay

1. Bovine nasal septum cartilage (local abbatoir).
2. Dulbecco's phosphate buffered saline (PBS) with 100 U/mL penicillin, 100 µg/mL streptomycin, and 20 U/mL nystatin (DPBS/antibiotics)
3. Explant culture medium: Dulbecco's Modified Eagle's Medium (DMEM) with 25 mM HEPES supplemented with 2 mM glutamine, 100 U/mL penicillin, 100 µg/mL streptomycin, and 20 U/mL nystatin.
4. Leather hole punch
5. 0.1 M Phosphate buffer (pH 6.5).
   a. Solution A: 0.1M sodium dihydrogen phosphate (9.36 g/600 mL distilled water).
   b. Solution B: 0.1M disodium hydrogen phosphate (7.10 g/500 mL distilled water).
   c. Add 137 mL of solution A to 63 mL of solution B and check that the pH is 6.5.
6. Papain solution (make fresh): dissolve 0.25 g in 10 mL phosphate buffer.
7. Cysteine-hydrochloride solution: dissolve 0.08 g in 10 mL phosphate buffer.
8. Ethylenediaminetetraacetic acid (EDTA) solution: dissolve 0.19 g in 10 mL phosphate buffer.
9. 4-(dimethylamino)benzaldehyde (DAB) reagent *(Caution: Highly Toxic)*: dissolve 20 g in 30 mL of 70% perchloric acid in a glass bottle and cover with foil and store at 4°C.
10. Acetate citrate buffer (pH 6.0): 57.0 g sodium acetate · 3 H_2O + 5.5 g citric acid (H_2O) + 385 mL propan-2-ol, make up to 1 L with distilled water and pH to 6.0.
11. Hydroxyproline standard: dry hydroxyproline extensively in a vacuum dessicator and dissolve at 1 mg/mL in water. Store at −20°C.
12. Chloramine T solution: Make up immediately prior to use. Dissolve 0.14 g chloramine T in 2 mL distilled water to make a 7% (w/v) chloramine T solution and dilute to 10 mL with acetate-citrate buffer; discard remainder after use as this reagent is not stable.
13. Sarstedt tubes (Sarstedt, Leicester).

14. Dimethyl-methylene blue (DMB) reagent: 3.04 g glycine, 2.37 g sodium chloride, 95 mL 0.1 M HCl make up to 1 L with distilled water and pH to 3.0 in a glass bottle. Then add 16 mg of DMB to dissolve and cover with foil. The absorbance at 530 nm should be approx 0.3. Store at room temperature.
15. Vacuum dessicator.

2.3. Immunolocalisation of Collagenases in Cartilage

1. 1 LPBS (pH 7.4) : 9.0 g sodium chloride, 0.2 g potassium chloride, 1.2 g disodium hydrogen phosphate, 0.2 g potassium dihydrogen phosphate, and 0.2 g sodium azide. Store at room temperature.
2. 10 L Tris buffered saline (TBS) (pH 7.6), 80 g sodium chloride, 6 g Tris (hydroxymethyl) methylamine, 38 mL of 1 M hydrochloric acid. Make up to 10 L with distilled water. Store at room temperature.
3. Sucrose.
4. 4% Paraformaldehyde in PBS (make fresh).
5. OCT (Tissue-Tek, Sakura).
6. 3-aminopropyltriethoxysilane (APES).
7. Control rabbit serum.
8. Biotinylated secondary antibody (rabbit anti-goat).
9. Avidin-biotin complex/Horse radish peroxidase (ABComplex/HRP) detection system (DakoCytomation Ltd., Cambridgeshire, UK).
10. Diaminobenzidine (DAB).
11. Industrial methylated spirits.
12. Ammonium hydroxide.
13. Hydrogen peroxide.
14. Ethanol.
15. Methanol.
16. Acetone.
17. Xylene.
18. Sodium azide.
19. Dry ice.
20. Anti-MMP antibody (sheep anti-MMP-1) *(9)*.
21. Anti-TIMP antibody *(10)*.
22. DECON.
23. Slides (Bios Europe Ltd., Edinburgh, UK).
24. Coverslips (Bios Europe Ltd., Edinburgh, UK).
25. 10-cm^2 Petri dish.
26. DPX.
27. Weigert's iron-haematoxylin. (Make just before use. May be kept for 2 d at room temperature or for 10–14 d at 4°C for multiple use.)
    a. Solution A: 5.0 g Haematoxylin (Bios Europe Ltd., Edinburgh, UK) + 95% ethanol.
    b. Solution B: 5.8 g ferric chloride (FeCl$_3$·6 H$_2$O), 485 mL distilled water, and 5 mL concentrated hydrochloric acid.
    c. Mix equal volumes of solutions A and B.

3. Methods

3.1. Assay of Collagenase Activity in Conditioned Culture Medium—Diffuse Fibril Assay

This method requires the preparation of radiolabelled collagen. Collagen forms a gel at neutral pH and body temperature and after digestion degraded products are separated from undegraded collagen fibrils by centrifugation. The radioactive fragments in the supernatant are measured and used to determine collagenase activity.

3.1.1. Preparation of Type I Collagen Substrate (see **Note 1**)

1. Freeze new-born calf skin (30 cm × 30 cm) onto a wooden tray. Remove hair first with scissors and then with a large scalpel blade. Allow to thaw, cut skin into small pieces (approx 50 mm × 50 mm) and maintain a temperature of 4°C throughout this procedure.
2. Grind skin in an electric mincer (Lynx Asco meat grinder) with chips of dry ice to prevent heating of the tissue during mincing. Allow minced skin to thaw and perform all subsequent steps at 0–4°C. Add toluene (0.03% [v/v]), or an alternative preservative, at each stage of the procedures below.
3. Extract minced skin three times with 1.5 L of 0.9% (w/v) NaCl for 30 min while stirring followed by 2 extractions with ice-cold water. Filter after each extraction through cheesecloth or through a fine mesh-beer making bag. Discard supernatants.
4. Resuspend the extracted skin in 1.5 L of 0.5 M acetic acid and stir slowly overnight at 4°C, filter and retain supernatant. Repeat this step.
5. Combine acetic acid extracts from **step 4** and centrifuge at 7500g for 2 h at 4°C, discard pellet.
6. Dialyze supernatant against 2 changes of 12 L of 5% *(w/v)* NaCl in 0.1 M acetic acid until collagen precipitates. Centrifuge at 7500g for 30 min and retain pellet.
7. Resuspend pellets in 2 to 3 L of 0.5 M acetic acid and stir. Dialyze against 0.5 M acetic acid (5 L) overnight (or until pellet dissolves).
8. Dialyze against at least 4 changes of 20 mM disodium hydrogen phosphate (20 L) until collagen precipitates and centrifuge at 7500g for 30 min. Retain pellets.
9. Resuspend pellets in 2 to 3 L of 0.5 M acetic acid. Stir until dissolved and then slowly add 5% *(w/v)* NaCl to precipitate the collagen. Centrifuge at 7500g for 30 min, wash pellet in 1 L of 20% *(w/v)* NaCl and collect the collagen by centrifugation (7500g for 30 min).
10. Resuspend pellets in 0.5 M acetic acid and stir slowly overnight until dissolved. Centrifuge at 30,000g for 1 h, retain supernatant and dialyze against 0.1 M acetic acid (10 L). Freeze-dry thoroughly and store desiccated at –20°C until required.

3.1.2. Labeling of Collagen Substrate With ^3H Acetic Anhydride
(see **Note 2**)

1. Dissolve 250 mg of freeze-dried collagen in 50 mL of 0.2 M acetic acid at 4°C and dialyze against 10 mM disodium tetraborate (pH 9.0) containing 0.2 M CaCl$_2$ (2 L, one change). If the collagen precipitates redialyze against acetic acid and repeat.
2. Place collagen in a conical flask with a large stir bar and stir very slowly.
3. Place the bottom of tube containing [^{3}H]-acetic anhydride (25 mCi) into dry ice. Open tube and add 1 mL dry dioxane into tube; it will freeze in the bottom of the tube. Thaw dioxane as quickly as possible and add dioxane containing [^{3}H] acetic anhydride into the collagen; wash tube with a further 1 mL of dry dioxane. Stir for 30 min at 4°C.
4. Dialyze collagen against 50 mM Tris-HCl (pH 7.6) containing 0.2 M NaCl, 5 mM calcium acetate and 0.03% (v/v) toluene until the radioactivity in the diffusate falls to background levels. Dilute collagen to a final concentration of 1 mg/mL and add unlabeled collagen (1 mg/mL) until the [^{3}H] content equals 200,000 dpm/mg. This solution is dialysed against 0.2 M acetic acid and stored at –20°C until required.
   Specific assays can be made for type II and III collagen if these substrates are radioactively labeled as described above.

3.1.3. Collagenase Assay

1. Thaw collagen at 1 mg/mL in 0.2 M acetic acid and dialyse against 50 mM Tris/HCl, buffer (pH 7.6) containing 0.2 M NaCl and 0.03% *(v/v)* toluene.
2. Set up control 400 µL microfuge tubes 1–2 with 100 µL of buffer A, tubes 3–4 with 10 µL of trypsin (100 µg/mL) in 1 mM HCl + 90 µL of buffer A; tubes 5–6 with 100 µL of bacterial collagenase at 100 µg/mL in buffer A. If samples contain appreciable amounts of salt then appropriate blanks should be used (*see* **Note 3**).
3. Add test samples in duplicate or triplicate to tube 7 onward and make up volume to 100 µL with buffer A. Add 100 µL buffer B to all tubes.
4. Add 100 µL of [^{3}H]-labelled collagen (1 mg/mL) to all tubes. Cap and incubate in water bath at 37°C for 1 to 20 h. At the end of the assay period centrifuge at 13,000g for 10 min to remove the undigested collagen. Remove 200 µL of the supernatant and combine with 3 mL of Ultima Gold scintillation fluid and count for [^{3}H] in a scintillation counter.
5. Subtract mean blank values (tubes 1–2) from all test results. If trypsin digests substantial amounts of collagen (tubes 3–4) it means the collagen is denatured and assay results should be discarded. The mean values obtained for tubes 5–6 represent the total lysis figure and corresponds to 100 µg of collagen.
6. Results are expressed as units/mL where one unit of activity represents the amount of enzyme that degrades 1 µg of collagen/min at 37°C. Thus, to obtain results in units/mL use the following formula:

$$\text{units/mL} = [\text{test mean} - \text{mean tubes } 1-2] \times \frac{100 \,(\mu g \text{ of collagen}) \times 1000}{[\text{mean tubes } 5-6 \times \text{time (min)} \times \text{volume of sample } (\mu L)]}$$

If test samples have been diluted then the dilution factor needs to be included in this formula.

3.1.4. Collagenase Assay in 96-Well Plate Format

1. Thaw collagen at 1 mg/mL in 0.2 *M* acetic acid and dialyze against 50 m*M* Tris/HCl buffer (pH 7.6) containing 0.2 *M* NaCl and 0.03% *(v/v)* toluene with one change.
2. In a 96-well V-bottomed plate set up control wells A1–2 with 50 µL of buffer A; wells A3–4 with 5 µL of trypsin (100 µg/mL) in 1 m*M* HCl + 45 µL of buffer A; wells A5–6 with 50 µL of bacterial collagenase at 100 µg/mL in buffer A. If samples contain appreciable amounts of salt then appropriate blanks should be used (*see* **Note 3**).
3. Add test samples in duplicate or triplicate to well A7 onwards and make up volume to 50 µL with Buffer A. Add 50 µL of Buffer B to all wells.
4. Add 50 µL of [^{3}H]-labeled collagen (1 mg/mL) to all wells. Cover with a plate sealer (Valeant Pharmaceuticals, Basingstoke, UK) and incubate in water bath at 37°C for 1–20 h. At the end of the assay period centrifuge at 1300*g* for 30 min (*see* **Note 4**) in a Mistral 3000I centrifuge using a four place swing out rotor (43124-129; Sanyo Gallenkamp PLC, Loughborough, UK). Remove 50 µL of each supernatant and combine with 200 µL of Supermix scintillation fluid in a flexible 96-well sample plate placed in a no-crosstalk cassette and count for [^{3}H] in a 1450 Microbeta Trilux liquid scintillation counter (Perkin Elmer LAS (UK) Ltd., Cambridge, UK) or similar instrument.
5. Subtract mean blank values (wells A1–2) from all other test results. If trypsin digests substantial amounts of collagen (wells A3–4) it means the collagen is denatured and assay results should be discarded (*see* **Note 5**). The mean total values obtained for wells A5–6 represent the total lysis figure and correspond to the total counts released from 50 µg of collagen.
6. Results are expressed as units/mL where one unit of activity represents the amount of enzyme that degrades 1 µg of collagen/min at 37°C. Thus to obtain results in units/mL use the following formula:

$$\text{units/mL} = [\text{test mean} - \text{mean wells A1} - 2] \times \frac{50 \,(\mu g \text{ of collagen}) \times 1000}{[\text{mean tubes A5} - 6 \times \text{time (min)} \times \text{volume of sample } (\mu L)]}$$

If test samples have been diluted then the dilution factor needs to be included in this formula (*see* **Note 6**). Other assays have been described for collagenases and these are reviewed in **refs.** *11–14* (*see* **Notes 7** and **8**).

3.1.5. TIMP Assay

The collagenase assay can easily be adapted to allow for the measurement of samples containing TIMPs. Extra control tubes are set up (tubes 7–8) containing a known amount (approx 0.06 units) of active interstitial collagenase, MMP-1, (bacterial collagenase is not suitable as it is not inhibited by TIMPs) which is known to digest approx 70 to 80% of the collagen over the assay period. This amount of collagenase is added to all subsequent tubes (tube 9 onwards) along with the test samples. TIMP activity is then measured as the reduction of released collagen fragments in test samples compared to the active enzyme control. The formula for expressing the results in units/mL becomes:

$$\text{units/mL} = \frac{\left[\text{active collagen mean (tubes 7 – 8)} - \text{text mean}\right] \times 100 \, (\mu g \text{ of collagen}) \times 1000}{\left[\text{mean tubes 5 – 6} \times \text{time (min)} \times \text{volume of sample } (\mu L)\right]}$$

3.1.6. Activation of proMMPs With APMA or Trypsin

Many samples of conditioned culture medium contain proMMPs that require activation and this can be accomplished with either APMA or trypsin (*see* **Note 9**).

3.1.6.1. APMA ACTIVATION.

Replace buffer B in the collagenase assay with a mixture of buffer B (4 parts) to 10 m*M* APMA (1 part). The inclusion of APMA throughout the assay period is sufficient to activate proMMPs. If short assays are required (less than 3 h) then trypsin activation (*see* below) should be used or the sample should be preincubated with APMA at 37°C for 1 h before the substrate is added. APMA can be made up at 10 m*M* by dissolving 35.2 mg of APMA in 200 mL of dimethyl sulfoxide (DMSO) and diluting to 10 mL with 20 m*M* Tris/HCl buffer (pH 8.0–9.5).

3.1.6.2. TRYPSIN ACTIVATION

Add an equal volume of trypsin (20 µg/mL) to each sample in the assay tube, mix and incubate at room temperature for 15 min. Then add the same volume of soyabean trypsin inhibitor (100 µg/mL), mix and make up to 100 mL with buffer A. Proceed with assay by adding buffer B etc.

3.2. Bovine Nasal Cartilage Assay

3.2.1. Cartilage Preparation and Experiment Setup

1. Bovine nasal septum cartilage is obtained from the abattoir and used immediately or can be held overnight at 4°C. The connective tissue sheath is removed

from the cartilage and the cartilage is cut into 2-mm slices.
2. Discs 2 mm in diameter are punched out of the cartilage slices using a sterilized leather punch. Care is taken to avoid cartilage with obvious vascular channels. Cartilage discs are washed twice with DPBS/antibiotics and once in explant culture medium. Three discs per well of a 24-well plate are incubated in 1 mL of serum free control medium at 37°C in 5% carbon dioxide/humidified air to allow the explants to equilibrate.
3. Medium is removed from each well and replenished with 600 µL fresh culture medium containing appropriate cytokines and test reagents (4 wells per condition). This is considered day 0.
4. Plates are incubated at 37°C for 7 d, supernatants harvested and the cartilage discs in each well replenished with identical test reagents to day 0. The experiment was continued for a further 7 d and at day 14, supernatants were removed and stored at –20°C until assayed.

3.2.2. Digestion of Cartilage

1. To determine total glycosaminoglycan (GAG) and hydroxyproline (OHPro) content of the cartilage fragments remaining, the 3 fragments in each well are transferred to a LP3 tube (Life Sciences International, Basingstoke, UK) and mixed with 350 µL phosphate buffer, 100 µL papain, 50 µL cysteine-HCl, and 50 µL EDTA solutions.
2. The tubes are capped and the caps pierced with a 21-gage needle and incubated at 65°C overnight until digestion is complete. Phosphate buffer (450 µL, 0.1 M) is then added to each tube. Sodium azide (0.02%) was then added to these digests and the day 7 and 14 conditioned media samples and they were stored at –20°C until assayed for GAG and OHPro. Conditioned media samples can also be assayed for collagenase activity.

3.2.3. Hydroxyproline Assay (see **Note 10**)

1. Add 200 µL sample (i.e., media or cartilage digest) + 200 µL conc. HCL (CARE) into 2 mL Sarstedt screw cap tubes (Sarstedt, Leicester), cap tightly and put on hotblock (105°C) overnight. Next morning remove tubes from hot block, allow to cool, and spin on pulse to 8000 rpm in bench top microfuge. Remove caps from tubes and remove HCl using speedvac (Savant, Life Sciences International), 2 to 2.5 h or until dry.
2. Remove tubes and resuspend in 200 µL distilled water, cap, mix, and store at room temperature. The standard curve (0–30 µg/mL) is made from the stock solution of hydroxyproline (1 mg/mL) as shown below using deionized water as diluent. These standards can be stored for up to 1 wk at 4°C.

Standard (µg/mL)	H_2O (µL)	Hydroxyproline (µL)
0	1000	0
5	995	5
10	990	10
15	985	15

20	980	20
25	975	25
30	970	30

3. Initially assay day 7 media samples neat only, and day 14 media samples and cartilage digests at neat and 1 in 10 dilution (diluted in distilled water). The standards are added in columns 1 and 2 of a 96-well plate as shown below adding 40 µL of standard or sample per well.

1	2	3	4	5	6	7	8	9	10	11	12
A 0 µg/mL		Samples →		Samples →		Samples →		Samples →		Samples →	
B 5 µg/mL		Samples →									
C 10 µg/mL		Samples →									
D 15 µg/mL		Samples →									
E 20 µg/mL		Samples →									
F 25 µg/mL		Samples →									
G 30 µg/mL		Samples →									
H 0 µg/mL		Samples →									

4. Starting at time zero and using a 12-channel pipet, apply 25 µL of freshly prepared chloramine T-reagent to each row working down the plate in a defined sequence at 10 s intervals. Wait until 4 min have elapsed from time zero and then apply 150 µL DAB reagent per well in the same sequence at 10 s intervals, working down the plate. Seal plate with plastic plate sealer (Valeant Pharmaceuticals, Basingstoke, UK) and incubate in an oven for 35 min at 65°C. Leave to cool for a few minutes and read at 560 nm.
5. These results are in µg/mL: to correct to the original volume of conditioned media harvested, 600 µL, multiply all day 7 and 14 results by 0.6. Do not multiply cartilage digest results by 0.6. The concentration of OHPro in the samples were calculated with reference to the standard curve and these data were used to calculate the % release of OHPro from cartilage on day 7 and 14, respectively.

3.2.4. DMB assay for proteoglycan content (see **Note 11**)

1. Make standards up as shown below using a 1 mg/mL chondroitin sulphate stock solution:

Standard (µg/mL)	Phosphate buffer (µL)	Chondroitin sulfate (µL)
0	1000	0
5	995	5
10	990	10
15	985	15
20	980	20
25	975	25
30	970	30
35	965	35
40	960	40

2. Samples are diluted in phosphate buffer and 40 µL of standard or sample is applied to each well as shown below:

1	2	3	4	5	6	7	8	9	10	11	12
A Std 0		Std 5		Std 10		Std 15		Std 20		Std 25	
B Std 30		Std 35		Std 40		Samples →					
C Samples →											
D Samples →											
E Samples →											
F Samples →											
G Samples →											
H Samples →											

3. 250 μL of the DMB solution is added to all wells and the absorbance read immediately at 530 nm. As for the hydroxyproline assay, all results for day 7 media samples and day 14 media samples should be multiplied by 0.6; cartilage digest results do not need correcting. The concentration of GAG in the samples were calculated with reference to the standard curve and these data were used to calculate the % release of GAG from cartilage on day 7 and 14, respectively.

3.3. Immunolocalisation of Collagenases in Cartilage

3.3.1. Preservation of Tissue Sample

Cartilage and synovial tissue samples should be fixed to preserve the cellular structure of the specimens. When harvesting and trimming joint specimens prior to fixation, avoid air-drying specimens (*see* **Note 12**). Specimens should be no thicker than 4 mm for good fixation. Avoid over-fixation which may make specimens difficult to cut and resistant to staining.

1. Specimens should be fixed in 4% paraformaldehyde using a 10:1 ratio of fixative to tissue. The fixative diffuses 1 to 4 mm every h. For a 2 mm^3 piece of cartilage, samples were fixed for 4 h.
2. Rinse specimen in PBS to remove the fixative.
3. Immerse specimen in PBS containing 30% sucrose and leave at 4°C for 48 h to cryoprotect the tissue.

3.3.2. Mounting Specimen and Freezing

The fixed tissue sample must next be mounted in OCT for ease of handling and cryosectioning.

1. Add a small amount of OCT to cover the bottom of a plastic mould big enough to hold the specimen and that can be snap frozen.
2. Place the specimen in the mould.
3. Add enough OCT around the sides of the mould to encase the specimen in OCT.
4. Put some dry ice into an ice tray and add enough methanol to cover the dry ice. Then place the mould in the mixture to freeze for a couple of minutes.
5. Store frozen specimen at −80°C until ready for cryosectioning.

3.3.3. APES-Coated Slides

1. Immerse slide in 1% DECON in distilled water for 30 min.
2. Wash in running tap water for 30 min.
3. Wash in distilled water twice for 5 min.
4. Wash in 95% industrial methylated spirits twice.
5. Dry in a hot air incubator for 10 min.
6. Coat slides in a freshly prepared 2% solution of APES in dry acetone for 5 min.
7. Wash twice in distilled water.
8. Air dry at 42°C overnight.
9. Store slides at room temperature, protected from dust.

3.3.4. Cryosectioning

The inclusion of calcified tissue or bone-tissue segments in specimens intended for cryosectioning should be avoided because they make it very difficult to section and may cause serious tissue disruption. Samples with calcifications or adherent bone material require a decalcification step (*see* **Note 13**). Sections must be placed on precoated slides so they will adhere during processing.

1. A cryostat (Leica) is needed for cryosectioning. Cryosections can be as thin as 5 µm. If you have no experience in cryosectioning you will have to send them to a lab with the appropriate facilities and expertise.
2. Place on an APES-coated slide (*see* **Note 14**) and air dry at room temperature for 2 h.
3. Store slides at −20°C.

3.3.5. Immunohistochemistry Staining of Frozen Sections of Cartilage

For immunolocalisation of antigen, it is recommended to use frozen sections because they are easier to work with than paraffin embedded sections (*see* **Note 15**). Immunolocalization on frozen sections can be performed with an antibody of low titre and result in less background staining problems. The following method for frozen sections is performed at room temperature in small tanks with the specified liquids, unless otherwise indicated. Antibody dilutions must be optimized individually (try 1:100, 1:500, and 1:1000). This method works for MMP and TIMP antibodies. The antibodies used for MMP-1 immunolocalisation have been described below.

1. Soak slide in 100% acetone for 10 min. Then air dry.
2. Rehydrate slide in 70% acetone for a couple of minutes and then transfer to TBS (pH 7.6) for a couple of minutes.
3. Block endogenous peroxidase activity by soaking slides in a solution made up of 200 mL of 0.02 M sodium azide in PBS plus 2 mL of hydrogen peroxide and rock for 10 min.

Collagenase Activity in Cartilage

4. Wash slide briefly in TBS.
5. Add 100 µL of blocking serum (150 µL of normal (rabbit) serum added to 10 mL TBS) to the slide covering the specimen on the slide. Place slide at 4°C for 1 h to incubate in a humidity chamber (*see* **Note 16**). Then wash twice in TBS for 5 min on a rocker.
6. Add 50 µL of primary antibody (1:500 sheep anti-MMP-1) diluted in TBS with normal serum (150 µL normal serum added to 10 mL TBS) to the slide as before. Place slide at 4°C to incubate overnight in a humidity chamber. Then wash twice in TBS for 5 min on a rocker.
7. Add 50 µL of secondary antibody (1:200 biotinylated rabbit anti-sheep) diluted in TBS with normal serum (150 µL normal serum added to 10 mL TBS) to the slide as before. Incubate for 1 h in a humidity chamber. Then wash twice in TBS for 5 min on a rocker.
8. Add 50 µL of ABComplex/HRP reagent to the slide as before (*see* **Note 17**). Incubate for 30 min in a humidity chamber. Then wash twice in TBS for 5 min on a rocker.
9. Add 50 µL of 0.67 mg/mL DAB in TBS to the slide as before. Incubate for up to 2 min or until you see the formation of a brown precipitate at the desired intensity when viewed by a microscope (*see* **Note 18**). Then wash twice in running tap water.
10. Counter stain with Weigert's iron-haematoxylin for 10 s to stain the nucleus.
11. Place slide in water with ammonia (add a few drops of ammonium hydroxide to tap water and mix well) until the stain turns blue.
12. Dehydrate by transferring the slide to a series of ethanol gradients (75, 95, and finally 100% ethanol).
13. Place slide in xylene, a lipid solvent, to clear or cause the tissue to become transparent.
14. Mount slide in DPX by adding a couple drops of DPX to the slide and placing a cover slip on top. Slides can be stored at room temperature.

4. Notes

1. Collagen preparations cannot be hurried and extra dialysis steps may be required if a flocculent white precipitate is not seen when expected (**steps 6**, **8**, and **9**). Because large diameter dialysis tubing is used it is essential to ensure that each dialysis step, especially the first, is fully equilibrated before proceeding with subsequent steps. The temperature *MUST* be maintained at 4°C throughout the procedure so large volumes of pre-cooled buffers are required throughout the preparation.
2. Calcium is included in the buffer when labelling collagen at high pH to prevent precipitation.
3. Blank values in the collagenase assay are affected by high salt, high serum, high calcium, other chaotrophic ions and some metal ions. High blank values can also be caused by some collagen preparations not forming good fibrils. There is often variation between collagen preparations in their ability to form fibrils and varia-

tion of the levels of calcium ions can often control these differences. MMPs require calcium for thermal stability so it must be included in the assay.
4. Collagen can be difficult to pellet in polypropylene tubes because the fibrils tend to adhere to the tube at the surface of the liquid. Lower centrifugation speeds are used in the 96-well plates because this problem does not appear to occur in polystyrene plates.
5. Trypsin blanks are higher in acetylated collagen because some labeled lysine groups are located in the telopeptide region that is susceptible to trypsin. Excessive labeling with acetic anhydride increases trypsin blanks and can retard fibril formation *(7)*. It has been reported that trypsin blanks should be no higher than 5% of the total counts above the blank. This fictional figure is only achieved if trypsin is stored for long periods at neutral pH when it degrades itself. Trypsin should be stored frozen in small aliquots in 1 mM HCl and used immediately upon thawing and any surplus discarded.
6. The linear portion of the collagenase assay lies between 10 and 80% lysis and results that fall outside this range should be repeated at higher or lower dilutions.
7. Other enzymes can cleave collagen and it is theoretically possible that esterases could remove the [^{3}H] from the labelled collagen. Confirmation of the 3/4 and 1/4 products produced by collagenase can be confirmed by incubating enzyme with collagen at 23°C in the presence of 1 M glucose (to prevent fibril formation) followed by SDS-PAGE to demonstrate the 3/4 and 1/4 cleavages *(2)*.
8. If blank values and trypsin values are high then the temperature can be reduced to say 35°C. However this will result in a substantial loss in sensitivity of the assay.
9. Cell culture medium with serum contains α_2-macroglobulin which inhibits MMPs. Activation of the proMMPs in the presence of α_2-macroglobulin leads to the formation of an enzyme:inhibitor complex such that activity cannot be detected. To avoid this problem serum should be treated by lowering the pH to pH 3.0 for 90 min and returning the pH to neutral by the addition of NaOH prior to adding to cells in culture. This destroys α_2-macroglobulin activity.
10. Hydroxyproline is used to measure release of collagen from cartilage. Free hydroxyproline is released from proteins and peptides by acid hydrolysis. The acid is then neutralized. The hydroxyproline is oxidized to a pyrrole with chloramine T. This intermediate then gives a reddish color with 4-dimethylaminobenzaldehyde (DAB) *(15)*.
11. The DMB assay measures the amount of glycosaminoglycan (GAG) in a sample. DMB is a strongly metachromatic dye and it forms a complex with sulfated GAG which absorbs at 530 nm. Chondroitin sulphate is used as the standard: stock solution stored at –20°C, at 1 mg/mL *(16)*.
12. It is necessary to avoid prolonged air exposure of joint specimens because it has been shown to result in glycosaminoglycan (GAG) loss. Speer et al have reported that when articular cartilage was allowed to dry for 1 h during a surgical procedure, GAG depletion was seen by loss of surface staining with toluidine blue *(17)*.

13. Samples that require decalcification can be immersed in a chelating agent, such as 0.1 M EDTA, after fixation. Chelating agents do not affect the structure of antigens and therefore do not interfere with antigen retrieval *(18)*.
14. If you have trouble with tissue adherence to the slide during processing, you may need to purchase Superfrost Plus slides (Fisher) which have superior adherence properties for immunohistochemistry.
15. Paraffin embedding may cause the tissue to undergo various chemical alterations that may affect antigen preservation, making it difficult to retrieve certain antigens. If paraffin sections are used, the slides must be dewaxed twice in xylene for 5 and 10 min, rehydrated in a series of industrial methylated spirits gradients (twice in 99, 95, 70, 50%) and then in distilled water. If no antigen is detected then the slide can be microwaved for antigen retrieval in 0.01 M citrate *(19)* for 3.5 to 4 min on full power until boiling, 10 min at 450 W, and then allowed to cool for 20 min. Other protocols can be used with trypsin or hyaluronidase digestion to reveal antigen epitopes (*see* **Note 18**). The protocol can be resumed at **step 4**.
16. A humidity chamber can be made by placing wet tissue on the bottom of a square Petri dish. The slide is placed on top of the tissue and the Petri dish cover is placed on top. The Petri dish is then placed into a plastic bag.
17. Avidin-Biotin Complex/Horse radish peroxidase reagent must be made at least 30 min before use depending on the manufacturer. This involves combining reagent A (avidin) and B (biotinylated enzyme) in buffer to form complexes.
18. If the desired antigen is not detected, then an enzyme digestion step *(20)* may be included before **step 4**. Treat the slide with 1 mg/mL hyaluronidase in PBS for 30 min at 37°C in a humidity chamber. Next wash in TBS three times for a couple of minutes, and then proceed with the method.

References

1. Gisslow, M. T. and McBride, B. C. (1975) A rapid sensitive collagenase assay. *Anal. Biochem.* **68,** 70–78.
2. Cawston, T. E. and Barrett, A. J. (1979) A rapid and reproducible assay for collagenase using [1-^{14}C]acetylated collagen. *Anal. Biochem.* **99,** 340–345.
3. Terato, K., Nagai, Y., Kawaninski, K., and Shinro, Y. (1976) A rapid assay method of collagenase activity using 14C-labelled soluble collagen as substrate. *Biochim. Biophys. Acta* **445,** 753–762.
4. Johnson-Wint, B. and Gross, J. (1980) A quantitative collagen film collagenase assay for large numbers of samples. *Anal. Biochem.* **104,** 175–181.
5. Lefebvre, Vaes, G. (1989) Enzymatic evaluation of procollagenase and collagenase inhibitors in crude biological media. *Biochim. Biophys. Acta* **992,** 355–361.
6. Murphy, G., Koklitis, P., and Carne, A. F. (1989) Dissociation of TIMP from enzyme complexes yields fully active inhibitor. *Biochem. J.* **261,** 1031–1034.
7. Catterall, J. B. and Cawston, T. E. (2003) Assays of media, serum and synovial fluid matrix metalloproteinases (MMPs) and MMP inhibitors: bioassays and

immunoassays applicable to cell culture medium, serum, and synovial fluid, in *Methods in Molecular Biology*; vol. 225 *Inflammation protocols* (Winyard, P. G. and Willoughby, D. A., eds.), Humana, Totowa, NJ, pp. 353–364.
8. Cawston, T. E., Ellis, A. J., Humm, G., Ward, D., and Curry, V. (1995) IL-1 and OSM in combination promote the release of collagen fragments from bovine nasal cartilage in culture. *Biochim. Biophys. Res. Comm.* **215**, 377–385.
9. Clark, I. M., Powell, L. K., Wright, J. K., Cawston, T. E., and Hazelman, B. L. (1992) Monoclonal antibodies against human fibroblast collagenase and the design of an enzyme-linked immunosorbent assay to measure total collagenase. *Matrix* **12**, 475–480.
10. Clark, I. M., Powell, L. K., Wright, J. K., and Cawston, T. E. (1991) Polyclonal and monoclonal antibodies against human tissue inhibitor of metalloproteinases (TIMP) and the design of an enzyme-linked immunosorbent assay to measure TIMP. *Matrix* **11**, 76–85.
11. Harris, E. D. and Vater, C. A. (1982) Vertebrate collagenases. *Meth. Enzymol.* **82**, 423–458.
12. Cawston, T. E. and Murphy, G. (1981) Mammalian collagenases. *Meth. Enzymol.* **80**, 711–722.
13. Dioszegi, M., Cannon, P., and Van Wart, H. E. (1995) Vertebrate collagenases. *Meth. Enzymol.* **248**, 413–431.
14. Fields, G. B. (2001) Using flourogenic peptide substrates to assay MMPs, in *Methods in Molecular Biology; vol. 151 Matrix metalloproteinase protocols,* Clark, I. M., ed., Humana Press, Totowa, NJ, pp. 389–397.
15. Bergman, I. and Loxley, R. (1963) Two improved and simplified methods for the spectrophotometric determination of hydroxyproline. *Anal. Chem.* **35**, 1961–1965.
16. Farndale, R. W., Buttle, D. J., and Barrett, A. J. (1986) Improved quantitation and discrimination of sulphated glycosaminoglycans by use of dimethylmethylene blue. *Biochim. Biophys. Acta* **883**, 173–177.
17. Speer, K. P., Callaghan, J. J., Seaber, A. U., et al (1990) The effects of exposure of articular cartilage to air. *J. Bone Joint Surg. Am.* **72**, 1442–1450.
18. Jonsson, R., Tarkowski, A., and Klareskog, L. (1986) A demineralization procedure for immunohistopathological use. EDTA treatment preserves lymphoid cell surface antigens. *J. Immunol. Methods* **88**, 109–114.
19. Brown, R. W. and Chirala, R. (1995) Utility of microwave-citrate antigen retrieval in diagnostic immunohistochemistry. *Mod. Pathol.* **8**, 515–520.
20. Roberts, S., Caterson, B., Evans, E. H., and Eisenstein, S. M. (1994) Proteoglycan components of the intervertebral disc and cartilage endplate: an immunolocalization study of animal and human tissues. *Histochem. J.* **26**, 402–411.

14

Assessment of Gelatinase Expression and Activity in Articular Cartilage

Rosalind M. Hembry, Susan J. Atkinson, and Gillian Murphy

Summary

Two methods for the assessment of the expression of gelatinases A and B, MMP-2 and MMP-9, in articular cartilage are described. Immunohistochemical analysis of tissue sections provides information about the precise localization of the enzymes within the tissue, pinpointing the cells that synthesize the proteinases, and zymography of cell/tissue conditioned culture media allows a semi-quantitative assessment of the gelatinases and their activation status.

Key Words: Gelatinase A; gelatinase B; MMP-2; MMP-9; cartilage; chondrocyte; immunolocalization; zymography.

1. Introduction

The assessment of metalloproteinase expression and function in articular cartilage presents a number of challenges. Cartilage is a dense, highly charged, and insoluble tissue with relatively few cells, the chondrocytes, embedded therein. These cells are responsible for the normal maintenance of the extracellular matrix (ECM) and can also be implicated in the excessive matrix degradation associated with the arthritic diseases. The production and secretion of matrix metalloproteinases (MMPs) represents a major contributor to their degradative capacity and there has been extensive investment in technologies to monitor these enzymes. The two "gelatinases," gelatinase A MMP-2 and gelatinase B MMP-9 (http://merops.sanger.ac.uk) have been identified in cartilage tissue. Originally they were implicated in the turnover of denatured collagen, but it has recently become clear that, as for other MMPs, they are involved in the mobilization of growth factors and the modulation of chemokines (1). It is anticipated that further work will reveal the true extent of their roles in both cell physiology and pathology.

The measurement of the gelatinases associated with cartilage has largely depended on the fact that they are secreted from chondrocytes into the extracellular milieu. Within the tissue they can become sequestered to ECM components, especially collagen, but they also may be endocytosed by the cell as inhibited or active forms. As they are secreted in a latent proenzyme form, the extent of their activation is critical to the interpretation of their role in diseased cartilage. They are also subject to inhibition by the natural inhibitors, the tissue inhibitors of metalloproteinases, TIMPs, which are secreted by the same cells.

The direct assessment of gelatinase A and B levels in intact cartilage tissue has largely been confined to *in situ* hybridization *(2)*, immunohistochemistry *(3)* (*see* **Subheading 1.1.**) or *in situ* zymography *(4)*, but the extraction and assay of mRNA is also effective. Studies by Bau et al. *(5)* and Kevorkian et al. *(6)* have described the extraction of mRNA from normal and osteoarthritic cartilage and the analysis of metalloproteinases and inhibitors by quantitative reverse transcriptase polymerase chain reaction (RT-PCR) methods. Kevorkian et al. *(6)* published data for MMP-2 and MMP-9 which showed that they are upregulated in the diseased tissue relative to normal.

The majority of MMP studies have preferred to put small explants of tissue or dissociated chondrocytes into short term culture in order to analyse the enzymes emerging into the medium as a consequence of *de novo* synthesis and secretion. Measurements of the gelatinases may be effected by gelatin zymography (*see* **Subheading 1.2.**), by enzyme assay *(7,8)*, by immunoblotting *(3)* or by commercial enzyme linked immunosorbent assay (ELISA) techniques *(9,10)*. Only the latter is truly quantitative, but an element of semiquantitation is possible using the other methods. Zymography is an extraordinarily sensitive method of detecting the gelatinases and can distinguish between the pro and active forms of both MMP-2 and MMP-9, but is a poor measure of the extent of MMP-TIMP complexes. It is recommended that the activities detected are confirmed as MMP-2 and MMP-9 by immunoblotting, but it is often necessary to extensively concentrate the conditioned medium to obtain a signal. This is easily achieved for the gelatinases as they bind to gelatin-Sepharose *(11)*. The level of combined gelatinolytic activity may be assayed using ^{14}C-labeled gelatin, with an element of specificity, since the other MMPs have very weak gelatinolytic capacities, but there are no totally specific activity assays for these enzymes *(7)*. The available ELISAs are sensitive and specific but need to be checked for the enzyme forms that will be detected.

1.1. Immunodetection of Gelatinases in Cartilage

A number of immunohistochemical methods have been used for the detection of gelatinases in cartilage. Because many MMPs and TIMPs are present in tissues in tiny amounts and secreted from the cell immediately after synthesis the more sensitive indirect method, using primary antibody followed by labeled secondary antibody, is the method of choice rather than the direct method in which the primary antibody is conjugated to a reporter molecule. Culture of the tissue with monensin may be required to allow synthesis but stop translocation of synthesized enzyme, thus increasing the amount of antigen within the cells available for detection.

Optimal preservation of tissue structure and antigen are crucial for accurate immunolocalization. Frozen sections retain the enzyme antigen in the native state allowing maximum antibody binding but tissue architecture may be less than optimal. Paraffin embedding preserves tissue structure but may denature or mask the antigen such that antibody binding is reduced or cannot take place, or nonspecific binding may be increased. Some fixatives, such as glutaraldehyde, crosslink proteins so extensively that the immunoreaction cannot take place. Thus, the choice between whether frozen sections or paraffin sections are used, and which fixative, will be a compromise for each antigen and antibody. The detection method chosen will depend on microscope availability and whether three -dimensional (3D) spatial resolution of signal is required. The avidin-biotin-peroxidase complex (ABC) and immunogold-silver (IGSS) methods for immunohistochemical staining of MMPs and TIMPs using monoclonal antibodies screened for reactivity on paraffin sections are comprehensively described in a previous volume *(12)*.

In this chapter the method for immunostaining frozen sections by indirect immunofluorescence using specific polyclonal antisera raised in sheep to MMPs and TIMPs is described in detail. Other commercially available antibodies may also be used with this method but the specificity of each antibody should first be rigorously checked. Tissue samples obtained at surgery are halved. One half is frozen directly to identify enzyme that has been synthesized and/or bound to matrix before excision from the joint. The other half is subjected to short term culture with monensin to inhibit the cellular secretion of the antigen under study before freezing, to identify cells actively engaged in synthesis *(13)*. The effect of monensin treatment on gelatinase synthesis and localization in rabbit chondrocyte monolayers has been illustrated previously *(14)*. Frozen sections are then cut, fixed, and incubated with primary antibody followed by secondary antibody conjugated with a fluorescent probe and viewed by fluorescence or confocal microscopy. Common problems that may occur during the procedure are discussed.

This method has been used extensively to successfully demonstrate gelatinase synthesis by cartilage from several species. During normal development of the rabbit growth plate intracellular gelatinase fluorescence was seen within chondrocytes of the resting zone and the proximal one half of the proliferative zone and was also visible in the articular and nonhypertrophic cells of

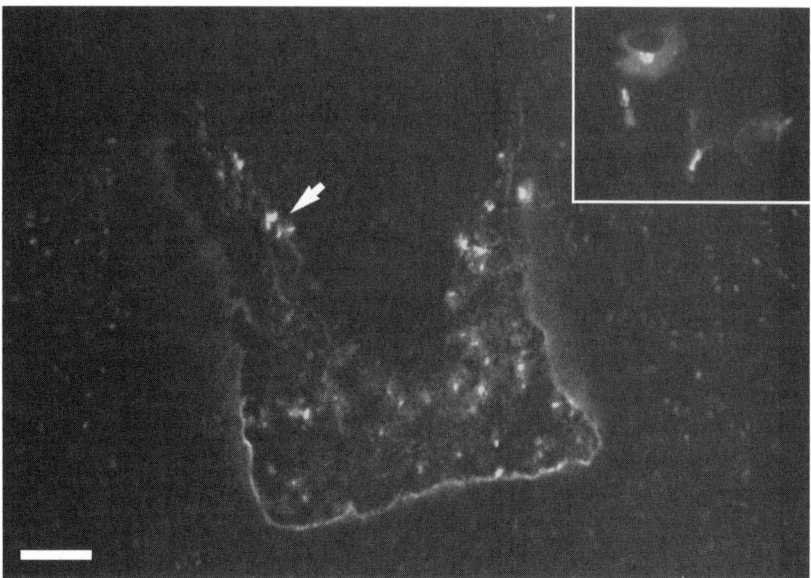

Fig. 1. Immunolocalization of MMP-9. Partial thickness defects were created in the medial femoral condylar cartilages of immature pigs. Defects were treated with chondroitinase AC, filled with transforming growth factor-β/matrix as described previously (*17*) and the joints closed. Defects and surrounding cartilage were excised 8 d later, cultured with monensin, frozen, sectioned and the sections stained by indirect immunofluorescence for MMP9. Bar = 250 μm.

There is strong staining for MMP-9 on the damaged cartilage matrix of the lateral and basal margins of the defect, extending throughout the zone of cell necrosis. MMP-9 positive cells within the defect and staining on defect contents are also visible. The insert is a higher magnification view of the cells marked with the arrow and shows a group of macrophage-like cells with juxtanuclear immunofluorescence, probably in the Golgi apparatus. (Reprinted with permission from **ref. *17*.**)

the epiphysis (*15*). MMP-2 was seen in human subarticular chondrocytes at gestational ages from 10 wk onward (*16*). MMP-9 was present in some chondrocytes in the upper third of epiphyseal cartilage from immature pigs as well as in terminal hypertrophic chondrocytes at the interface with the secondary ossification centre (*17*). The importance of MMP-9 during the process of endochondral ossification has been well documented (*4,18,19*).

Gelatinase involvement in joint pathology has also been studied using this immunostaining method. In a study of the early in vivo repair of partial thickness defects in pig articular cartilage, macrophages infiltrating the defects at 2 and 8 d after defect formation were shown to synthesize and deposit MMP-9 onto the damaged cartilage matrix, the zone of necrosis (*see* **Fig. 1**). Analysis of the zone of necrosis matrix at 8 d and 6 wk after defect formation showed loss of matrix metachromasia and the presence of the MMP-derived DIPEN$_{341}$ neo-epitope, indicating aggrecan degradation *(17)*. Active gelatinases have been shown to cleave cartilage aggrecan *(20)* and type XI collagen *(21)*, whereas experiments with either monocytes or macrophages cocultured with human articular chondrocytes demonstrated that articular chondrocytes provide factors that activate macrophage-derived proMMP-9 *(22)*.

1.2. Analysis of Gelatinases in Cell or Tissue Conditioned Culture Media by Gelatin Zymography

The presence of both MMP-2 and MMP-9 secreted into the media by short term chondrocyte monolayer cell cultures can be readily assessed by substrate zymography *(23)*. The technique utilizes nonreducing sodium dodecyl sulfate polyacrylamide gel electrophoresis (SDS-PAGE) in which the substrate, gelatin, is incorporated into the polymerized acrylamide gel. Both MMPs can be detected in latent as well as active forms because SDS causes conformational changes that expose the active site without cleavage of the propeptide. Removal of SDS with the use of the nonionic detergent, Triton X-100, allows degradation of the gelatin at the points in the gel where the active site is exposed. After a suitable incubation time (*see* **Note 7**) the remaining gelatin is stained with Coomassie Brilliant Blue. MMPs appear as white zones of lysis on a blue background (*see* **Fig. 2**). As cell culture media are also likely to contain non-MMP gelatin degrading enzymes (e.g., plasmin), gels may be incubated in the presence or absence of MMP inhibitors (ethylene diamine tetraacetic acid [EDTA], 1,10 phenanthroline). Any white bands revealed on gels incubated in the presence of the inhibitors are not of MMP origin and therefore can be discounted. Alternatively, gelatinases can be separated from the conditioned media, prior to electrophoresis, by binding to gelatin agarose and elution with 10% dimethyl sulfoxide (DMSO). It is advisable that cells are cultured, if possible, in the absence of serum for the experimental period, as both MMP-2 and TIMPs are present in foetal bovine serum (FBS). Suitable serum substitutes can be used (e.g., insulin/transferrin/selenium [ITS] supplement, bovine serum albumin [BSA], lactalbumin hydrolysate [LH]). However, if serum is absolutely necessary then the concentration can be lowered to 1 or 2%. Media from control cultures without cells should be run for comparison. Gelatin zymography is not strictly quantitative but a degree of comparison can be achieved by the titration of known amounts of purified MMPs. A standard curve of concentrations below the level of complete degradation of the incorporated gelatin should be run alongside the unknown samples *(24)*.

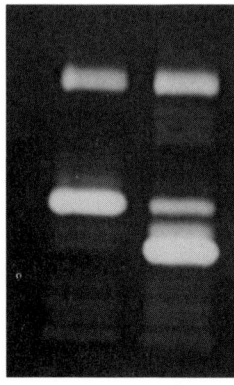

Pro MMP-9

Pro MMP-2
Int MMP-2
Act MMP-2

−   +   IL1β, oncostatin M, Con A

Fig. 2. Zymography of gelatinases. Human chondrosarcoma cells (SW1353) were incubated for 24 h in serum free medium containing IL1β (10 ng/mL) and oncostatin M (50 ng/mL). Concanavalin A (50 µg/mL) was added to the cultures for a further 24 h and then the culture supernatants were analyzed by zymography to assess gelatinase activity. Unstimulated cells produce both proMMP-9 and proMMP-2 which can be detected as single bands. Under the activation conditions described the cells process the proMMP-2 propeptide via an intermediate form (Int MMP-2) to the fully active form (Act MMP-2). The zymography technique shows the sequential fall in molecular weight as the propeptide is processed. Zymogram courtesy of Dr. S. Cowell.

2. Materials
2.1. Immunodetection of Gelatinases in Cartilage

1. Holding medium such as L15 air-buffered medium (Sigma), sterile.
2. 7% Gelatin prepared by heating 7.0 g gelatin and 0.9 g NaCl in 100 mL distilled H_2O until dissolved. After cooling add sodium azide to final concentration of 0.02%. Store at 4°C and warm to 37°C before use.
3. Stock solution of 10 mM monensin (Sigma) dissolved in 100% ethanol may be stored at 4°C for 2 to 3 mo.
4. Dulbecco's modified Eagle's medium (DMEM) with 10% foetal bovine serum (FBS).
5. Coated glass slides prepared by immersing slides in poly-L-lysine solution (Sigma; cat. no. P8920) (diluted 1:10 in deionized water and filtered) for 5 min, drain and dry at room temperature overnight. The poly-L-lysine solution may be stored at 4°C.
6. 4% Formaldehyde freshly prepared from paraformaldehyde dissolved in PBS (pH 7.4).
7. 0.1% Triton X-100 in PBS.
8. Primary antibody IgGs and control sera IgGs in PBS.

9. Secondary antibody conjugated with either fluorescein isothiocyanate (FITC) or Alexa Fluor 488. We now use either F(ab')$_2$ fragments of donkey anti-sheep IgG conjugated with FITC (Jackson Immunoresearch Laboratories Inc. ML; cat no. 713-096-147) or donkey anti-sheep IgG conjugated with Alexa Fluor 488 (Molecular Probes; cat. no. A-11015) as commercially available alternatives to the Pig anti-Sheep Fab'-FITC described in *(13)*. Optimal working concentrations need to be established for each batch.
10. Vectashield mountant (Vector Laboratories Inc., CA) or other mountant for fluorescence containing antifadants (e.g., Citifluor; University of Kent, Canterbury, UK).

2.2. Analysis of Gelatinases in Cell or Tissue Conditioned Culture Media by Gelatin Zymography

1. 40% Acrylamide/Bis solution 29:1 (3.3% C). *Toxic!* Wear protective gloves and goggles, polymerize before disposal, and store at 4°C.
2. Gelatin, type A from porcine skin, 2 or 4 mg/mL (*see* **Note 1**) in H$_2$O heated to 50°C for 20 min. Stable for 4 wk at 4°C.
3. Lower gel buffer (4X): 1.5 M Tris-HCl (pH 8.8) and 4% SDS. Stable at 4°C for 12 mo.
4. Upper gel buffer (4X): 500 mM Tris-HCl (pH 6.8) and 4% SDS. Stable at 4°C for 12 mo.
5. 10% Ammonium persulphate (APS), stable at 4°C for 1 wk.
6. *N,N,N',N'*-Tetramethylethylenediamine (TEMED), store at 4°C.
7. Laemmli sample buffer (5X), *(25)*; 0.625 M Tris-HCl (pH 6.8), 2% SDS, 0.2% bromophenol blue, and 10% glycerol. Stable at room temperature for 3 mo.
8. Laemmli reducing sample buffer (4X): Add β-mercaptoethanol to 5X buffer. Stable at room temperature for 2 wk.
9. Electrophoresis running buffer (5X): 125 mM Tris, 960 mM glycine, and 0.5% SDS. Stable at room temperature for 3 mo.
10. Coomassie Brilliant Blue G 250 stain (CBBG): 0.25% (w/v) in 50% (v/v) methanol, and 10% (v/v) acetic acid. Stable at room temperature.
11. Destain: 30% (v/v) methanol, 1% (v/v) acetic acid. Stable at room temperature.
12. Triton X-100 (2.5% [v/v] in H$_2$O). Stable at room temperature for 1 mo.
13. Assay buffer (TCAB): 100 mM Tris-HCl (pH 7.9), 30 mM CaCl$_2$, and 0.02% sodium azide. Stable at 4°C for 2 mo.

3. Methods

3.1. Immunodetection of Gelatinases in Cartilage

1. Excise cartilage from the joint under sterile conditions, documenting the position within the joint from which each sample is taken and keeping each piece separate. To preserve good structure tissues should be handled carefully and processed within 1 to 3 h of surgical removal. Divide into pieces approximately 5 mm^2 to give paired samples from each site; place in holding medium.

2. Transfer one piece from each pair to a labeled plastic tube, embed tissue in either 7% gelatin or OCT compound (see **Note 1**) and freeze in liquid nitrogen for 90 s. Cap tube and store at −80°C. This sample will show enzyme that has been synthesized and/or bound to matrix before excision from the joint. When it is important to preserve tissue orientation the tissue can be placed before embedding onto Millipore paper (Millipore, Cat.no. SMWP0190R cut to fit tube; this does not cause damage to cryostat knives).
3. Transfer the second piece from each pair to a 3-cm plastic tissue culture dish and culture in DMEM with 10% FBS with 5 µM monensin for 6 to 24 h at 37°C to allow synthesis but stop translocation of synthesized enzyme, thus increasing the amount of antigen within the cells available for detection to show cells actively engaged in antigen synthesis (see **Note 2**).
4. At the end of the culture period gently rinse tissues briefly with PBS at room temperature to remove serum proteins, freeze and store as in **step 2**.
5. Section tissue at 6 to 10 µm using a cryostat, take up sections onto poly-L-lysine coated glass slides and air dry briefly. Fix sections with 4% formaldehyde freshly prepared from paraformaldehyde in PBS (pH 7.4), 5 min at room temperature (see **Note 3**). Wash 3 times in PBS for 5 min each at room temperature.
6. Incubate sections with 0.1% Triton X-100 in PBS for 5 min at room temperature to permeabilize cells and allow penetration of IgG. Wash 3 times in PBS for 5 min each at room temperature.
7. Incubate with primary antiserum or normal serum IgG. Usually 50 µg/mL in PBS for 30 to 60 min works well but IgG concentration can be reduced to 5 µg/mL or less and incubate overnight at 4°C (see **Note 4**). Appropriate control slides must always be included with each set of samples (e.g., no primary antibody, nonimmune IgG preferably from the same animal as the antibody, antibody absorbed with antigen, and an irrelevant antibody raised in the same species). After incubation, wash 3 times in PBS for 5 min each at room temperature.
8. Incubate with secondary antiserum IgG conjugated with FITC diluted in PBS, 30 min at room temperature. Wash 3 times in PBS for 5 min each.
9. To aid tissue identification, cell nuclei may be counterstained with either methyl green (100 µg/mL, 2 min, red fluorescence), propidium iodide (1 µg/mL, 2 min, red fluorescence) or DAPI (1 µg/mL, 5 min, blue fluorescence) according to the fluorescence filter sets available.
10. Coverslip sections using Vectashield. Seal coverslip to slides, for example with nail varnish.
11. Observe by epifluorescence microscopy using a wide band FITC filter and ×40 or ×60 fluorescence objectives with a numerical aperture of 1.2 or 1.4 (see **Note 5**).

3.2. Analysis of Gelatinases in Cell or Tissue Conditioned Culture Media by Gelatin Zymography

1. Prepare SDS 7% poyacrylamide resolving gels with gelatin incorporated to a final concentration of 0.5 or 1.0 mg/mL (see **Note 7**). For one gel (0.75 mm

thickness), using the BioRad mini-protean 3 apparatus, mix: 1.05 mL acrylamide/bis, 1.5 mL gelatin (2 or 4 mg/mL), 1.5 mL 4X lower gel buffer, 1.95 mL H_2O, 36 µL ammonium persulphate (10%) and 7 µL TEMED. Pour, overlay with a few drops H_2O and leave at room temperature until polymerized (approx 20 min).
2. When set, drain off overlaying water and insert a 15-well comb.
3. Prepare SDS 5% polyacrylamide stacking gel. Mix 0.5 mL acrylamide/bis, 1.0 mL 4X upper gel buffer, 1.9 mL H_2O, 600 µL 1% ammonium persulphate, and 7 µL TEMED. Pour and leave to polymerize at room temperature (5–10 min).
4. Prepare samples: 20 µL cell supernatants, 5 µL nonreducing Laemmli sample buffer (5X). Prepare molecular weight markers in reducing sample buffer: 10 µL protein standards, 5 µL H_2O, 5 µL 4X reducing sample buffer, boil for 2 min.
5. Load samples: 12 µL per track.
6. Run gel in the cold room at constant current (20 mA) for approx 1 h until the tracking dye reaches the bottom (*see* **Note 6**).
7. Remove gel from glass plates and incubate in 2.5% Triton X-100 at room temperature (2 times for 15 min).
8. Rinse gel briefly in H_2O.
9. Incubate in assay buffer (TCAB) overnight at room temperature or 37°C (*see* **Note 7**).
10. Stain gel for 20 min at room temperature in CBBG. Destain at room temperature using several changes of destain solution until all the blue colour has been removed from the stacking gel.

4. Notes

1. Seven percent gelatin has the advantage over OCT compound (e.g., TissueTek®) in that when cut it forms inert "tissue" surrounding each cartilage section that is retained on the slide during staining; this improves adhesion of the cartilage to the slide, aids retention of antibody solutions over the section, and reduces edge effects. Once cut, blocks can be removed from the cryostat chucks, replaced in capped tubes, and again stored without drying out. Blocks may be stored in this way for at least 10 yr without reduction in MMP signal.
2. The length of time in culture with monensin required for positive immunolocalization depends upon the antigen, the cell type, activity of the tissue, and age of patient or animal. Cartilage from elderly humans usually requires 24 h for most MMPs, whereas 6 h is sufficient for synovial samples *(13)* and colorectal tumors can be stained without pretreatment *(26)*. Care should be taken with embryonic and perinatal tissues since most cell synthesis is affected and cell necrosis may occur; a maximum of 3 h is recommended.
3. When compared with 4% formaldehyde, fixatives such as acetone and methanol markedly reduce binding of our sheep antibodies to MMPs. Glutaraldehyde crosslinks proteins so extensively that the antigenic sites become obscured, signal is reduced and tissue autofluorescence is increased.
4. Sheep antisera are very "sticky" and may cause high background staining particularly where inflammatory infiltrates contain cells with Fc receptors; this is

not usually a problem with intact cartilage. This can easily be overcome by addition of 1% BSA and/or 1 to 5% serum of the species in which the secondary antiserum is raised to both antibody incubations. Many immunohistochemical methods include a 1 h blocking step with 1% serum prior to antibody incubations but we find that addition of serum proteins to both antibody incubations is as, or more, effective and saves time. Sections likely to contain polymorphonuclear leucocytes, such as those from inflamed joints, should be incubated with 4-chloro-1-naphthol (2.8 mM in methanol/PBS with 0.01% H_2O_2 for 10–20 min) prior to the addition of the primary antibody to block non-specific binding of fluorescein to eosinophilic granules.
5. Articular cartilage matrix from elderly humans has significant yellowish autofluorescence, probably resulting from the high degree of collagen crosslinking, at the wavelengths commonly used for fluorescence microscopy that can obscure genuine matrix staining. This can be overcome using a Zeiss 510 meta confocal microscope to identify the excitation and emission spectra of both specific fluorescence (fluorescent probe [e.g., FITC]) and autofluorescence (unstained cartilage matrix) separately, then digitally remove the autofluorescent component from the stained specimen fluorescence.
6. For good separation of the bands, particularly the pro, intermediate, and fully active forms of MMP-2 the tracking dye can be run off the bottom of the gel for 5 to 10 min.
7. The temperature at which gelatin zymograms are developed depends on the levels of enzymes present. For very low levels of MMP-2 or MMP-9 it may be necessary to incubate the gels at 37°C for an extended period (e.g., 24 or even 48 h). In this case it is advisable to increase the final concentration of gelatin incorporated from 0.5 to 1.0 mg/mL. Some nonspecific degradation of the gelatin may occur at 37°C and lower concentrations of gelatin result in a background stain that is too pale, thus making identification of the zones of lysis as a result of MMPs difficult. However, for unknown samples a starting point of 0.5 mg/mL gelatin and overnight incubation at room temperature is advised. The volumes of samples can be reduced or the samples diluted, if enzyme levels are too high, until clearly defined bands are achieved.

Acknowledgments

We thank the Medical Research Council UK and the Arthritis Research Campaign for financial support.

References

1. Murphy, G., Knäuper, V., Atkinson, S., et al. (2002) Matrix metalloproteinases in arthritic disease. *Arthritis Res.* **4,** Suppl 3, S39–S49.
2. Flannelly, J., Chambers, M. G., Dudhia, J., et al. (2002) Metalloproteinase and tissue inhibitor of metalloproteinase expression in the murine STR/ort model of osteoarthritis. *Osteoarthritis Cartilage* **10,** 722–733.

3. Tetlow, L. C., Adlam, D. J., and Woolley, D. E. (2001) Matrix metalloproteinase and proinflammatory cytokine production by chondrocytes of human osteoarthritic cartilage: associations with degenerative changes. *Arthritis Rheum.* **44,** 585–594.
4. Lee, E. R., Murphy, G., El-Alfy, M., et al. (1999) Active gelatinase B is identified by histozymography in the cartilage resorption sites of developing long bones. *Dev. Dyn.* **215,** 190–205.
5. Bau, B., Gebhard, P. M., Haag, J., Knorr, T., Bartnik, E., and Aigner, T. (2002) Relative messenger RNA expression profiling of collagenases and aggrecanases in human articular chondrocytes in vivo and in vitro. *Arthritis Rheum.* **46,** 2648–2657.
6. Kevorkian, L., Young, D. A., Darrah, C., et al. (2004) Expression profiling of metalloproteinases and their inhibitors in cartilage. *Arthritis Rheum.* **50,** 131–141.
7. Murphy, G. and Crabbe, T. (1995) Gelatinases A and B. *Methods Enzymol.* **248,** 470–484.
8. Clegg, P. D. and Carter, S. D. (1999) Matrix metalloproteinase-2 and -9 are activated in joint diseases. *Equine Vet. J.* **31,** 324–330.
9. Imai, K., Ohta, S., Matsumoto, T., Fujimoto, N., Sato, H., Seiki, M., and Okada, Y. (1997) Expression of membrane-type 1 matrix metalloproteinase and activation of progelatinase A in human osteoarthritic cartilage. *Am. J. Pathol.* **151,** 245–256.
10. Fujimoto, N. and Iwata, K. (2001) Use of EIA to measure MMPs and TIMPs, in *Methods in Molecular Biology* vol. 151, *Matrix Metalloproteinase Protocols*, Clark, I. M., ed., Humana Press, Totowa, NJ, pp. 347–358.
11. Overall, C. M., Wrana, J. L., and Sodek, J. (1989) Independent regulation of collagenase, 72-kDa progelatinase, and metalloendoproteinase inhibitor expression in human fibroblasts by transforming growth factor-beta. *J. Biol. Chem.* **264,** 1860–1869.
12. Okada, Y. (2001) Immunohistochemistry of MMPs and TIMPs, in *Methods in Molecular Biology* vol. 151, *Matrix Metalloproteinase Protocols,* Clark I. M., ed., Humana, Totowa, NJ, pp. 359–365.
13. Hembry, R. M., Murphy, G., and Reynolds, J. J. (1985) Immunolocalization of tissue inhibitor of metalloproteinases (TIMP) in human cells. Characterization and use of a specific antiserum. *J. Cell Sci.* **73,** 105–119.
14. Reynolds, J. J. and Hembry, R. M. (1992) Immunolocalization of metalloproteinases and TIMP in normal and pathological tissues. *Matrix.* **Suppl. 1,** 375–382.
15. Brown, C. C., Hembry, R. M., and Reynolds, J. J. (1989) Immunolocalization of metalloproteinases and their inhibitor in the rabbit growth plate. *J. Bone Joint Surg. Am.* **71,** 580–593.
16. Edwards, J. C., Wilkinson, L. S., Soothill, P., Hembry, R. M., Murphy, G., and Reynolds, J. J. (1996) Matrix metalloproteinases in the formation of human synovial joint cavities. *J. Anat.* **188,** 355–360.
17. Hembry, R. M., Dyce, J., Driesang, I., et al. (2001) Immunolocalization of matrix metalloproteinases in partial-thickness defects in pig articular cartilage. A preliminary report. *J. Bone Joint Surg.* **83-A,** 826–838.

18. Vu, T. H., Shipley, J. M., Bergers, G., et al. (1998) MMP-9/gelatinase B is a key regulator of growth plate angiogenesis and apoptosis of hypertrophic chondrocytes. *Cell* **93,** 411–422.
19. Davoli, M. A., Lamplugh, L., Beauchemin, A., et al. (2001) Enzymes active in the areas undergoing cartilage resorption during the development of the secondary ossification center in the tibiae of rats aged 0-21 days: II. Two proteinases, gelatinase B and collagenase-3, are implicated in the lysis of collagen fibrils. *Dev. Dyn.* **222,** 71–88.
20. Fosang, A. J., Neame, P. J., Last, K., Hardingham, T. E., Murphy, G., and Hamilton, J. A. (1992) The interglobular domain of cartilage aggrecan is cleaved by PUMP, gelatinases, and cathepsin B. *J. Biol. Chem.* **267,** 19,470–19,474.
21. Smith, G. N., Jr, Hasty, K. A., Yu, L. P., Jr, Lamberson K. S., Mickler E. A., and Brandt K. D. (1991) Cleavage of type XI collagen fibers by gelatinase and by extracts of osteoarthritic canine cartilage. *Matrix* **11,** 36–42.
22. Dreier, R., Wallace, S., Fuchs, S., Bruckner, P., and Grassel, S. (2001) Paracrine interactions of chondrocytes and macrophages in cartilage degradation: articular chondrocytes provide factors that activate macrophage-derived pro-gelatinase B (pro-MMP-9). *J. Cell Sci.* **114,** 3813–3822.
23. Heussen, C. and Dowdle, E. B. (1980) Electrophoretic analysis of plasminogen activators in polyacrylamide gels containing sodium dodecyl sulfate and copolymerized substrates. *Anal. Biochem.* **102,** 196–202.
24. Kleiner, D. E. and Stetler-Stevenson, W. G. (1994) Quantitative zymography: detection of picogram quantities of gelatinases. *Anal. Biochem.* **218,** 325–329.
25. Laemmli, U. K. (1970) Cleavage of structural proteins during the assembly of the head of bacteriophage T4. *Nature* **227,** 680–685.
26. Gallegos, N. C., Smales, C., Savage, F. J., Hembry, R. M., and Boulos, P. B. (1995) The distribution of matrix metalloproteinases and tissue inhibitor of metalloproteinases in colorectal cancer. *Surg. Oncol.* **4,** 111–119.

15

Analyses of MT1-MMP Activity in Cells

Richard D. Evans and Yoshifumi Itoh

Summary

Membrane-type 1 matrix metalloproteinase (MT1-MMP) is a type I transmembrane protein which exhibits biological activities on the cell surface. One of the characteristic functions of this enzyme is activation of proMMP-2 on the cell surface *(1)*. This process can be monitored using gelatin zymography to detect the pro- and active forms of MMP-2. Cellular MT1-MMP activity can also be directly detected by *in situ* degradation of layers of either gelatin or collagen. By combining inhibitors, western blotting, and immunohistochemistry, one can detect and conclude the presence of MT1-MMP activity in the cells.

Key Words: MT1-MMP; proMMP-2; activation; Zymography; TIMP-2.

1. Introduction

In general, detection of MMP activities in biological samples such as tissues or cells requires fractionation, concentration, and treatment of the samples to remove inhibitors, concentrate enzyme to a detectable level and to activate the precursor form of the enzymes. In the case of MT1-MMP, however, one can detect the activity without pretreatments of the sample as it is expressed on the cell surface in the active form. MT1-MMP expresses a broad range of extracellular matrix (ECM) degrading activity including fibrillar collagens (I, II, and III), laminin (type I and V), fibrin, fibronectin, vitronectin, and aggrecan *(2,3)*. It is also capable of shedding a range of cell surface proteins such as CD44, α-v integrin, transglutaminase, low-density lipoprotein receptor related protein, and syndecan-1 *(4–10)*. It also activates other MMPs, namely proMMP-2 and proMMP-13 *(1)*. MT1-MMP is localised at the cell surface and can degrade the ECM beneath the cell. MT1-MMP activity can be inhibited by tissue inhibitor of metalloproteinase-2 (TIMP-2), TIMP-3, and TIMP-4 as well as by the general MMP inhibitor GM6001, but not by TIMP-1. Using these characteristic properties, one can detect the activity of MT1-MMP in a relatively specific manner. Because some of these biological activities and properties are shared by other MT-MMPs, it is necessary to show the expression of MT1-MMP using specific antibodies either by immunofluorescence (**Fig. 1**) or by western blot (**Fig. 1**) in order to confirm the activities detected are derived from MT1-MMP.

1.1. Detection of ProMMP-2 Activation

MT1-MMP activates proMMP-2 at the cell surface. The activation steps involve formation of a trimolecular complex of MT1-MMP, TIMP-2, and proMMP-2 and homophilic complex formation of two MT1-MMP molecules through their hemopexin-like domains *(11)*. The relative molecular weight of proMMP-2 in sodium dodecyl sulfate polyacrylamide gel electrophoresis (SDS-PAGE) is 72-kDa in reducing conditions and 68-kDa in nonreducing conditions. Upon activation, the propeptide of MMP-2 is removed, and the molecular weight shifts to 65-kDa in reducing conditions or 62-kDa in nonreducing conditions. Thus, one can analyze the activation of proMMP-2 by looking at the shift of the molecular weight (**Fig. 1**). The most sensitive and easiest way to analyze this is a gelatin zymography as it can detect as little as 1 ng of MMP-2 and requires only SDS-PAGE apparatus (**Fig. 1**). Many isolated cells produce MMP-2 by themselves, thus one may just analyse their conditioned medium if it is the case. If the cells in question do not produce proMMP-2, one can use fibroblast-conditioned culture medium that contains proMMP-2 to test their ability to activate proMMP-2.

1.2. In Situ *Gelatin Zymography*

When cells express MT1-MMP, they degrade the ECM beneath them. Thus, by using fluorescence-labeled gelatin as a substrate, one can detect this degradation (**Fig. 2**). The area degraded will be detected as a dark area where fluorescence is lost against a bright background of undegraded gelatin. By combination with TIMP-1, TIMP-2, and GM6001, one can study MT1-MMP derived matrix degrading activity by the cells (**Fig. 2**). This experiment provides the information not only about the presence of activity, but also about the localization of MT1-MMP on the cell surface at ECM attachment sites.

1.3. In Situ *Collagen Degradation*

MT1-MMP is capable of directly cleaving type-I collagen in its native form, cleaving the molecule into 3/4 and 1/4 length fragments *(12,13)*. These fragments can then be further degraded by the enzyme, allowing cells to invade into collagen gels. This activity can be assayed in a similar manner to *in situ* gelatin zymography. In this technique the matrix is visualised by coomassie blue staining and the degradation generated by the cells are seen as light areas against a dark background (**Fig. 3**).

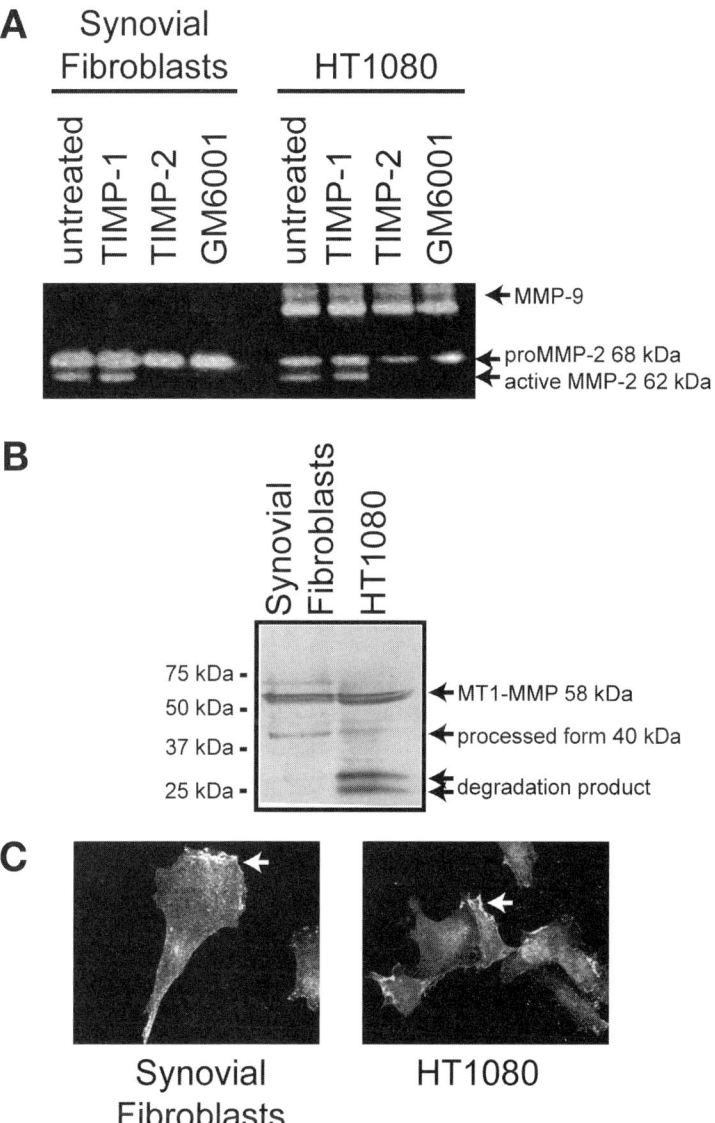

Fig. 1. Activation of pro-MMP-2 by MT1-MMP expressed in human rheumatoid synovial fibroblasts or HT1080 cells. (**A**) Addition of TIMP-1 to the culture medium has no effect on MMP-2 activation, TIMP-2, or GM6001 inhibit the conversion to the active form. (**B**) Both HT1080 and synovial fibroblasts express MT1-MMP as detected by an antibody to the hemopexin like domain (anti-MT1-MMP hemopexin domain antibody is commercially available from Oncogene Science Inc.). (**C**) Immunostaining with an anti-MT1-MMP antibody shows expression of the enyzme in both HT1080 and human rheumatoid synovial cells. MT1-MMP localises to membrane ruffles and the basal layer of the cell.

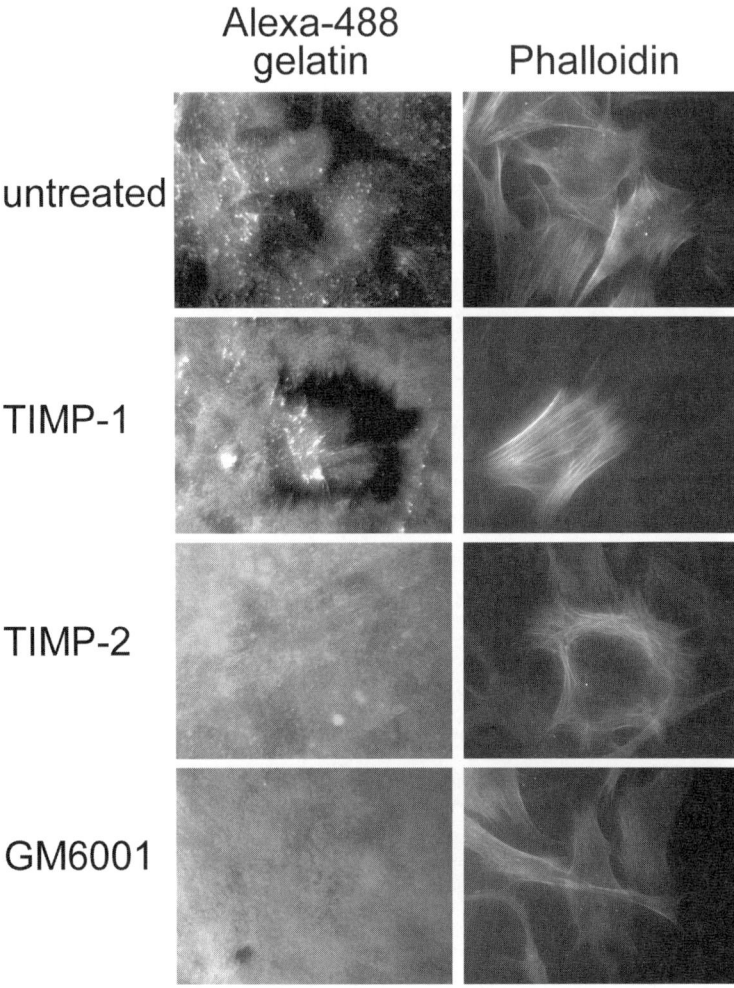

Fig. 2. Degradation of Alexa-488 labeled gelatin by human rheumatoid synovial cells. TIMP-1 does not inhibit MT1-MMP degradation but does block MMP-2. This removes the diffuse degradation patterns seen in the untreated cells leaving just the sharper MT1-MMP degraded areas. TIMP-2 and GM6001 inhibit gelatin degradation completely.

Fig. 3. *(opposite page)* Degradation of collagen by human rheumatoid synovial cells. Degradation is visible as bleached areas. This collagen degradation activity is TIMP-2 and GM6001 sensitive but TIMP-1 insensitive indicating that it is MT1-MMP dependent.

Synovial Fibroblasts

untreated

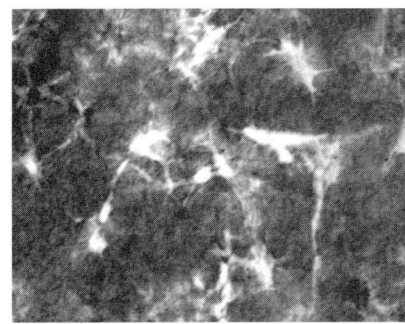

TIMP-1

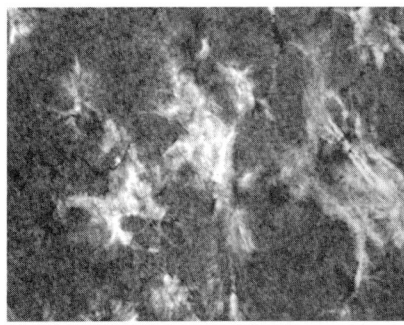

TIMP-2

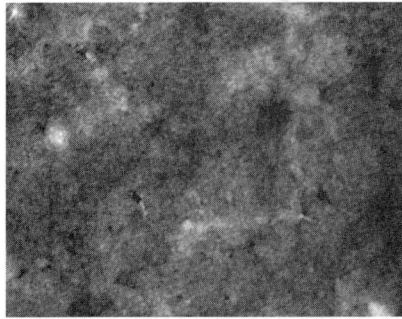

GM6001

2. Materials
2.1. Detection of ProMMP-2 Activation (Fig. 1)
1. Cell culture facility.
2. Culture ware.
3. Growth medium.
4. Serum free medium.
5. SDS-PAGE mini-gel apparatus.
6. Gelatin (Sigma).
7. SDS-PAGE reagents .
8. Washing buffer: 2.5% Triton-X100, 100 mM Tris-HCl (pH 7.5), 5 mM CaCl$_2$, 10 μM ZnCl$_2$, and 0.02 % NaN$_3$.
9. Incubation buffer: 100 mM Tris-HCl (pH 7.5), 5 mM CaCl$_2$, 10 μM ZnCl$_2$, and 0.02 % NaN$_3$.
10. Human fibroblast conditioned medium. (*Note:* this is an option for the cells that do not express proMMP-2.) Most of fibroblasts produce proMMP-2 constitutively. Thus conditioned media from fibroblasts can be used as a source of proMMP-2.)

2.2. In Situ *Zymography (Fig. 2)*
1. Cell culture facility.
2. 4-Well chamber slides (Nunc).
3. 1% Gelatin in phosphate buffered saline (PBS).
4. Alexa 488 labelling kit (Molecular probes).
5. 1% Glutaraldehyde in dH$_2$O (freshly prepared).
6. 1 M Ammonium chloride in dH$_2$O.
7. 70% EtOH.
8. Growth culture medium.
9. 3% Para-formaldehyde in PBS.
10. 80% Glycerol in PBS.
11. Cover slip (22 × 50 mm).
12. Nail varnish.
13. Fluorescence microscope with CCD camera and standard fluorescent filtersets.

2.3 Immunostaining of MT1-MMP on Cell Layers (Fig. 1)
1. Cell culture facility.
2. Sterile glass coverslips or chamber slides (Falcon/Becton Dickenson).
3. Goat Serum (Sigma).
4. BSA (Sigma).
5. PBS.
6. Rabbit Anti-human MT1-MMP antibody (Calbiochem).
7. Anti-rabbit fluorescent labeled secondary antibody (Molecular Probes).
8. Nail varnish.
9. 80 % Glycerol in PBS.
10. Fluorescence microscope with CCD camera and standard fluorescent filtersets.

2.4. In Situ Collagen Degradation (Fig. 3)

1. Cell culture facility.
2. 6-Well culture dishes (Nunc).
3. Collagen (Vitrogen).
4. 1 M NaOH.
5. 3% Para-formaldehyde in PBS.

3. Methods

3.1. Detection of ProMMP-2 Activation (Fig. 1)

1. Seed cells in appropriate culture plates or dish (i. e., 12-well plates) and culture them until near confluent in growth medium.
2. Remove growth medium and wash cells with serum free medium once, and culture cells further in serum free medium for next 24 to 48 h (*see* **Notes 1** and **2**).
3. Harvest culture medium and remove and cell debris by centrifugation. Samples can be frozen and stored for later analyses.
4. Mix 20 µL portion of medium with 20 µL of nonreducing SDS-PAGE loading buffer. Do not boil the sample or add ethylene diamine tetraacetic acid (EDTA).
5. Apply the samples to SDS-PAGE on 10 to 7.5% polyacrylamide gel containing 0.1% gelatin. Apply prestained molecular weight marker.
6. After electrophoresis, wash gels in washing buffer 4 times for 15 min each.
7. Incubate gels in the incubation buffer at 37°C for 4 h overnight (*see* **Note 3**).
8. Stain the gels with Coomassie Brilliant Blue R250.

3.2. In Situ Zymography (Fig. 2)

3.2.1. Preparation of Alexa488-Labeled Gelatin

1. Dissolve gelatin in PBS by heating the solution adjusting its concentration to 3 mg/mL. Do not boil. Amine containing buffers must not be used as these interfere with the labeling reaction.
2. Transfer a 1 mL portion of gelatin solution to reaction vial of Alexa488 reactive dye, stir at room temperature for 1 h (*see* **Note 4**).
3. Stop reaction by adding 100 µL of 2M Tris-HCl (pH 7.5).
4. Transfer gelatin-Alexa488 solution to a dialysis tube (molecular weight cut off 6000–8000) and dialyze against 1 L cold PBS over night to remove free Alexa488 dye.
5. After dialysis, mix with gelatin to give a final concentration of 50% and keep at –20°C until use. Labeled gelatin is stable for several months.

3.2.2. Preparation of Alexa-gelatin Coated Slide Chamber

1. Dilute Alexa488-gelatin to a concentration of 50 µg/mL in PBS.
2. Add 250 µL/well in 4-well chambers and incubate for 1 h (*see* **Note 5**).
3. Remove the solution and wash with PBS once.
4. Add 1% glutaraldehyde in H_2O 250 mL/well and incubate for 5 min (*see* **Note 6**).

5. Remove glutaraldehyde, add 1 M NH$_4$Cl 250 mL/well, and incubate 15 min to block glutaraldehyde.
6. Wash each wells with 70% EtOH.
7. Unused slides can be stored at 4°C for several days in 70% ethanol.

3.2.3. Culture and Visualization

1. Wash each well of the coated chamber with culture medium to remove ethanol.
2. Seed cells on it and culture for 4 to 18 h as desired (see **Notes 1** and **2**).
3. Fix cells with 3% para-formaldehyde for 15 min at room temperature.
4. Counter stain for F-actin with Alexa 568-Phalloidin (Molecular Probes).
5. Remove plastic divider, and mount with cover slip (22 × 50 mm) in 80% glycerol/PBS or appropriate mounting media.
6. Seal the edge of cover slip with nail polish.
7. Visualize using a fluorescent microscope with fluorescein isothiocyanate (FITC)/green fluorescent protein (GFP) and rhodamine filtersets (see **Note 7**).

3.3. Immunofluorescent Visualisation of MT1-MMP (Fig. 1)

3.3.1. Staining of Cell Layers for MT1-MMP

1. Culture cells on chamber slides or on glass coverslips placed in tissue culture wells. Slides may be uncoated or previously coated with matrixes such as collagen, fibronectin, or gelatin, or with serum. Cells should be allowed to attach and spread on the slides for between 1 h and overnight. The number of cells, time to attach and matrix used will vary for different cell types.
2. After the cells are attached wash the coverslips or slides with PBS, then fix in 3% paraformaldehyde in PBS for 10 min. Do not exceed 10 min as this will result in partial permeabilization of the cell membrane.
3. Wash once with PBS and then block the samples in 5% goat serum, 1% BSA in PBS for 90 min.
4. Wash once with PBS and incubate with 2 µg/mL primary antibody (anti-MT1-MMP, Calbiochem) for 90 min in block solution.
5. Wash twice with PBS and incubate with appropriate secondary antibody for 90 min in block solution. (Alexa dye labeled secondary antibodies are available from Molecular Probes).
6. Wash three times with PBS, and mount coverslips in 80% glycerol/PBS onto microscope slides and seal with nail varnish. Chamber slides should be sealed with an appropriate sized coverslip in a similar manner.
7. Visualize staining using a fluorescent microscope with appropriate filter sets or confocal microscope.

3.4. In Situ Collagen Degradation (Fig. 3)

3.4.1. Preparation of the Collagen Layer

1. Add 330 µL of 10X RPMI to 3 mL of 3 mg/mL Collagen (Vitrogen) on ice and mix briefly.

2. Add 1 M NaOH dropwise to the collagen preparation until neutralized.
3. Cover one well of a 6-well dish with 1 mL of collagen, swirl briefly, and remove 800 µL to leave a thin layer on the well. Repeat for each well (*see* **Note 8**).
4. Leave dish at 37°C for 1 h to allow collagen to polymerize.

3.4.2. Assay

1. Wash wells gently with culture medium and plate cells onto collagen layer.
2. Incubate cells on collagen for between 1 and 3 d.
3. After incubation remove the cells from the wells using trypsin, overnight incubation may be required (*see* **Note 9**).
4. Fix the dish with 3% para-formaldehyde for 30 min then stain the remaining collagen using with Coomassie Brilliant Blue R250 (*see* **Note 10**)
5. Wash dish briefly with dH_2O and dry overnight at room temperature.
6. Visualize degradation using brightfield microscopy.

4. Notes

1. TIMP-1 and TIMP-2 were added to the cultures medium at 0.5 µM. GM6001 at 10 µM. TIMP-1 and TIMP-2 are available from Calbiochem, GM6001 from Elastin Products Co.
2. The degree of MMP-2 activation or gelatin degradation observed will be dependent on the level of expression of MT1-MMP in the cells. If low levels of activation/degradation are observed increasing the length of incubation may provide more useful results.
3. The length of incubation required is dependent on the level of MMP-2 present in the medium, not on the amount of MT1-MMP present.
4. Alexa-488 reactive dye vials are available commercially.
5. The sensitivity of the *in situ* gelatin degradation assay can be adjusted by increasing or decreasing the concentration of gelatin used to coat the slide, a thinner gelatin layer allows more rapid detection of lower levels of activity but results in decreased brightness from the undigested areas.
6. Extended fixation of the gelatin layer with glutaraldehyde can result in a gelatin layer which is too highly crosslinked to be degraded efficiently.
7. Upon microscopic analysis it often appears that the position of the cells as indicated by the phalloidin counterstain do not match with the position of the degradations. This is because the cells will move around the slide during the incubation in a random manner and at varying speeds. As a result the degradation may be visible in areas where cells are absent as the cell has now moved to a different position. Also varying degrees of degradation will be seen depending on the length of time the cell remained in that position.
8. The thickness of the layer is critical in this assay, too thick a layer will result in degradation not being clearly visible. Titration to around pH 8.0 is also vital. Addition of excess NaOH will result in the collagen becoming damaged and the assay will not perform as intended.

9. Whereas addition of a protease to a degradation assay may seem counterintuitive the polymerised collagen is highly resistant to trypsin. The trypsin will only be able to digest collagen that has already had its triple helical structure damaged by the action of MT1-MMP.
10. Fixation of the collagen with paraformaldehyde prior to coomassie staining is essential to prevent the acidity of the stain dissolving the fibrillar collagen.

References

1. Sato, H., Takino, T., Okada, Y., Cao, J., Shinagawa, A., Yamamoto, E., and Seiki, M. (1994) A matrix metalloproteinase expressed on the surface of invasive tumour cells. *Nature* **370,** 61–65.
2. Zucker, S., Pei, D., Cao, J., and Lopez-Otin, C. (2003) Membrane type-matrix metalloproteinases, (MT-MMP). *Curr. Top Dev. Biol.* **54,** 1–74.
3. Itoh, Y. and Nagase, H. (2002) in *Proteases in Biology and Medicine,* vol. 38, Hooper, N. M., ed., , Portland Press, London, pp. 21–35.
4. Seiki, M., Mori, H., Kajita, M., Uekita, T., and Itoh, Y. (2003) Membrane-type 1 matrix metalloproteinase and cell migration. *Biochem. Soc. Symp.* 253–262.
5. Kajita, M., Itoh, Y., Chiba, T., Mori, H., Okada, A., Kinoh, H., and Seiki, M. (2001) Membrane-type 1 matrix metalloproteinase cleaves CD44 and promotes cell migration. *J. Cell Biol.* **153,** 893–904.
6. Endo, K., Takino, T., Miyamori, H., et al. (2003) Cleavage of syndecan-1 by membrane type matrix metalloproteinase-1 stimulates cell migration. *J. Biol. Chem.* **278,** 40,764–40,770.
7. Deryugina, E. I., Ratnikov, B. I., Postnova, T. I., Rozanov, D. V., and Strongin, A. Y. (2002) Processing of integrin alpha(v) subunit by membrane type 1 matrix metalloproteinase stimulates migration of breast carcinoma cells on vitronectin and enhances tyrosine phosphorylation of focal adhesion kinase. *J. Biol. Chem.* **277,** 9749–9756.
8. Deryugina, E. I., Bourdon, M. A., Jungwirth, K., Smith, J. W., and Strongin, A. Y. (2000) Functional activation of integrin alpha V beta 3 in tumor cells expressing membrane-type 1 matrix metalloproteinase. *Int. J. Cancer* **86,** 15–23.
9. Belkin, A. M., Akimov, S. S., Zaritskaya, L. S., Ratnikov, B. I., Deryugina, E. I., and Strongin, A. Y. (2001) Matrix-dependent proteolysis of surface transglutaminase by membrane-type metalloproteinase regulates cancer cell adhesion and locomotion. *J. Biol. Chem.* **276,** 18,415–18,422.
10. Rozanov, D. V., Hahn-Dantona, E., Strickland, D. K., and Strongin, A. Y. (2004) The low density lipoprotein receptor-related protein LRP is regulated by membrane type-1 matrix metalloproteinase (MT1-MMP) proteolysis in malignant cells. *J. Biol. Chem.* **279,** 4260–4268.
11. Itoh, Y., Takamura, A., Ito, N., et al. (2001) Homophilic complex formation of MT1-MMP facilitates proMMP-2 activation on the cell surface and promotes tumor cell invasion. *EMBO J.* **20,** 4782–4793.

12. Ohuchi, E., Imai, K., Fujii, Y., Sato, H., Seiki, M., and Okada, Y. (1997) Membrane type 1 matrix metalloproteinase digests interstitial collagens and other extracellular matrix macromolecules. *J. Biol. Chem.* **272,** 2446–2451.
13. Visse, R. and Nagase, H. (2003) Matrix metalloproteinases and tissue inhibitors of metalloproteinases: structure, function, and biochemistry. *Circ. Res.* **92,** 827–839.

16

Analysis of TIMP Expression and Activity

Linda Troeberg and Hideaki Nagase

Summary

This chapter provides practical information on the assay of tissue inhibitor of metalloproteinase (TIMP) activity and background information enabling meaningful interpretation of the data. Protocols are given for assessing the presence of TIMPs in biological samples by immunoblotting and by virtue of their ability to inhibit matrix metalloproteinase (MMP) hydrolysis of protein substrates (reverse zymography) and synthetic fluorogenic substrates.

Key Words: MMP; titration; inhibition kinetics; zymography; immunobotting; inactivation of TIMPs; SDS-PAGE; MMP assay; fluorogenic substrate.

1. Introduction

TIMPs are 22 to 30 kDa proteins that inhibit MMPs by forming a 1:1 complex with a target MMP. There are four structurally homologous TIMPs in vertebrates (**Table 1**). The activity of TIMP-1 was first recognized in plasma and the conditioned medium of connective tissue cells as a collagenase inhibitor. It was subsequently purified as a 29 to 30 kDa protein *(1,2)* and later shown to inhibit almost all MMPs except some membrane-type MMPs (MT-MMPs) *(3)*. TIMP-2 was found as a molecule tightly bound to proMMP-2 in the medium of human melanoma cells *(4)*, whereas TIMP-3 was originally recognized as a protein secreted from Rous sarcoma virus-transfected chicken embryonic fibroblasts *(5)*. The protein binds tightly to the extracellular matrix (ECM), so unlike with other TIMPs, where inhibitory activity is detected in the conditioned medium, TIMP-3 activity is only observed when the matrix is extracted with sodium dodecyl sulfate (SDS) and subjected to reverse zymography *(6)*. TIMP-4 was first identified by cDNA cloning *(7)* and is expressed in restricted tissues (e.g., heart, ovary and brain). These TIMPs inhibit all MMPs so far tested, except that TIMP-1 is a poor inhibitor of MMP-

Table 1
Properties of TIMPs

	TIMP-1	TIMP-2	TIMP-3	TIMP-4
Molecular mass	28.5 kDa	21.5 kDa	21.6 kDa	22.3 kDa
Tissue distribution	Wide distribution	Wide distribution	Wide distribution	Most restricted, heart, brain, and ovary
Glycosylation	yes	no	yes	no
Interaction with zymogens	proMMP-9	proMMP-2	proMMP-2	proMMP-2
Specificity	MMPs (but not MT1, 2, 3 or 5 or MMP-19), ADAM-10	MMPs	MMPs and some member of the ADAM and ADAMTS families	MMPs
Extinction coefficient ($M^{-1} \cdot cm^{-1}$)	24 750	31 720	43 240	36 840
A_{280} for 1 mg/mL	1.195	1.458	1.994	1.642

19, MT1-MMP (MMP-14), MT2-MMP (MMP-15), MT3-MMP (MMP-16) and MT5-MMP (MMP-24) *(3)*. TIMP-3 also inhibits some of the ADAM and ADAMTS metalloproteinases *(8)*, including tumor necrosis factor (TNF)-α converting enzyme (ADAM-17), ADAM-10, ADAM-12 and aggrecanases (ADAMTS-1, -4 and -5).

2. Materials

2.1. Titration of TIMPs Against MMPs Using Fluorogenic Substrate Assay

1. TNC buffer: 50 mM Tris-HCl (pH 7.5) 10 mM CaCl$_2$, 150 mM NaCl, 0.05% Brij 35 (Sigma, St. Louis MO), 0.02% NaN$_3$. Store at 4°C, stable for at least 1 mo.
2. Fluorogenic substrate stock solution: Prepare a 10 mM stock solution of Mca-Pro-Leu-Gly-Leu-Dpa-Ala-Arg-NH$_2$ *(9)* (Bachem, Bubendorf, Switzerland or Peptides International, Louisville, KY) in dimethyl sulfoxide (store at –20°C, stable for 6 mo). Dilute to give a 3 µM working solution as required (prepare freshly). *See* **Note 1** on alternative substrates.
3. MMP. MMPs are commercially available from various sources (Chemicon, Temecula, CA; Sigma, St. Louis MO; and R&D Systems, Minneapolis, MN).

Analysis of TIMP Expression and Activity

MMP-1, -3 and -7 are ideal enzymes for this purpose as they are stable. MMP-2 is not recommended as it autolyzes easily, hampering accurate assessment of the amount of enzyme used. Store as recommended by suppliers.

4. TIMP. These are commercially available from sources such as Chemicon and R&D. Store as recommended by suppliers.
5. Stop solution: 100 mM disodium ethylene diamine tetraacetic acid (Na$_2$EDTA). Store at room temperature, stable for at least 6 mo.
6. Fluorometer, such as models from Sequoia-Turner, Perkin-Elmer, or Molecular Devices. *See* **Note 2** on choice of fluorometers.
7. 37°C incubator.

2.2. Kinetic analysis of TIMP Interactions With MMPs

As for **Subheading 2.1.**

2.3. Reverse Zymography

1. 10 mg/mL Porcine skin gelatin (Sigma-Aldrich). Dissolve gelatin in H$_2$O by brief microwaving or heating to 37°C. Vortex the solution to ensure that the gelatin is fully dissolved to ensure a uniform background. The solution can be stored at –20°C for up to 1 mo, but must be reheated to 37°C before each use.
2. 2.5 µg/mL MMP-2: MMP-2 is available from Chemicon, Sigma, or R&D Systems. *See* **Note 7** on the use of conditioned medium a source of gelatinase.
3. 30% (w/v) Acrylamide and 0.8% (w/v) N,N'-methylene-*bis*-acrylamide: **Caution!** Unpolymerized acrylamide is highly toxic—consult your local laboratory procedures for safe use and disposal.
4. Lower gel buffer: 110 mM amediol, 47 mM HCl, and 0.02% NaN$_3$ (pH 8.96). Store at 4°C, stable for at least 1 mo.
5. Upper gel buffer: 84 mM amediol, 62 mM HCl, and 0.02% NaN$_3$ (pH 8.37). Store at 4°C, stable for at least 1 mo.
6. Sucrose solution: 50% (w/v) sucrose, 0.03% (v/v) toluene, and 0.02% NaN$_3$. Store at 4°C, stable for at least 1 mo.
7. TEMED (N,N,N',N'-tetramethyl ethylene diamine).
8. Ammonium persulfate: 10% (w/v) ammonium persulfate in H$_2$O. Stable at –20°C for at least 2 wk.
9. 4X Upper reservoir buffer: 10 mM amediol, 10 mM glycine, 0.1% (w/v) SDS (pH 9.39). Store at 4°C, stable for at least 1 mo.
10. 4X Lower reservoir buffer: 62.5 mM amediol, 50 mM HCl (pH 8.23). Store at 4°C, stable for at least 1 mo.
11. 2X Sample loading solution: 5% (w/v) SDS, 0.1% (w/v) bromophenol blue, and 40% (v/v) glycerol. Store at room temperature, stable for at least 3 mo.
12. Prestained molecular mass standards, for example Precision Plus All Blue protein standards from Bio-Rad (Hercules, CA).
13. Staining solution: 0.125% (w/v) Coomassie brilliant blue R-250, 62.5% (v/v) methanol, and 25% (v/v) acetic acid. Store at room temperature, stable.

14. Destaining solution: 30% (v/v) methanol and 1% (v/v) formic acid. Store at room temperature, stable.
15. Zymogram developing buffer: 50 mM Tris-HCl (pH 7.5), 5 mM $CaCl_2$, 5 μM $ZnCl_2$, and 0.02% NaN_3. Store at 4°C, stable for at least 1 mo.
16. Zymogram renaturing solution: 2.5% (w/v) Triton X-100 in zymogram developing buffer. Store at 4°C, stable for at least 1 mo.
17. Electrophoresis apparatus (e.g., mini-gel system, with gel dimensions approx 9 × 6 × 0.75 mm).
18. Constant voltage power supply (200 V, 500 mA).
19. Sealed plastic container for incubating gels.
20. 37°C Incubator.

2.4. Inactivation of TIMPs

1. Reducing agent: 100 mM Dithiothreitol (DTT) dissolved in water. Store at –70°C, stable for at least 2 wk.
2. Alkylating agent: 500 mM Iodoacetamide (IAN) dissolved in water. Store at –20°C, stable for at least 1 mo.

2.5. Immunoblotting

As per **Subheading 2.3.**, plus

1. Pre-stained molecular mass markers (e. g., Precision Plus Protein All Blue Standards, Bio-Rad).
2. Blotting buffer: 25 mM Tris, 192 mM glycine (pH 8.3), 20% (v/v) methanol, and 0.1% (m/v) SDS.
3. Tris-buffered saline (TBS): 25 mM Tris-HCl (pH 7.4) and 200 mM NaCl.
4. Blocking solution: 5% low fat milk powder in TBS.
5. Primary antibody: Antibodies against TIMPs are commercially available from various companies (Chemicon, Sigma, Santa Cruz Biotechnology; and R&D Systems).
6. TBS-Tween: 0.1% (v/v) Tween-20 (Sigma) in TBS.
7. Secondary antibody: Secondary (anti-species) antibodies are commercially available from various companies (Sigma and Santa Cruz Biotechnology).
8. Substrate solution: 5-Bromo-4-chloro-3-indolylphosphate/nitroblue tetrazolim (BCIP/NBT) (e.g., Western Blue Stabilized substrate [Promega]).
9. Electroblotting cell (e.g., Novex XCell SureLock Mini-Cell [Invitrogen, Carlsbad, CA] or Bio-Rad Mini Trans-Blot Cell, [Bio-Rad, Hercules, CA]).

3. Methods

3.1. Titration of TIMPs Against MMPs Using Fluorogenic Substrate Assay

3.1.1. Principle of Assay

As a first step to demonstrating the presence of TIMPs in a sample, assays based on their ability to inhibit MMP activity can be used. MMPs readily degrade various synthetic substrates and this activity can be inhibited by TIMPs. In the following section, we first describe analysis of MMP activity against synthetic fluorogenic substrates and then describe how these assays can be used to investigate the presence of TIMPs and the kinetics of their interactions with MMPs.

MMP activity is easily measured in vitro by following their hydrolysis of fluorogenic peptide substrates, which consist of an MMP-cleavable peptide sequence (8–12 amino acid residues long), coupled to a fluorophore and a quenching group. In the intact substrate, the fluorophore and the quencher are in close proximity and so the energy emitted by the fluorophore is absorbed by the quencher, as a result of the fluorescence resonance energy transfer (FRET). Once the susceptible peptide bond is cleaved, the fluorophore and the quencher are separated and the fluorophore emits energy at its characteristic emission wavelength. Mca-Pro-Leu-Gly-Leu-Dpa-Ala-Arg-NH_2 is a commonly used fluorogenic MMP substrate, where Mca is methylcoumarin and Dpa is dinitrophenyl diaminopropionate *(9)*. MMPs cleave the Gly-Leu bond of the substrate.

TIMPs and MMPs bind reversibly to form tight complexes with 1:1 stoichiometry. The concentration of partially purified or purified TIMPs can thus be determined by titration against an MMP solution of known concentration. A fixed, known concentration of MMP is mixed with dilutions of the TIMP solution and the mixture incubated to allow formation of stable 1:1 complexes. The amount of MMP remaining uninhibited after incubation is measured using a fluorogenic substrate. The TIMP concentration can then be determined from the maximum dilution (lowest concentration of TIMP) required to give complete inhibition of MMP activity. This method can also be used to determine the concentration of MMPs, provided that a TIMP solution of known concentration is available.

3.1.2. Determination of TIMP Concentration in Biological Samples by Titration

1. Prepare a 50 n*M* solution of MMP in TNC buffer. *See* **Note 4** for information on choosing an appropriate enzyme concentration and determining the linear range of the assay.
2. The sample can be heated to 90°C for 10 min to inactivate MMPs, if these are likely to be present. TIMPs will not be affected by this treatment, as they retain activity after incubation at 100°C for 30 min or exposure to pH 2.0 *(10)*.
3. Prepare various dilutions of the TIMP-containing sample (at least 5 different concentrations, that give partial inhibition of MMP activity). It may be necessary to concentrate the sample first. We routinely concentrate conditioned medium from various cell types 10-fold before titration. If available, also make dilutions of a purified or partially purified TIMP to serve as a positive control for inhibition.

The concentration of purified or partially purified TIMPs can be estimated from their absorbance at 280 nm, using the extinction coefficients in **Table 1**.

4. Incubate 50 μL of the 50 nM MMP solution with 50 μL of the various TIMP dilutions for 30 min at room temperature. This gives an enzyme concentration of 25 nM during the incubation step.
5. Add 100 μl Mca-Pro-Leu-Gly-Leu-Dpa-Ala-Arg-NH$_2$ (3 μM) *(9)* and allow the assay to proceed at 37°C for 30 min. This gives a final enzyme concentration of 12.5 nM and a final substrate concentration of 1.5 μM.
6. Stop the assay by adding 20 μL of 100 mM Na$_2$EDTA. This works by chelating Zn^{2+}, which is required for MMP activity.
7. Measure the product fluorescence using an excitation wavelength of 325 nm and an emission wavelength of 393 nm.
8. For purified or partially purified TIMP samples, plot the estimated TIMP concentration (*x*-axis) against change in fluorescence (*y*-axis). Draw a straight line through the points. The *x*-intercept gives the estimated TIMP concentration equivalent to 50 nM MMP. In **Fig. 1A**, a purified TIMP-2 solution with an estimated concentration of 1 μM was diluted to concentrations between 5 and 100 nM and titrated against 50 nM MMP-1 and MMP-3 to accurately determine its active concentration. Here, the *x*-intercept of 55 nM indicates that an estimated TIMP concentration of 55 nM is required to fully inhibit 50 nM MMP-1. The TIMP solution is thus slightly less concentrated than was estimated, as more TIMP than expected is required for full MMP inhibition. The concentration of the stock is thus 50/55 × 1 μM, or 910 nM.
9. For unpurified biological samples, plot the dilution factor (*x*-axis) against the change in fluorescence (*y*-axis). Draw a line through the points. The *x*-intercept gives the sample dilution required to completely inhibit the MMP. **Fig. 1B** shows titration of various dilutions of 10-fold concentrated fibroblast conditioned medium (CM) with 50 nM MMP-1. Here, the intercept on the *x*-axis indicates that 50 nM MMP-1 is completely inhibited by 6 μL of 10X concentrated conditioned medium, which is a 8.4X dilution of the 10X CM. The TIMP concentration in the sample is thus 50 nM × 8.4/10, or 42 nM (equivalent to 1.05 μg/mL, using the molecular mass of TIMP as 25 kDa).

3.2. Kinetic analysis of TIMP Interactions With MMPs

3.2.1 Theoretical Background

Because TIMPs bind to MMPs in tight, reversible 1:1 complexes, their binding can be analysed and described using techniques developed for tight-binding inhibitors *(11)*.

$$[E]+[I] \underset{k_{off}}{\overset{k_{on}}{\rightleftharpoons}} [EI] \tag{1}$$

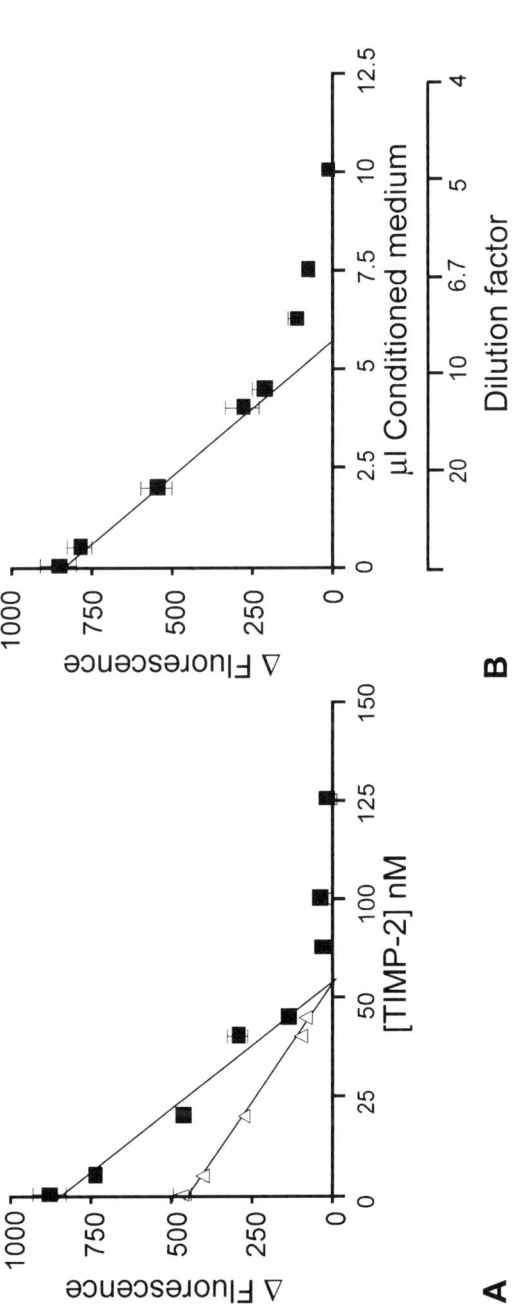

Fig. 1. (**A**) Titration of MMP-1 and MMP-3 with TIMP-2. Titration was used to accurately determine the concentration of a TIMP-2 solution which, from its absorbance at 280 nm, had an estimated concentration of 1 μM. Various dilutions of the TIMP-2 solution (50 μL of dilutions with estimated concentrations between 5 and 125 nM) were incubated with MMP-1 (■) or MMP-3 (△) (50 μL of 50 nM) for 30 min at room temperature to allow for the formation of stable enzyme/inhibitor complexes. Fluorogenic substrate (Mca-Pro-Leu-Gly-Leu-Dpa-Ala-Arg-NH$_2$, 100 μL of 3 μM) was added and the reaction allowed to proceed for 30 min at 37°C before being stopped with Na$_2$EDTA. Product fluorescence was read using a SPECTRAmax microwell fluorometer (Molecular Devices). The data indicate that 50 nM MMP-1 and MMP-3 are completely inhibited by 55 nM TIMP-2, indicating that the actual TIMP-2 concentration of the stock is 50/55 × 1 μM, which is 990 nM. (**B**) Titration of MMP-1 with fibroblast conditioned medium (CM). MMP-1 (50 μL of 50 nM) was incubated with various volumes (2.5–10 μL) of a 10-fold concentrated solution of fibroblast CM (all made up to 50 μL) for 30 min at room temperature. Fluorogenic substrate (Mca-Pro-Leu-Gly-Leu-Dpa-Ala-Arg-NH$_2$, 100 μL of 3 μM) was added and the reaction allowed to proceed for 30 min at 37°C before being stopped with Na$_2$EDTA. Product fluorescence was read using a SPECTRAmax microwell fluorometer. The graph shows that 50 nM MMP-1 is completely inhibited by 6 μL of concentrated CM, which is a 8.4-fold dilution of the 10 × CM. The TIMP concentration in the CM is thus 50 × 8.4/10, which is 42 nM, or 1.05 μg/mL.

The association rate constant (k_{on}) is the rate constant for the forward reaction and has units of $M^{-1}\cdot s^{-1}$. The dissociation rate constant for the reverse reaction (k_{off}) has units of s^{-1}. K_i, the equilibrium dissociation constant, describes the equilibrium position of the interaction (affinity between the inhibitor and the target enzyme), and is defined as k_{off}/k_{on}, with units of M.

In order to determine K_i accurately, various experimental conditions must be met:

1. No more than 10% of substrate should be hydrolyzed during the assay.
2. The enzyme concentration must be below the K_i value. For MMPs and TIMPs, K_i is usually in the low nanomolar range, so it follows that sub nanomolar amounts of MMP must be used.
3. Inhibitor concentrations of at least 10× the enzyme concentration must be used, and only partial inhibition must be analyzed.

    These conditions differ from those used for titrations (*see* **Subheading 2.1.**), where both the enzyme and inhibitor are used at concentrations at least $50 \times K_i$, ensuring that the reverse reaction is minimal, thus forming a stable enzyme/inhibitor complex. Here, by using enzyme concentrations below K_i, the association is slowed down to a rate that can be measured more accurately. We recommend reading of Bieth *(12)* before determining these kinetic parameters.

    K_i can be determined from analysis of equilibrium levels of MMP activity (*see* **Subheading 3.2.2.**) or from progress curve analysis (*see* **Subheading 3.2.3.**), which monitors the formation of the enzyme/inhibitor complex more closely and also allows for determination of k_{on}.

3.2.2. Protocol for Equilibrium Determination of K_i

1. Incubate MMP (90 µL of 2.2 n*M*) with at least 5 different concentrations of TIMP (90 µL of 0.022–2.2 µ*M*) in TNC buffer for 1 h at 37°C. See **Note 5** on determining enzyme concentration and assay time and **Note 6** on establishing a suitable incubation time.
2. Add substrate solution (20 µL of 15 µ*M*, warmed to 37°C) to start the reaction. (Final enzyme concentration = 1 n*M*, final inhibitor concentration = 10–100 n*M*, final substrate concentration = 1.5 µ*M*).
3. Allow reaction to proceed for 5 h at 37°C and stop by addition of 20 µL of 100 m*M* Na$_2$EDTA.
4. Determine v_0, the rate of substrate hydrolysis in the absence of TIMP ($M\cdot s^{-1}$).
5. Determine v_s, the equilibrium rate of substrate hydrolysis reached after inhibition (rate of hydrolysis in the inhibited steady state) ($M\cdot s^{-1}$) for each TIMP concentration, [*I*].
6. Calculate $a = v_s/v_0$ and plot $1/a$ (*x*-axis) versus $[I]/(1-a)$ (*y*-axis).
7. Determine K_i, which is equal to the gradient (slope) of this line, by linear regression.

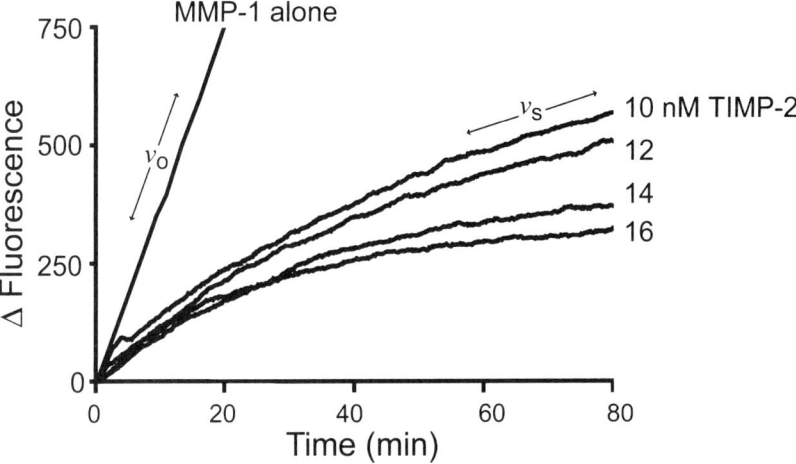

Fig. 2. Progress curve analysis of MMP-1 inhibition by TIMP-2. MMP-1 (1 nM) was incubated with various concentrations of TIMP-2 (10–16 nM) and hydrolysis of Mca-Pro-Leu-Gly-Leu-Dpa-Ala-Arg-NH$_2$ (1.5 μM) at 37°C monitored using an LS-50B fluorometer. v_s and v_0 were determined by linear regression of the indicated regions of the curve, and k at various TIMP concentrations determined using the integrated rate equation. k_{on} is equal to the gradient of a plot of TIMP concentration versus k, whereas K_i is equal to the gradient of a plot of $(v_0/v_s - 1)$ against TIMP concentration.

3.2.3. Protocol for Progress Curve Determination of K_i and k_{on}

1. Prepare a 20 nM stock solution of MMP in TNC buffer and keep on ice.
2. Prepare TIMP solutions of at least 5 different concentrations between 100 and 250 nM. Warm to 37°C.
3. In a 1-mL cuvet, warm:
   - 500 μL of 3 μM substrate (final concentration of 1.5 μM).
   - 100 μL of 100 nM TIMP (final concentration of 10–25 nM).
   - 350 μL TNC.
4. Add 50 μL enzyme solution (final concentration of 1 nM) to start the reaction.
5. Monitor the reaction until the rate of substrate hydrolysis is constant, indicating that the equilibrium rate of substrate hydrolysis has been reached. **Figure 2** shows a typical progress curve.
6. Estimate v_0 from initial rate of substrate hydrolysis in the absence of inhibitor.
7. Estimate v_s from the equilibrated rate of substrate hydrolysis with inhibitor.
8. Use nonlinear regression to calculate k from these estimates using the integrated rate equation:

$$P = v_s t + (v_0 - v_s)(1 - e^{-kt})/k$$

where P is product fluorescence (arbitrary units), v_s is the equilibrium rate of substrate hydrolysis (M·s^{-1}) (from **step 7**), v_0 is the initial rate of substrate hydrolysis (M·s^{-1}) (from **step 6**), t is time (s), and k is the pseudo-first order rate constant (M^{-1}·s^{-1}).

9. Determine k at several inhibitor concentrations and plot inhibitor concentration (x-axis) against k (y-axis). For a simple bimolecular interaction between enzyme and inhibitor, this plot should be linear, with k_{on} equal to the gradient.
10. Plot $(v_0/v_s - 1)$ (x-axis) against inhibitor concentration (y-axis). K_i is equal to the gradient of this straight line. Because $K_i = k_{off}/k_{on}$, k_{off} can be determined indirectly by multiplying k_{on} by K_i. The k_{off} value can also be determined directly using α_2M *(13)*.

3.2.4. K_i Vs IC_{50}

Many investigators describe enzyme inhibition in terms of IC$_{50}$ values or percentage inhibition. These parameters can be useful if understood and applied correctly, but are often used erroneously. IC$_{50}$ values are particularly misleading, because the inhibition observed is influenced by the enzyme and inhibitor concentration, the time of incubation, K_m values for the substrate and the type of inhibition. A low percentage inhibition can be increased markedly simply by increasing the inhibitor concentration or the time of incubation *(12)*. Thus, at enzyme concentrations far above K_i ($[E]>20 \cdot K_i$), IC$_{50}$ is a measure of the enzyme concentration, not the effectiveness (affinity) of the inhibitor. At lower enzyme concentrations, closer to or below K_i, the IC$_{50}$ value is affected by the presence and the concentration of substrate. IC$_{50}$ is only a meaningful parameter if determined at low enzyme concentrations, with substrate concentrations below K_m. It is not meaningful to compare IC$_{50}$ values of a particular inhibitor (e.g., TIMP) or a synthetic MMP inhibitor against different MMPs, because the K_m values for the substrate will be different for the different MMPs and often different enzyme concentrations are used.

3.3. Reverse Zymography

In much the same way as zymography is useful for analyzing MMP activity in complex biological samples *(14)*, reverse zymography is a method for analyzing the presence of TIMPs in biological samples. A sodium dodecyl sulfate-polyacrylamide gel electrophoresis (SDS-PAGE) gel is copolymerized with a gelatinase (MMP-2 or -9) and gelatin, a substrate for the enzyme. Samples thought to contain TIMPs are electrophoresed on the gel without reduction or boiling and the proteins then renatured by washing the gel in a non-ionic detergent such as Triton X-100. The renatured proteinase digests the substrate protein throughout the gel, except at positions in the gel to which TIMPs have electrophoresed. TIMPs are thus visualized as areas of MMP inhibition, seen as darkly staining bands against a clear, digested background. A plain SDS-

Table 2
Composition of Lower Reverse Zymogram Gel (Per Gel)

2 mL acrylamide/bisacrylamide stock solution
1.50 mL lower gel buffer
1.29 mL sucrose solution
0.61 mL H$_2$O
0.60 mL gelatin solution
400 µL of 2.5 µg/mL MMP-2 (= 250 ng/mL)
21 µL ammonium persulfate
2.25 µL TEMED

Table 3
Composition of Upper Reverse Zymogram Gel (Per Gel)

0.32 mL acrylamide/bisacrylamide stock solution
0.64 mL upper gel buffer
0.64 mL sucrose solution
0.96 mL H$_2$O
31.5 µL ammonium persulfate
6.75 µL TEMED

PAGE gel, without the substrate protein or the proteinase, must always be run to confirm that the "inhibitor" bands are actually zones of inhibition and not normally stained proteins. Picogram quantities of TIMP can be vizualized using this method *(15)* (*see* **Note 7** on factors affecting sensitivity). Reverse zymography can be used to quantify the amount of TIMP present in a sample, provided that careful calibration with a known standard is performed in parallel with the test sample. We use the amediol-based SDS-PAGE method of Bury *(16)*, but Tris-glycine-based methods can also be used *(15)*.

1. Cast the separating gel. Assemble the gel casting apparatus according to the manufacturer's instructions and cast a 10% gel according to **Table 2**. Overlay the gel with water to exclude atmospheric oxygen and leave to polymerize (about 2 h).
2. Cast the stacking gel. Decant the water from the top of the separating gel and cast a stacking gel according to **Table 3**. Insert a well comb, and leave to polymerize (about 30 min).
3. Prepare samples and TIMP standards (if available) by mixing 5 to 20 µL of sample with an equal volume of sample loading buffer. To maintain the activity of the MMPs and TIMPs, do not add reducing agents or EDTA, or boil the sample.
4. Electrophorese. Assemble the gel apparatus according to the manufacturer's instructions. Add upper and lower reservoir buffers to the appropriate chambers. Load samples using either a Hamilton syringe or disposable tips. Electrophorese

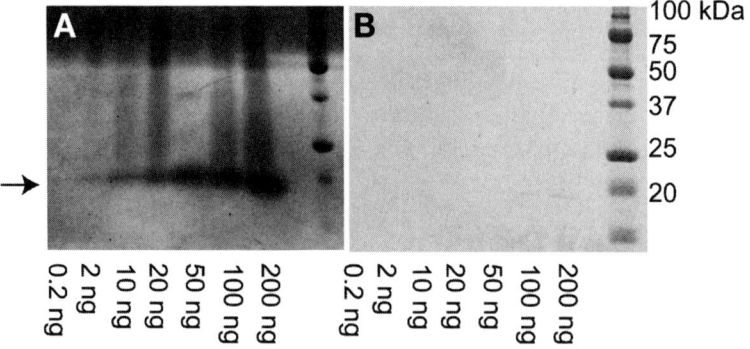

Fig. 3. Reverse zymogram showing TIMP-2 activity. (**A**) TIMP-2 (0.2 – 200 ng) was analyzed on a reverse zymogram with 1 mg/mL gelatin and 250 ng/mL MMP-2. After electrophoresis, the gel was washed for 4 × 15 min in a buffer containing 1% Triton X-100 to renature MMP-2 and TIMP samples, and the gel then incubated at 37°C overnight. When stained with Coomassie Brilliant Blue, bands of TIMP activity were visible as dark areas against the lightly stained, digested background (*arrow*). The limit of detection for this reverse zymogram is 2 ng TIMP-2. (**B**) A companion SDS-PAGE gel, with the same samples as in A, but cast without gelatin or MMP-2, shows that the bands on A are the results of MMP-2 inhibition.

at 150 to 200 V until the bromophenol blue tracker dye reaches a few mm from the bottom of the gel (about 45 min to 1 h).

5. Wash the gels. Disassemble the apparatus and place the gels in a plastic or glass container. Wash the gels in zymogram renaturing buffer for 15 min on a shaker. Repeat this 3 more times. This buffer renatures electrophoresed MMPs and TIMPs by replacing SDS with Triton X-100.
6. Incubate the gels in zymogram developing buffer overnight at 37°C. During this time, the MMPs will digest the copolymerized gelatin throughout the gel, except at positions to which TIMPs have been electrophoresed.
7. Stain the gel in staining solution for 20 min and destain thoroughly using several changes of destain solution. **Figure 3** shows a typical reverse zymogram.

3.4. Inactivation of TIMPs

TIMPs all contain 6 conserved disulfide bonds, three in the N-terminal domain and three in the C-terminal domain. These disulfide bonds are crucial for maintaining the three-dimensional structure of the molecules, so TIMPs are sensitive to reducing agents. Reduction and alkylation can thus be used to inactivate TIMPs *(17)*.

1. Add DTT to TIMP solution to a final concentration of 2 m*M*.
2. Incubate at 37°C for 30 min.

3. Add IAN to a final concentration of 5 mM.
4. Incubate at 37°C for 30 min.
5. Remove reducing and alkylating agents by dialysis against 10 volumes of TNC buffer for 5 h at 4°C.

3.5. Immunoblotting

Immunoblotting, or Western blotting, is a technique for identifying proteins on the basis of their size and reactivity with an antibody. Tissue samples are solubilized in SDS, separated on an SDS-PAGE gel, and electroblotted onto a nitrocellulose or polyvinylidene fluoride (PVDF) membrane. The blot is incubated with an antibody specific for the target protein and immunoreactive bands then visualized using an enzyme-labeled secondary detection antibody. The main advantage of this technique over enzyme linked immunosorbent assay (ELISA) is that it allows identification of the mass of the target protein, indicating for example, whether it is in a glycosylated form. This technique has been used to investigate the involvement of TIMPs and MMPs in angiogenesis in the inflammatory pannus *(18)* and to investigate whether the anti-rheumatic effects of heparinoid are TIMP-related *(19)*. The sensitivity of the technique depends largely on the antibody used, but in general, high nanogram quantities of TIMP can be detected.

1. Cast a 10% acrylamide gel as described in **Subheading 3.3.2.** and **Tables 2** and **3**, but replacing the gelatin and MMP-2 solutions with water.
2. Load samples, including prestained molecular mass markers and positive controls (*see* **Subheading 2.1.**)
3. Electrophorese as described in **Subheading 3.3.2.**
4. Wet a sheet of PVDF in methanol and then wash in blotting buffer. The membrane must not be allowed to dry after wetting.
5. Assemble the blotting apparatus according to the manufacturer's instructions, with the gel at the cathode (negative) side and the PVDF at the anode (positive) side.
6. Blot for 180 min at 200 mA (unlimiting current).
7. Disassemble the apparatus.
8. Incubate the PVDF in blocking solution for an hour at room temperature (or overnight at 4°C if convenient) to block remaining protein-binding sites.
9. Incubate with primary antibody for 3 h at room temperature, preferably on a gel rocker. Use a dilution recommended by the antibody manufacturer.
10. Wash three times for 5 min each in TBS-Tween.
11. Incubate with secondary antibody for 1 h at room temperature, preferably on a gel rocker. Use a dilution recommended by the antibody manufacturer.
12. Wash three times for 5 min each in TBS-Tween.
13. Add substrate solution.
14. Once bands have developed sufficiently, wash the blot in distilled H_2O and leave to dry on a tissue.

3.6. Enzyme-Linked Immunosorbent Assays

ELISAs provide a rapid and convenient way to quantify the presence of TIMPs in biological samples. The sample is adsorbed onto a 96-well plastic plate and a specific antibody added to identify and quantify TIMPs present. Binding of the primary antibody is usually detected using a labeled secondary antibody, linked to an enzyme that generates a coloured product. The main advantage of ELISA is that many samples can be processed simultaneously. Kits are commercially available from various companies, such as Chemicon and R&D Systems, with the sensitivity of the assays varying for different TIMPs and manufacturers.

4. Notes

1. Other fluorogenic substrates can be used to monitor MMP activity. Some show a degree of selectivity for particular MMPs, but none is specific for any one enzyme. The substrate described here, Mca-Pro-Leu-Gly-Leu-Dpa-Ala-Arg-NH$_2$ *(9)* is a commonly used general MMP substrate, whereas Mca-Arg-Pro-Lys-Pro-Val-Glu-Nva-Trp-Arg-Lys(Dnp)-NH$_2$ is selective for MMP-3 *(20)*. Other enzymes (e.g., chymotrypsin-like proteinases) also hydrolyze both of these substrates. Synthetic fluorogenic substrates are thus not suited to assaying MMP activity in complex biological fluids.
2. Fluorometers. Various kinds of fluorometer are available and most machines have or can be fitted with filters suitable for these assays. Manufacturers include Sequoia-Turner (Mountain View, CA), Perkin Elmer (Wellesley, MA), and Molecular Devices (Sunnyvale, CA). Some machines allow for reading of one sample at a time in a cuvet, whereas others read microwell plates and are suitable for rapid assay of a large number of samples. This protocol gives volumes suitable for microwell assays or small cuvettes. Volumes can be increased proportionally if required.
3. Determine the linear range of the assay. When establishing an assay, it is necessary to determine the conditions for which the amount of fluorescent product increases linearly with the amount of enzyme present. This is affected by the substrate concentration, the enzyme concentration and the time of the assay. If substrate concentration becomes limiting, then the fluorescence detected is not an accurate reflection of the amount of enzyme present. For example, **Fig. 4** shows that for hydrolysis of 1.5 µ*M* Mca-Pro-Leu-Gly-Leu-Dpa-Ala-Arg-NH$_2$ by 12.5 or 25 nM MMP-1, the amount of product fluorescence increases linearly over the chosen 30 min assay time, while for 50 nM MMP-1, substrate becomes limiting and fluorescence does not increase linearly over 30 min. For a 30 min stopped-time assay, a final concentration of 12.5 n*M* MMP-1 is thus appropriate. Such determination of the linear range should always be carried when establishing a stopped-time enzyme assay.
4. Source of gelatinase. Many authors use conditioned medium from cells secreting MMP-2 and/or MMP-9 as a source of gelatinase activity. Although such media

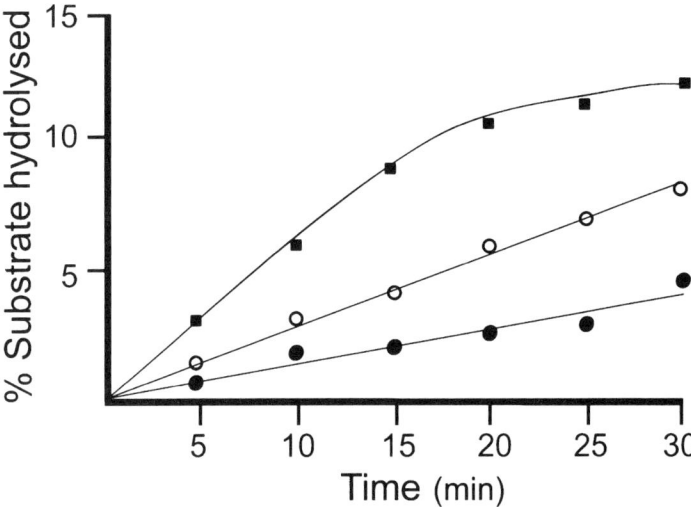

Fig. 4. Determination of linear range of enzyme assay. Hydrolysis of Mca-Pro-Leu-Gly-Leu-Dpa-Ala-Arg-NH$_2$ (1.5 μM) by various concentrations of MMP-1 was monitored using stopped-time assays. For 12.5 (●) and 25 nM (○) MMP-1, the amount of product fluorescence increases linearly over 30 min. For 50 nM MMP-1 (■), however, the rate of substrate hydrolysis is linear only up to 15 min, after which the rate of hydrolysis decreases (above 10% hydrolysis) as substrate becomes limiting. A single 30 min stopped-time assay for 50 nM MMP-1 would thus not give an accurate representation of the amount of enzyme present. For a 30 min stopped-time assay with this substrate, 25 nM MMP-1 is more appropriate.

are useful, they contain a variable amount of gelatinase activity depending on the cells used to prepare the medium and the culture conditions. Reverse zymograms using purified enzyme are generally more sensitive and reproducible than those prepared using conditioned media *(15)*. Additionally, the longer the incubation time chosen, the lower the concentration of gelatinase that can be used and hence the lower the amount of TIMP that can be detected.

5. Enzyme concentration for K_i determinations. To determine K_i accurately, the enzyme concentration must be well below K_i. At such concentrations (<1 nM), many MMPs have to be incubated with fluorogenic substrates for several hours to generate sufficient signal for reliable measurement. Test various enzyme concentrations and assay times to determine the lowest enzyme concentration that can be reliably measured, and ensure that the enzyme is stable over the chosen assay period by confirming the linearity of the assay.
6. Suitable incubation time for K_i determinations. For equilibrium determination of K_i values, lower concentrations of enzyme and inhibitor are used than when performing titrations, so it takes longer for equilibrium between the enzyme and inhibitor to be reached. When performing such analyses, it is necessary to opti-

mize the incubation time, to ensure that a true equilibrium rate of substrate hydrolysis is being measured. Assay enzyme activity after various periods of incubation with the inhibitor. If the rate of substrate hydrolysis decreases with time, the reaction of the enzyme and the inhibitor has not reached equilibrium. In this case, increase the time for which the enzyme and inhibitor are preincubated, until hydrolysis of the substrate is linear with time.
7. Sensitivity of reverse zymograms. MMP-2 reverse zymograms can detect as little as 1 pg of TIMP-2 and 40 pg of TIMP-1. MMP-9 reverse zymograms can detect 60 pg of TIMP-2 and 40 pg of TIMP-1 *(15)*. The sensitivity of reverse zymograms depends largely on the source and concentration of enzyme used. For optimal sensitivity, a reverse zymogram must contain sufficiently active gelatinase activity to clear the gelatin background in the incubation time chosen. If too little gelatinase is present, the background will be too dark for visualization of TIMP inhibitory activity. If too much gelatinase is used, however, higher concentrations of TIMP are required to inhibit gelatinolysis and the sensitivity of the zymogram is consequently reduced.

Acknowledgments

This work is supported by a Wellcome Trust grant no. 057508.

References

1. Cawston, T. E., Galloway, W. A., Mercer, E., Murphy, G., and Reynolds, J. J. (1981) Purification of rabbit bone inhibitor of collagenase. *Biochem. J.* **195**, 159–165.
2. Valle, K. and Bauer, E. A. (1979) Biosynthesis of collagenase by human skin fibroblasts in monolayer culture. *J. Biol. Chem.* **254**, 10,115–10,122.
3. Baker, A. H., Edwards, D. R., and Murphy, G. (2002) Metalloproteinase inhibitors: biological actions and therapeutic opportunities. *J. Cell Sci.* **115**, 3719–3727.
4. Stetler-Stevenson, W. G., Krutzsch, H. C., and Liotta, L. A. (1989) Tissue inhibitor of metalloproteinase (TIMP-2). A new member of the metalloproteinase inhibitor family. *J. Biol. Chem.* **264**, 17,374–17,378.
5. Blenis, J. and Hawkes, S. P. (1984) Characterization of a transformation-sensitive protein in the extracellular matrix of chicken embryo fibroblasts. *J. Biol. Chem.* **259**, 11,563–11,570.
6. Pavloff, N., Staskus, P. W., Kishnani, N. S., and Hawkes, S. P. (1992) A new inhibitor of metalloproteinases from chicken: ChIMP-3. A third member of the TIMP family. *J. Biol Chem.* **267**, 17,321–17,326.
7. Greene, J., Wang, M., Liu, Y. E., Raymond, L. A., Rosen, C., and Shi, Y. E. (1996) Molecular cloning and characterization of human tissue inhibitor of metalloproteinase 4. *J. Biol Chem.* **271**, 30,375–30,380.
8. Nagase, H. and Brew, K. (2002) Engineering of tissue inhibitor of metalloproteinases mutants as potential therapeutics. *Arthritis Res.* **4**, S51–S61.

9. Knight, C. G., Willenbrock, F., and Murphy, G. (1992) A novel coumarin-labelled peptide for sensitive continuous assays of the matrix metalloproteinases. *FEBS Lett.* **296,** 263–266.
10. Osthues, A., Knäuper, V., Oberhoff, R., Reinke, H., and Tschesche, H. (1992) Isolation and characterization of tissue inhibitors of metalloproteinases (TIMP-1 and TIMP-2) from human rheumatoid synovial fluid. *FEBS Lett.* **296,** 16–20.
11. Morrison, J. F. and Walsh, C. T. (1988) The behaviour and significance of slow-binding enzyme inhibitors. *Adv. Enzymol. Relat. Areas Mol. Biol.* **61,** 201–301.
12. Bieth, J. G. (1995) Theoretical and practical aspects of proteinase inhibition kinetics. *Meth. Enzymol.* **248,** 59–84.
13. Troeberg, L., Tanaka, M., Wait, R., Shi, Y. E., Brew, K., and Nagase, H. (2002) *E. coli* expression of TIMP-4 and comparative kinetic studies with TIMP-1 and TIMP-2: insights into the interactions of TIMPs and matrix metalloproteinase 2 (gelatinase A). *Biochemistry* **41,** 15,025–15,035.
14. Troeberg, L. and Nagase, H. (2003) Measurement of Matrix Metalloproteinase activities in the medium of cultured synoviocytes using zymography. In: *Inflammation Protocols*, Winyard, P. and Willoughby, D. A., eds., Humana Press, Totowa, NJ, pp. 77–87.
15. Oliver, G. W., Leferson, J. D., Stetler-Stevenson, W. G., and Kleiner, D. E. (1997) Quantitative reverse zymography: analysis of picogram amounts of metalloproteinase inhibitors using gelatinase A and B reverse zymograms. *Anal. Biochem.* **244,** 161–166.
16. Bury, A. F. (1981) Analysis of protein and peptide mixtures. Evaluation of three sodium dodecyl sulfate-polyacrylamide gels electrophoresis buffer systems. *J. Chromatog.* **213,** 491–500.
17. Woessner, J. F., Jr. (1995) Quantification of Matrix Metalloproteinases in Tissue Samples. *Meth. Enzymol.* **248,** 510–528.
18. Jackson, C. J., Arkell, J., and Nguyen, M. (1998) Rheumatoid synovial endothelial cells secrete decreased levels of tissue inhibitor of MMP (TIMP1). *Ann. Rheum. Dis.* **57,** 158–161.
19. Watanabe, H., Wada, H., Itoh, M., Kataoka, M., Kido, H., and Naruse, T. (2002) Effect of heparinoid on the production of tissue inhibitor of metalloproteinases (TIMP)-3 in rheumatoid synovial fibroblasts. *J. Pharm. Pharmacol.* **54,** 699–705.
20. Nagase, H., Fields, C. G., and Fields, G. B. (1994) Design and characterization of a fluorogenic substrate selectively hydrolyzed by stromelysin 1 (matrix metalloproteinase-3). *J. Biol. Chem.* **269,** 20,952–20,957.

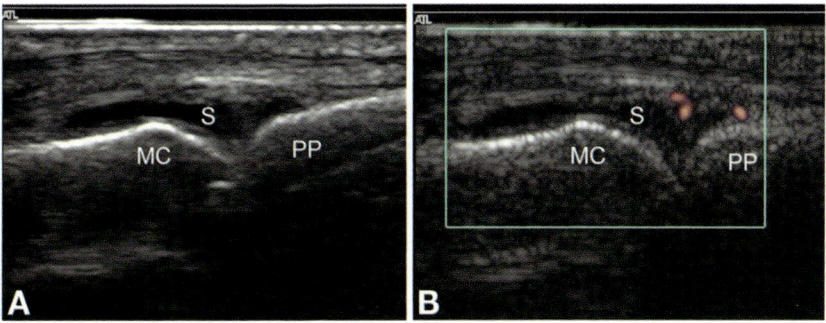

Fig. 4. Ultrasonography of a metacarpophalangeal joint. (*See* discussion in Ch. 1 on pp. 15–16 and complete caption on p. 16.)

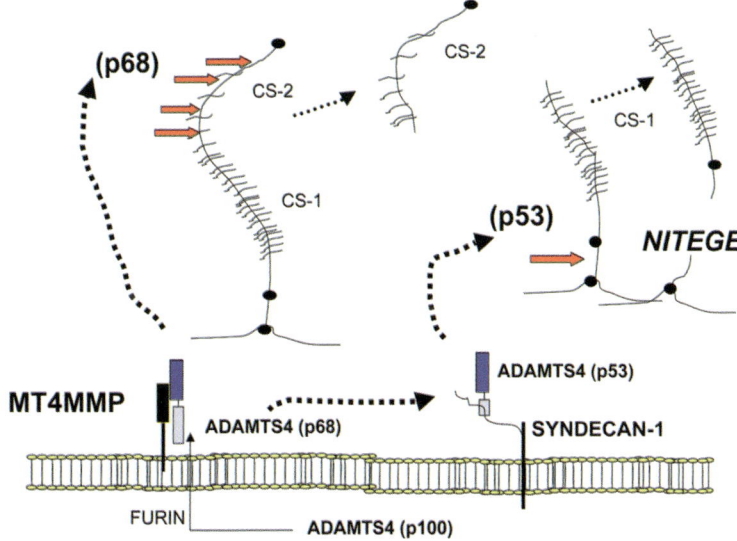

Fig. 1. Chondrocyte-mediated processing of ADAMTS-4. (*See* discussion in Ch. 10 on p. 168 and complete caption on p. 169.)

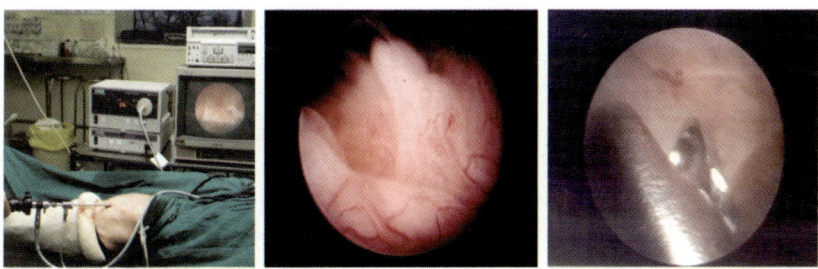

Fig. 1. Demonstration of a 2.7-mm-diameter arthroscope. (*See* discussion in Ch. 22 on p. 343 and complete caption on p. 346.)

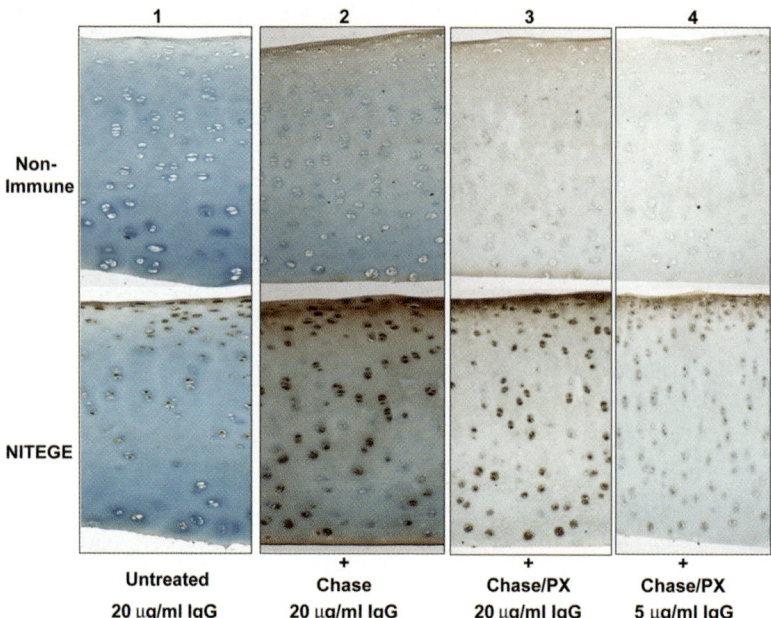

Fig. 2. IHC with non-immune IgG and anti-NITEGE of freshly excised bovine articular cartilage. (*See* discussion in Ch. 10 on p. 173 and complete caption on p. 174.)

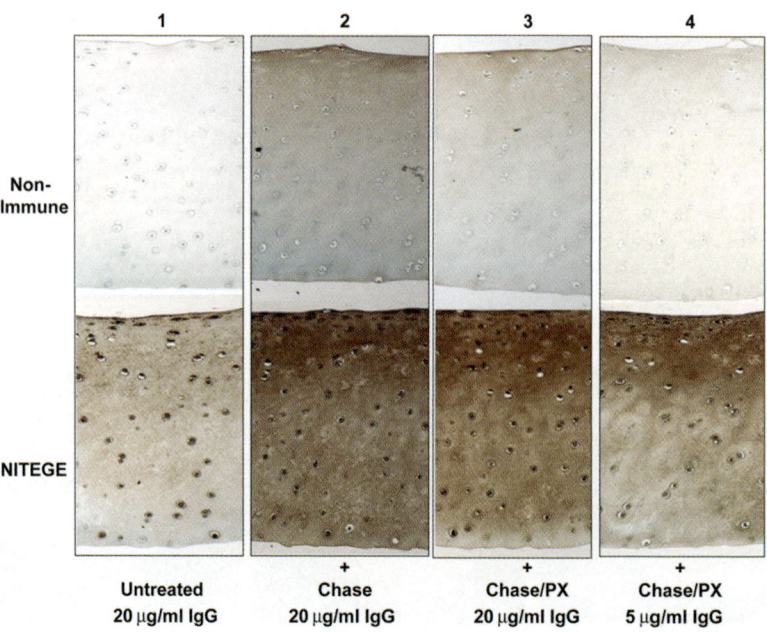

Fig. 3. IHC with non-immune IgG and anti-NITEGE of cartilage after explant culture in RA. (*See* discussion in Ch. 10 on pp. 173, 177 and complete caption on p. 175.)

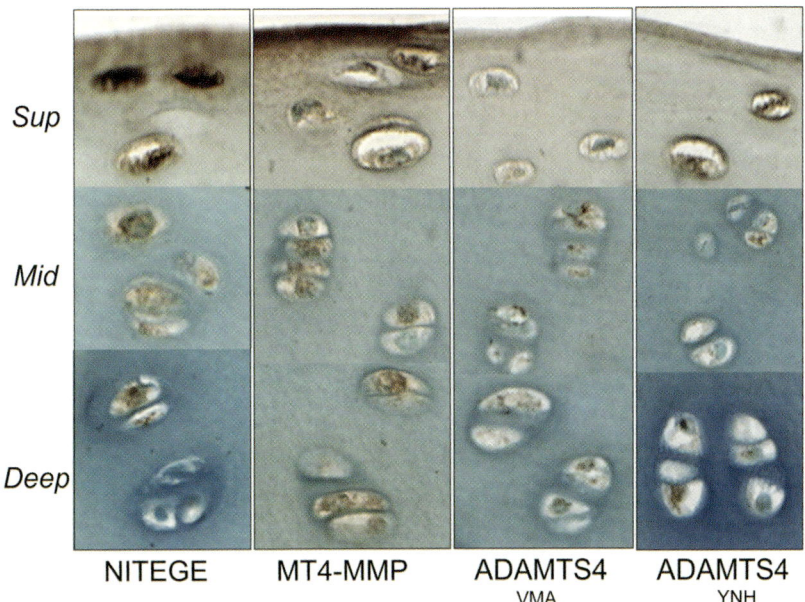

Fig. 4. IHC of freshly excised bovine articular cartilage using anti-NITEGE, anti-MT4MMP, and anti-ADAMTS4. (*See* discussion in Ch. 10 on pp. 173, 175, 177 and complete caption on p. 176.)

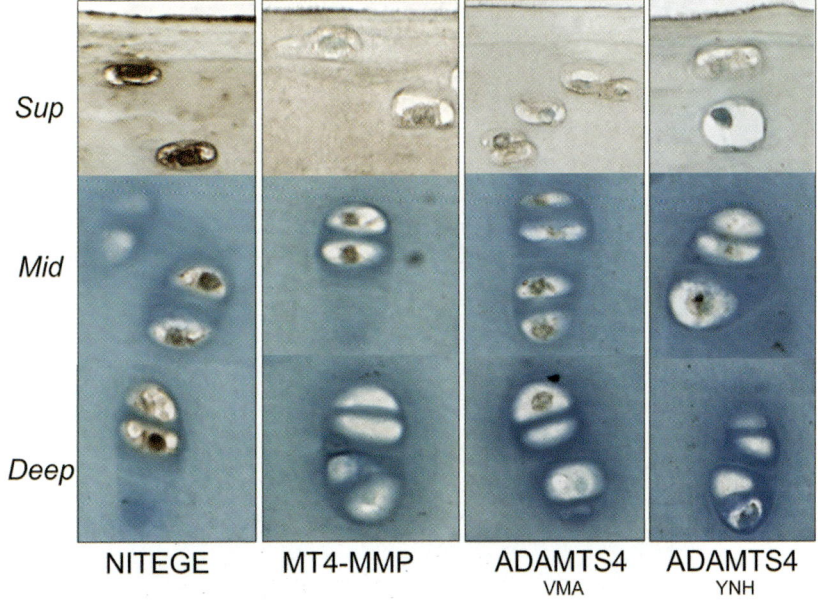

Fig. 5. IHC of bovine articular crtilage after 4 d of serum-free culture. (*See* discussion in Ch. 10 on pp. 173, 175 and complete caption on p. 177.)

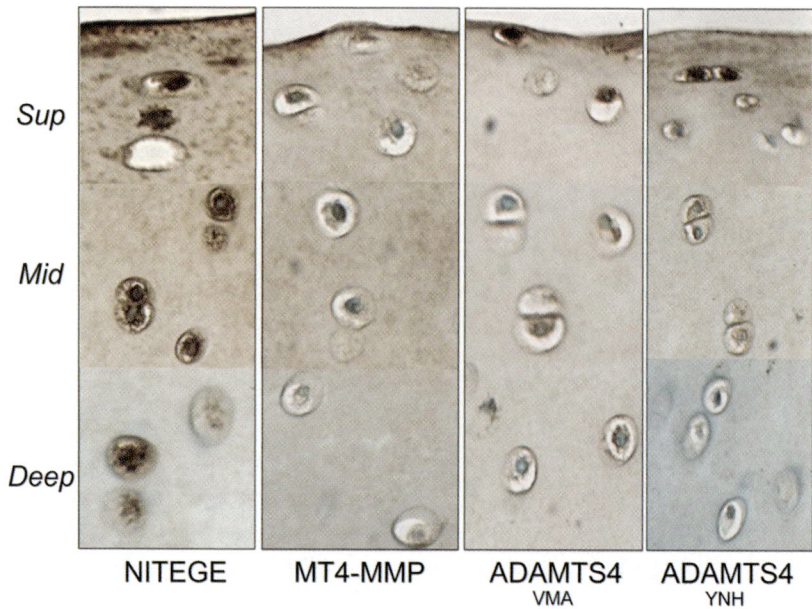

Fig. 6. IHC of bovine articular cartilage after 4 d of serum-free culture with 3 µ*M* RA. (*See* discussion in Ch. 10 on pp. 173, 178 and complete caption on p. 178.)

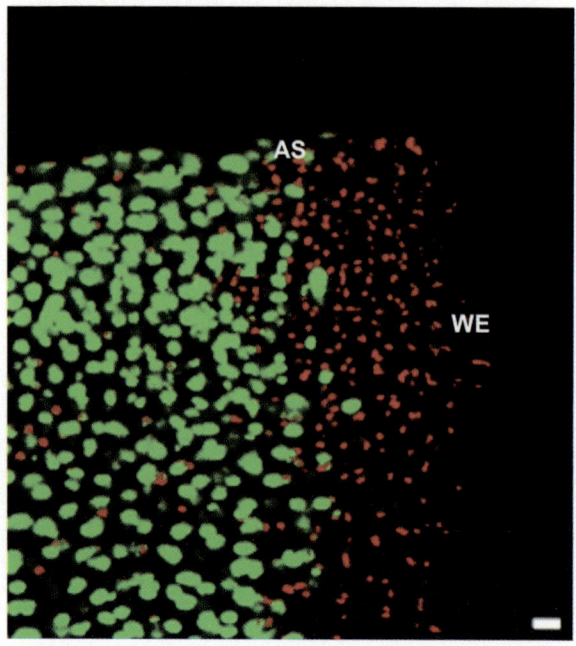

Fig. 5. Use of Live/Dea kit to detect live and dead cells. (*See* discussion and complete caption in Ch. 11 on p. 195.)

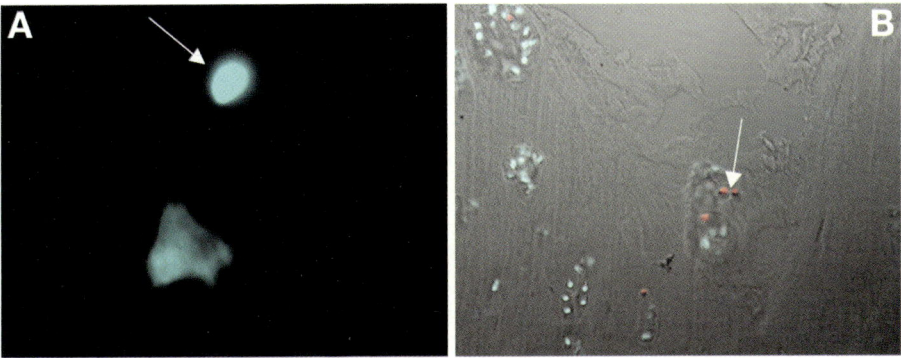

Fig. 4. Use of DAPI to detect apoptotic nuclei. (*See* discussion in Ch. 11 on p. 192 and complete caption on p. 193.)

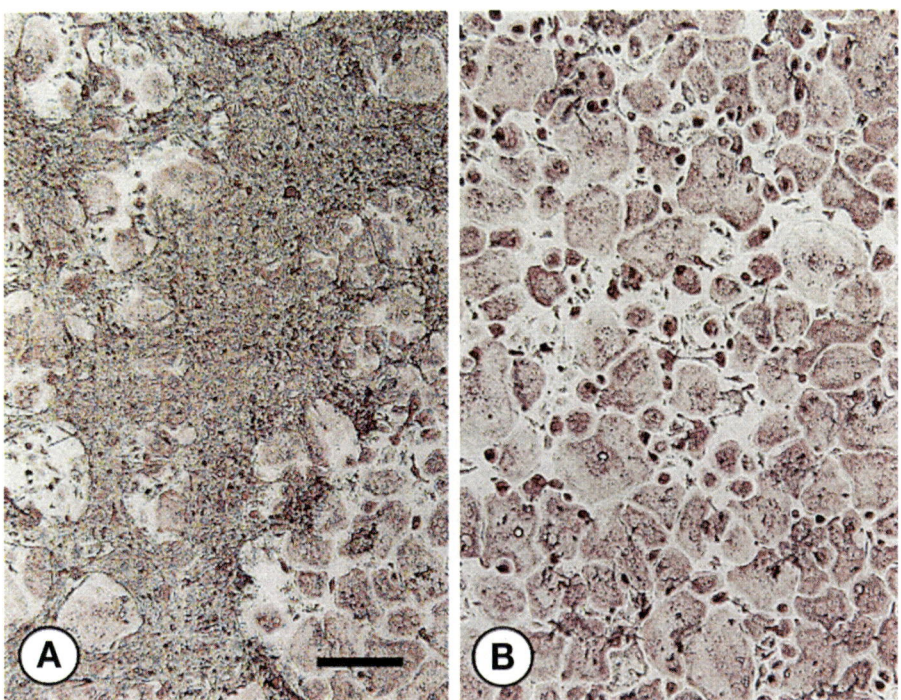

Fig. 2. Purified TRAP-positive osteoclasts. (*See* discussion in Ch. 18 on p. 292 and complete caption on p. 293.)

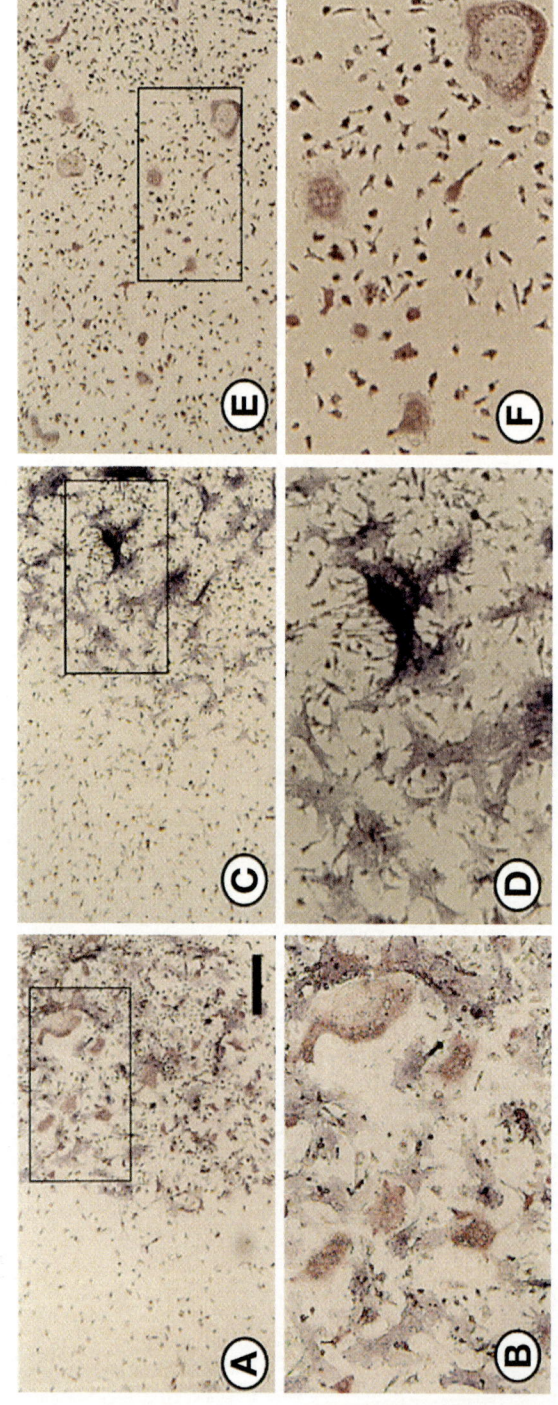

Fig. 1. Enzyme histochemistry for TRAP and ALP in mouse bone marrow cultures. (*See* discussion in Ch. 18 on p. 288 and complete caption on p. 289.)

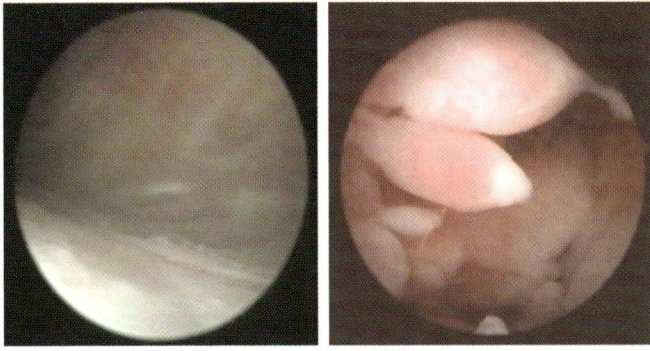

Fig. 2. Arthroscopic images of chondropathy and synovitis in RA. (*See* discussion and full caption in Ch. 22 on p. 348.)

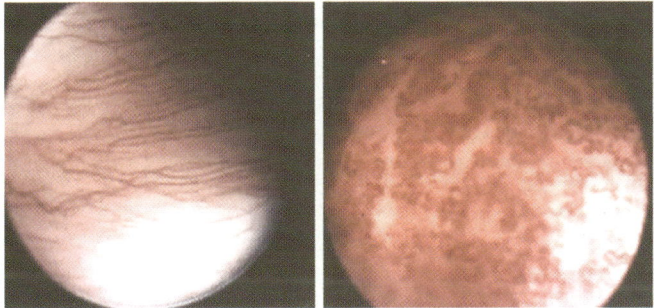

Fig. 3. Arthroscopic images of straight and tortuous vessels. (*See* discussion in Ch. 22 on p. 348 and complete caption on p. 349.)

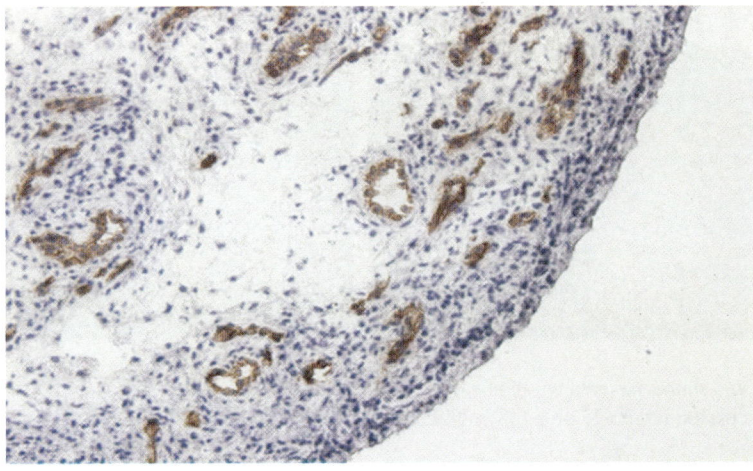

Fig. 4. CD31 staining in early PsA synovial tissue section. (*See* discussion and caption in Ch. 22 on p. 350.)

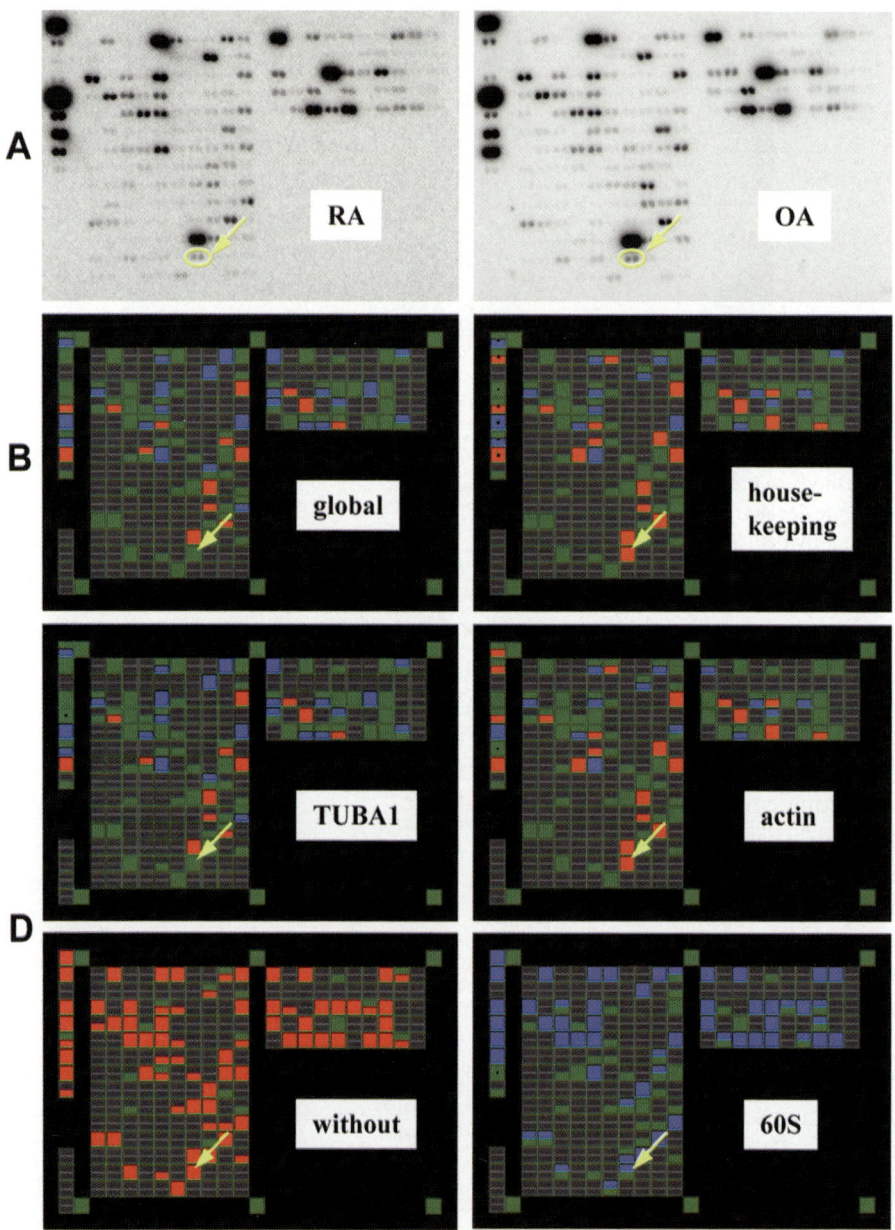

Fig. 4. Comparison views using different normalization settings of two cDNA arrays. (*See* discussion in Ch. 25 on p. 387 and complete caption on p. 388.)

17

Bone Histomorphometry in Arthritis Models

Georg Schett and Birgit Tuerk

Summary

Rodent arthritis models are tools to study joint destruction. This chapter describes methods to prepare and analyze histological sections from inflamed paws. Protocols to create decalcified paraffin-embedded as well as undecalcified plastic-embedded sections are provided. Strengths and limitations of these two histological techniques are described. Major staining techniques on hard tissue sections focusing on the evaluation of joint destruction are indicated. Finally, evaluation of arthritic paw sections by histomorphometry, a semiautomated computerized method to measure areas of interest, is described.

Key Words: Arthritis; histology; decalcified paraffin-embedded sections; undecalcified plastic-embedded sections; and histomorphometry.

1. Introduction

Chronic arthritis usually leads to structural changes of bone and cartilage. In human diseases, like rheumatoid arthritis (RA), the assessment of structural articular changes is dominated by conventional X-ray and other imaging methods. Direct histological analysis of bone and cartilage is usually not possible because of the scarcity of human material. Typically specimens from blind or arthroscopic synovial biopsies are analyzed. These are usually "soft tissue" of the synovial membrane but do not stem from the adjacent cartilage or bone.

Animal models of arthritis allow a direct histological analysis of joint destruction. Each animal model has its peculiarities, which are important determinants for choosing the optimal localization for an adequate histological analysis. This chapter describes a method that allows a detailed analysis of joint destruction. This description comprises the generation of paraffin-embedded joint sections, the generation of undecalcified plastic-embedded joint sections and histomorphometric evaluation of sections. Histomorphometry is a semiautomated computerized method to measure areas of interest on histo-

Table 1
Strengths and Limitations of Paraffin- and Plastic-Embedded Joint Sections

Intended analysis	Paraffin-embedded joint section	Plastic-embedded joint section
Duration of sample preparation	Shorter (d)	Longer (d)
Analysis of soft tissue (e.g., synovia)	Excellent	Moderate
Analysis of hard tissue (e.g., bone)	Moderate	Excellent
Analysis of synovial inflammation	Excellent	Moderate
Analysis of proteoglycan loss	Excellent	Good
Analysis of bone erosion	Excellent	Excellent
Analysis of new bone formation	Poor	Excellent
Analysis of mineralization	Poor	Excellent
Analysis of osteoclasts	Excellent	Excellent
Analysis of osteoblasts	Poor–good*	Eexcellent
Immunohistochemistry:	Yes	No
Histochemistry (e.g., TRAP)	Yes	Yes
In situ hybridization	Yes	No

Note: *Good if osteoblasts are detected by *in situ* hybridization

pathological sections *(1–3)*. It is widely used to characterize systemic bone changes, such as osteoporosis, but is also useful for many other applications such as measuring inflammatory joint destruction *(4)*.

Because most animal models of arthritis are systemic diseases, which involve more than one single joint, it is possible that both techniques are used in combination. Thus, the joint of one side can analyzed by standard histology, whereas the other side can be used for histomorphometry. However, such procedure requires that the disease similarly affect both sides. If only one side is affected the respective technique, which yields best answers for the questions to be addressed, needs to be chosen. **Table 1** lists the limitations and strengths of each technique are listed, which can facilitate this decision.

2. Materials

2.1. Preparation of Sections

1. 4.5% Paraformaldehyde (, zink-free, neutral-buffered (pH 7.2).
2. 14% Ehtylene diamine tetraacetic acid (EDTA) decalcification solution (pH 7.2): mix 140 g EDTA free acid with 850 mL distilled water, add 90 mL ammonium hydroxide (NH_4OH), and adjust pH to 7.2 with ammonium hydroxide.
3. Make up to 1 L with distilled water.
4. Ethanol in different concentrations (70, 80, 96, 100%).
5. 100% Xylol.

Bone Histomorphometry in Arthritis Models 271

6. 100% Methanol.
7. 100% Methylmethacryl acid or K-Plast embedding System (Medim, Buseck, Germany).
8. Benzoylperoxide or K-Plast embedding medium.
9. Dibutylphthalate or K-Plast embedding medium.
10. Paraffin 52°C and 56°C melting temperature.

2.2. Staining of Sections

1. Azophloxine (0.4 g ponceau de xylidine and 0.1 g acid fuchsine dissolved in 0.6 mL acetic acid and 300 mL distilled water.
2. Phosphomolybdic acid orange G: 3 to 5g phosphomolybdic acid dissolved in 100 mL distilled water, add 2 g orange G.
3. Sodium formol solution: 5 g sodium carbonate water free in 25 mL 37% formaldehyde and 75 mL distilled water.
4. Light Green Solution: 0.2g to 0.3g Light Green SF solved in 0.2 mL acetic acid and 100 mL distilled water.
5. Tartrate-resistant acid phosphatase staining kit (Sigma Diagnostics; cat. no. 387-A).
6. 100% *n*-butylacetate.
7. 10% Silver nitrate solution.
8. 0.1% Toluidine blue solution.
9. 5% Sodium thiosulfate.
10. 1% Acetic acid .
11. Weigert's iron hematoxylin.
12. Mayer's hematoxylin.

2.3. Hardware and Other Items

1. Vacuum infiltration processor (for automated paraffin embedding).
2. Tissue embedding console (for paraffin embedding).
3. Glass or polypropylene containers (for plastic embedding).
4. (Airtight) embedding moulds (for plastic embedding).
5. Embedding rings (allow a better positioning of specimen in the microtome; for plastic embedding).
6. Sliding microtome (for paraffin sections).
7. Stainless steel microtome blade holders (for paraffin sections).
8. Disposable microtome blades (for paraffin sections).
9. Rotation microtome with motor equipped with a carbide knife (for plastic sections).
10. Cooling plate (to cool specimen before cutting; for paraffin sections).
11. Silan coated glass slides (for paraffin sections).
12. 1% Glue coated slides (for plastic sections).
13. Water bath (to stretch sections).
14. Incubator (to dry sections).
15. Polyethylene foil (for plastic sections).
16. Image analysis system (microscope, camera, computer and software, either OsteoMeasure or BioQuant).

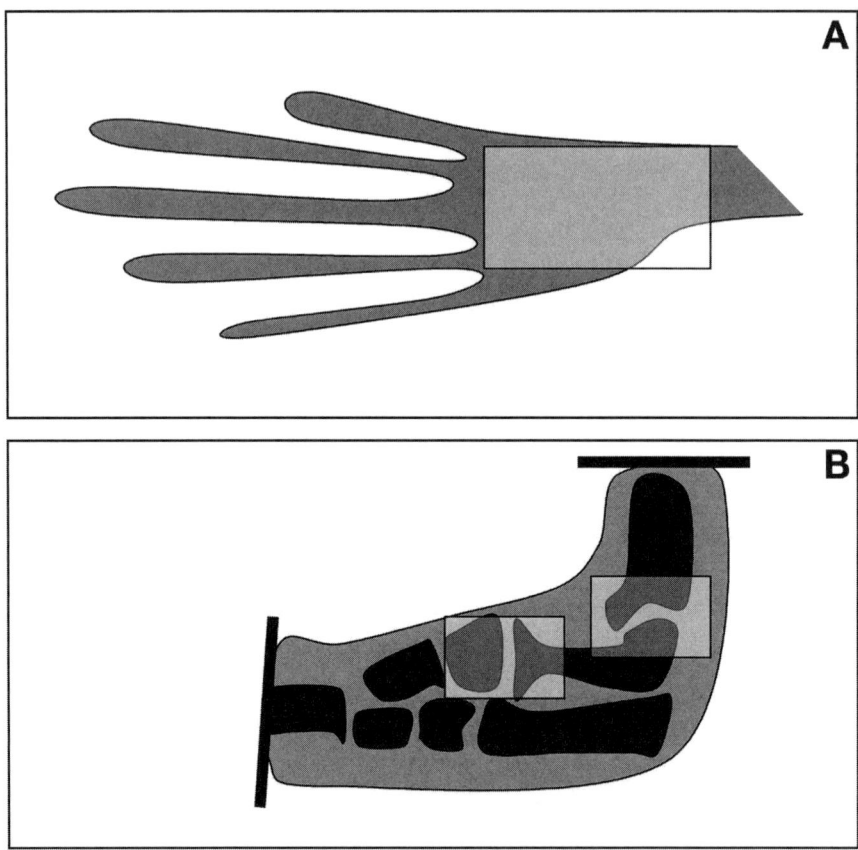

Fig. 1. Sections of mouse (**A**) and rat (**B**) paws. Mouse paws can be dissected in a frontal direction. The region of interest is the carpal or tarsal area in the front or hind paw, respectively. Rat paws are cut sagitally into two pices and then cut in sagittal direction. Toes have been removed. The region of interest is the tarsotibial (TT) and the tarsonavicular (TN) joint of the hind paw.

3. Methods

3.1. Preparation of Decalcified Paraffin-Embedded Paw Sections

3.1.1. Preparation of Paws

Animals are deeply anaesthetised for necropsy. Paws will be removed using sharp rongeurs, with the cut placed at the fur line of the distal limbs (just proximal to the carpometacarpal joints [wrist] or tibiotarsal joint [ankle]). Skin, major tendons and muscles as well as the plantar aponeurosis are removed. Rat paws are then cut sagitally into two pieces (*see* **Fig. 1** and **Note 1**).

3.1.2. Fixation

Paws will be immersed in fixative for 5 to 8 h (mice) or up to 72 h (rats). Proper fixation requires the following practices (*see* **Note 2**):

1. Choice of fixative: 4.5% neutral-buffered zink-free paraformaldehyde (pH 7.2).
2. Quantity of fixative: Ten volumes of fixative (in mL) per volume of tissue (in g).
3. Fixation temperature: room temperature (approx 21–22°C).
4. Fixation schedule: fix freshly harvested paws for 24 h, then change the fixative and fix for up to additional 48 h (rats). For mice no change of fixative necessary.
5. Maximum length of fixation: 8 h (mice) to 72 h (rats).
6. Conditions for long-term storage of adequately fixed tissue:
   a. Preferred fixative: 70% ethanol.
   b. Temperature: room temperature (approx 21–22°C)
   c. Quantity of fixative: cover tissue with fixative.
   Alternative fixation techniques are described elsewhere in detail *(5)*.

3.1.3. Decalcification

Proper decalcification requires the following practices:

1. Choice of decalcifying method: Prepare the following decalcification solution containing 14% of EDTA.
2. Quantity of decalcifying solution: ten volumes of solution (in mL) per volume of tissue (in g) or no more than 20 paws for each large histology container filled to the brim with decalcifying solution.
3. Fixation temperature: 4°C on a shaker.
4. Length of decalcification: 5 to 7 d (*see* **Note 3**).
5. Fixation schedule: six to eight serial changes, at intervals of 24 to 36 h.

Alternative decalcification techniques are described elsewhere in detail *(6)*.

3.1.4. Confirmation of Adequate Decalcification

After four to six changes of solution, bone hardness is tested in a control bone sample, which will not be used for any further analysis. A fresh razor blade is used to cut the bone near border formed by the rongeur cut. Alternatively, also a needle may be used. Adequately decalcified bone trims easily, with no evidence of firmness (crunching).

3.1.5. Paraffin Embedding

Table 2 shows an embedding procedure that is usually performed by an automated system.

The specimens are then transferred into a tissue embedding console and embedded in paraffin (56°C melting temperature) with the site down, which is later used for cutting (= plantar site down in mice). Alternative embedding techniques are described elsewhere in detail *(7)*.

Table 2

	Time	Temperature
4.5% neutral-buffered zinc-free paraformaldehyde (pH 7.2)	1 h	40°C
70% ethanol	1 h	40°C
80% ethanol	1 h	40°C
96% ethanol	1 h	40°C
96% ethanol	1 h	40°C
100% ethanol	0.5 h	40°C
100% ethanol	1 h	40°C
100% ethanol	1 h	40°C
100% xylene	1 h	40°C
100% xylene	1.5 h	40°C
Paraffin (52°C melting temperature)	1 h	56°C
Paraffin (52°C melting temperature)	1 h	56°C
Paraffin (52°C melting temperature)	1 h	56°C
Paraffin (52°C melting temperature)	1.5 h	56°C

3.1.6. Cutting of Sections

For cutting paraffin sections use a sliding microtome equipped with a microtome blade holder and disposable blade (*see* **Note 4**).

1. Make sections of 2 µm thickness.
2. Stretch section in a water bath (50°C).
3. Mount section onto a silan-coated slide (*see* **Note 5**).

3.2. Processing of Decalcified Paraffin-Embedded Paw Sections

Paraffin sections can be used for a variety of different detection techniques. The above-mentioned protocol is suitable for application of various staining methods. **Table 3** summarizes some of the most useful histological staining techniques, which can be applied on paraffin sections and their specific value in arthritis models.

Some of these techniques can be used simultaneously on the same section. Examples are:

1. Combination of TRAP with ISH (*see* **Fig. 2**).
2. Combination of TRAP with ICH.
3. Combination of IHC with ISH.

A more detailed description of basic staining techniques and immunohistochemistry on decalcified paraffin-embedded bone sections is found elsewhere *(8)*.

Table 3
Staining Techniques on Paraffin Sections of Relevance for Arthritis Studies

Method	Major aim of application
Haematoxylin & Eosin	Overview of inflammation and structural damage.
Toluidine blue	Cartilage proteoglycan loss. Detection of mast cells.
Safranin O	*Cartilage proteoglycan loss.*
Van Gieson	*Collagen fibers.*
Naphthol AS-D chloraceta testerase	Detection of granulocytes.
Tartrate-resistant acid phosphatase (TRAP)	Osteoclasts and osteoclast precursors, bone erosion.
Immunohistochemistry (IHC)	Characterization of cell types and/or synovial protein expression.
In situ hybridization (ISH)	Synovial mRNA expression (*see* **Note 6**)

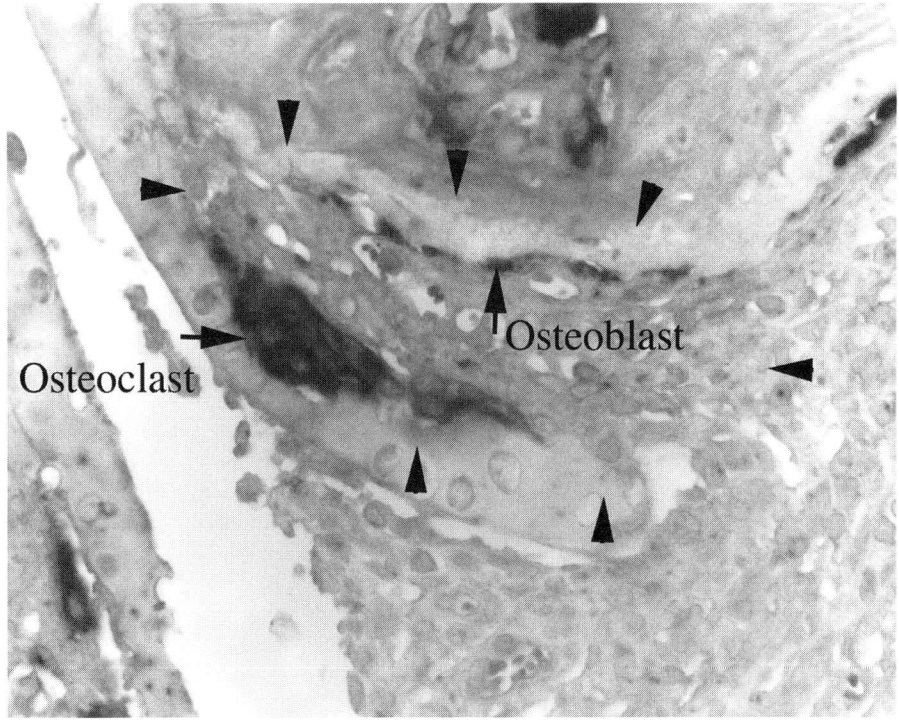

Fig. 2. Decalcified paraffin-embedded section of an arthritic mouse paw showing double labeling of osteoclasts (tartrate-resistant acid phosphatase) and osteocalcin (ISH). Arrowheads indicate the subchondral bone erosion.

3.3. Preparation of Undecalcified Plastic-Embedded Paw Sections

3.3.1. Preparation of Paws

Refer to protocol for paraffin sections.

3.3.2. Fixation

Paws can be fixed and stored in 70% alcohol before embedding the material in plastic. Additional fixation by other agents is not generally necessary, however, can sometimes improve the quality of tissue sections and also allows staining of osteoclasts by TRAP. We therefore highly recommend a short fixation with 4.5% neutral-buffered zink-free formalin (pH 7.2) for up to 12 h at room temperature.

3.3.3. Plastic Embedding

Plastic embedding allows conservation of the mineral structure of bones, which is necessary for the morphometric and functional assessment. The combination of polymethylmethacrylates with plastisisers makes the material sliceable on special rotation microtome with carbide knives. To induce polymerization catalysts (e.g., benzoylperoxide) are added and the temperature is carefully increased. It takes exercise to find the right methacrylate—mixture and polymerization temperature depending on the specimen size.

The following protocol is a standard protocol for generating plastic sections:

1. Dehydration: 70% ethanol for 24 h at room temperature followed by 100% methanol for 2 × 24 h at room temperature.
2. Infiltration: Bone samples are put in 85 mL methylmethacrylate +15 mL dibutylphthalate placed under vacuum for 3 h at room temperature. Then the bone samples are moved from vacuum and put at room temperature under fume hood for totally 3 d with a stirrer. Bone samples are then placed in 85 mL methylmethacrylate + 15 mL dibutylphthalate + 0.5 g benzoylperoxide under vacuum for 3 h at room temperature. Then the bone samples are moved from vacuum and put at room temperature under fume hood for totally 3 d with a stirrer (*see* **Note 7**).
3. Polymerization: Bone samples are faced down on the bottom of precoated vials with 12 mL of embedding solution (85 mL methylmethacrylate + 15 mL dibutylphthalate + 2.5 g benzoylperoxide). The vials are capped tightly and then put in water bath with 40°C for 24 to 48 h.
   For smaller specimens commercial embedding sets have been developed (e.g., K-Plast® System FA Medim, Buseck, Germany), which work very well for generating plastic sections of arthritic paws. These sets follow a similar protocol with dehydration (*see* above), infiltration (for 2 × 24 h at room temperature) and polymerization (in air tight in embedding moulds at 37°C overnight).

Table 4
Staining Techniques on Plastic Sections of Relevance for Arthritis Studies

Method	Major aim of application
Goldner-Trichrome	Green/blue: mineralized bone.
	Intense red: osteoid, osteoblasts, and osteoclasts.
	Reddish: soft tissue.
Von Kossa	Black: mineralised bone.
	Red: other tissue.
Toluidine blue	Blue: Cartilage proteoglycan.
TRAP	Purple: Osteoclasts.

3.3.4. Cutting of Sections

For cutting plastic sections use rotation microtome equipped with a motor and a carbide knife (*see* **Note 8**).

1. Make sections of 2 to 3 µm thickness.
2. Stretch section in a waterbath (80°C).
3. Mount section onto a slide coated with 1% glue.
4. Cover section with polyethylen foil avoiding air bubbles.
5. Compress section and polyethylen foil.
6. Place it in an incubator (60°C) overnight (maintain compression).

A detailed description of the cutting procedure of plastic-embedded tissue specimen is found elsewhere *(9)*.

3.4. Processing of Undecalcified Plastic-Embedded Paw Sections

Plastic sections can be used for a limited number of detection techniques *(10)*. **Table 4** summarizes some of the most useful histological staining techniques that can be applied on plastic sections and their specific value in arthritis models.

3.4.1. Goldner-Trichrome Stain

1. Remove plastic: 100% methylmetarylic acid 30 min at room temperature.
2. Rinse briefly in 100% ethanol followed by rising in 96% ethanol.
3. Air dry.
4. Dip in distilled water.
5. Weigert's iron hematoxylin at room temperature for 10 min.
6. Differentiate in 1% HCl alcohol.
7. Rinse in tap water 10min.
8. Azophloxine 5min.

9. Rinse in 1% acetic acid.
10. Differentiate in phosphomolybdic acid orange G 5 for15 s.
11. Rinse in 1% acetic acid.
12. Place into light green solution 5min.
13. Rinse in 1% acetic acid.
14. Dehydrate rapidly with 96 and then 100% ethanol.
15. Clear with *n*-butylacetate.
16. Mount with resinous medium.

3.4.2. Von Kossa stain (see **Fig. 3**)

1. Remove plastic: 100% methylmetarylic acid.
2. Rinse briefly in 100% ethanol followed by rising in 96% ethanol.
3. Air dry.
4. Dip in distilled water.
5. Place slides in 10% silver nitrate solution 10 to 30 s.
6. Two changes distilled water 5 min each.
7. Bring slides into sodium formol solution until the calcified structures appear black 2to 5 min.
8. Fix in 5 % sodium thiosulfate 5 min.
9. Two changes distilled water 5 min each.
10. Bring slides into azophloxine 5min.
11. Rinse in 1% acetic acid.
12. Differentiate in phosphomolybdic acid orange G 5 for15 s.
13. Rinse in 1% acetic acid.
14. Dehydrate rapidly with 96 and then 100% alcohol.
15. Clear with *n*-butylacetate.
16. Mount with resinous medium.

3.4.3. Toluidine Blue Stain

1. Remove plastic: 100% methylmetarylic acid.
2. Rinse briefly in 100% ethanol followed by 96% ethanol.
3. Air dry.
4. Dip in distilled water.
5. Bring slides in 0.1% toluidine blue solution for 10 min.
6. Rinse briefly in distilled water.
7. Differentiate in 96% alcohol.
8. Rinse in 100% alcohol.
9. Clear with *n*-butylacetate.
10. Mount with resinous medium.

3.4.4. TRAP Stain

Activity of tartrate-resistant acid phosphatase can be detected by commercial kits (e.g., Sigma; cat. no. 387-A). Staining procedure is according to the following protocol:

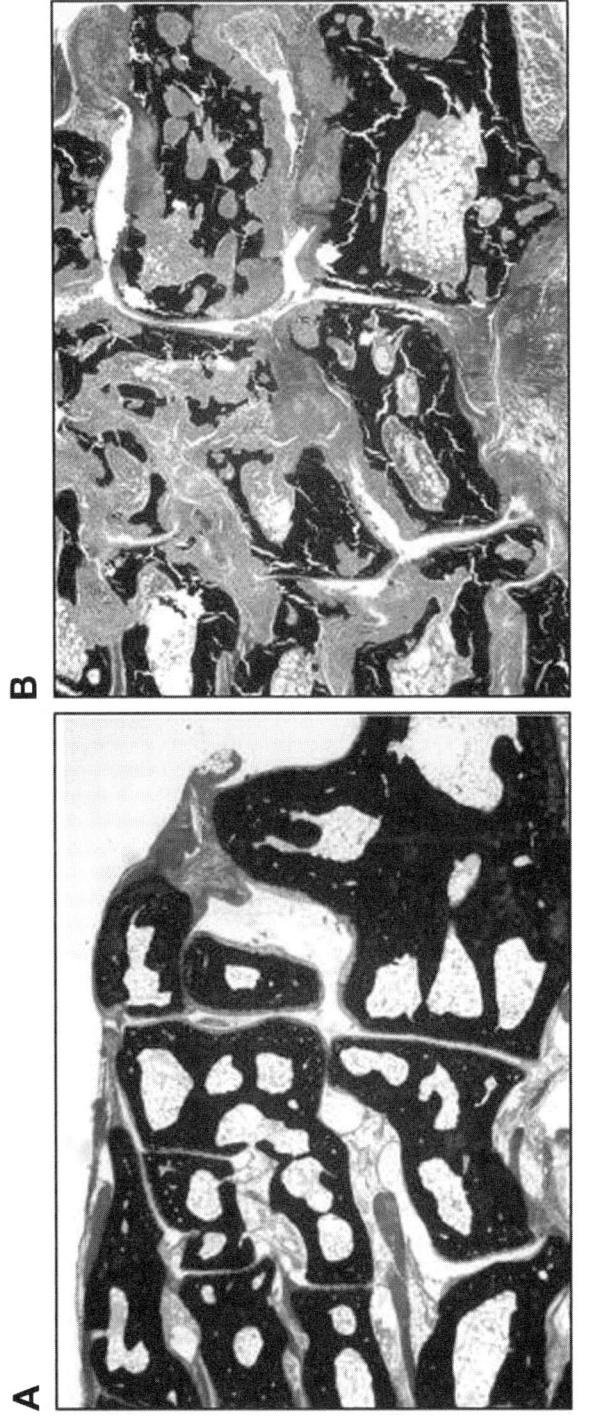

Fig. 3. Undecalcified plastic-embedded section of a normal (**A**) and arthritic (**B**) mouse paw. Sections were stained by von Kossa. Mineralized bone appears black, soft tissue and cartilage is reddish. Note the profound bone loss in arthritic mice evident from a fragmentation of black areas.

1. Remove plastic: 100% methylmetacrylic acid.
2. Rinse briefly in 100% ethanol followed by rising in 96% ethanol.
3. Air dry.
4. Dip in distilled water.
5. Heat 45 mL distilled water up to 37°C.
6. Mix 0.5 mL naphthol AS-BI phosphoric acid, 2 mL acetate solution,1 m: tartrate solution with 45 m: distilled water under continuous shacking.
7. Incubate at 37°C in the dark for 1 h.
8. Generate the substrate by gently mixing 0.5 mL fast garnet GBC base solution and 0.5 mL sodium nitrite solution and leave them for 2 min at room temperature.
9. Incubate the substrate for 2 min at room temperature.
10. Add substrate and incubate for 8 to 15min at 37°C.
11. Perform counter stain with hematoxylin for 10 to 20 s.
12. Rinse with tap water to blue nuclei.
13. Embed slide with aqueous mounting media.

3.5. Histomorphometry

Histomorphometry is defined as the quantitative measurement and characterization of microscopy images using a computer. A manual or automated digital image analysis typically involves measurements and comparisons of selected geometric areas, perimeters, length angle of orientation, form factors, center of gravity coordinates, as well as image enhancement. Static histomorphometry involves the identification of cellular and tissue components for the measurement of lengths (mm), areas (mm^2) or cell counts (#/mm or $\#/mm^2$). Dynamic histomorphometry, in contrast, makes use of fluorochromes, such as calcein, that are incorporated into bone at the front of calcification (*see* **Note 9**). These labeled sites will fluoresce and can be viewed with ultraviolet (UV) microscopy. When fluorochromes are administered over a time interval, the rates of formation and mineralization can be calculated from measurements of tissue growth between the labels.

Analysis of sections by histomorphometry requires an image analysis system, which consists of a camera linking the microscope to a computer equipped with special software, allowing histomorphometric evaluation of slides (**Table 5**). The most frequently used software systems are OsteoMeasure and BioQuant. Parameters of histomorphometry are standardized and included in these programmes *(11)*. The investigator needs to be able to define bone, osteoid, osteoclasts and osteoblasts on the respective section. After defining and marking these structures histomorphometry parameters are calculated automatically (*see* **Notes 10** and **11**).

Table 5
Most Useful Histomorphometry Parameters for Arthritis Studies

BV/TV[a]	Bone volume per tissue volume.
	This parameter gives on overview of inflammatory bone loss.
ES/BS[a]	Eroded surface per bone surface.
	This parameter describes the extent of inflammatory bone loss.
N. OC/B.Pm[b]	Numbers of osteoclasts per bone perimeter and
OcS/BS[2]	osteoclast surface per bone surface.
	These parameters depict the activity of inflammatory bone resorption.
N. OB/B. Pm[b]	Numbers of osteoblasts per bone perimeter.
ObS/BS[b]	Osteoblast surface per bone surface.
OV/TV[b]	Osteoid volume per tissue volume.
	These parameters evaluate the activity of bone repair.
MAR[b]	Mineral apposition rate.
	These parameters demonstrate the dynamics of bone repair.

[a]Parameter refers to the whole paw section.
[2]parameter may refer to inflammatory bone erosions only. However, the parameter may be measured at other sites too (e.g., periarticular bone).

4. Notes

1. Careful dissection of paws is important for the quality of sections. Fibrous tissues like skin, tendons, and muscles need to be removed to prevent fragmentation of sections during cutting. This includes also the plantar aponeurosis, which is plate-like fibrous structure. Also claws should be removed, since decalcification does not improve pliability of these structures.
2. Fixation is a critical step and should be as short as possible. If applied to long, paraformaldehyde interferes with immunohistochemistry due to masking of antigens, requires harsher pretreatment with proteinase K for ISH and abolishes TRAP enzyme activity, which is an important tool to detect osteoclasts in inflammatory tissue.
3. If decalcification time is too short, cutting is difficult and quality of sections decreases. In contrast, longer decalcification is usually not a problem.
4. Quality of paraffin section depends on the microtome blade used. There are disposable blades, which work very good for cutting paraffin sections (e.g., Feather Microtome Blade A35).
5. Thorough drying of sections (in an incubator) is very important to prevent detaching.
6. If ISH is planned, all materials and procedures should be performed in RNase free conditions.
7. Plastic embedding of long bones is facilitated if the marrow space is opened through cutting away one end of the bone to allow better penetration of embedding solution.

8. To prevent tearing of plastic embedded tissue cutting should be started carefully with low speed and feed.
9. If dynamic histomorphometry should be used the investigator has to inject a fluorochrome before sacrificing the animal. Two injections are needed and the time interval depends on the amount of bone turnover in an animal. Younger animals have higher bone turnover than older animals. A standard procedure, for example, is the subcutaneous injection of 0.3 mg calcein 7 and 1 d before necropsy.
10. Results obtained from histomorphometry depend on the orientation of the sections. Exactly matching orientation of sections is very important to achieve comparable results.
11. Histomorphometric measurements can be also performed on articular cartilage. For example, measurement of the fraction of cartilage area with proteoglycan loss (toluidine blue negative) compared to total cartilage area (toluidine blue negative and positive) yields a quantitative measurement of cartilage damage *(12)*.

References

1. Recker, R. (ed.) (1983) *Bone histomorphometry*: Techniques and Interpretation. CRC, Boca Raton, FL.
2. Vedi, S. and Compston, J. (2003) Bone Histomorphometry, in *Bone Research Protocols*, Ralston, SH, Helfrich, MH, eds., Humana Press, Totowa, NJ, pp. 283–298.
3. Aaron, J. E. and Shore., P. A. (2003) Bone histomorphometry: Concepts and common techniques, in *Handbook for Histology Methods for Bone and Cartilage*, An, Y. H., Martin, K. L., eds., Humana Press, Totowa, NJ, pp. 331–352.
4. Redlich, K., Gortz, B., Hayer, S., et al. (2004) Repair of local bone erosions and reversal of systemic bone loss upon therapy with anti-tumor necrosis factor in combination with osteoprotegerin or parathyroid hormone in tumor necrosis factor-mediated arthritis. *Am. J. Pathol.* **164**, 543–555.
5. Scarano, A., Moreira, P. L., Kang, Q. K., and Gruber, H. E. (2003) Common fixatives in hard tissue technology, in *Handbook for Histology Methods for Bone and Cartilage*, An, Y. H., Martin, K. L., eds., Humana Press, Totowa, NJ, pp. 159–166.
6. Skinner, R. A. (2003) Decalcification of bone tissue, in *Handbook for Histology Methods for Bone and Cartilage*, An, Y. H., Martin, K. L., eds., Humana Press, Totowa, NJ, pp. 167–184.
7. An, Y. H., Moreira, P. L., Kang, Q. K., and Gruber, H. E. (2003) Principles of embedding and common protocols, in *Handbook for Histology Methods for Bone and Cartilage*, An, Y. H., Martin, K. L., eds., Humana Press, Totowa, NJ, pp. 185–198.
8. Gruber, H. E. and Ingram, J. A. (2003) Basic staining and histochemical techniques and immunohistochemical localization using bone sections, in *Handbook for Histology Methods for Bone and Cartilage*, An, Y. H., Martin, K. L., eds., Humana Press, Totowa, NJ, pp. 281–286.

9. Ries, W. L. (2003) Techniques for Sectioning undecalcified bone tissue using microtomes, in *Handbook for Histology Methods for Bone and Cartilage,* An, Y. H., Martin, K. L., eds., Humana Press, Totowa, NJ, pp. 221–232.
10. Scarano, A., Petrone, G., and Piattelli, A. (2003) Staining techniques of plastic embedded tissue specimen, in *Handbook for Histology Methods for Bone and Cartilage,* An, Y. H., Martin, K. L., eds., Humana Press, Totowa, NJ, pp. 315–320.
11. Parfitt, A. M., Drezner, M. K., Glorieux, F. H., et al. (1987) Bone Histomorphometry: Standardization of Nomenclature, Symbols and Units. *J. Bone Miner. Res.* **6,** 595–610.
12. Zwerina, J., Hayer, S., Tohidast-Akrad, M., et al. (2004) Single and combined inhibition of tumor necrosis factor, interleukin-1, and RANKL pathways in tumor necrosis factor-induced arthritis: Effects on synovial inflammation, bone erosion, and cartilage destruction. *Arthritis Rheum.* **50,** 277–290.

18

Generation of Osteoclasts In Vitro, and Assay of Osteoclast Activity

Naoyuki Takahashi, Nobuyuki Udagawa, Yasuhiro Kobayashi, and Tatsuo Suda

Summary

Osteoclasts are bone-resorbing multinucleated cells derived from the monocyte-macrophage lineage. The authors have developed a mouse marrow culture system and a coculture system of mouse osteoblasts and hemopoietic cells, in which osteoclasts are formed in response to various osteotropic factors such as 1α,25-dihydroxyvitamin D_3, parathyroid hormone, prostaglandin E_2, and interleukin -11. Recent studies have revealed that osteoblasts express two cytokines essential for osteoclastogenesis: receptor activator of nuclear factor κB ligand (RANKL) and macrophage colony-stimulating factor (M-CSF). Using RANKL and M-CSF, we can induce osteoclasts from monocyte-macrophage lineage cells even in the absence of osteoblasts. This chapter describes the methods for osteoclast formation in vitro in the presence and absence of osteoblasts, and for pit-formation assay using dentine slices and osteoclasts formed in vitro. These culture systems have made it possible to investigate each step of osteoclast development and function separately.

Key Words: Osteoclast; osteoblast; RANKL; RANK; M-CSF; bone marrow cell; RAW264.7 cell; collagen gel-culture; 1α,25-$(OH)_2D_3$; TRAP; calcitonin; bisphosphonate; pit assay.

1. Introduction

Osteoclasts, the multinucleated giant cells that resorb bone, originate from hemopoietic cells of the monocyte–macrophage lineage *(1–6)*. We have developed a mouse bone marrow culture system in which osteoclasts are formed in response to several bone-resorbing factors such as 1α,25-dihydroxyvitamin D_3 [1α,25-$(OH)_2D_3$], parathyroid hormone (PTH), prostaglandin E_2 (PGE_2) and interleukin-11 (IL-11) *(4,5)*. We also developed a mouse coculture system of primary osteoblasts and hemopoietic cells to examine the regulatory mechanism of osteoclastogenesis *(4–6)*. A series of experiments using the coculture system have established the concept that osteoblasts or bone marrow-derived stromal cells have a key role in regulating osteoclast differentiation. Macrophage colony-stimulating factor (M-CSF, also called CSF-1) produced by osteoblasts/stromal cells was shown to be an essential factor for differentiation of osteoclasts from the progenitor cells *(7–9)* (*see* **Note 1**). Receptor activator of nuclear factor κB ligand (RANKL), a new member of the tumor necrosis factor (TNF) family, was also identified as another essential factor for osteoclastogenesis *(10–12)*. Osteoblasts/stromal cells express RANKL as a membrane-associated factor in response to various bone resorbing factors (*see* **Note 2**). Osteoclast precursors possess RANK, a TNF receptor family member, recognize RANKL through cell–cell interaction with osteoblasts/stromal cells, and differentiate into osteoclasts in the presence of M-CSF *(12,13)*. Recent studies have shown that mouse macrophage-like RAW264.7 cells can differentiate into osteoclasts in response to RANKL even in the absence of M-CSF *(14)*. We have also developed a method for obtaining a large number of functionally active osteoclasts from cocultures grown on collagen gel-coated dishes *(15)*. Using osteoclasts recovered from the collagen gel culture, a reliable pit-formation assay was established to investigate the regulatory mechanisms of osteoclast function *(16)*. The methods for osteoclast formation in vitro and for the pit-formation assay using dentine slices and osteoclasts formed in vitro are described here.

2. Materials
2.1. Mice and Cell Lines

1. ddY mice (*see* **Note 3**).
2. Mouse bone marrow-derived stromal cell lines, ST2 and MC3T3-G2/PA6, and a mouse macrophage cell line, RAW264.7 (RIKEN Cell Bank, Tsukuba, Japan).

2.2. Reagents

1. Recombinant human M-CSF (Leukoprol; Kyowa Hakko Kogyo Co. Tokyo, Japan, or R &D Systems, Minneapolis, MN) (*see* **Note 4**).
2. 1α,25-$(OH)_2D_3$ and PGE_2 (Wako Pure Chemical Industries, Ltd., Osaka, Japan),
3. PTH (Peptide Institute, Inc., Osaka) and IL-11 (Pepro Tech EC Ltd., London, UK).
4. Human osteoprotegerin (OPG) and a soluble form of human RANKL (Pepro Tech EC Ltd., London).
5. Synthetic analogue of eel calcitonin (Elcatonin, Asahi Kasei Pharma, Tokyo).
6. Type I collagen gel solution (cell matrix type IA; Nitta Gelatin Co., Osaka) (*see* **Note 5**).
7. Bacterial collagenase (Wako Pure Chemical Industries, Ltd.).
8. Tissue culture plastic ware (Corning).

Osteoclast Generation and Function

9. α-Modification of minimum essential medium (α-MEM), RPMI-1640 and Ca^{2+}- and Mg^{2+}-free phosphate-buffered saline [PBS(–)] (Sigma Chemical Co., St. Louis, MO).
10. Fetal bovine serum (FBS) (JRH Biosciences, Lenexa, KS) (see **Note 6**).
11. A monoclonal antibody against mouse cathepsin K (Serotec, Oxford, UK).
12. Biotinylated second antibodies, avidin-biotin conjugated peroxidase, and an AEC substrate kit (Histofine, Nichirei Co., Tokyo, Japan).
13. Sterile instruments, syringes, and needles.
14. Other chemicals and reagents are of analytical grade.

2.3. Culture Media and Buffer Solutions

1. α-MEM containing 10% FBS for cultures of osteoclast differentiation and function.
2. RPMI-1640 containing 10% FBS for maintenance of RAW264.7 cells.
3. PBS(–) for washing cells.
4. a-MEM containing 0.2% bacterial collagenase for detachment of cells cultured on collagen gel-coated dishes.
5. Trypsin–ethylene diamine tetraacetic acid (EDTA) solution (Sigma, St. Louis, MO): PBS(–) containing 0.05% trypsin and 0.5 mM EDTA for detachment of cells from culture plates.
6. Pronase–EDTA solution: PBS(–) containing 0.001% pronase and 0.02% EDTA for removal of osteoblasts from cocultures. Pronase is dissolved in PBS(–) containing 0.02% EDTA just before use.
7. 3.7% Formaldehyde in PBS(–) for fixation of cells.
8. 0.1% Triton X-100 in PBS(–) for permeabilization of cells fixed with 3.7% formaldehyde in PBS(–).
9. Tartrate-resistant acid phosphatase (TRAP) staining solution: Five milligrams of naphthol AS–MX phosphate is dissolved in 0.5 mL of N,N'-dimethyl formamide in a glass container. Thirty milligrams of fast red violet LB salt and 50 mL of 0.1 M sodium acetate buffer (pH 5.0) containing 50 mM sodium tartrate are added to the mixture. TRAP-staining solution can be stored for 1 mo in the refrigerator.
10. Type I collagen mixture for preparing collagen gel-coated dishes: Type I collagen solution (see **Subheading 3.5.**), 5X α-MEM, and 200 mM N-2-hydroxyethyl-piperazine-N'-2-ethanesulfonic acid (HEPES) buffer (pH 7.4) containing 2.2% $NaHCO_3$ (7:2:1, by vol) are quickly mixed at 4°C just before use.

3. Methods
3.1. Bone Marrow Cultures

The mouse bone marrow culture system was developed for examining the effects of bone-resorbing factors on osteoclast formation *(6)*. The discovery of the RANKL–RANK interaction for osteoclastogenesis indicated that the growth of stromal cells is an essential step for osteoclast development in bone marrow cultures *(5)* (*see* **Note 2**).

1. Tibiae are removed aseptically from 7- to 9-wk-old male mice and the bone ends are cut off with scissors. The marrow cavities are flushed by injecting 1 mL of α-MEM at one end of the bone using a sterile 27-gage needle.
2. Bone marrow cells are washed once with α-MEM, suspended in α-MEM containing 10% FBS, and cultured at 1×10^6 cells/0.5 mL/well in 48-well plates (Corning, Corning, NY) in a humidified atmosphere of 5% CO_2. Cultures are fed every 2 to 3 d by replacing 0.4 mL of spent medium with fresh medium.
3. Osteotropic factors such as $10^{-8}\,M$ 1α,25-$(OH)_2D_3$, 100 ng/mL of PTH, $10^{-6}\,M$ PGE_2, and 10 ng/mL of IL-11 induce osteoclast formation in this marrow culture. These factors are usually added at the beginning of the culture and at each time the medium is changed.
4. Cells are fixed and stained for TRAP (a marker enzyme of osteoclasts) as described in **Subheading 3.7.1.**

TRAP-positive mononuclear cells appear on days 3 to 4 and multinucleated cells on days 4 to 5 in the presence of bone-resorbing factors *(6)*. The number of TRAP-positive multinucleated cells reaches a maximum on days 6 to 8. TRAP-positive osteoclasts are formed only near the colonies of alkaline phosphatase (ALP)-positive osteoblasts in the cultures treated with PTH *(6)* (*see* **Fig. 1A, B**; *see* **Note 7**). OPG completely inhibits the TRAP-positive cell formation induced by PTH in bone marrow cultures (*see* **Fig. 1C,D**). Osteoclasts are also formed when mouse bone marrow cultures are treated with 50 ng/mL M-CSF and 100 ng/mL RANKL (*see* **Fig. 1E,F**). In this culture, osteoclasts are formed uniformly throughout the culture dish.

3.2. Cultures of Bone Marrow-Derived Macrophages

Macrophages appearing in bone marrow cultures are the precursors of osteoclasts *(17)*. The mouse bone marrow culture system was modified to prepare highly purified osteoclast precursors *(18,19)*.

1. First incubation: Bone marrow cells suspended in α-MEM containing 10% FBS are cultured with 100 ng/mL M-CSF in 6-cm culture dishes for 16 h (2×10^7 cells/10 mL/dish) (*see* **Note 8**).
2. Second incubation: Nonadherent cells were gently collected and further cultured for 2 d with 100 ng/mL M-CSF. Then, non-adherent cells are removed by pipetting. Adherent cells strongly express macrophage specific antigens such as Mac-1, Moma-2, and F4/80. Therefore, adherent cells are called "M-CSF-dependent bone marrow macrophages (BMMϕ)." Typically, 1×10^4 BMMϕ are obtained, when 1×10^5 bone marrow cells are cultured for 3 d in the presence of M-CSF.
3. Adherent BMMϕ are scraped off with a rubber scraper, and collected by centrifugation. BMMϕ are further cultured for 3 d in 24-well plates (1×10^4 cells/well) with 50 ng/mL RANKL and 50 ng/mL of M-CSF (*see* **Note 9**).
4. Cells are fixed and stained for TRAP.

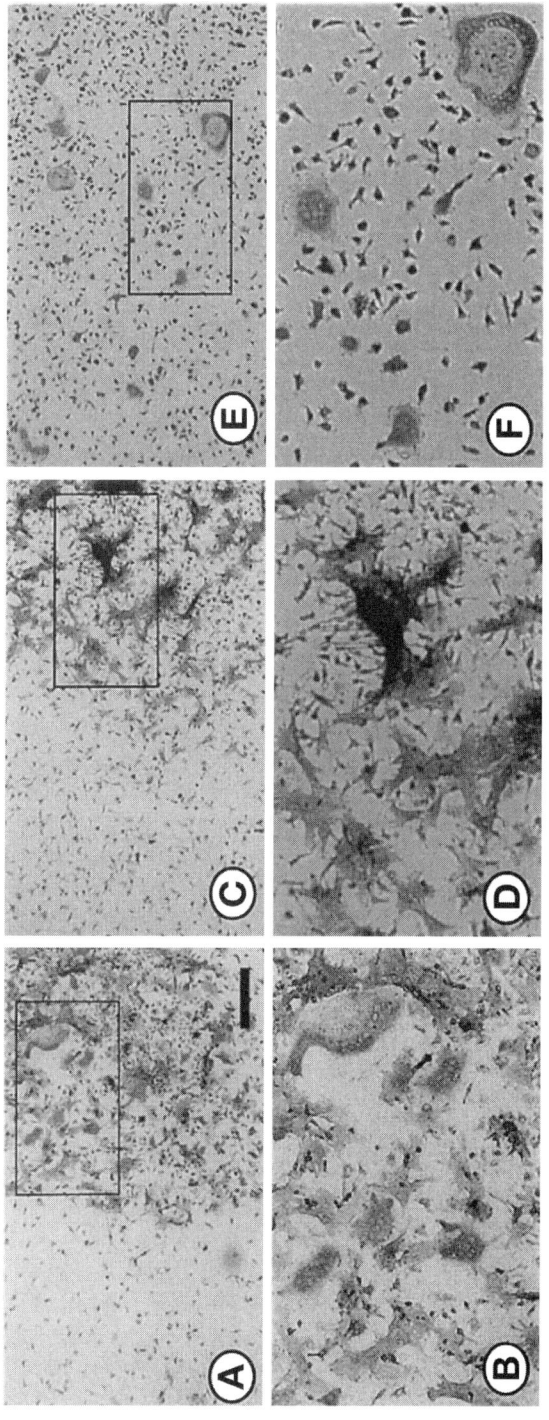

Fig. 1. Enzyme histochemistry for TRAP and ALP in mouse bone marrow cultures. Bone marrow cells of ddY were cultured for 7 d with 100 ng/mL PTH (**A**), 100 ng/mL PTH plus 100 ng/mL OPG (**C**), or 100 ng/mL RANKL plus 50 ng/mL of M-CSF (**E**). Marrow cultures were then fixed and double-stained for TRAP and ALP. TRAP-positive cells appeared as red cells and ALP-positive cells as blue cells. **B**, **D**, and **F** show high power views of the boxed portions in **A**, **C**, and **E**. respectively. Note that TRAP-positive cells formed in the culture are observed near or within the colonies of ALP-positive cells in the presence of PTH (**B**). In contrast, TRAP-positive cells are distributed uniformly on the culture dish in the presence of both RANKL and M-CSF (**F**). Adding OPG completely suppressed the formation of TRAP-positive cells induced by PTH (**D**). Scale bar = 200 mm. (Color illustration in insert following p. 268.)

Table 1
Characteristics of Culture Systems for Mouse Osteoclast Formation

Culture sysem[a]	Number of osteoclasts formed[b] (cells/cm^2)	Culture period required for oseoclast formation (day)
Bone marrow culture	25–50	6–8
Coculture on plastic dishes primary osteoblasts + bone marrow cells	400–700	6–8
Coculture on collagen gel primary osteoblasts + bone marrow cells	1000–1300	6–8

[a]Each culture system is performed according to methods described in the text.
[b]The number of osteoclasts formated in each culture treated with $1\alpha,25(-(OH)_2D_3$.

3.3. Cultures of RAW264.7 Cells

RAW264.7, a murine macrophage cell line, has the capability to differentiate into osteoclasts in response to RANKL (14,20). RAW264.7 cells are used as osteoclast progenitors.

1. A frozen vial of RAW264.7 cells is thawed in a 37°C water bath. Cells are suspended in RPMI-1640 containing 10% FBS, and collected by centrifugation. RAW264.7 cells are resuspended in RPMI-1640 containing 10% FBS, and cultured for 3 to 5 d to confluence (*see* **Note 10**). RAW264.7 cells are treated with trypsin–EDTA solution, scraped off with a rubber scraper, and collected by centrifugation.
2. RAW264.7 cells are suspended in α-MEM containing 10% FBS, and cultured in a 24-well plate (3×10^4 cells/well) in the presence of RANKL (100 ng/mL) (*see* **Note 11**).
3. The culture is fed every 2 to 3 d with fresh medium.
4. Cells are fixed and stained for TRAP.

3.4. Osteoclast Formation in Cocultures

Cocultures of primary osteoblasts with bone marrow cells or spleen cells produce more osteoclasts than bone marrow cultures do *(9)* (**Table 1**). Bone marrow-derived stromal cell lines such as ST2 and MC3T3-G2PA6 can be used instead of primary osteoblasts in coculture with bone marrow cells *(20)* (*see* **Note 12**).

3.4.1. Preparation of Primary Osteoblasts

1. Thirty to fifty newborn ddY mice (1- to 3-d-old) are used for one preparation of osteoblasts. Mice are anaesthetized with ether, and sacrificed by decapitation. Heads are placed in a Petri dish with PBS(–). Calvariae are taken out, and the attached skin and muscle are removed carefully.
2. Calvariae are put into a 50-mL centrifuge tube and washed twice with α-MEM to remove blood cells. Then, calvaria are incubated in 10 mL of α-MEM containing 0.1% bacterial collagenase (Wako) and 0.2% dispase (Godo Shusei Co., Tokyo) for 10 min at 37°C in a shaking water bath (120 cycles/min).
3. The collagenase solution is recovered, and a fresh collagenase solution (10 mL) is added. Calvariae are further incubated for another 10 min at 37°C in a shaking water bath (120 cycles/min). The incubation of calvariae with collagenase solution is repeated 5 times.
4. Primary osteoblasts isolated in Fractions 2 to 5 are collected by centrifugation, and cultured for 3 days in α-MEM containing 10% FCS in 10-cm culture dishes (Corning) (cells obtained from 10 calvariae/dish) (*see* **Note 13**).
5. Primary osteoblasts are detached from the dish by treating with trypsin-EDTA, centrifuged ($250g$, 5 min) and suspended in α-MEM containing 20% FCS and 15% dimethyl sulfoxide, and stored at –80°C (1×10^6 cells/mL/freezing vial). The osteoclast formation-supporting activity of primary osteoblasts is not destroyed in the freezer for at least 3 mo.

3.4.2. Cocultures of Osteoblasts and Bone Marrow Cells

1. Primary osteoblasts (1×10^4 cells/well) are cocultured with bone marrow cells (1×10^5 cells/well) in 48-well plates (0.5 ml/well) in α-MEM containing 10% FCS. Cultures are fed every 3 d by replacing 0.4 mL of spent medium with fresh medium.
2. Osteotropic factors such as $1\alpha,25\text{-}(OH)_2D_3$ (10^{-8} M), PTH (100 ng/mL), PGE_2 (10^{-6} M) and IL-11 (10 ng/mL) induce osteoclast formation in the coculture. These factors are usually added at the beginning of the culture and each time the medium is changed.
3. Cells are fixed and stained for TRAP.

3.5. Collagen-Gel Cultures

Osteoclasts formed on plastic culture dishes are very difficult to detach by the treatment with either trypsin–EDTA or bacterial collagenase. To obtain functionally active osteoclasts formed in cocultures with osteoblasts, a collagen gel culture is recommended *(15,21)* (**Table 1**).

1. A 10-cm culture dish (Corning) is coated with 5 mL of the type I collagen mixture on ice. The dish is put in a CO_2 incubator for 10 min to make the aqueous type I collagen gelatinous at 37°C.

2. Primary osteoblasts (2×10^6 cells) and bone marrow cells (2×10^7 cells) are cocultured on a collagen gel-coated dish in 15 mL of α-MEM containing 10% FBS and 10^{-8} M 1α,25-$(OH)_2D_3$. The medium is changed every 2 to 3 d.
3. After the cells are cultured for 7 d, the dish is treated with 4 mL of 0.2% collagenase solution for 20 min at 37°C in a shaking water bath (60 cycles/min). The culture dishes are carefully placed on a sheet of aluminum foil put on the water surface of the water bath to maintain the sterile condition of the dishes.
4. The cells released from the dish are collected by centrifugation at 250g for 5 min and suspended in 10 mL of α-MEM containing 10% FBS (the crude osteoclast preparation). Usually, 4×10^4–1×10^5 osteoclasts are recovered from a 10-cm collagen gel-coated dish, and the purity of osteoclasts is 2–3% in this crude preparation.
5. The crude osteoclast preparation is used for biological and biochemical studies of osteoclasts.

3.6. Purification of Osteoclasts Formed In Vitro

Because the purity of osteoclasts in the crude osteoclast preparation is only 2 to 3%, further purification is essential for biochemical studies of osteoclasts (*see* **Fig. 2**). Osteoclasts are easily purified from the crude osteoclast preparation placed on plastic dishes by treatment with pronase–EDTA solution *(22,23)*.

1. Five milliliters of the crude osteoclast preparation is placed on a 6-cm culture dish (Corning) for 6 to 15 h in the presence of 10% FBS.
2. Adherent cells are washed with α-MEM, and treated with 4 mL of pronase-EDTA solution for 10 min.
3. Osteoblasts are then removed by gentle pipetting. More than 90% of the adherent cells on the dishes are TRAP-positive mononuclear and multinucleated cells.

The purified osteoclasts rapidly die as a result of apoptosis within 48 h *(22–24)*. Therefore, biochemical experiments on osteoclasts should be performed within 24 hr after purification of osteoclasts.

3.7. Identification of Osteoclasts Formed In Vitro

3.7.1. TRAP Staining

Cytochemical staining for TRAP is widely used for identification of osteoclasts in vivo and in vitro.

1. Cells are fixed with 3.7% (v/v) formaldehyde in PBS(–) for 10 min, fixed again with ethanol–acetone (50:50 [v/v]) for 1 min, and incubated with the TRAP-staining solution for 10 min at room temperature (*see* **Note 14**).
2. TRAP-positive osteoclasts appear as dark-red cells within 10 min. An incubation period longer than 10 min should be avoided, because cells other than osteoclasts become weakly positive for TRAP with time.
3. After staining, cells are washed with distilled water, and TRAP-positive multinucleated cells having three or more nuclei are counted as osteoclasts under a microscope.

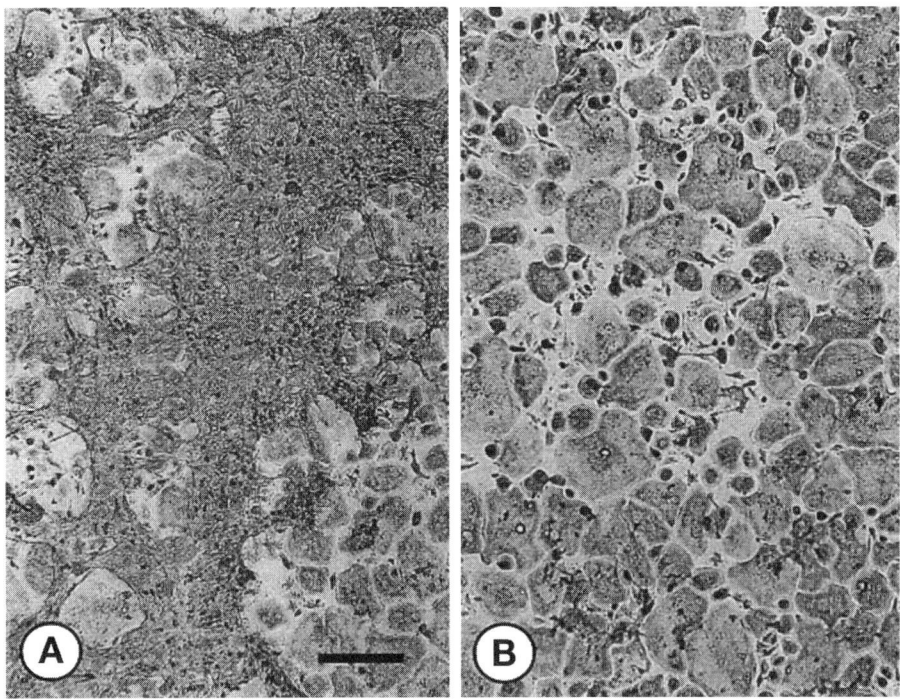

Fig. 2. Purified TRAP-positive osteoclasts formed in cocultures of mouse osteoblasts and bone marrow cells. Primary osteoblasts (2×10^6 cells) and bone marrow cells (2×10^7 cells) were cocultured for 7 d on a collagen gel-coated dish in the presence of $1\alpha,25\text{-}(OH)_2D_3$. The dish was then treated with 0.2% collagenase solution to recover all the cells from the dish. The cells released from the dish were collected by centrifugation and suspended in 10 mL of α-MEM containing 10% FBS (the crude osteoclast preparation). The crude osteoclast preparation was placed on a 10-cm culture dish for 10 h in the presence of 10% FBS (**A**). The purity of osteoclasts in this crude preparation was only 2 to 3%. Adherent cells were washed with α-MEM, then treated for 10 min with 8 mL of trypsin–EDTA solution. Osteoblasts were then removed by gentle pipetting. More than 90% of the adherent cells on the dish were TRAP-positive mononuclear and multinucleated cells (**B**). Scale bar = 200 μm. (Color illustration in insert following p. 268.)

3.7.2. Immunostaining for Cathepsin K

Osteoclasts specifically express cathepsin K, carbonic anhydrase II, and vacuolar proton ATPase *(5,25)*. Immunohistochemical staining of these markers is used for identification of osteoclasts formed in vitro.

1. Cells in a 24-well plate are fixed with 0.5 mL of cold methanol-acetone (50:50 [v/v]) for 10 min on ice.
2. Cells are treated for 10 min with 0.5 mL of 3% H_2O_2 solution to inactivate intrinsic peroxidase on ice. H_2O_2 solution is removed by aspiration.
3. One milliliter of PBS is added to each well, and the plate is incubated for 5 min on ice. This washing step is repeated three times.
4. Cells are then incubated for 10 min with 200 µL of 10% rabbit serum to block nonspecific binding. The rabbit serum solution is removed by aspiration, followed by washing with PBS three times (5 min incubation for each wash).
5. Cells are incubated with polyclonal rabbit antibodies against mouse cathepsin K (1:500 dilution with PBS) or with nonimmune rabbit serum (200 µL/well) at room temperature. After incubation for 1 h, the serum solution is removed by aspiration, and cells are washed with PBS (5 min, three times).
6. Cells are incubated for 10 min with biotinylated second antibodies (200 µL) at room temperature. Cells are washed with PBS (5 min, three times).
7. The bound antibodies are visualized with avidin-biotin conjugated peroxidase, and an AEC substrate kit (Histofine, Nichirei Co., Tokyo, Japan).

Immunohistochemical staining of osteoclasts with antibodies against other specific markers such as carbonic anhydrase II and vacuolar proton ATPase are utilized for identification of osteoclasts. The pattern of immunohistochemical staining with antibodies against carbonic anhydrase II and vacuolar proton ATPase is essentially similar to that with anticathepsin K antibodies *(25)*.

3.8. Pit Formation Assay

When osteoclasts are placed on dentine slices, they form resorption pits within 24 h in the presence of osteoblasts. A reliable pit formation assay was established using the crude osteoclast preparation and dentine slices.

1. Dentine blocks (ivory) are obtained through donation from a local zoo. Dentine slices (φ 4 mm, 200 µm thick) are prepared from ivory blocks using a band saw (BS-3000, Exakt, Germany) and a cutting punch (*see* **Note 15**).
2. Dentine slices are cleaned by ultrasonication in distilled water, sterilized using 70% ethanol, and dried under ultraviolet light. Dentine slices are stored at room temperature.
3. Dentine slices are placed in 96-well plates containing 0.1 mL/well of α-MEM with 10% FBS (1 slice/well). A 0.1-mL aliquot of the crude osteoclast preparation is transferred onto the slices (*see* **Note 16**).
4. After the plates are allowed to stand for 60 min at 37°C, the slices are removed and placed onto 24-well plates containing α-MEM with 10% FBS (0.5 mL/slice/well).
5. After incubation for 24–48 h, the dentine slices are recovered from the culture. The surface of dentine slices is rubbed strongly with a cotton bud to remove all cells on the slices.

6. Ten microliters of Mayer's hematoxylin (Wako Pure Chemical Industries) is placed on the surface of each dentine slice using the surface tension for 35 to 45 s. The dentine slices are washed with distilled water, and rubbed with a cotton bud.
7. Resorption pits are clearly visualized with Mayer's hematoxylin under transmitted light.
8. The number of resorption pits formed on dentine slices is counted under a light microscope. Alternatively, the resorbed area is measured using an image analysis system linked to a light microscope.

Resorption pits are first observed on dentine slices after culturing for 6 to 8 h, and the resorbed areas increase with time up to 72 h. Many resorption pits are observed on a dentine slice recovered from the culture for 48 h (see **Fig. 3A**). When calcitonin at $10^{-9} M$ was added to the culture, the pit-forming activity of osteoclasts was completely inhibited (see **Fig. 3B**). Bisphosphonates also strongly inhibited the pit-forming activity of osteoclasts.

3.9. Actin Ring Formation

Osteoclasts adhere to the bone surface through specialized discrete structures called "podosomes" in the clear zone, which consist mainly of dots containing F-actin *(26,27)*. Therefore, the ringed structure of podosomes (actin ring) formed in osteoclasts is a characteristic of polarized osteoclasts *(28)*. The actin rings are visualized by staining F-actin with rhodamine-conjugated phalloidin.

1. Rhodamine-conjugated phalloidin (Sigma) is dissolved in a small volume of methanol, diluted with PBS(–) to be the final concentration of 0.3 mM, and stored at 4°C in the dark.
2. Cells cultured on dentine slices in 48-well plates are fixed with 0.4 mL of 3.7% formaldehyde in PBS(–) for 10 min and washed with PBS(–).
3. Dentine slices are treated with 0.4 mL of 0.1% Triton X-100 in PBS(–) for 1 min.
4. Dentine slices are incubated for 3 h with the rhodamine-conjugated phalloidin solution in a refrigerator. The rhodamine-conjugated phalloidin solution is recovered, and the slices are washed with water. The rhodamine-conjugated phalloidin solution can be used repeatedly unless F-actin staining becomes weak.
5. Actin rings formed by osteoclasts on dentine slices are detected with a fluorescence microscope (Olympus BX-FLA, Osaka).

Actin rings formed in osteoclasts are disrupted by adding several inhibitors of bone resorption such as calcitonin *(29)* and bisphosphonates *(30)* (see **Fig. 4A,B**).

4. Notes

1. Osteopetrosis is an inherited disorder characterized by an increase in bone mass resulting from reduced bone resorption. Experiments on the osteopetrotic op/op

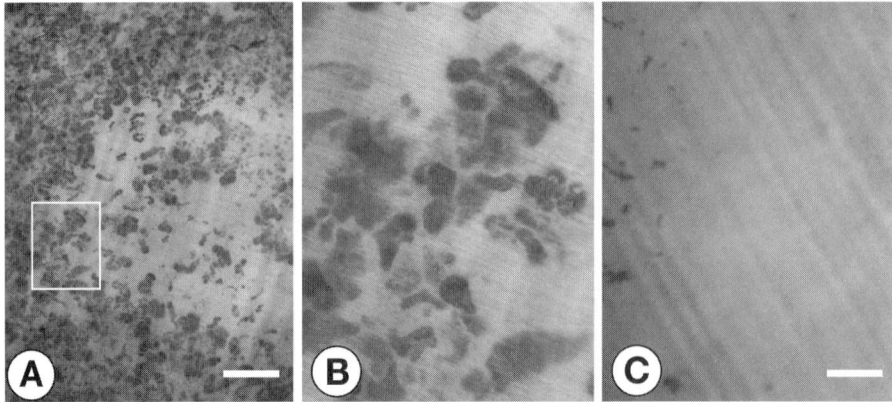

Fig. 3. Resorption pits formed by osteoclasts on dentine slices. The osteoclast preparation recovered from a collagen-gel culture was placed on a dentine slice (φ 4 mm), and cultured for 48 h in the presence (**C**) or absence (**A**) of eel calcitonin ($10^{-9}\,M$) After culturing for 48 h, cells were removed from the dentine slice. The slice was then stained with Mayer's hematoxylin to visualize resorption pits. Many resorption pits are observed on the dentine recovered from the control culture (**A**). **B** shows a high power view of the boxed portion in **A**. Calcitonin strongly suppressed pit-forming activity of osteoclasts (**C**). **B** shows an enlarged portion of **A**. Scale bar = 200 μm.

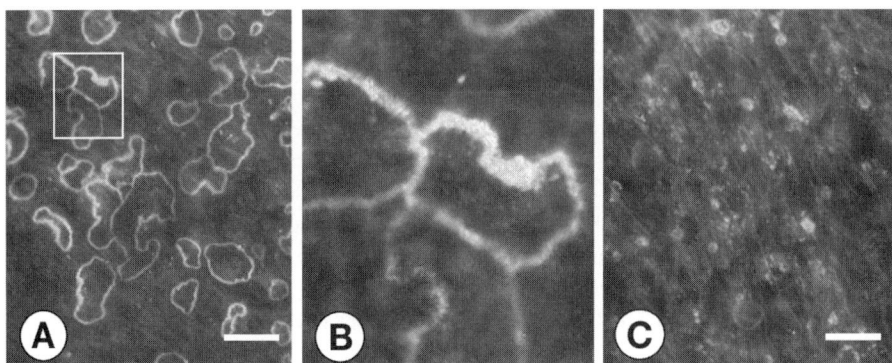

Fig. 4. Actin ring formation by osteoclasts. The osteoclast preparation recovered from a collagen-gel culture was placed on a dentine slice (φ 4 mm), and cultured for 24 h in the presence (**C**) or absence (**A**) of eel calcitonin ($10^{-9}\,M$). Cells were then fixed, and incubated with rhodamine-conjugated phalloidin to visualize the distribution of F-actin. Many actin rings are observed on the dentine slices recovered from the control culture (**A**). **B** shows a high power view of the boxed portion in (**A**). **C** shows that calcitonin strongly suppressed actin ring formation by osteoclasts (**C**). Bars = 50 mm.

mouse model have established that an osteoblast/stromal cell product, M-CSF, is a crucial factor for osteoclast formation *(7–9)*. The M-CSF gene of osteopetrotic op/op mice dose not encode functionally active M-CSF protein because of insertion of extra thymidine in the coding region of the M-CSF gene *(7)*. Administration of recombinant human M-CSF restored impaired bone resorption in op/op mice *(8)*. Calvarial osteoblasts obtained from op/op mice failed to support osteoclast development in cocultures with normal spleen cells, but the addition of M-CSF to cocultures induced osteoclast formation in response to $1\alpha,25\text{-}(OH)_2D_3$ *(9)*. These findings indicate that M-CSF produced by osteoblasts/ stromal cells plays an essential role in osteoclast development.

2. In 1997, osteoprotegerin (OPG) and osteoclastogenesis inhibitory factor (OCIF), that inhibit osteoclast development in vivo and in vitro, respectively, were cloned independently by two different research groups *(31,32)*. It was later demonstrated that OPG and OCIF are the same protein molecule. OPG/OCIF is a member of the TNF receptor family, but it does not have a transmembrane domain, suggesting that OPG/OCIF functions as a circulating soluble factor. Subsequently, the cDNA encoding a binding molecule of OPG/OCIF was isolated from an expression library of the mouse stromal cell line ST2 and was named as osteoclast differentiation factor (ODF) *(10)*. A ligand for OPG/OCIF was also cloned from an expression library of the mouse myelomonocytic cell line 32D, and was named as OPG ligand (OPGL) *(11)*. OPGL was found to be identical to ODF. The binding molecule of OPG/OCIF is a membrane-associated protein of the TNF ligand family. ODF/OPGL is also identical to TNF-related activation-induced cytokine (TRANCE) *(33)* and RANKL *(34)*, which were independently cloned from mouse T-cell hybridomas and mouse dendritic cells, respectively. RANK is the transmembrane receptor of ODF/OPGL/TRANCE/RANKL, which is expressed by osteoclast precursors and mature osteoclasts *(12,13,34)*. OPG/OCIF is a decoy receptor of ODF/OPGL/TRANCE/RANKL. Thus, ODF, OPGL, TRANCE, and RANKL are different names for the same protein, which is essential for the development and function of osteoclasts. The American Society for Bone and Mineral Research (ASBMR) President's Committee on Nomenclature has recommended using the terms "RANKL," "RANK," and "OPG" for these factors with different names *(35)* (*see* **Fig. 5**).

3. Other strains of mice such as BALB/c, C57BL, and ICR can also be used for mouse osteoclast formation.

4. Human M-CSF is effective in both human and murine cells, whereas murine MCSF is effective in murine cells but not in human cells.

5. Only this type of collagen (cell matrix type IA; Nitta Gelatin Co.) is suitable for this procedure. Nitta Gelatin Co. supplies cell matrix type IA to scientists with an import permit in foreign countries.

6. Because FBS is one of the important factors that affect osteoclast formation, careful batch testing of FBS is recommended.

7. In bone marrow cultures, bone marrow stromal cells support osteoclast formation from osteoclast progenitor cells. The target cells of osteotropic factors for

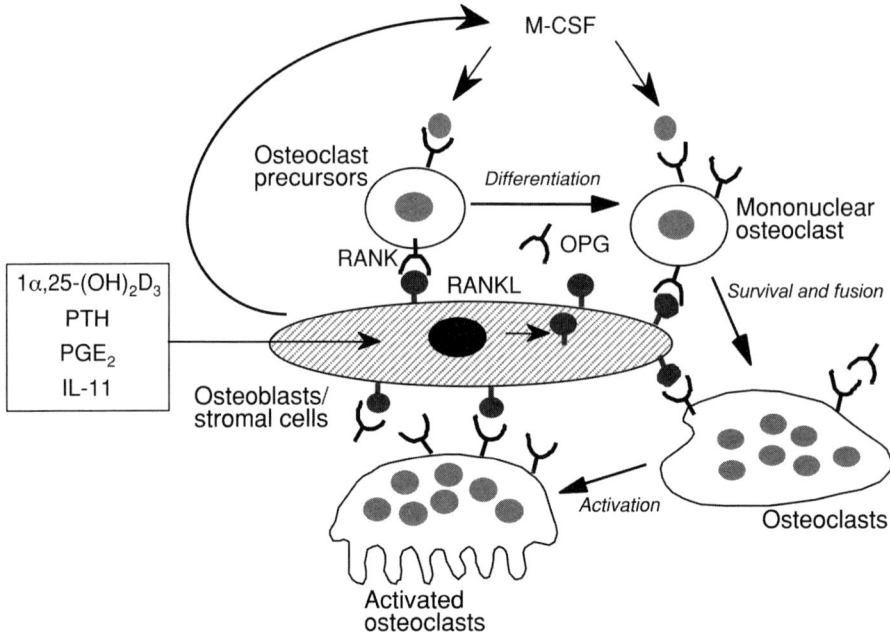

Fig. 5. Schematic representation of osteoclast differentiation regulated by osteoblasts/stromal cells. RANKL expressed by osteoblasts/stromal cells is a key molecule that induces osteoclast differentiation and function. Osteotropic factors such as $1\alpha,25$-$(OH)_2D_3$, PTH, PGE_2, and IL-11 stimulate expression of RANKL as a membrane-associated factor in osteoblasts/stromal cells. In addition, osteoblasts/stromal cells constitutively produce M-CSF, another cytokine essential for osteoclast formation. Osteoclast progenitors express RANKL and differentiate into osteoclasts though cell–cell interaction with osteoblasts/stromal cells in the presence of M-CSF. Mature osteoclasts also express RANK. RANKL directly stimulates fusion and activation of osteoclasts. OPG, a soluble decoy receptor of RANKL, is mainly produced by osteoblasts/stromal cells. OPG strongly inhibits the entire differentiation, fusion and activation processes of osteoclasts induced by RANKL.

inducing osteoclast formation are bone marrow stromal cells, not osteoclast progenitors. Therefore, the growth of stromal cells is one of the determinants of osteoclast formation in bone marrow cultures.
8. Bone marrow stromal cells adhere to culture dishes during the first incubation for 10-20 h. During the first incubation, bone marrow macrophages and their precursors can survive as nonadherent cells in the presence of M-CSF.
9. No TRAP-positive cells are formed even in the presence of RANKL, when M-CSF is not added to the culture. Instead of RANKL, mouse TNF-α (20–100 ng/mL) stimulates formation of osteoclasts from BMMϕ in the presence of M-

CSF *(18,19)*. Osteotropic hormones and cytokines including $1\alpha,25\text{-}(OH)_2D_3$, PTH, PGE_2 and IL-11 fail to induce osteoclast formation in BMMϕ cultures.
10. RAW264.7 cells proliferate and retain their potency to differentiate into osteoclasts in RPMI-1640 containing 10% FBS. In the osteoclast differentiation assay, RAW264.7 cells are cultured in α-MEM containing 10% FBS.
11. Primary BMMϕ require both M-CSF and RANKL to differentiate into osteoclasts. However, RAW264.7 cells can differentiate into osteoclasts even in the absence of M-CSF.
12. ST2 and MC3T3-G2PA6 have been shown to support osteoclast differentiation in cocultures with mouse bone marrow cells *(20)*. However, the number of osteoclasts formed in the cocultures with primary osteoblasts is much higher than in those with ST2 cells or MC3T3-G2PA6 cells. Therefore, we usually use primary osteoblasts in the coculture for osteoclast formation.
13. Fraction 1 contains considerable numbers of hemopoietic cells. Cells recovered in Fractions 2–5 show strong ability to support osteoclast differentiation in coculture with bone marrow cells.
14. TRAP is intracellular enzyme. Treatment of cells with ethanol–acetone (50:50 [v/v]) is an essential step for TRAP staining. Instead of ethanol–acetone (50:50 [v/v]), PBS(–) containing 0.1% Triton X-100 can be used in this step for TRAP staining.
15. Bone slices are often used for the pit-formation assay. However, we prefer dentine slices, because dentine has a homogeneous structure and is free of vascular canals and osteocyte lacunae, which are present in bone slices.
16. A mark is put on one surface of each dentine slice with the pencil in order to check the orientation of the slice during the pit formation assay. The dentine slices are placed into wells of 96-well plates so that the surface without the mark faces up in the well.

References

1. Chambers, T. J. (2000) Regulation of the differentiation and function of osteoclasts. *J. Pathol.* **192,** 4–13.
2. Roodman, G. D. (1996) Advances in bone biology: the osteoclast. Endocr. Rev. **17,** 308–332.
3. Teitelbaum, S. L. and Ross, F. P. (2003) Genetic regulation of osteoclast development and function. *Nat. Rev. Genet.* **4,** 638–649.
4. Takahashi, N., Yamana, H., Yoshiki, S., et al. (1988) Osteoclast-like cell formation and its regulation by osteotropic hormones in mouse bone marrow cultures. *Endocrinology* **122,** 1373–1382.
5. Suda, T., Takahashi, N., and Martin, T. J. (1992) Modulation of osteoclast differentiation. *Endocr. Rev.* **13,** 66–80.
6. Takahashi, N., Akatsu, T., Udagawa, N., et al. (1988) Osteoblastic cells are involved in osteoclast formation. *Endocrinology* **123,** 2600–2602.
7. Yoshida, H., Hayashi, S., Kunisada, T., et al. (1990) The murine mutation osteopetrosis is in the coding region of the macrophage colony stimulating factor gene. *Nature* **345,** 442–444.

8. Felix, R., Cecchini, M. G., and Fleisch, H. (1990) Macrophage colony stimulating factor restores in vivo bone resorption in the op/op osteopetrotic mouse. *Endocrinology* **127,** 2592–2594.
9. Takahashi, N., Udagawa, N., Akatsu, T., Tanaka, H., Isogai, Y., and Suda, T. (1991) Deficiency of osteoclasts in osteopetrotic mice is due to a defect in the local microenvironment provided by osteoblastic cells. *Endocrinology* **128,** 1792–1796.
10. Yasuda, H., Shima, N., Nakagawa, N., et al. (1998) Osteoclast differentiation factor is a ligand for osteoprotegerin/osteoclastogenesis-inhibitory factor and is identical to TRANCE/RANKL. *Proc. Natl. Acad. Sci. USA* **95,** 3597–3602.
11. Lacey, D. L., Timms, E., Tan, H. L., et al. (1998) Osteoprotegerin ligand is a cytokine that regulates osteoclast differentiation and activation. *Cell* **93,** 165–176.
12. Suda, T., Takahashi, N., Udagawa, N., Jimi, E., Gillespie, M. T., and Martin, T. J. (1999) Modulation of osteoclast differentiation and function by the new members of the tumor necrosis factor receptor and ligand families. *Endocr. Rev.* **20,** 345-357.
13. Boyle, W. J., Simonet, W. S., and Lacey, D. L. (2003) Osteoclast differentiation and activation. *Nature* **423,** 337–342.
14. Hsu, H., Lacey, D. L., Dunstan, C. R., et al. (1999) Tumor necrosis factor receptor family member RANK mediates osteoclast differentiation and activation induced by osteoprotegerin ligand. *Proc. Natl. Acad. Sci. USA* **96,** 3540–3545.
15. Akatsu, T., Tamura, T., Takahashi, N et al. (1992) Preparation and characterization of a mouse multinucleated cell population. *J. Bone Miner. Res.* **7,** 1297–1306.
16. Tamura, T., Takahashi, N., Akatsu, T., et al. (1993) A new resorption assay with mouse osteoclast-like multinucleated cells formed in vitro. *J. Bone Miner. Res.* **8,** 953–960.
17. Udagawa, N., Takahashi, N., Akatsu, T., et al. (1990) Origin of osteoclasts: Mature monocytes and macrophages are capable of differentiating into osteoclasts under a suitable microenvironment prepared by bone marrow-derived stromal cells. *Proc. Natl. Acad. Sci. USA* **87,** 7260–7264.
18. Kobayashi, K., Takahashi, N., Jimi, E., et al. (2000) Tumor necrosis factor a stimulates osteoclast differentiation by a mechanism independent of the ODF/RANKL-RANK interaction. *J. Exp. Med.* **191,** 275–286.
19. Fuller, K., Murphy, C., Kirstein, B., Fox, S. W., and Chambers, T. J. (2002) TNFα potently activates osteoclasts, through a direct action independent of and strongly synergistic with RANKL. *Endocrinology* **143,** 1108–1118.
20. Udagawa, N., Takahashi, N., Akatsu, T., et al. (1989) The bone marrow-derived stromal cell lines MC3T3-G2/PA6 and ST2 support osteoclast-like cell differentiation in cocultures with mouse spleen cells. *Endocrinology* **125,** 1805–1813.
21. Suda, T., Jimi, E., Nakamura, I., and Takahashi, N. (1997) Role of 1α,25-dihydroxyvitamin D_3 in osteoclast differentiation and function. *Methods Enzymol.* **282,** 223–235.
22. Jimi, E., Ikebe, T., Takahashi, N., Hirata, N., Suda, T., and Koga, T. (1996) Interleukin-1b activates an NF-κB-like factor in osteoclast-like cells. *J. Biol. Chem.* **271,** 4605–4608.

23. Nakamura, I., Jimi, E., Duong, L. T., et al. (1998), Tyrosine phosphorylation of p130^{Cas} is involved in actin organization in osteoclasts. *J. Biol. Chem.* **273,** 11,144–11,149.
24. Suda, T., Nakamura, I., Jimi, E., and Takahashi, N. (1997) Regulation of osteoclast function. *J. Bone Miner. Res.* **12,** 869–879.
25. Udagawa, N., Takahashi, N., Sasaki, T., et al. (1992) Failure of bone resorption in osteosclerotic (oc/oc) mice is due to a microenvironment. In: *Calcium Regulating Hormones and Bone Metabolism*, Cohn, D. V., Gennari C, Tashjian, Jr., A. H. eds., Elsevier Science Publishers, pp. 151–156.
26. Zambonin-Zallone, A., Teti, A., Carano, A., and Marchisio, P. C. (1988) The distribution of podosomes in osteoclasts cultured on bone laminae, effect of retinol. *J. Bone Miner. Res.* **3,** 517–523.
27. Chellaiah, M. A., Soga, N., Swanson, S., et al. (2000) Rho-A is critical for osteoclast podosome organization, motility, and bone resorption. *J. Biol. Chem.* **275,** 11,993–12,002.
28. Nakamura, I., Takahashi, N., Sasaki, T., Jimi, E., Kurokawa, T., and Suda, T. (1996) Chemical and physical properties of the extracellular matrix are required for the actin ring formation in osteoclasts. *J. Bone Miner. Res.* **11,** 1873–1879.
29. Suzuki, H., Nakamura, I., Takahashi, N., et al. (1996) Calcitonin-induced changes in cytoskeleton are mediated by a signal pathway associated with protein kinase A in osteoclasts. *Endocrinology* **137,** 4685–4690.
30. Murakami, H., Takahashi, N., Sasaki, T., et al. (2995) A possible mechanism of the specific action of bisphosphonates on osteoclasts: Tiludronate preferentially affects polarized osteoclasts having ruffled borders. *Bone* **17,** 137–144.
31. Simonet, W. S., Lacey, D. L., Dunstan, C. R., et al. (1997) Osteoprotegerin, a novel secreted protein involved in the regulation of bone density. *Cell* **89,** 309–319.
32. Tsuda, E., Goto, M., Mochizuki, S., et al. (1997) Isolation of a novel cytokine from human fibroblasts that specifically inhibits osteoclastogenesis. *Biochem. Biophys. Res. Commun.* **234,** 137–142.
33. Wong, B. R., Rho, J., Arron, J., et al. (1997) TRANCE is a novel ligand of the tumor necrosis factor receptor family that activates c-Jun N-terminal kinase in T cells. *J. Biol. Chem.* **272,** 25,190–25,194.
34. Anderson, D. M., Maraskovsky, E., Billingsley, W. L., et al. (1997) A homologue of the TNF receptor and its ligand enhance T-cell growth and dendritic-cell function. *Nature* **390,** 175–179.
35. The American Society for Bone and Mineral Research President's Committee on Nomenclature (2000) Proposed standard nomenclature for new tumor necrosis factor family members involved in the regulation of bone resorption. *J. Bone Miner. Res.* **15,** 2293–2296.

III

CELL TRAFFICKING, MIGRATION, AND INVASION

19

Isolation and Analysis of Large and Small Vessel Endothelial Cells

Justin C. Mason, Elaine A. Lidington, and Helen Yarwood

Summary

The in vitro isolation, propagation and study of endothelial cells (EC) is an invaluable means by which the function of the vascular endothelium in physiology and pathophysiology can be explored. In recent years heterogeneity between large and small vessel EC, between arteries and veins, and between microvascular EC derived from different organs has become increasingly apparent. This has led to the development of protocols for the isolation of these different EC. In addition, the data emerging on vascular EC function in gene-targeted mice has highlighted the need for reliable methods of isolation of murine EC. This chapter describes methods for the isolation, characterization and culture of macro- and microvascular EC from a variety of species and introduces simple approaches to investigating their surface antigen expression.

Key Words: Human umbilical vein endothelial cells; dermal microvascular endothelial cells; murine endothelial cells; porcine endothelial cells; flow-cytometry; enzyme-linked immunosorbent assay.

1. Introduction

The in vitro isolation, propagation and study of endothelial cells (EC) remains an invaluable method by which the role of the vascular endothelium in a variety of physiological and pathophysiological processes can be studied. This approach has resulted in a dramatic increase in our understanding of endothelial function. The culture of EC in vitro has been in routine laboratory use for over 30 yr *(1)*, and during this time methodology for the isolation of EC from specific vascular beds has constantly evolved. In particular, methods for the isolation of EC from microvascular beds have been developed *(2–9)*. In addition, the development of genetically modified mice provides the opportunity to study the role of individual endothelial genes not only in vivo, but also in vitro through isolation of endothelial cells from different murine organs *(10–14)*. This approach is likely to be of specific importance in view of previous studies, in both human and murine models, in which the presence of functional heterogeneity between endothelial cells derived from different vascular beds has been demonstrated *(14–18)*. The aim of this chapter is to describe methods for the isolation, culture and study of endothelial cells from large and small vessels in a variety of species.

2. Materials
2.1. Isolation of Endothelial Cells

1. Penicillin/streptomycin (Sigma; cat no. P0781). Store in aliquots at –20°C.
2. Hanks balanced salt solution (HBSS) (Sigma; cat no. H9269).
3. Hanks balanced salt solution without calcium and magnesium (HBSS w/o Ca/Mg) (Sigma; cat no. H9394).
4. Sodium pyruvate (Invitrogen; cat no. 11360-039). Store at 4°C.
5. Gentamicin (Invitrogen; cat no. 15710-049). Store at room temperature.
6. Amphotericin B (Fungizone) (Invitrogen; cat no. 15290-018). Aliquot and store at –20°C.
7. Dispase (Boehringer Mannheim). Store at –20°C and make up fresh on each occasion.
8. Medium 199 (MP Biomedicals; cat. No. 1220254). Store at 4°C.
9. Heat-inactivated human AB positive serum (Sigma; cat. no. S7418). To heat inactivate, incubate in a water bath at 56°C for 45 min. Store in aliquots at –20°C.
10. L-glutamine (Sigma; cat no. G7513). Store in aliquots at –20°C.
11. Preservative free heparin 1000 iu/mL (CP Pharmaceuticals, Wrexham UK). Store at 4°C.
12. Endothelial cell growth factor (ECGF) (Sigma; cat no. E2759). Reconstitute each vial with 5 mL of M199 and pass through a 0.2 μM filter and collect into a sterile universal container. Divide into 1-mL aliquots and freeze at –20°C. Reconstituted ECGF is stable at 4°C for 2 wk.
13. Bovine fibronectin (Sigma; cat no. F-1141). Reconstitute in sterile water, aliquot and store at –20°C.
14. Trypsin/ethylene diamine tetraacetic acid (EDTA) (ICN; cat no. 16-891-49). Aliquot and store at –20°C.
15. MiniMacs beads (cat no. 130-048-502 or 130-047-202), preseparation filters (cat no. 130-041-407), and columns (cat no. 130-042-401 or 130-042-201) (Miltenyi Biotec).
16. 2% Gelatin (Sigma; cat no. G1393). Prewarm at 37°C for 30 min and dilute to 1% in sterile water or phosphate buffered saline (PBS). Aliquot and store at 4°C.
17. Collagenase A (Roche Diagnostics). Store at –20°C and make up fresh on each occasion in HBSS.
18. Three-way tap (Becton Dickinson; cat no. 394601).
19. RPMI 1640 (Sigma; cat no. R0883).
20. 70 µm filters (Becton Dickinson; cat no. 352350).

21. Rat anti-mouse endoglin antibody (MJ7/18) (BD Pharmingen or hybridoma from developmental studies hybridoma bank, University of Iowa).
22. DMEM (Invitrogen; cat no. 41965-039).
23. Fetal calf serum (FCS) batch tested for EC growth properties and heat inactivated to 56°C for 30 min. 30% Bovine serum albumin solution (BSA) (Sigma; cat no. A-3424).

2.2. Storage of Endothelial Cells in Liquid Nitrogen

1. Cryogenic vials (Corning Inc.; cat no. 430489).
2. Dimethyl sulfoxide (DMSO). Use tissue culture grade and store at room temperature. (Sigma; cat no. D2650).

2.3. Characterization of Cultured Endothelial Cells

1. Matrigel (Becton Dickinson; cat no. 35 40234). Store at −20°C.
2. Acetylated low-density lipoprotein (Intracel; cat no. RP-078). Store at 4°C.
3. FITC-anti-von Willebrand Factor polyclonal Ab (Serotec; cat no. AHPO62F). Store at 4°C.
4. Biotinylated *Ulex europaeus* agglutinin-I (UEA-I) (Vector Laboratories; cat no. B-1065).
5. Biotinylated *Griffonia simplicifolia* isolectin B4 (Vector Laboratories; cat no. B-1205).

2.4. Analysis of Endothelial Cell Activation

1. Paraformaldehyde (BDH Laboratory supplies; cat no. 294474L).
2. L-lysine (Sigma; cat no. L-5626)
3. Sodium *m*-periodate (Sigma; cat no. S-1147).
4. Biotinylated rabbit anti-mouse Ig F(ab')$_2$ (DAKO; cat no. E0413).
5. FITC- rabbit anti-mouse Ig F(ab')$_2$ (DAKO; cat no. E0313).
6. RPE- rabbit anti-mouse Ig F(ab')$_2$ (DAKO; cat no. R0439)
7. Alexa Fluor 488-labeled anti-mouse Ig (Molecular Probes; cat no. A-11011 or A-11017).
8. Streptavidin-biotin-peroxidase (DAKO; cat no. P0347)
9. *O*-phenylenediamine (Sigma; cat no. P-5412).
10. Hydrogen peroxidase (Sigma; cat no. H-1009).
11. Crystal violet (Sigma; cat no. C-0775).
12. FACS tubes (Becton Dickinson; cat no. 62052).
13. V-bottomed 96-well plate (Nunc; cat no. 245128).

3. Methods
3.1. Isolation of Human Dermal Microvascular Endothelial Cells

1. Human skin should be collected into HBSS, supplemented with 100 IU/mL penicillin, 100 µg/mL streptomycin and 20 µg/mL amphotericin B (*see* **Note 1**).

2. In a tissue culture hood, the skin is cut into strips (2 × 0.5 cm), washed twice in HBSS and incubated overnight in 2 µg/mL Dispase at 4°C. Following enzymatic digestion the epidermis should be easily removed with the use of sterile fine forceps. The remaining tissue is cut into 0.5 cm^2 pieces, washed twice in HBSS and transferred into Medium 199 (M199).
3. Skin segments are transferred individually into a single well of a 6-well tissue culture plate containing 1 mL M199. Microvascular segments are expressed from the edge of the skin, into the M199, by firm downward pressure with the curved edge of a sterile stitch cutter *(19)* (*see* **Note 2**). Once all the tissue segments have been used, carefully discard them (following local regulations for disposal of human tissue). Collect the M199 containing expressed microvascular fragments and pass through a sterile 100 µm filter to remove contaminating cell debris.
4. Following centrifugation of the cell suspension at 200g for 5 min, aspirate the M199, and gently resuspend the cell pellet in DMEC growth medium (M199 supplemented with 20% heat-inactivated human AB positive serum, 100 IU/mL penicillin, 100 µg/mL streptomycin and 5 µg/mL amphotericin B, 2mM L-glutamine, 10 U/mL preservative free heparin, and 50 µg/mL endothelial cell growth factor).
5. The initial isolate is plated out onto a single well of a 6-well tissue culture plate, precoated for 30 min at 37°C with human fibronectin (10 µg/mL) and cultured for 72 h. The cell monolayer is then washed twice in calcium and magnesium free HBSS and the cells detached by incubation for 1 to 2 min with 0.125% trypsin/EDTA at 37°C. The harvested EC are transferred to a 25cm^2 fibronectin-coated tissue culture flask containing prewarmed (37°C) and pregassed (5% CO_2) DMEC medium and grown to confluence.
6. The initial isolate consists of approx 70% DMEC and therefore requires further purification by positive selection. EC are harvested with trypsin/EDTA, as described in **Subheading 3.1.5.**, and resuspended in HBSS/5% FCS containing a mouse anti-human endoglin (CD105) MAb (final concentration 10 µg/mL) (*see* **Note 3**). Following incubation for 20 min at 4°C, the DMEC are centrifuged at 200g for 5 min, washed twice in HBSS/1% FCS and resuspended with anti-mouse IgG-coated MiniMacs beads and incubated for 20 mins at 4°C (*see* **Note 4**).
7. The DMEC suspension is centrifuged at 200g for 5 min, washed thrice in HBSS/1% FCS, and passed over a MS+ miniMac column suspended in a miniMac magnet. Following washing in 3-column volumes of HBSS/1% FCS, the purified DMEC are eluted from the column following removal of the column from the magnetic field.
8. DMEC are collected by centrifugation at 200g for 5 min and resuspended in DMEC growth medium and transferred into a 25 cm^2 fibronectin-coated tissue culture flask and grown to confluence prior to detachment with trypsin/EDTA and transfer to a 75 cm^2 fibronectin-coated tissue culture flask. The DMEC can be sub-cultured up to passage 6 in 75 cm^2 fibronectin-coated tissue culture flask (*see* **Note 5**). DMEC require feeding every 2 to 3 d by exchanging half the growth medium.

3.2. Isolation of Human Umbilical Vein Endothelial Cells

1. Collect cords into HBSS supplemented with 1 mM sodium pyruvate, 100 IU/mL penicillin, 100 μg/mL streptomycin, and 100 μg/mL gentamicin. Place umbilical cord onto a silver foil covered tray, sprayed with 70% ethanol and air dried in a tissue culture hood *(1)*.
2. Wipe cord with tissues soaked in 70% ethanol to remove blood.
3. Prepare collagenase A solution (0.5 mg/mL) in HBSS (without calcium and magnesium) allowing 10 to 15 mL per umbilical cord. Filter collagenase through 0.2 μm filter and prewarm to 37°C. Add 2 mL of 1% gelatin to one 25 cm^2 tissue culture flask for each umbilical cord and incubate for 30 min at 37°C.
4. Cut approx 1 cm off each end of the cord and identify the two umbilical arteries and one umbilical vein. Dissect 0.5 to 1 cm of the vein at each end of the cord and dilate with a pair of blunt forceps. Cannulate one end with a three-way tap, secure firmly with fine thread.
5. Attach a sterile 20-mL syringe to the three way tap and repeatedly flush the umbilical vein with HBSS (without calcium and magnesium) until all blood and blood clots are removed. Ensure that no holes exist in the vein and if found use a fine artery clamp or a crocodile clip to occlude them. Finally, push 15 mL of air through the umbilical vein to remove residual HBSS.
6. Cannulate the other end of the umbilical vein with a three-way tap, secure, and close the three-way tap. Attach an empty 20-mL syringe to this end and a second 20-mL syringe containing the prewarmed collagenase to the opposite end.
7. Place umbilical cord on saran wrap on a small tray. Distend the cord with collagenase. Wrap the cord in Saran™ wrap and place in an incubator at 37°C for 8 to 10 min (*see* **Note 6**).
8. Remove cord from incubator and relieve pressure by drawing 5 mL of collagenase into the previously empty syringe. Massage the cord firmly along the whole length between thumb and forefinger for 1 min. Push any remaining collagenase from the original syringe through the cord into the collecting syringe and then push 15 mL of air through the umbilical vein to remove residual collagenase.
9. Remove syringe containing collagenase and empty contents into a 50-mL centrifuge tube. Replace the syringe with one containing 20 mL HBSS. Pass 10 mL into the cord and repeat the massage step and collect into an empty syringe attached to the other end of the cord. Repeat this step 4 times adding each 10 mL of HBSS to the 50-mL centrifuge tube (*see* **Note 7**). Pellet human umbilical vein endothelial cells (HUVEC) by centrifugation at 200*g* for 5 min.
10. Carefully dispose of umbilical cord in a sealed container following local regulations for disposal of human tissue.
11. Aspirate gelatin from 25 cm^2 tissue culture flask and add 4 mL of prewarmed and pregassed (37°C in 5% CO_2 incubator) HUVEC growth medium (M199 supplemented with 20% heat-inactivated FCS [*see* **Note 8**], 100 IU/mL penicillin, 100 μg/mL streptomycin, 2 mM L-glutamine, 10 U/mL preservative free heparin, and 30 μg/mL endothelial cell growth factor).

12. Resuspend HUVEC pellet gently in 1 mL of growth medium and transfer to 25 cm² tissue culture flask and leave EC to attach overnight.
13. The following day prewarm and pregas 5 mL of HUVEC growth medium for 30 min. Aspirate medium from tissue culture flask and wash HUVEC monolayer in HBSS 3 times to remove all residual blood and cell debris. Add prewarmed HUVEC growth medium.
14. HUVEC will reach confluence in 1 to 5 d and require feeding every 2 to 3 d by exchanging half the growth medium. At confluence HUVEC should be harvested and transferred to a gelatin-coated 75 cm² containing 12 mL of HUVEC growth medium. For harvesting, wash the monolayer twice in calcium and magnesium free HBSS and detach cells by incubation for 1 to 2 min in 0.125% trypsin/EDTA at 37°C. HUVEC can be subcultured up to passage 6 in 75 cm² gelatin-coated tissue culture flask (*see* **Notes 9** and **10**).

3.3. Isolation of Porcine Arterial Endothelial Cells

1. Collect porcine aortas into HBSS supplemented with 1 mM sodium pyruvate, 100 IU/mL penicillin, 100 µg/mL streptomycin, 100 µg/mL gentamicin, 20 µg/mL, and amphotericin B (*see* **Note 11**).
2. Place aorta onto a silver foil covered tray, sprayed with 70% ethanol and air dry in a tissue culture hood. Cut off excess fat and connective tissue and carefully wash vessel lumen with HBSS supplemented with 1 mM sodium pyruvate, 100 IU/mL penicillin, 100 µg/mL streptomycin, 100 µg/mL gentamicin, and 20 µg/mL amphotericin B.
3. Prepare collagenase A solution (0.25 mg/mL) in HBSS (without calcium and magnesium) allowing 20 to 25 mL per aorta. Filter collagenase through 0.2 µM filter and prewarm to 37°C. Add 2 mL of 1% gelatin to one 25 cm² tissue culture flask for each aorta and incubate for 30 min at 37°C.
4. Any vessels arising from the aorta should be carefully clipped using fine artery clips and one end of the aorta closed with an arterial clamp. The aorta should then be filled with HBSS to check for leaks (*see* **Note 12**).
5. Remove HBSS and fill aorta with collagenase, clamp the open end, wrap aorta in Saran wrap, and incubate for 15 min at 37°C.
6. The porcine arterial endothelial cells (PAEC) are harvested by careful collection of the collagenase into a 50-mL centrifuge tube and repeated gentle massage of the aorta and washing of the lumen with 10 mL HBSS supplemented with antiobiotics as above. Pellet PAEC by centrifugation at 200g for 5 min.
7. Carefully dispose of each aorta in a sealed container following local regulations for disposal of animal tissue.
8. Aspirate gelatin from 25 cm² tissue culture flask and add 4 mL of PAEC growth medium (RPMI 1640, supplemented with 20% heat-inactivated FCS, 100 IU/mL penicillin, 100 µg/mL streptomycin, 2 mM L-glutamine, 10 U/mL preservative free heparin, and 30 µg/mL endothelial cell growth factor).
9. Resuspend PAEC pellet gently in 1 mL of growth medium and transfer to the 25 cm² tissue culture flask and leave EC to attach overnight.

10. The following day prewarm 5 mL of PAEC growth medium for 30 min. Aspirate medium from tissue culture flask and wash PAEC monolayer in HBSS 3 times to remove all residual blood and cell debris. Add prewarmed PAEC growth medium.
11. PAEC will reach confluence in 1 to 5 d and require feeding every 2 to 3 d by exchanging half the growth medium. At confluence PAEC should be harvested and transferred to a gelatin-coated 75 cm^2 containing 12 mL of growth medium. For harvesting, wash the monolayer twice in calcium and magnesium free HBSS and detach cells by incubation for 1 to 2 min in 0.125% trypsin/EDTA at 37°C. PAEC can be subcultured up to passage 6 in 75 cm^2 gelatin-coated tissue culture flask (*see* **Notes 9** and **10**).

3.4. Isolation of Murine Cardiac Endothelial Cells

1. Prepare collagenase A solution (1 mg/mL) in HBSS, allowing 5 mL per isolation. Filter collagenase through 0.2 µM filter and prewarm to 37°C. In addition, prewarm 5 mL 0.125% trypsin/EDTA to 37°C and one 25 cm^2 tissue culture flask containing 2 mL of 1% gelatin per preparation (*see* **Note 13** on preparation of control EC).
2. Use 6 mice per isolation, killed via an approved schedule 1 method. Immerse each mouse briefly in 70% ethanol and pin-out onto a silver foil covered polystyrene board, in a tissue culture hood.
3. Using sterile scissors and fine forceps carefully remove skin from the chest, making incisions across the abdomen and up the right flank without piercing the peritoneum. Reflect the skin to the left and rinse loose hairs off with HBSS. Flame instruments and once cool, hold the sternum and make a horizontal incision just beneath the sternum and above the diaphragm followed by a vertical incision along the right border of the sternum and reflect the ribcage to expose the heart. Flame instruments again, remove the heart and transfer into 3 mL HBSS supplemented with 100 IU/mL penicillin, 100 µg/mL streptomycin, and 100 µg/mL gentamicin in a 60-mm Petri-dish. Repeat for all six mice. EC can also be obtained from lung and brain tissue (*see* **Note 14**).
4. Using a sterile scalpel cut hearts in half, remove any connective tissue, blood clots, and the base of the aorta. Collect tissue into a fresh 60-mm Petri-dish containing HBSS (with antibiotics), dice-up tissue into pieces of about 1 to 2 mm^3. Remove HBSS using a P1000 pipette and wash tissue in 3 ml HBSS removing the HBSS again using a P1000 pipette.
5. Place tissue into 5 mL of prewarmed collagenase A (0.5 mg/mL) in a Universal tube and incubate for 30 min at 37°C, shaking tube at regular intervals. After a final vigorous shake, allow the undigested tissue to fall to the bottom of the tube. Collect the cell suspension into a 50-mL Falcon tube. Wash the undigested tissue in 10 mL calcium and magnesium free HBSS, shake vigorously and once settled add suspension to the previously collected suspension. Repeat wash twice more and centrifuge total cell suspension at 700*g* for 8 min at 4°C.
6. Aspirate supernatant and wash pellet in 30 mL calcium and magnesium free HBSS, centrifuging at 700*g* for 8 min at 4°C. Aspirate the supernatant and incu-

bate the pellet with 5 mL Trypsin/EDTA at 37°C for 10 min. Pipette up and down several times before passing suspension through a 70 µm filter placed over a 50-mL Falcon tube, wash through with 25 mL cold HBSS (Ca/Mg) 1% BSA. Centrifuge cell suspension at 700g for 8 min at 4°C.
7. Aspirate the supernatant and incubate the pellet with 5 µg irrelevant MAb (see **Note 15**) in 500 µL at 4°C for 20 min, to block Fc receptors on macrophages (optional wash in HBSS/1%BSA). Add 5 µg rat anti-murine endoglin MAb MJ7/18 (10 µg/mL) and incubate at 4°C for 20 min. Add 30 mL HBSS/1% FCS and centrifuge cell suspension at 700g for 8 min.
8. Aspirate supernatant and incubate pellet with 50 µL goat anti-rat IgG microbeads in 200 µL HBSS/1%BSA for 15 min at 4°C. Add 30 mL HBSS/1%BSA and centrifuge cell suspension at 700g for 8 min. Meanwhile put a miniMAC MS+ column (or midiMAC) onto its magnet with a pre-separation filter on top and wash through with 500 µL HBSS/1%BSA (see **Note 16**). Aspirate pellet and resuspend in 500 µL HBSS/1%BSA and pipette up and down several times to disperse clumps. Add to column and once suspension has passed through, wash column 3 times with 500 µL HBSS/1%BSA. Finally, wash through with 500 µL complete medium (5 mL DMEM, 10% FCS, 100 IU/mL penicillin, 100 µg/mL streptomycin, 2 m*M* L-glutamine, 10 U/mL preservative free heparin, and 30 µg/mL endothelial cell growth factor).
9. Remove column from the magnet and place in a 15-mL tube. Add 4 mL complete medium and apply plunger to collect bead-bound cells. Transfer cell suspension into gelatin-coated flasks (once excess gelatin has been removed).
10. After 24 h replace culture media with fresh media. The monolayers should become confluent in 7 d. Use the EC as a primary culture or after one passage into a 75-cm^2 flask. Use of EC at early passage is recommended as murine cardiac EC are liable to lose their differentiated morphology. However, murine lung EC can be used up to passage 4. The purity of each isolate should be confirmed by flow-cytometric analysis of 1×10^5 cells with an anti-endoglin antibody (see **Subheadings 3.6.** and **3.7.3.**).

3.5. Storage of Endothelial Cells in Liquid Nitrogen

Human and porcine EC at passages 1 to 3 and murine EC at passage 1 can be frozen and stored in liquid nitrogen and subsequently recovered without significance loss of function.

1. Add 1 mL of tissue grade DMSO to 9 mL heat-inactivated FCS, mix carefully and store at 4°C.
2. Wash the EC monolayer in a 75-cm^2 tissue culture flask twice in calcium and magnesium free HBSS and detach cells by incubation for 1 to 2 min in 0.125% trypsin/EDTA at 37°C. Add 5 mL of appropriate EC growth medium and transfer cell suspension to a 50-mL centrifuge tube. Pellet EC by centrifugation at 200g for 5 min.
3. Aspirate supernatant and gently resuspend the pellet in 3 mL of freezing medium.

4. Transfer EC in 1-mL aliquots to freezing vials. Label tubes with cell type, passage number, and date. Suspend cells in a liquid nitrogen tank freezing rack and gradually wind down into liquid nitrogen over 3 h. Transfer vials to liquid nitrogen storage box.
5. For retrieval of EC from liquid nitrogen, prepare a 50-mL centrifuge tube containing 40 mL HBSS. Prepare a gelatin coated 75-cm^2 tissue culture flask containing 12 mL of prewarmed and pregassed EC growth medium.
6. Thaw frozen EC by suspension in a water bath for 2 to 3 min.
7. Ensuring sterility, carefully transfer cells from the freezing vial into the 50-mL centrifuge tube. Pellet EC by centrifugation at 200g for 5 min.
8. Aspirate HBSS and resuspend EC in 1 mL of EC growth medium and transfer to the pre-prepared 75-cm^2 tissue culture flask.

3.6. Characterization of Endothelial Cells

When establishing the culture of endothelial cells in a laboratory it is essential to carefully characterize the cells to ensure their endothelial lineage, and to exclude the presence of significant numbers of contaminating cells, such as vascular smooth muscle cells and fibroblasts *(7,9)*. This requires the use of a variety of different assays in order to definitively demonstrate that the culture represents pure EC.

3.6.1. Characterization of Human EC

3.6.1.1. MORPHOLOGY

Viewed by phase contrast microscopy endothelial cells form contact-inhibited cobblestone-like monolayers of polygonal cells with prominent nuclei and nucleoli (*see* **Fig. 1A,B**).

3.6.1.2. MICROTUBULE FORMATION

ECs plated on Matrigel, a solubilized basement membrane preparation from a murine tumour, will spontaneously form microtubules with a clear lumen *(20)*.

1. Matrigel should be thawed overnight at 4°C prior to plating in the wells of a 6-well plate, at a volume sufficient to just cover the surface of the well.
2. After a 30 min incubation at 37°C, 3 × 10^5 EC, in 2 mL prewarmed growth medium, are added per well and cultured at 37°C. Within 4 to 6 h (DMEC) and 6 to 12 h (HUVEC) the EC will spontaneously form microtubules on the surface of the Matrigel.

3.6.1.3. UPTAKE OF ACETYLATED LOW-DENSITY LIPOPROTEIN

1. EC (2 × 10^5) are cultured overnight in gelatin-coated wells of a 24-well plate. The EC monolayer is washed in pre-warmed HBSS, prior to addition of 500 µL

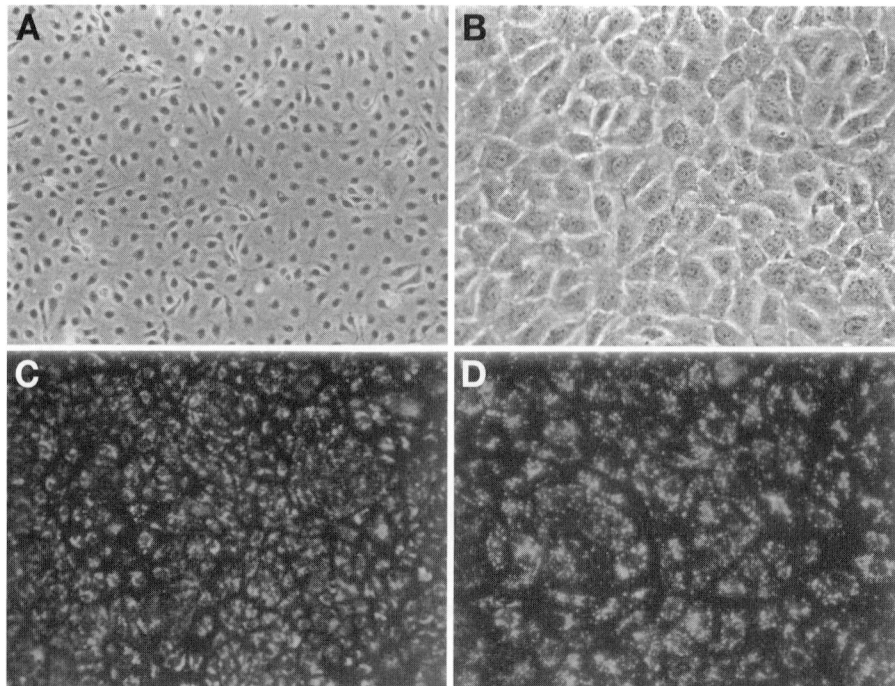

Fig. 1. Endothelial cell monolayers in culture imaged by phase contrast microscopy showing the characteristic contact-inhibited cobblestone monolayers of (**A**) human dermal microvascular EC (DMEC) and (**B**) human umbilical vein EC (HUVEC). (**C**) and (**D**) show the uptake of acetylated low-density lipoprotein labelled with 1,1'-dioctadecyl-3,3,3'3'-teramethy-indocarbocyanine perchlorate by these cells following 4 h incubation at 37°C. (**C**) DMEC and (**D**) HUVEC.

of prewarmed growth medium containing 20 µg/mL of acetylated low-density lipoprotein labelled with 1,1'-dioctadecyl-3,3,3'3'-teramethyl-indocarbocyanine perchlorate, and culture for 4 h at 37°C *(21)*.

2. EC are washed 3 times in HBSS, fixed in 2% paraformaldehyde, and visualized by fluorescent microscopy with rhodamine excitation (*see* **Fig. 1C,D**). Fluorescent cells are counted and expressed as a percentage of the total cells, counted under light microscopy using phase-contrast microscopy in the same field.

3.6.1.4. Staining for von Willebrand Factor

1. EC (2×10^5) are cultured overnight in gelatin-coated wells of a 24-well plate. The EC monolayer is washed in PBS and fixed in ice-cold 1:1 acetone-methanol for 20 min at 4°C.
2. Following aspiration of fixative, 500 µL of blocking solution (100 m*M* glycine, 1% BSA [w/v] in HBSS) is added for 30 min at 4°C.

3. The monolayer is washed 2 times in HBSS prior to addition of fluorescein isothiocyanate (FITC)-labeled goat anti-human vWF Ab diluted 1:50 in PBS for 60 min at room temperature.
4. Following washing 3 times in PBS, visualize by fluorescent microscopy with fluorescein excitation. Fluorescent cells are counted and expressed as a percentage of the total cells counted under light microscopy using phase-contrast microscopy in the same field.

3.6.1.5. STAINING WITH ULEX EUROPAEUS AGGLUTININ I (UEA-I)

1. EC (2×10^5) are cultured overnight in gelatin-coated wells of a 24-well plate. The EC monolayer is washed in PBS and fixed in ice-cold 1:1 acetone-methanol for 20 min at 4°C.
2. Following aspiration of fixative, 500 µL of blocking solution (100 mM glycine, 1% BSA [w/v] in HBSS) is added for 30 min at 4°C.
3. The monolayer is washed 2 times in HBSS prior to addition of biotinylated UEA-I (1:50) for 60 min at room temperature.
4. Following washing 3 times in PBS, streptavidin-FITC is added (1:50) for 30 min at room temperature, prior to 3 further washes in PBS.
5. EC are visualized by fluorescent microscopy with fluorescein excitation. Fluorescent cells are counted and expressed as a percentage of the total cells counted under light microscopy using phase-contrast microscopy in the same field (*see* **Note 17**).

3.6.1.6. SURFACE ANTIGEN EXPRESSION

Flow-cytometric analysis (*see* **Subheading 3.7.3.**) provides an excellent method for rapid analysis of EC surface antigen expression.

1. A variety of surface proteins should be analyzed using appropriate MAbs. Suitable target antigens include PECAM-1 (CD31), ICAM-2, endoglin (CD105) and VE-cadherin.
2. EC treated with the proinflammatory cytokine tumor necrosis factor (TNF)-α for 4 to 6 h will express the endothelial-specific antigen E-selectin. Significant upregulation of the expression of vascular cell adhesion molecule-1 (VCAM-1) and intercellular adhesion molecule-1 (ICAM-1) is seen following treatment of EC with TNFα for 12 to 24 h *(22–24)*.
3. The data from flow cytometric analysis will indicate the percentage of cells expressing each surface antigen.

3.6.2. Characterisation of Porcine Endothelial Cells

The methods described for human EC characterization in **Subheading 3.6.1.** also apply to the analysis of porcine EC. However, PAEC will not bind UEA-I and the staining of surface antigens is limited to some extent by the availability of mAbs binding the porcine antigen.

3.6.3. Characterization of Murine Endothelial Cells

The methods described for human EC characterization in **Subheading 3.6.1.** also apply to the analysis of murine EC *(10,12)*. However, murine EC will not bind UEA-I and biotinylated *Griffonia simplicifolia* isolectin B4 should be used in its place. Heterogeneity in ECs from different murine vascular beds has also been described with variable expression of CD31 and VCAM-1 *(12–14)*.

3.7. Analysis of Endothelial Cell Activation
3.7.1. Treatment of Endothelial Cells With Cytokines

The conditions under which the activation of EC is investigated will vary according to the area of study and below are some general guidelines.

1. Culture plates should be prepared by pre-coating with an appropriate substrate such as gelatin or fibronectin and pre-warmed culture medium.
2. For the majority of assays, EC are studied at confluence. This can be achieved by plating EC 12 to 24 h prior to the start of the assay in growth medium with the concentration of endothelial cell growth supplement reduced by 50%.
3. To achieve 80 to 100% confluence HUVEC should be plated as follows: $3–4 \times 10^4$ per well of 96-well plates, 2×10^5 per well in 24-well plates, and $5–6 \times 10^5$ per well of a 6-well plate.
4. Activation of EC is performed by adding the appropriate volume of freshly prepared 10X stock of the cytokine to each test well, followed by culture at 37°C for the desired time.

3.7.2. Analysis of Surface Antigen Expression by 96-Well Plate Cell Based Enzyne Linked Immunosorbent Assay

1. EC monolayers, in 96-well plates, are fixed by incubation with an aqueous solution of 2% paraformaldehyde, with 100 mM L-lysine monohydrochloride and 2.1 mg/mL sodium *m*-periodate, for 10 min at room temperature.
2. Following aspiration of fixative, blocking solution (100 mM glycine, 1% BSA w/v in HBSS) is added for 30 min at 4°C.
3. Aspirate blocking solution and add 50 µL of test or control mAb (*see* **Note 18**). Incubate for 1 h at room temperature.
4. Aspirate mAb and wash each well 2 times with 200 µL PBS (*see* **Note 19**).
5. Add 50 µL of affinity purified biotinylated rabbit anti-mouse Ig F(ab')$_2$ and incubate for 1 h at room temperature (*see* **Note 20**).
6. Aspirate secondary Ab and wash each well 2 times with 200 µL PBS.
7. Add to each well 50 µL of a high-molecular-weight complex of streptavidin-biotin-peroxidase and incubate at room temperature for 30 min.
8. Aspirate and wash each well 3 times with 200 µL PBS.
9. To develop the enzyme linked immunosorbent assay (ELISA), add 0.5 mg/mL *o*-phenylenediamine, 0.03% hydrogen peroxidase (v/v) in a pH 5.0 citrate-phosphate buffer (200 µL/well). Colour development is stopped after 2 to 5 min with 2 M sulfuric acid 50 µL/well.

Assessment of Endothelial Cell Activation

10. The optical density (OD) is measured at 491 nm using an ELISA plate reader. Specific mAb binding is calculated by subtracting the background as represented by the mean OD of triplicate wells incubated with an irrelevant isotype-matched primary MAb.
11. If required, protein estimation can be performed using crystal violet to exclude differential cell loss. Aspirate and wash each well times with 200 µL PBS.
12. Add 100 µL of crystal violet (0.1% [w/v] in distilled water) for 10 min. Wash wells carefully with 200 µL of distilled water 2 times and add 100 µL of 33% (v/v) acetic acid and measure OD at 620 nm.

3.7.3. Analysis by Flow Cytometry

Following harvesting of EC, flow-cytometric analysis should be performed throughout at 4°C. Two alternative methods are described and are equally effective.

3.7.3.1. PROTOCOL 1: TUBE METHOD

1. The treated EC monolayers are washed in calcium and magnesium free HBSS and cells detached by incubation for 1 to 2 min with 0.125% trypsin/EDTA at 37°C. Collect the cell suspension into labeled FACS tubes in 0.5 to 2 mL HBSS + 1% FCS per tube to give $1-2 \times 10^5$ cells/tube. For each treatment have a separate negative control (*see* **Note 21**). Centrifuge the tubes at $360g$ at 4°C for 5 min, carefully aspirate medium leaving the cell pellet in 50 µL.
2. Add 50 µL ice-cold HBSS/1% FCS containing 1 µg primary antibody (or 50 µL neat hybridoma supernatant). Resuspend the cell pellets by gentle vortexing and leave for 20 to 30 min at 4°C in the dark.
3. Add 2 mL ice-cold HBSS/1% FCS to each tube. Centrifuge the cells at $360g$ for 5 min, aspirate medium leaving 50 µL.
4. Add 50 µL ice-cold HBSS/1% FCS with 1 µg FITC- Alexa Fluor 488 or PE-conjugated secondary antibody (raised against the species of primary antibody) and resuspend the pellets by gentle vortexing. Leave for 20 to 30 min at 4°C in the dark.
5. Add 2 mL cold PBS to each tube. Centrifuge the tubes $360g$ for 5 min and aspirate medium leaving 50 µL and then add 450 µL of cold PBS.
6. For analysis run samples through the flow cytometer using an appropriate negative control to set-up the cytosettings. Initially set a gate on the whole endothelial cell population. Then to ensure analysis of live EC, add propidium iodide to each tube (final concentration 1–5 µg/mL) and using the control EC, set a separate gate on PI-negative cells. For FITC- or Alexa Fluor 488-labeled cells, measure log FL1 (FL2 for PE-labeled cells) of the gated cells and plot this as a histogram against the cell count. Set the gain and volts of FL1 so that the peak of the negative control is visible on the left edge of the histogram. Count 5000 to 10,000 events per tube and do not change the cytosettings for the remainder of the experiment. Note the geometric mean fluorescent intensity for each tube and use this to calculate the relative fluorescence intensity (RFI) for each treatment (*see* **Notes 22 and 23**).

3.7.3.2 PROTOCOL 2: PLATE METHOD

1. Release cells using trypsin/EDTA (as per protocol 1) and put $1–2 \times 10^5$ cells into Universal tubes in 2 to 5 mL HBSS/10% FCS. Centrifuge the tubes $350g$ for 5 min, aspirate medium and resuspend the pellet in 200 µL ice-cold HBSS/1% FCS.
2. Add 100 µL of cell suspension per well of a 96-well V-bottomed plate, spin down the plate at $350g$ for 5 min and then flick off supernatant.
3. Add 50 µL ice-cold HBSS/1% FCS containing 1 µg primary antibody (or 50 µL neat hybridoma supernatant) and resuspend pellet by gentle pipetting. Leave the plate for 30 min on a rocking table at 4°C.
4. Centrifuge plate at $350g$ at 4°C for 5 min, flick off supernatant and add 200 µL ice-cold HBSS/1% FCS to each well, centrifuge again and flick off supernatant. Repeat wash 1 time.
5. Add 50 µL ice-cold HBSS/1% FCS with 1 µg of appropriate FITC-, Alexa Fluor- or PE-conjugated secondary antibody and resuspend pellets by gentle pipetting. Leave for 20 to 30 min on rocking table at 4°C wrapped in foil.
6. Add 200 µL ice-cold HBSS/1% FCS to each well, spin down the plate at $350g$ for 5 minutes then flick off supernatant. Repeat wash 2 times.
7. Add 100 µL of ice-cold PBS to each well and gently resuspend the cells prior to transfer in to FACS tubes containing 400 µL of ice cold PBS.
8. Analyze the samples as described above in **Subheading 3.7.3.1.**

4. Notes

1. Before handling human tissues ensure adequate vaccination against Hepatitis B. In addition, consider all tissue as potentially infected with human immunodeficiency virus and wear double gloves and avoid the use of needles or other sharp objects wherever possible.
2. Be sure to express microvascular segments from the edge of the tissue and not to scrape across the surface of the tissue. This will reduce the risk of harvesting large numbers of cutaneous fibroblasts.
3. A mouse anti-human CD31 (PECAM-1) mAb is a suitable alternative here.
4. As an alternative to the two-step procedure, anti-human endoglin (CD105) coated MiniMacs beads can be used at this stage.
5. During the propagation of purified DMEC, precoating of tissue culture flasks with 0.1% gelatin is a suitable alternative to fibronectin.
6. Collagenase may leak out at this point from the three-way tap insertion or through a hole in the vein. In this case resecure the tap or occlude the hole(s) where possible and add further collagenase. Do not be tempted to incubate the cord with collagenase for more than 10 min. This simply results in poor quality EC and increases the risk of isolating smooth muscle cells.
7. Diligent repeated massage and flushing of umbilical vein greatly increases the yield of HUVEC.
8. To heat-inactivate serum warm to 56°C for 30 to 45 min.

Assessment of Endothelial Cell Activation

9. Confluent endothelial cells in 75-cm^2 tissue-culture flasks are best sub-cultured into 2 or at the most 3 further 75-cm^2 flasks.
10. The most common reasons for the failure of endothelial cells to proliferate in culture are (1) failure to precoat tissue culture flasks with gelatin or fibronectin and (2) failure to add endothelial cell growth factor.
11. This method can also be applied to the isolation of bovine aortic endothelial cells.
12. All instruments should be sterilized by autoclaving prior to use.
13. The genetic backgound of the mice effects the rate of EC proliferation, responsiveness to growth factors may lead to minor differences in expression of surface molecules. Therefore always generate wild-type EC controls at the same time as preparations from genetically modified animals (preferably using littermate controls).
14. Isolation may also be performed using murine lung tissue. Only 2 or 3 mice are required and after collagenase treatment tissue floats to the surface rather than to the bottom. Subsequent rounds of antibody/bead purification using MiniMAC column may be required to get a pure population.
15. An irrelevant mAb will bind and block Fc receptors on monocytes/macrophages and thus reduce nonspecific binding to the anti-endoglin antibody.
16. We recommend use of the MiniMAC system from Miltenyi as opposed to larger magnetic beads. It is not recommended to incubate the MiniMAC beads with the antibody prior to isolation. EC isolation using positive selection with MiniMAC columns is only effective when there are significantly more EC than contaminating cells in the preparation.
17. As an alternative, this method can be performed on an EC suspension and analyzed by flow cytometry to estimate the percentage of cells binding UEA-I. *See* **Subheading 3.7.3.** for the details of surface antigen expression by flow-cytometric analysis.
18. To minimize background staining affinity-purified mAbs should be used.
19. Wash wells using a multichannel pipette, expelling PBS carefully down the side of each well so as to avoid disturbing the EC monolayer.
20. Ensure that an appropriate secondary antibody is used as dictated by the species of origin of the primary mAb.
21. EC are autofluorescent and the degree of autofluorescence may be altered by treatment of the cells. Therefore, always compare EC treated with an antibody of interest to those treated with an isotype-matched negative control antibody.
22. As an alternative to analyzing FACS samples immediately, the EC pellet can be fixed by adding 400 µL of 1% paraformaldehyde (PFA) in PBS and vortexing gently. The samples can be stored, wrapped in sliver foil, at 4°C for a maximum of 5 d before analysis. For analysis use an appropriate negative control to set-up the cytosettings. Gate intact EC by studying a dot plot of forward scatter vs log side scatter (dead cells/debri are smaller than the intact EC population and can be excluded from the gate).
23. Relative fluorescence intensity (RFI) is calculated as the mean fluorescent intensity (MFI) for cells treated with test antibody divided by the MFI for the EC treated with an isotype-matched negative Ab.

References

1. Jaffe, E. A., Nachman, R. L., Becker, C. G., and Minick, C. R. (1973) Culture of human endothelial cells derived from umbilical veins. Identification by morphologic and immunologic criteria. *J. Clin. Invest.* **52(11),** 2745–2756.
2. Jackson, C. J., Garbett, P. K., Marks, R. M., et al. (1989) Isolation and propagation of endothelial cells derived from rheumatoid synovial microvasculature. *Ann. Rheum. Dis.* **48(9),** 733–736.
3. Jackson, C. J., Garbett, P. K., Nissen, B., and Schrieber, L. (1990) Binding of human endothelium to Ulex europaeus I-coated Dynabeads: application to the isolation of microvascular endothelium. *J. Cell Sci.* **96,** 257–262.
4. Abbot, S. E., Kaul, A., Stevens, C. R., and Blake, D. R. (1992) Isolation and Culture of Synovial Microvascular Endothelial Cells - Characterization and Assessment of Adhesion Molecule Expression. *Arthritis Rheum.* **35(4),** 401–406.
5. Hewett, P. W., Murray, J. C., Price, E. A., Watts, M. E., and Woodcock, M. (1993) Isolation and Characterization of Microvessel Endothelial Cells from Human Mammary Adipose Tissue. *In Vitro Cell Dev. Biol. Animal* **29a,** 325–331.
6. Hewett, P. W. and Murray, J. C. (1993) Human Lung Microvessel Endothelial Cells - Isolation, Culture, and Characterization. *Microvascular Res.* **46(1),** 89–102.
7. Hewett, P. W. and Murray, J. C. (1993) Human Microvessel Endothelial Cells - Isolation, Culture and Characterization. *In Vitro Cell Dev. Biol. Animal* **29A(11),** 823–830.
8. Hewett, P. W. and Murray, J. C. (1993) Immunomagnetic Purification of Human Microvessel Endothelial Cells Using Dynabeads Coated with Monoclonal Antibodies to PECAM-1. *Eur. J. Cell Biol.* **62(2),** 451–454.
9. Scott, P. A. E. and Bicknell, R. (1993) The Isolation and Culture of Microvascular Endothelium. *J. Cell Sci.* **105,** 269–273.
10. Gerritsen, M. E., Shen, C. P., McHugh, M. C., et al. (1995) Activation-dependent isolation and culture of murine pulmonary microvascular endothelium. *Microcirculation* **2(2),** 151–163.
11. Dong, Q. G., Bernasconi, S., Lostaglio, S., et al. (1997) A general strategy for isolation of endothelial cells from murine tissues. Characterization of two endothelial cell lines from the murine lung and subcutaneous sponge implants. *Arterioscler. Thromb. Vasc. Biol.* **17(8),** 1599–1604.
12. Lidington, E. A., Rao, R. M., Marelli-Berg, F. M., Jat, P. S., Haskard, D. O., and Mason, J. C. (2002) Conditional immortalization of growth factor-responsive cardiac endothelial cells from H-2Kb-tsA58 mice. *Am. J. Physiol. Cell Physiol.* **282(1),** C67–C74.
13. Marelli-Berg, F. M., Peek, E., Lidington, E. A., Stauss, H. J., and Lechler, R. I. (2000) Isolation of endothelial cells from murine tissue. *J. Immunol. Meth.* **244,** 205–215.
14. Lim, Y.-C., Garcia-Cardena, G., Allport, J. R., et al. (2003) Heterogeneity of Endothelial Cells from Different Organ Sites in T-Cell Subset Recruitment. *Am. J. Pathol.* **162(5),** 1591–1601.

15. Gerritsen, M. E. (1987) Functional heterogeneity of vascular endothelial cells. *Biochem. Pharmacol.* **36(17),** 2701–2711.
16. Page, C., Rose, M., Yacoub, M., and Pigott, R. (1992) Antigenic heterogeneity of vascular endothelium. *Am. J. Pathol.* **141(3),** 673–683.
17. Petzelbauer, P., Bender, J. R., Wilson, J., and Pober, J. S. (1993) Heterogeneity of Dermal Microvascular Endothelial Cell Antigen Expression and Cytokine Responsiveness *In situ* and in Cell Culture. *J. Immunol.* **151(9),** 5062–5072.
18. Swerlick, R. A. and Lawley, T. J. (1993) Role of Microvascular Endothelial Cells in Inflammation. *J. Invest. Dermatol.* **100(1),** S111–S115.
19. Mason, J. C., Yarwood, H., Tárnok, A., et al. (1996) Human Thy-1 is cytokine inducible on vascular endothelial cells and is a signaling molecule regulated by protein kinase C. *J. Immunol.* **157,** 874–883.
20. Kubota, Y., Kleinman, H. K., Martin, G. R., and Lawley, T. J. (1988) Role of laminin and basement membrane in the morphological differentiation of human endothelial cells into capillary-like structures. *J. Cell Biol.* **107(4),** 1589–1598.
21. Voyta, J. C., Via, D. P., Butterfield, C. E., and Zetter, B. R. (1984) Identification and isolation of endothelial cells based on their increased uptake of acetylated-low density lipoprotein. *J. Cell Biol.* **99,** 2034–2040.
22. Wellicome, S. M., Thornhill, M. H., Pitzalis, C., et al. (1990) A monoclonal antibody that detects a novel antigen on endothelial cells that is induced by tumor necrosis factor, IL-1, or lipopolysaccharide. *J. Immunol.* **(7),** 2558–2565.
23. Thornhill, M. H., Wellicome, S. M., Mahiouz, D. L., Lanchbury, J. S., Kyan-Aung, U., and Haskard, D. O. (1991) Tumor necrosis factor combines with IL-4 or IFN-gamma to selectively enhance endothelial cell adhesiveness for T cells. The contribution of vascular cell adhesion molecule-1-dependent and -independent binding mechanisms. *J. Immunol.* **146(2),** 592–598.
24. Mason, J. C. and Haskard, D. O. (1994) The clinical importance of leucocyte and endothelial adhesion molecules in inflammation. *Vasc. Med. Rev.* **5(3),** 249–275.

20

Analysis of Flow-Based Adhesion In Vitro

Oliver Florey and Dorian O. Haskard

Summary

To be able to visualize real time leukocyte – endothelial cell interactions in vitro opens up the possibility of exploring the complex cascade of events that culminate in leukocyte recruitment and diapedesis in a much more detailed and controlled way. Techniques have been developed whereby fluorescently labeled leukocytes are perfused over an endothelial substrate in a controlled manner. Interactions can then be visualized and, using motion tracking software, the movement of cells characterized. Dynamic flow based adhesion assay protocols build on previous static assays of leukocyte adhesion in better modeling the environment in which these interactions actually take place in vivo.

Key Words: Flow chamber; leukocyte recruitment; in vitro; rolling; adhesion; selectin.

1. Introduction

Leukocyte adhesion and subsequent extravazation through the endothelium is important in the pathogenesis of inflammatory disorders such as rheumatoid arthritis *(1)*. Understanding how leukocyte accumulation within the inflamed synovium is likely to provide new therapeutic approaches to resolve the inflammatory process *(2)*. As such, new techniques to better model the inflammatory environment are needed. With this in mind flow-based adhesion assays have been developed.

The recruitment of leukocytes to inflamed endothelium follows a series of interactions between the leukocyte and endothelium *(3)* which results in leukocytes tethering and rolling on the endothelium thus allowing activation of the leukocyte by factors such at interleukin (IL)-8 and MCP-1, leading to enhanced integrin binding to their counter receptors on the endothelium *(4)*. In the past, leukocyte adhesion to the endothelium was measured under static conditions whereby leukocytes were allowed to settle on adhesive substrates and amounts of adhesion measured through fluorescently labeling *(5)* or radiolabeling *(6)*

From: *Methods in Molecular Medicine, Vol. 135: Arthritis Research, Volume 1*
Edited by: A. P. Cope © Humana Press Inc., Totowa, NJ

the leukocytes. This has now been superseded by dynamic-flow-based adhesion assays where the effect of laminar flow on the cascade of events leading to adhesion can be both visualized and measured. This is a more complex adhesion assay, which has a greater physiological relevance than previous static assays and allows measurement of selectin mediated rolling interactions previously undetectable under static assays *(7,8)*. The key factor in this flow adhesion assay is the direct visualization of leukocyte endothelial interactions. To achieve this, a specific instrument set up is required in which leukocytes may be perfused over a slide coated with an adhesive substrate in a closed chamber and interactions captured to video. Leukocytes are first fluorescently labeled so as to be able to differentiate the flowing leukocytes from the endothelial background, something that has been shown not to influence their adhesive interactions *(9)*. All of this is carried out at 37°C in a heated microscope box. After the experiment, video footage can be analyzed using software programs that measure parameters such as velocity and cell motion characteristics (e.g., rolling, stationary, or free flowing). Automated analysis is what makes this assay more readily accessible. Variations in this adhesion assay allow for a range of questions to be investigated; for example, different isolated leukocyte cell types or whole blood may be used depending on the area of interest. Also, endothelial cells activated by different stimuli and their subsequent ability to recruit leukocytes may be investigated *(10)*. Immobilization of the adhesion molecules of choice on plastic *(11)* or the production of cell lines expressing adhesion molecules *(12)* can be used to investigate a more specific role these proteins have in adhesion. Recently a system in which adhesion molecules are introduced into resting endothelial cells through adenovirus infection has been used to reconstitute a specific model inflammatory environment *(13)*.

This chapter outlines the methods of a number of ways with which to analyze leukocyte-endothelial cell interactions under flow conditions. Each has their benefits and drawbacks.

2. Methods
2.1. Slide Preparation

1. Tissue culture facilities, incubators, hoods, and pipetes.
2. Nunc slide flaskettes (Invitrogen, Carlsbad, CA), or relevant slides for flow chamber.
3. Recombinant adhesion molecules (R&D systems, Minneapolis, MN).
4. Purified and titred adenoviral constructs encoding for adhesion molecules.
5. Phosphate buffered saline (PBS) and bovine serum albumin (BSA), 30% BSA (Sigma, St. Louis, MO).
6. Human umbilical vein endothelial cella (HUVEC) culture media M199 (Invitrogen, Carlsbad, CA): 10% fetal bovine serum (FBS), penicillin/streptomycin, heparin, endothelial cell growth factor (Sigma, St. Louis, MO).

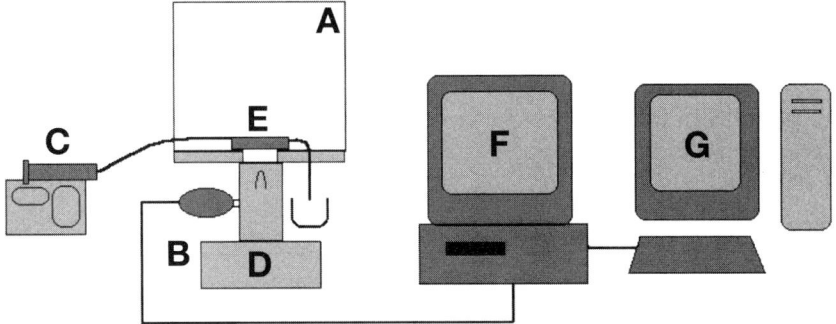

Fig. 1. (**A**) Heat box, (**B**) camera, (**C**) harvard perfusion pump, (**D**) inverted fluorescent microscope, (**E**) flow chamber, (**F**) video recorder, (**G**) analysis computer.

7. Cell culture reagents.
8. 1% Gelatin
9. Trypsin/EGTA.
10. Hank's balanced salt solution (HBSS) with or without calcium and magnesium (Sigma, St. Louis, MO).

2.2. Leukocyte Isolation, Cell Line Culture, and Fluorescent Labeling

1. Blood donations from healthy volunteers in anticoagulated syringes.
2. Ficoll Hypaque.
3. HBSS with and without calcium and magnesium.
4. Sterile culture water.
5. 2% Dextran in HBSS without calcium and magnesium.
6. Calcein AM, (Molecular Probes, Eugene, OR). Made up in dimethly sulfoxide (DMSO), this light sensitive and should be kept at 4°C.

2.3. Parallel Plate Flow Adhesion Assay

The details of the basic setup may differ, but the main features are listed here (*see* **Fig. 1**).

1. Inverted fluorescent microscope with heated stage box with a camera and video recording system attached. We use an inverted Nikon Diaphot 300 florescence microscope connected to a JVC TK-C1360 color video camera, and the images are recorded on a Panasonic AG-6730 S-VHS video recorder
2. Harvard perfusion pump.
3. Parallel plate flow chamber (built in-house, but commercial chambers available).
4. Connecting tubing (Zygon).
5. Perfusion buffer, HBSS with calcium and magnesium (Sigma, St. Louis, MO).

2.4. Motion Analysis

1. Computer linked to video recorder.
2. Motion analysis software. Many are available; we use Cell Motion from Ed Marcus Laboratories, MA.

3. Methods
3.1. Slide Preparation

There are a variety of adhesive substrates that can be used in flow adhesion assays depending on the aim of the investigation. Here we outline the preparation protocols of a few of the options available.

3.1.1. HUVEC Isolation and Culture/Adenovirus Infected cells

1. Isolate HUVEC as per protocol *(14)*, seed out cells to achieve confluent monolayer in 2 to 3 d. Stimulate with pro inflammatory cytokines at required time points (*see* **Note 1**).
2. For adenovirus infection seed cells to obtain approximately 70–80% confluency by the next day.
3. Replace growth media with infection media (growth media but with no serum).
4. Add desired dilution of adenovirus and leave for 3 h.
5. Change media to growth media and leave for 48 h (*see* **Note 2**).

3.1.2. Immobilized Protein

1. Slides are coated with 50 µL of 5 to 10 µg/mL soluble adhesion molecule and incubated overnight at 4°C under humid conditions.
2. To inhibit non-specific interactions of leukocytes to plastic, slides are incubated with PBS + 1% BSA for 2 h at 37°C (*see* **Note 3**).
3. Slides are washed twice with PBS + 0.05% Tween-20 prior to use.

3.2. Leukocyte Isolation, Cell line Culture and Fluorescent Labeling

Any leukocyte population can be studied as long as enough cells are purified. Here, we outline a method for neutrophil isolation.

3.2.1. Neutrophil Isolation and Cell Line Culture

1. Take 30 mL venous blood into 50 mL tube containing 3 mL sodium citrate, top up to 50 mL with HBSS and mix gently by inverting.
2. Carefully layer 25 mL onto 10 mL Ficoll. Spin at 400g for 30 min at room temperature with no brake.
3. Carefully aspirate entire peripheral blood mononuclear cells (PBMC) layer at interface. Aspirate further top layer until there is 1 to 2 mL above the red pellet.
4. Resuspend pellet with equal volume of HBSS, then make a 1:1 dilution with 2% dextran. Mix tubes by inverting.
5. Allow to stand for 30 min at room temperature to allow the red blood cells (RBC) to sediment.
6. Pool top layers from tubes leaving behind the RBC pellet.
7. Top up pooled layers with HBSS, spin at 1400 rpm for 5 min.
8. Aspirate supernatant, and then add 3 mL cold cell culture water resuspending the pellet gently. Swirl tube for 30 s to lyse the RBC then fill to top with HBSS.

9. Spin at 1400 rpm for 5 min, aspirate off HBSS, and resuspend cells at 2×10^6/mL in M199 media.
10. Add 1 µL of calcein AM (1 mg/mL) per 1 mL neutrophils, leave at 37°C for 30 min. Final concentration of calcein should be 1 µg/mL.
11. Wash cells and resuspend to desired concentration in perfusion buffer
12. For cell lines culture as per protocol and label as above.

3.2.2. Whole Blood

1. Take 5 mL venous blood in low-molecular-weight Heparin 25 U/mL to prevent clotting and interference with selectin interactions *(15)*. Split into 2 tubes, add 10 µL of calcein AM (1 mg/mL) to each tube and incubate at 37°C for 25 min.
2. Resuspend blood 1:18 with HBSS (with Ca^{2+}/Mg^{2+}) prior to use and keep blood at room temperature (*see* **Note 4**).

3.3. Parallel Plate Flow Chamber Adhesion Assay

Leukocytes or endothelial substrates are treated according to experimental design prior to use. This may include pretreatment with blocking antibodies (*see* **Note 5**) or specific compounds that will result in altered adhesion.

1. Prewarm perfusion buffer (HBSS with calcium and magnesium) and microscope stage to 37°C. Ultraviolet (UV) light switched on with filters to visualize green wavelength of calcein AM.
2. Leukocytes are loaded into syringe at desired concentration; this may vary depending on amount of interaction visualized (*see* **Note 6**). All air bubbles are removed from the syringe and connecting tubing. The syringe driver is programmed to correct setting to obtain shear stress required. This depends on various parameters of the flow chamber used (*see* **Note 7**). Shear stresses are calculated using the modified Poiseuille equation:

$$Tw = \frac{6Q\mu}{H^2W}$$

where:
- Tw is shear stress (dynes/cm²),
- Q is flow rate (mL/s),
- μ is viscocity at 37°C (Poise),
- H is height of channel (cm),

and
- W is width of channel (cm).

3. Slides are washed and assembled into flow chamber. Air inside the system is removed by flushing through with HBSS buffer. Once assembled chamber may be connected to the syringe again making sure no air bubbles are introduced into the system.
4. Flow is initiated using a Harvard syringe driver and perfusion is allowed for 2 min before any recording as this allows time for interactions to start. Video recordings of ten random fields of view using a 10X objective are taken each 15 s long. Fields are kept relatively central to the slide, as areas close to the edge have reduced laminar flow dynamics. The flow chamber size will determine the volume of cells required to be perfused per slide. Smaller chambers require only a few milliliters of cells and greatly save on resources.

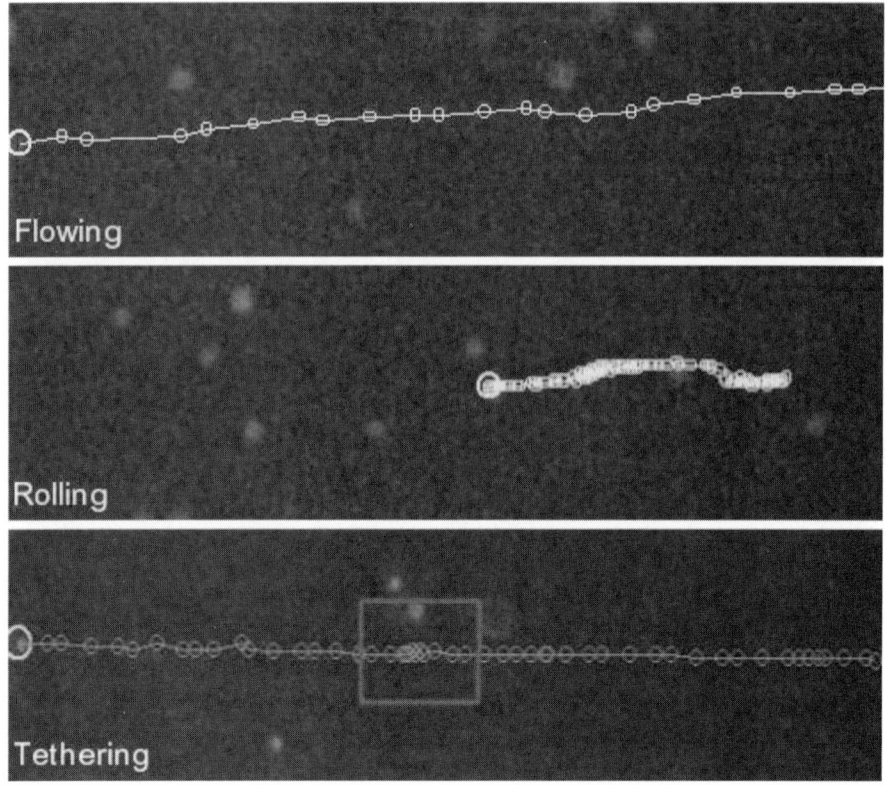

Fig. 2. Cell motion trajectories marked every 0.07 s. Pictures show representative tracks for a free flowing (noninteracting) cell, rolling cell and a cell that has tethered during it's pass.

5. After the recordings the slide is removed and chamber washed through with HBSS to remove any cells remaining in the inlet/outlet tubes. The next slide may now be assembled and experimental procedure repeated. Once all slides have been processed analysis of recordings may be carried out.

3.4. Motion Analysis

Recorded videos are transferred to digital video file (*.avi) in 10-s clips. Software is able to detect objects that fall within size exclusion limits and by highlighting areas of high light intensity (only the leukocytes are fluorescently stained), and track their movement frame-by-frame building up cell trajectory pathways (*see* **Fig. 2**). From this, mean velocity of individual cells as well as the whole population can be calculated. By setting predefined velocity parameters at which cells are considered to be rolling, the numbers of cells rolling/field, stationary/field and in free flow can be measured. A cell is usually defined as rolling if it travels between set velocities for at least 1 s. These parameters need to be optimized for specific interactions in the experimental setup as they vary between protein interactions *(16)*.

4. Notes

1. A varied range of responses by HUVEC to stimuli can be expected, as there are variations from individual to individual culture.
2. Adenovirus infection of cells has been shown in the past to activate endothelial cells increasing ICAM-1 and VCAM-1 level *(17)*. Chemokine production may also occur post infection. This should be investigated beforehand as it could affect the result of the assay.
3. Complete blocking of the slides is important as cells, especially neutrophils, can bind plastic through integrins increasing the amount of background adhesion thus obscuring any increases though the experiment. Blocking protocols may have to be optimized. Varying the amount of BSA or length incubation time are some options.
4. The use of whole blood in these assays is not common, as it doesn't allow for specific effects to be investigated. There will be many secondary interactions occurring such as platelet–leukocyte binding and leukocyte–leukocyte attachments. However, it does allow for a more physiological representation of an inflammatory response.
5. Using blocking antibodies to specific adhesion molecules allows assessment of the role they play in the interactions. For leukocytes preincubation of cells at 1×10^6/mL with 25 µg/mL of a blocking antibody for 15 min at 37°C is sufficient to block the interactions. Cells are not washed before perfusion but are resuspended to desired concentration for flow. Addition of specific antibodies to endothelial monolayers at 30 to 50 µg/mL for 20 min is also sufficient.
6. The usual perfusion density of leukocytes is between 250,000 and 1,000,000 cells/mL. Therefore, the study of small sub sets of cells may require a large amount of blood to start with in order to obtain enough cells for the experiment. The density depends on the level of interactions per field likely to be seen. Often this is determined by the number of interactions the analysis software is able to cope with in one field of view.
7. For the majority of experiments the viscosity of the perfusion media should remain the same, but when using whole blood there may be differences even when diluted 1:18. These need to be measured and compensated for in the shear equations.

References

1. Jalkanen, S., Steere, A. C., Fox, R. I., and Butcher, E. C. (1986) A distinct endothelial cell recognition system that controls lymphocyte traffic into inflamed synovium. *Science* **233**, 556–558.
2. Kavanaugh, A. F., Davis, L. S., Nichols, L. A., et al. (1994) Treatment of refractory rheumatoid arthritis with a monoclonal antibody to intercellular adhesion molecule 1. *Arthritis Rheum.* **37**, 992–999.

3. Springer, T. A. (1994) Traffic signals for lymphocyte recirculation and leukocyte emigration: the multistep paradigm. *Cell* **76,** 301–314.
4. Seo, S. M., McIntire, L. V., and Smith, C. W. (2001) Effects of IL-8, Gro-alpha, and LTB(4) on the adhesive kinetics of LFA-1 and Mac-1 on human neutrophils. *Am. J. Physiol. Cell Physiol.* **281,** C1568–C1578.
5. Price, E. A., Coombe, D. R., and Murray, J. C. (1995) A simple fluorometric assay for quantifying the adhesion of tumour cells to endothelial monolayers. *Clin. Exp. Metastasis* **13,** 155–164.
6. Arnaout, M. A., Lanier, L. L., and Faller, D. V. (1988) Relative contribution of the leukocyte molecules Mo1, LFA-1, and p150,95 (LeuM5) in adhesion of granulocytes and monocytes to vascular endothelium is tissue- and stimulus-specific. *J. Cell Physiol.* **137,** 305–309.
7. Brunk, D. K. and Hammer, D. A. (1997) Quantifying rolling adhesion with a cell-free assay: E-selectin and its carbohydrate ligands. *Biophys. J.* 72, 2820–2833.
8. Lawrence, M. B. and Springer, T. A. (1991) Leukocytes roll on a selectin at physiologic flow rates: distinction from and prerequisite for adhesion through integrins. *Cell* **65,** 859–873.
9. Abbitt, K. B., Rainger, G. E., and Nash, G. B. (2000) Effects of fluorescent dyes on selectin and integrin-mediated stages of adhesion and migration of flowing leukocytes. *J. Immunol. Methods* **239,** 109–119.
10. Kalogeris, T. J., Kevil, C. G., Laroux, F. S., Coe, L. L., Phifer, T. J., and Alexander, J. S. (1999) Differential monocyte adhesion and adhesion molecule expression in venous and arterial endothelial cells. *Am. J. Physiol.* **276,** L9–L19.
11. DiVietro, J. A., Smith, M. J., Smith, B. R., Petruzzelli, L., Larson, R. S., and Lawrence, M. B. (2001) Immobilized IL-8 triggers progressive activation of neutrophils rolling in vitro on P-selectin and intercellular adhesion molecule-1. *J. Immunol.* **167,** 4017–4025.
12. Rao, R. M., Haskard, D. O., and Landis, R. C. (2002) Enhanced recruitment of Th2 and CLA-negative lymphocytes by the S128R polymorphism of E-selectin. *J. Immunol.* **169,** 5860–5865.
13. Gersten, R. E., Garcia-Zepeda, E. A., Lim, Y. C., et al. (1999) MCP-1 and IL-8 trigger firm adhesion of monocytes to vascular endothelium under flow conditions. *Nature* **398,** 718–723.
14. Jaffe, E. A., Nachman, R. L., Becker, C. G., and Minick, C. R. (1973) Culture of human endothelial cells derived from umbilical veins. Identification by morphologic and immunologic criteria. *J. Clin. Invest.* **52,** 2745–2756.
15. Koenig, A., Norgard-Sumnicht, K., Linhardt, R., and Varki, A. (1998) Differential interactions of heparin and heparan sulfate glycosaminoglycans with the selectins. Implications for the use of unfractionated and low molecular weight heparins as therapeutic agents. *J. Clin. Invest.* **101,** 877–889.
16. Puri, K. D., Finger, E. B., and Springer, T. A. (1997) The faster kinetics of L-selectin than of E-selectin and P-selectin rolling at comparable binding strength. *J. Immunol.* **158,** 405–413.

17. Newman, K. D., Dunn, P. F., Owens, J. W., et al. (1995) Adenovirus-mediated gene transfer into normal rabbit arteries results in prolonged vascular cell activation, inflammation, and neointimal hyperplasia. *J. Clin. Invest.* **96,** 2955–2965.

21

Analysis of Leukocyte Recruitment in Synovial Microcirculation by Intravital Microscopy

Gabriela Constantin

Summary

A complete pattern of adhesion molecules and chemokines involved in leukocyte migration in different tissues and in homeostatic vs inflammatory conditions is still lacking. This chapter describes how to characterize the mechanisms of leukocyte recruitment in synovial vessels in vivo by using epifluorescence videomicroscopy. It has been proposed that inflamed endothelium expresses a combination of adhesion ligands and activating factor(s) for G_i-linked receptors that together orchestrate leukocyte recruitment in vivo. Thus, the combination of molecules involved in the adhesion cascade in synovial vessels may favor the arrest of specific leukocyte subpopulations during different phases of joint inflammation. The approach presented here may provide a useful tool for further investigations of physiologic and pathologic events that occur in normal synovial microcirculation as well as during arthritis.

Key Words: Intravital microscopy; synovial venules; endothelium; leukocyte recruitment; adhesion molecules; inflammation; arthritis.

1. Introduction

Rheumatoid arthritis (RA) is an autoimmune disease characterized by chronic inflammation and destruction of cartilage and bone in the joints *(1)*. Leukocyte recruitment into synovial vessels is a crucial step in the pathogenesis of RA. Extravazation of leukocytes during inflammation is a concurrent process leading to finely regulated steps controlled by adhesion molecules and activating factors *(2)*. It involves: (1) initial contact (tethering or capture) and rolling along the vessel wall mediated by selectins and integrins, and their ligands. The slow motion of rolling leukocytes during inflammation then facilitates sensing of chemoattractants exposed on the endothelial surface and leukocyte activation by (2) chemoattractant-induced heterotrimeric G_i protein-dependent intracellular biochemical changes leading to integrin activation; (3)

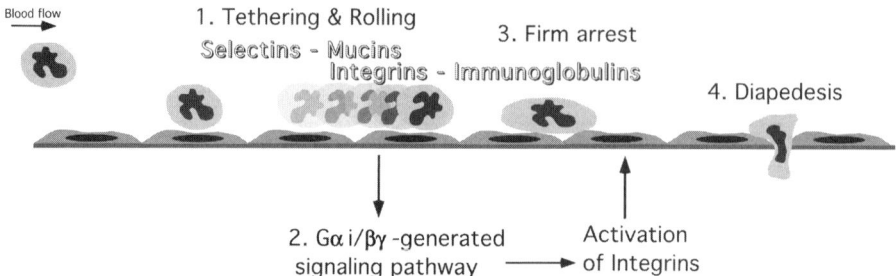

Fig. 1. The multistep model of leukocyte migration through the endothelium. It involves: (1) Tethering (capture) and rolling along the vessel wall mediated mainly by selectins and mucins (e.g., endothelial E- and P-selectin and leukocyte PSGL-1). Integrins and their endothelial ligands (e.g., VLA-4-VCAM-1) may also contribute to this step; (2) integrin activation by insight-out signaling genereted by chemoattractant receptors connected to heterotrimeric G_i proteins; (3) integrin-dependent firm arrest to immunoglobulin-like family endothelial ligands (e.g., LFA-1-ICAM-1; VLA-4-VCAM-1); and (4) transmigration/diapedesis.

integrin-dependent firm arrest to immunoglobulin-like family endothelial ligands; and (4) transmigration *(3–5)* (*see* **Fig. 1**).

Intravital microscopic visualization of these dynamic events at the microcirculatory level provides a powerful research tool coupling in vitro observations with events in vivo, for understanding complex biological interactions and disease mechanisms, and to develop and test novel therapeutic approaches for autoimmune diseases *(6)*. Intravital microscopy may be exploited to study the molecular mechanisms of blood cell recruitment in normal, acute, subacute, and chronically inflamed synovial vessels. As described below, the analysis may be performed by using different leukocyte subpopulations injected intraarterialy. Moreover, the approach takes advantage of the fact that human adhesion molecules expressed by leukocytes are able to efficiently interact with their endothelial counterligands expressed by mouse endothelium in the intravital microscopy setting *(7–10)*. Consequently, the adhesive capabilities of human leukocyte subpopulations derived from healthy donors or patients with RA may be eventually studied under pathophysiological conditions in murine synovial microcirculation as previously described for other autoimmune diseases *(10)*.

2. Materials and Instruments
2.1. Mice

Young mice (8–12 wk old) should be used. Arthritis may be induced using different protocols described in other chapters of this book. The expression of adhesion molecules and activating factors by the synovial venules might differ between various arthritis models, especially from the quantitative point of view. However, the expression of adhesion molecules in more general model of subacute or chronic vasculitis (*see* **Subheading 3.1.**) should be independent of the mouse strain used.

2.2. Reagents

1. Phosphate buffer saline (PBS).
2. DMEM without sodium bicarbonate (Sigma).
3. Hepes.
4. Green CMFDA (5-chloromethylfluorescein diacetate) (Molecular probes, Eugene, OR).
5. Orange CMTMR (5-(and-6)-(chloromethyl)benzoyl)amino)tetramethylrhodamine) (Molecular probes, Eugene, OR).
6. Fluorescein isothiocyanate (FITC) Dextran (MW 150 kDa) (Sigma).
7. Tetramethyl-rhodamine isothiocyanate (TRITC) Dextran (MW 160kDa) (Sigma).
8. High vacuum grease (Dow Corning GmbH, Wiesbaden, Germany).
9. 3-mL Plastic Pasteur pipets.
10. Schwabs, rounded cotton tip (Amarillo, TX).
11. Surgical sutures, 6-0 and 5-0 silk (Denkel, MA).
12. PE10 Polyethylene tube, I.D. 0.28 mm, aptical density (OD) 0.61 mm (Clay Adams/Becton Dickinson, MD)
13. Ketamine (Sigma).
14. Xylazine hydrochloride (Sigma).
15. Halothane (Sigma).
16. Square 24 × 24 mm glass coverslips.
17. Heparin grade I-A (Sigma).
18. Rhodamine 6G chloride (Sigma).
19. Lipopolysaccharide *Escherichia coli* 026:B6 (Sigma).
20. 1-mL Sterile syringes.
21. Alexa 568/488 protein labeling kit (Molecular Probes, Eugene, OR).
22. Rubber O-rings, internal diameter 11mm.
23. Tumor necrosis factor (TNF)-α (R&D, Minneapolis, MN).
24. Mycobacterium butyricum (Difco/BD).
25. Anti-adhesion molecules mAbs: anti-ICAM-1 (Y.N.1.7, ATCC), anti-VCAM-1 (MK 2.7, ATCC), anti-α4 integrins (PS/2, ATCC), anti-LFA-1 (TIB213, ATCC), anti-L-selectin (MEL-14, ATCC), anti-PSGL-1 (4RA10, Pharmingen/BD), anti-P-selectin (RB40.34, Pharmingen/BD).

2.3. Instruments

2.3.1. Microsurgery

1. Fiber optic light source.
2. Dissection microscope.

3. Forceps: biology tip 0.05 × 0.02 mm, 11-cm length; forceps 45°, 0.10 × 0.06 mm, 11-cm length; eye dressing forceps serrated, tip width 0.8 to 0.5 mm.
4. Scissors: fine iris scissors with large finger loops; extra fine spring scissors for incising vessel walls for catheter insertion.
5. Retractor, maximum spread 1.5 cm, 3 to 4 cm length.
6. Vascular clamps: microserrefines with delicate, atraumatic serrations and 2-mm spring width, 4-mm length, 0.75 width, pressure 125 g.
7. Forcep-type clip applicator.
8. Linkam stage heater CO102 (Olympus).

2.3.2. Videomicroscopy

1. Silicon-intensified target videocamera (VE-1000 SIT, Dage MTI, Michigan, IL).
2. Epifluorescence microscope with large stage adapted for intravital microscopy studies.
3. Microscope objectives: 10X (water immersion, focal distance 3,3, NA 0.3×); 20X (water immersion, focal distance 3.3 mm, NA 0.5), 40X (water immersion, focal distance 3.3 mm, NA 0.8).
3. Monitor (Sony SSM-125CE).
4. Digital VCR (Panasonic NV-DV10000).
5. SP100i digital syringe pump (World Precision Instruments, Sarasota, FL).

2.3.3. Data Analysis

1. Computer.
2. NIH image software.

3. Methods

3.1. Activation of Synovial Endothelium

1. Mice must be kept in pathogen free conditions (*see* **Note 1**).
2. Mice receive no treatment if the study requires a normal synovial endothelium.
3. In the case of a subacutely inflamed endothelium, animals may be injected intraperitoneally with 15 µg LPS or with 1 µg TNF-α, 5 to 6 h or 3 to 4 h respectively, before starting the intravital experiment *(11)*.
4. Chronic inflammation may be studied in mice with RA or by inducing a chronic vasculitis by immunizing Mycobacterium butyricum 4 to 8 d before performing the intravital microscopy experiments *(12)*.

3.2. Leukocyte Labeling

1. Leukocytes (e.g., neutrophils, monocytes or lymphocyte subpopulations) derived from donor syngenic animals are suspended at a concentration of 5×10^6/mL in DMEM without sodium bicarbonate supplemented with 20 mM Hepes, 5% fetal calf serum (FCS) (pH 7.1).
2. Then label with green CMFDA or orange CMTMR for 20 to 25 min at 37°C.
3. Alternatively, total blood leukocytes may be studied by in vivo labeling by injecting rhodamine 6G (0.3 mg/kg intravenous [iv], Sigma-Aldrich), immedi-

ately before microscopic visualization (rhodamine 6G will label leukocytes and platelets).

3.3. Cannulation of Carotid Artery

1. When the study specifies a particular leukocyte subtype (e.g., monocytes, Th1 or Th2 cells, B, T lymphocytes, $CD4^+CD25^+$ T-cells), cells must be injected intraarterialy
2. Animals are anesthetized by intraperitoneal injection (10 mL/kg) of physiologic saline containing ketamine (5 mg/mL) and xylazine (1 mg/mL).
3. The recipient is maintained at 37°C by a stage mounted strip heater Linkam CO102.
4. A heparinized PE-10 polyethylene catheter is inserted into the left common carotid artery toward the aortic arch for the injection of fluorescent cells (*see* **Note 2**) *(11)*.
5. When the number of the cells is small ($<1 \times 10^6$), cells could be injected through a thin catheter inserted into the femoral or inguinal artery.

3.4. Preparation of the Knee Joint

1. Expose the mouse knee joint on a stage (the hindlimb should be in slight in flexion). A dissection microscope should be used.
2. The skin covering the joint is removed and the ligamentum patellæ (*anterior ligament*) is cut from its insertion on the tibial tuberosity and carefully removed (*see* **Note 3**).
3. The infrapatellar pad which is attached to the articular capsule is then exposed. The fat pad has a capillary network resembling a honeycomb *(13)* (*see* **Fig. 2**).
4. The exposed tissue preparation needs to be bathed with sterile physiologic saline, and a 24 × 24 mm glass coverslip is then applied and fixed with silicon grease.
5. A round chamber with 11mm internal diameter is attached to the coverslip and filled with water.

3.5. Visualization of the Synovial Vessels

The preparation is placed on a microscope with epifluorescence equipped with large stage and water immersion objectives (*see* **Fig. 3**). Blood vessels are visualized by using fluorescent dextrans: 5 mg of FITC-dextran and/or 8 mg of TRITC-dextran are diluted in 0.5-mL sterile physiologic saline and centrifuged for 5 min at 14,000g (each mouse receives 0.05 mL supernatant) (*see* **Fig. 2**). Inflammation of the synovial vessels may lead to increased fluorescence (around the vessels and in the vessel wall) because of the presence of inflammatory cells and calcifications. $3–10 \times 10^6$ fluorescent labeled cells in 0.3 to 0.5 mL DMEM are injected into the carotid artery for each condition (control cells versus treated cells/endothelium) using the SP100 digital pump. Images are visualized by using a high performance silicon-intensified target videocamera and a monitor. Recordings are digitalized and stored on videotapes employing a digital VCR.

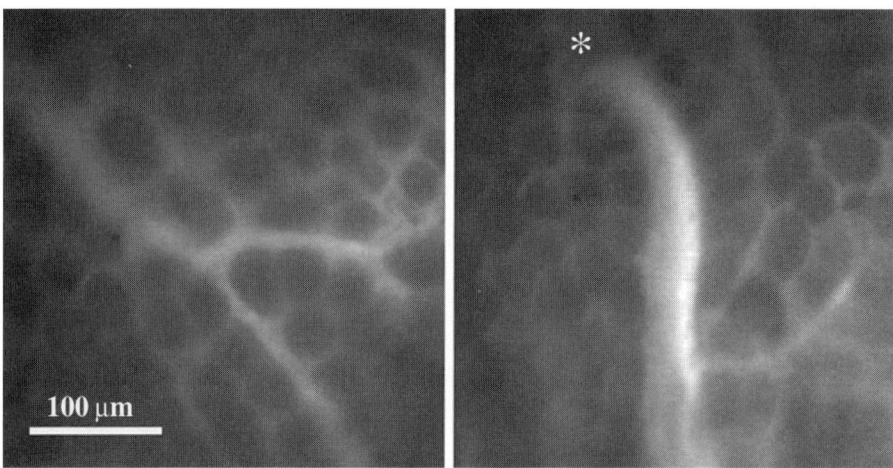

Fig. 2. Visualization of synovial vessels. The skin was reflected and the connective tissue adhered to the ligamentum patellæ was scraped away. The infrapatellar pad beneath the Ligamentum was carefully removed. The pad was bathed with sterile physiologic saline, and a 2 4 × 24 mm coverslip was then applied and fixed with silicon grease. A round chamber with 11-mm internal diameter was attached on the coverslip and filled with water. The preparation was placed on a microscope with epifluorescence equipped with a large stage and water immersion objectives. Synovial vessels were visualized by using FITC-dextran. The fat pad has a rich capillary network resembling a honeycomb. Superficial collecting synovial venules may have a convex origin due to the emergence on the pad surface from the more profound layers (*).

3.6. Analysis of the Interactions Between Leukocytes and Synovial Endothelium

1. Video analyse are performed by playback of digital videotapes in real time or at reduced speed, and frame by frame *(14)*. Vessel diameter (D), hemodynamic parameters and the velocities of rolling are determined by using a computer-based system and the National Institutes of Health (NIH) Image software.
2. The velocities of 20 consecutive freely flowing cells/venule are calculated, and from the velocity of the fastest cell in each venule (V_{fast}), the mean blood flow velocities (V_m) is calculated: $V_m = V_{\text{fast}}/(2 - \varepsilon^2)$ where ε is the ratio of the lymphocyte diameter to vessel diameter.
3. Alternatively, blood flow velocity may be calculated from the centerline velocity of the the red blood cells measured with a Doppler velocimeter *(12)*.
4. The wall shear rate (γ) is calculated from $\gamma = 8 \times V_m/D$ (s^{-1}), and the shear stress (τ) acting on rolling cells is approximated to $\gamma \times 0.025$ (dyn/cm^2), assuming a blood viscosity of 0.025 Poise (*see* **Note 4**) *(11)*.
5. Leukocytes are considered as rolling if they travel at velocities below V_{crit} ($V_{\text{crit}} = V_m \times \varepsilon \times (2 - \varepsilon)$) *(11)*. Leukocytes that remained stationary on venular wall for

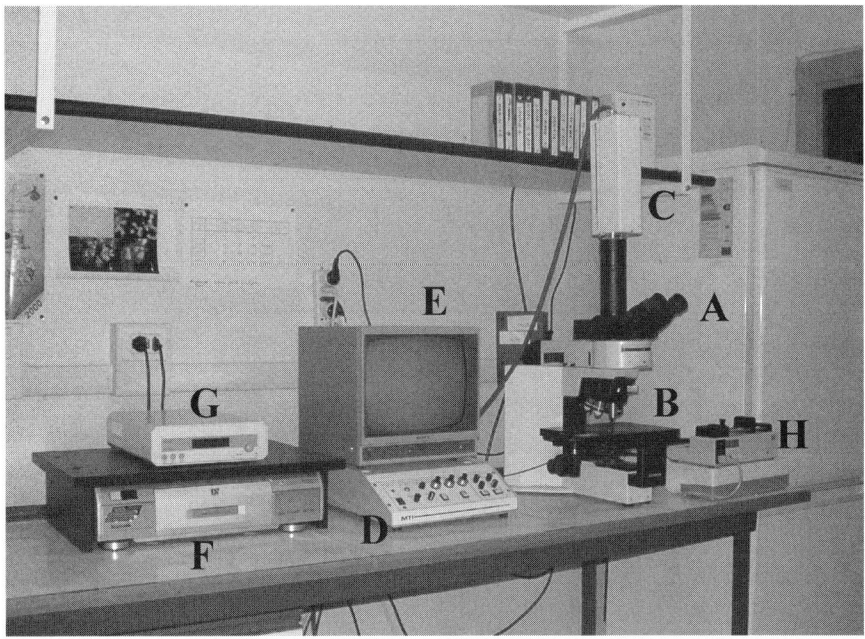

Fig. 3. Intravital microscopy equipment. (**A**) Epifluorescence microscope; (**B**) Water immersion objective; (**C**) SIT videocamera (DAGE MTI VE1000); (**D**) control board of the videocamera; (**E**), monitor; (**F**) digital VCR; (**G**) stage heater; and (**H**) digital pump for the injection of fluorescent cells.

    30 s are considered adherent. Transient tethering, "stop-and-go" very brief interactions with static binding are interactions of 1 s.
6. At least 100 consecutive cells/venule need to be examined.
7. In the case of the cells injected through a catheter, rolling and firm arrest fractions are determined as the percentage of cells that rolled or firmly arrested within a given venule in the total number of fluorescent cells that enter that venule during the same period *(11)*.
8. When total blood leukocytes are considered (Rhodamine-6G labeled), the rolling flux is calculated as the number of rolling cells that rolled past a fixed point in the venule during 1-min interval. Fifteen to twenty consecutive rolling cells should be analyzed to determine the rolling velocity (usually along a 100-µm venular segment).

3.7. In Vivo Staining of Endothelial Adhesion Molecules

1. MAbs to anti-adhesion molecules expressed by the endothelium (e.g., E-, and P-selectin, anti- VCAM-1, anti-ICAM-1) are fluoresceinated using Alexa Fluor 488 labeling kit. Control MAb may be rhodaminated using Alexa Fluor 566 kit.
2. 50 µg fluorescently labeled mAb is injected intravenously.

3. Twenty minutes later, the animal is anesthesized, the vena cava is cut, and the mouse is perfused through the left ventricle with cold PBS to remove the unbound MAb.
4. The synovial vessels are visualized using the intravital microscopy setting *(11)*.

3.8. Statistics

Statistical analysis of the results is performed by using SPSS or GraphPad software. A two-tailed Student's *t*-test may be employed for statistical comparison of two samples. Multiple comparisons are performed employing Kruskall-Wallis test with the Bonferroni correction of P. Velocity histograms are compared using Mann-Whitney U-test and Kolmogorov-Smirnov test. Linear regressions are analyzed employing the Spearman rank correlation test. Differences are regarded significant with a value of $p < 0.05$.

4. Notes

1. Endothelium from mice that are not kept in pathogen-free conditions and that are chronically exposed to infections mediates interactions with leukocytes.
2. The carotid artery lies medial to and below the jugular vein and is exposed by blunt dissection between the sternohyoid, sternomastoid, and omohyoid muscles. Care should be taken that the adjacent vagus nerve is dissected away from the artery. An occluding ligature (thread or staple) is placed on the left artery as anteriorly as possible (toward the head) and a loose posterior ligature is positioned several millimeters away. A small clamp is placed on the most distal portion of the carotid artery to stop the blood flow. Then a small incision is made using microscissors. The posterior ligature is lightly tightened round the artery and catheter and the catheter is pushed toward the heart while the clump is released, until the tip (blunt-ended) lies near/in the aortic arch. The posterior ligature is then tied round the artery and the catheter can be further secured by the free ends of the anterior ligature.
3. Before cutting the ligamentum, the transparent connective tissue that is placed above it needs to be removed. The ligamentum patellæ is the central portion of the common tendon of the quadriceps femoris, which is continued from the patella to the tuberosity of the tibia. It is a strong, flat, ligamentous band attached: *above,* to the apex and adjoining margins of the patella and the rough depression on its posterior surface; *below,* to the tuberosity of the tibia; its superficial fibers are continuous over the front of the patella with those of the tendon of the quadriceps femoris. The medial and lateral portions of the tendon of the quadriceps pass down on either side of the patella, to be inserted into the upper extremity of the tibia on either side of the tuberosity; these portions merge into the capsule, as stated above, forming the medial and lateral patellar retinacula. The posterior surface of the ligamentum patellæ is separated from the synovial membrane of the joint by a large infrapatellar pad of fat.
4. An example of how we calculate the hemodynamic parameters is as follows: blood is considered a Newtonian fluid with laminar flow in the venules (there is a parabolic distribution of velocities across the vessel). The diameter of the

leukocytes is between 8 and 10 μm. Considering a venule with $D = 16$ μm, ε is 0.5. If the fastest velocity (obtained from the velocities of 20 consecutive freely flowing cells/venule) is 2500 μm/s, V_m is 1428 μm/s. The wall shear rate (γ) in the vessel is 714 (s^{-1}), and the shear stress (τ) acting on rolling cells is 17.85 assuming a blood viscosity of 0.025 Poise.

References

1. Harris, Jr., E. D. (1990) Rheumatoid arthritis. Pathophysiology and implications for therapy. *N. Engl. J. Med.* **322,** 1277–1289.
2. D'Ambrosio, D., Lecca, P., Constantin G., Priami C., and Laudanna, C. (2004) Concurrency in leukocyte recruitment: the way to a predictive computer modelling. *Trends Immunol.* **25,** 411–416.
3. Butcher, E. C. (1991) Leukocyte-endothelial cell recognition: three (or more) steps to specificity and diversity. *Cell* **67,** 1033–1036.
4. Springer, T. A. (1994) Traffic signals for lymphoid recirculation and leukocyte emigration: the multi-step paradigm. *Cell* **76,** 301–314.
5. Butcher, E. C., Williams, M., Youngman K, Rott, L., Briskin, M. (1999) Lymphocyte trafficking and regional immunity. *Adv. Immunol.* **72,** 209–253.
6. Menger, D. M, and Lehr, H. A. (1993) Scope and perspectives of Intravital microscopy – bridge over from in vitro to in vivo. *Immunol. Today* **14,** 519–522.
7. Huo, Y., Hafezi-Moghadam, A., and Ley, K. (2000) Role of vascular cell adhesion molecule-1 and fibronectin connecting segment-1 in monocyte rolling and adhesion on early atherosclerotic lesions. *Circ. Res.* **87,** 153–159.
8. Ramos, C. L., Huo, Y., Jung, U., Ghosh, S., Manka, D. R., Sarembock, I. J., and Ley, K. (1999) Direct demonstration of P-selectin- and VCAM-1-dependent mononuclear cell rolling in early atherosclerotic lesions of apolipoprotein E-deficient mice. *Circ. Res.* **84,** 1237–1244.
9. Norman, K. E., Katopodis, A. G., Thoma, G., et al. (2000) P-selectin glycoprotein ligand-1 supports rolling on E- and P-selectin in vivo. *Blood* **96,** 3585–3591.
10. Battistini, L., Piccio, L., Rossi, B., et al. (2003). CD8+ lymphocytes from acute multiple sclerosis patients display selective increase of adesiveness in brain venules: a critical role for P-selectin-glycoprotein ligand-1. *Blood* **101,** 4775–4782.
11. Piccio, L, Rossi, B., Scarpini, E., et al. (2002) Molecular Mechanisms Involved in Lymphocyte Recruitment in Inflamed Brain Microvessels: Critical Roles for P-Selectin Glycoprotein Ligand-1 and Heterotrimeric G(i)-Linked Receptors. *J. Immunol.* **168,** 1940–1949.
12. Johnston, B., Walter, U. M., Issekutz, A. C., Issekutz, T. B., Anderson, D. C., Kubes, P. (1999) Differential roles of selectins and the alpha4-integrin in acute, subacute, and chronic leukocyte recruitment in vivo. *J. Immunol.* **159,** 4514–4523.
13. Veihelmann, A., Szczesny, G., Nolte, D., Krombach, F., Refior, H. J., and Messmer., K. (1998) A novel model for the study of synovial microcirculation in the mouse knee joint in vivo. *Res. Exp. Med. (Berl).* **198,** 43–54.
14. Constantin G., Majeed, M., Giagulli, C., Piccio, L., Kim, J. Y., and Laudanna C. (2000) Chemokines trigger immediate beta2 Integrin affinity and mobility Changes: differential regulation and roles in lymphocyte arrest under flow. *Immunity* **16,** 759–769.

22

Angiogenesis in Arthritis

Methodological and Analytical Details

Ursula Fearon and Douglas J. Veale

Summary

Angiogenesis is the formation of new blood vessels from existing vessels. The formation of new vessels appears to be an early and fundamental process for the evolution of the inflammatory response in synovial joints affected by arthritis. The propagation of new vessels in the synovial membrane allows the invasion of this tissue over the intra-articular cartilage in an adherent fashion. This process appears to support the active infiltration of synovial membrane into cartilage and results in erosion and destruction of the cartilage. This process results in joint damage and ultimately in deformity, as the normal joint architecture and balance of tendons becomes disrupted. Angiogenesis may be assessed in vivo by direct visualization through the introduction of a needle arthroscope using local anesthesia, differential patterns of vascular morphology have been described in seropositive rheumatoid arthritis and seronegative arthritides such as psoriatic and reactive arthritis. At a microscopic level, angiogenesis may be examined in the tissue sections using immunohistochemistry or immunofluorescence. Endothelial cells may also be studied in vitro in culture to examine production of angiogenic growth factors, cell activation, migration, and tubule formation. Finally, synovial biopsy explants may be *cultured ex vivo* to provide a model simulating the intra-articular milieu.

Key Words: Angiogenesis; blood vessels; endothelial cells; macroscopic; arthroscopy; synovial biopsy; microscopic; immunohistochemistry; migration; tubule formation; explant culture; mature and immature blood vessels.

1. Introduction

Angiogenesis is the formation of new vessels by sprouting of capillaries from existing vessels *(1–3)*. It is usually inactive in adults except in the female reproductive process and under pathologic conditions such as in diabetes, cancer and inflammatory arthritides *(4–6)*. Angiogenesis, in normal conditions, is tightly controlled by a balance of pro- and antiangiogenic stimuli, which promote/inhibit generation and proliferation of new endothelial cells (EC) *(6–13)*.

Angiogenesis is fundamental in inflammatory arthritis, such as rheumatoid (RA) and psoriatic arthritis (PsA)—the most common clinical types *(14)*; however, they have distinct pathogenic features *(14–16)*. A number of studies have described discrete micro- and macroscopic vascular changes in PsA and RA synovial membrane (SM), suggesting a possible basis for a differential pathogenesis *(14–17)*. The most striking difference appears at macroscopic assessment, at which we and others have described distinct blood vessel morphology using direct visualization at arthroscopy *(15–16)*. Blood vessels morphology/maturity and expression of angiogenic factors such as growth factors, adhesion molecules, and chemokines can be assessed using standard immunohistochemistry and confocal microscopy techniques. At a microscopic level several studies have demonstrated differential adhesion molecule and growth factor expression in PsA SM and psoriatic skin *(18–24)*. Macroscopic vascular morphology has also been correlated with increased expression of growth factors *(24)*. Finally the angiogenic capacity of the synovial joint can be assessed using joint tissue and fluid in standard angiogenic assays (Matrigel tubule formation, adhesion and migration).

2. Materials

2.1. Arthroscopy

1. Endoscopy Stack - VDU, Light source, (video or digital image capture, optional).
2. Camera.
3. 2.7- or 1.9-mm diameter arthroscope (Storz, Tuttlingen, Germany).
4. Introducer (Storz, Tuttlingen, Germany).
5. Outflow needle (Storz, Tuttlingen, Germany).
6. Grasping forceps (Storz, Tuttlingen, Germany).
7. Drip stand.
8. Saline for infusion.
9. Local anaesthetic: 2 % lignocaine.
10. Steristrips.
11. Crepe bandage.

2.2. Macroscopic Assessment of Angiogenesis Within the Joint

A comparison of the synovial findings between large and small joints demonstrated that similar changes are found at these sites, the procedure however differs and is outlined below *(25)*.

2.3. Microscopic Assessment of Angiogenesis Within the Joint

2.3.1. Preparation of Synovial Tissue Sections

Synovial fluid (SF) may be obtained by aspiration as the needle telescope is being introduced prior to full visualisation and SM biopsy. SF is incubated with hyalauronidase (Sigma) 30 min at 37°C and then centrifuged for 10 min at 3000 rpm, supernatants (reflecting that of the joint environment) are aspirated off and stored at −70°C for further assessment as outlined in **Subheading3**. Synovial tissue biopsies may be placed on a sterile moistened gauze and snap frozen in a cryomould containing OCT. Sample is removed from the cryomould, wrapped in tin foil, and stored in cryovials at −70°C in liquid nitrogen for further immunohistolchemical anlaysis. To assess the ex vivo environment of the joint, biopsies may be cultured in RPMI medium containing 10% fetal calf serum (FCS), penicillin, fugizone, Hepes and can be used for ex vivo mechanistic studies as outlined in **Subheading 3**.

1. Tissue Tek, OCT TM Compound (OCT), (Sakura, Netherlands).
2. Tissue-Tek, Cryomould Biopsy (10x10x5), (Sakura, Netherlands).
3. Haemotoxylin and eosin stain.
4. Tin foil.
5. Monoclonal mouse anti-human CD34 (Dako,Cytomation, UK).
6. Monoclonal mouse anti-human CD31 (Dako,Cytomation, UK).
7. Monoclonal mouse anti- human VIII (Dako,Cytomation, UK).
8. Monoclonal mouse anti-Human α smooth muscle actin (aSMA), Dako, Cytomation, UK).
9. Polyclonal rabbit anti-human VIII, (Dako, UK).
10. Mouse anti-human IgG, (Dako, UK).
11. Rabbit anti-human IgG, (Dako,Cytomation, UK).
12. Acetone (BDH, England).
13. Dako pen (Dako, UK).
14. Phosphate buffered saline (PBS) (Sigma).
15. Dako ABC kit (Dako, UK).
16. 3,3, Diaminobenzidine (DAB), (Sigma).
17. Mayers Haemaluam (BDH).
18. 30% Hydrogen peroxide solution (BDH).
19. Industrial methylated spirits T100 (IMS), (Lennox Chemical, Ireland).
20. Xylene (BDH, England).
21. DPX mountant for microscopy (BDH, England).
22. DAPI (Vector laboratories, UK).

2.3.2. Blood Vessel Maturity: Dual Immunoflourescent Staining

1. 1% Paraformaldehyde.
2. Polyclonal rabbit anti-human VIII polyclonal antibody (Dako, UK).

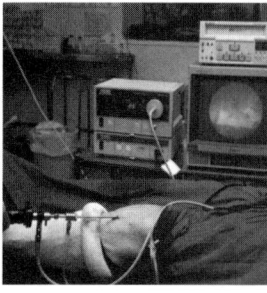 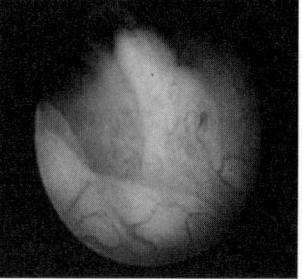 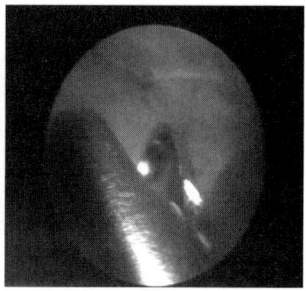

Fig. 1. Demonstrates a 2.7-mm diameter needle arthroscope inserted into the knee joint, with the light source and VDU screen in background (**A**); macroscopic visualisation of the inflamed synovial villi with blood vessels clearly visible (**B**); and demonstrates a biopsy being taken by the grasping forceps (**C**). (Color illustration in insert following p. 268.)

3. Monoclonal mouse anti-human αSMA monoclonal antibody (Dako).
4. Fluorescence labeled goat anti-mouse (Molecular Probes Inc).
5. Tetramethylrhodamine labeled goat anti-rabbit (Molecular Probes Inc).
6. Control Irrelevant Isotype matched IgG control (Dako).

2.4. Ex Vivo Assessment of the Joint Environment

1. RPMI 1640 Medium (Gibco BRL).
2. Penecillin (Gibco BRL).
3. Fugizone (Gibco BRL).
4. Hepes (Sigma).
5. 48 Well plates (Becton Dickinson).
6. Feotal Calf Serum (Gibco BRL).
7. Fibronectin (Roche Biochemicals).

2.5. Ex Vivo Assessment of Cell Migration

1. Human dermal microvascular endothelial cells (Clonetics™ Products. Biosciences Walkersville; cat. no. CC-2543).
2. Endothelial cell growth medium (EGM) + bullet kit (Clonetics Products; cat. no. CC3125).
3. Innocyte Migration Assay 96-well (Oncogene Research Products; cat. no. CBA010).
5. Ten to one thousand microliter pipets.
6. Twenty-four-well Transculture chambers with 3.0-μm pore size polycarbonate membrane insert (Corning; cat. no. 3415).
7. BD Matrigel™ Basement membrane Matrix (BD Biosciences; cat. no. 354234)
8. Phase contrast microscope.
9. Forty-eight-well and 25-mm Flasks (Falcon).

2.6. Synovial Fluid Supernatants

Synovial fluid (SF) may be collected at arthrocentesis prior to arthroscopy, and stored at –70°C until in vitro angiogenic assay assessment. Supernatants can also be assessed by enzyme linked immunosorbent assay (ELISA) for angiogenic factors such as VEGF, Ang1, Ang2, tumor growth factor (TGF)-β-1 and platelet derived growth factor (PDGF)-1 using routine ELISA kits (R&D Systems).

3. Methods
3.1. Arthroscopy
3.1.1. Large Joint Arthroscopy

The arthroscopy procedure is performed in the inflamed knee, although synovitis may be found prior to clinical manifestations have developed *(26)*, using a 2.7-mm bore arthroscope under local anesthesia. Most commonly a lateral inferopatellar portal is used for macroscopic examination of the synovium and a second lateral suprapatellar portal for the fluid outlet and for biopsy procedure *(15,17)*. The portal skin is infiltrated with a 21-gage needle and 2% lignocaine is delivered to the dermis and subcutaneous tissues up to and including the capsule and the synovial lining of the joint (*see* **Notes 1** and **2**). At each arthroscopy, a full, direct visual assessment is undertaken of all the synovial and the cartilage structures within the joint (*see* **Note 3**). In particular assessment of the synovial vascularity is most reliable and reproducible *(15)* (*see* **Note 4**). Synovial biopsy samples are potentially taken from a number of areas including the suprapatellar pouch, the cartilage pannus junction, the patellar gutters, and the tibiofemoral junction using a grasping forceps.

3.1.2. Small Joint Arthroscopy

The arthroscopy of the wrist joint was performed using a small bore 1.9-mm arthroscope under local anesthesia through a radial and an ulnar skin portal at the dorsum of the wrist at either the proximal and/or distal carpal row. Both portals were used for macroscopic examination of the synovium and for the biopsy procedure. At each arthroscopy, synovial biopsies were obtained with a grasping forceps.

The arthroscopy of the metacarpophalangeal (MCP) joint was performed using a small bore 1.9-mm arthroscope under local anesthesia through a medial and a lateral skin portal at the dorsum of the finger. Both portals were used for macroscopic examination of the synovium and for the biopsy procedure. At each arthroscopy, synovial biopsies were taken with a grasping forceps.

Arthroscopic analysis is a validated technique *(15,17,25)* and has advantages over the traditional blind needle biopsy in that macroscopic assessment of the

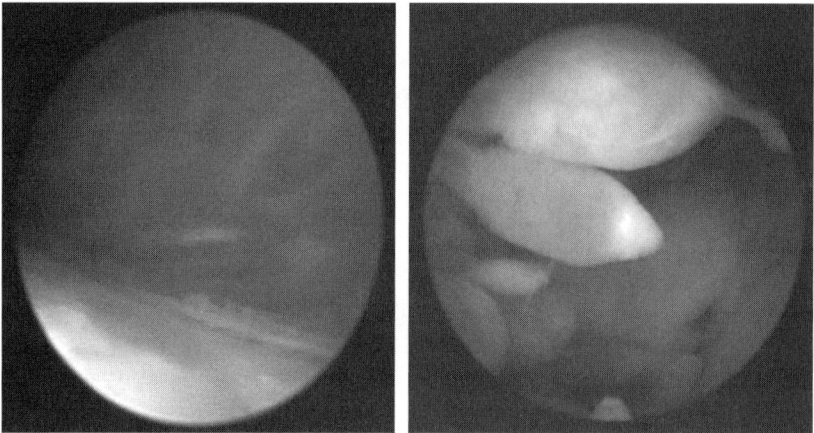

Fig. 2. Arthroscopic images of chondropathy (**A**) and synovitis (**B**) in RA. The vascular markings are further defined by their morphological appearence with two distinct vascular patterns recognized: straight, regular branching vessels, or tortuous, bushy vessels (*see* **Fig. 3**). This vascular pattern may be quantified by assigning a value of 0 to straight/branching vessels and a score of 1 to tortuous/bushy vessels. (Color illustration in insert following p. 268.)

joint and obtaining synovial tissue biopsies can be taken under direct visualization (*see* **Fig 2**). In particular, synovial biopsies can be specifically selected from different regions of the joint such as the suprapatellar pouch, medial and lateral walls and the cartilage pannus junction ensuring maximum representative sampling. Synovial biopsies may be obtained for immunohistochemical analysis, explant culture, and snap frozen for polymerase chain reaction (PCR) analysis.

3.2. Macroscopic Assessment of Angiogenesis and Blood Vessel Patterns in the Joint at Arthroscopy

Macroscopic examination under direct visualisation at arthroscopy may reveal information relating to capillary hyperaemia (redness), increased vascularity (blood vessels), granularity (ruffling of the synovial lining), villous hypertrophy (finger-like projections of the synovial tissue), vascularity, and chondropathy and synovitis (*see* **Fig. 2**) are scored using a 1 to 100-mm visual analogue scale (1 = minimal and 100 = maximal vascularity or synovitis).

3.3. Microscopic Assessment of Angiogenesis Within the Joint

3.3.1. Preparation of Synovial Tissue Sections and Analysis of Microscopic Vascularity

For microscopic analysis tissue biopsies are removed from storage and placed in a refrigerated microtome. Tissue sections 4 to 7 µm are cut and serial

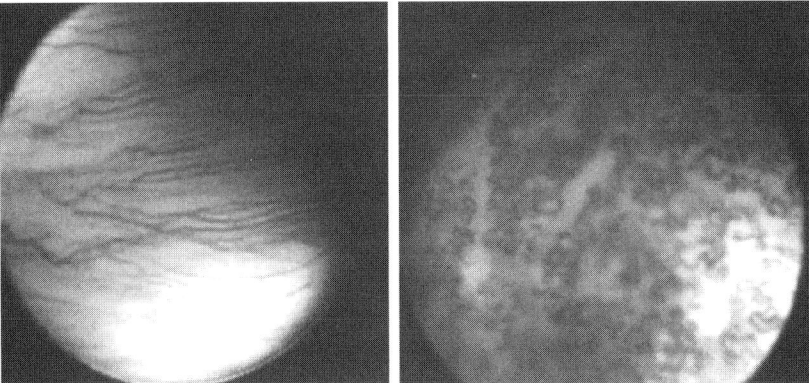

Fig. 3. Arthroscopic images of straight, regularly branching vessels (**A**) as seen in RA and tortuous, bushy vessels (**B**) as found in seronegative arthritides. (Color illustration in insert following p. 268.)

sections are placed on glass slides coated with 2% 3-amino-propyl-triethoixy-silane. A haemotoxylin and eosin stain may be rapidly performed to ensure tissue morphology is intact, only tissue sections with an intact lining layer are used. Tissue sections are allowed to dry overnight at room temperature and then wrapped in tin foil and stored at −70°C.

Three-step immunoperoxiadse staining is employed to assess the microsocpic vascularity and dual-labelled immunoflourescence is used to determine the percent of immature vs mature blood vessels within the joint.

To quantify microvascular density in the synovial tissue in terms of number and size of microvessel profiles, serial tissue sections are stained with a panel of antibodies directed against endothelial cell epitopes for CD31, CD34, and factor VIII using a three-step immunoperoxidase technique. The microscopic quantification can then be correlated with the macroscopic assessment of the joint quantified at arthroscopy as outlined in **Subheading 1.**

1. Prior to staining, sections are thawed at room temperature for 20 min, and then fixed in acetone for 10 min and allowed to air dry (*see* **Notes 5** and **6**).
2. Tissue sections are circled with a Dako wax pen to highlight the section position on the slide (.**Note 7**)
3. Slides are placed in a humidified chamber at room temperature and blocked with 1% normal blocking serum in phosphate buffered saline (PBS) for 20 min, to suppress nonspecific binding of IgG. Slides are then drained by tapping the slide on its side on blotting paper. The blocking serum used must be from the same species from which the secondary antibody is raised.
4. Serial sections are incubated for one hour at 37°C with primary antibodies against CD34, CD31, and Factor VIII at room temperature (*see* **Note 8**). For negative control, sections are incubated with a corresponding amount of isotype matched non-immune IgG control.

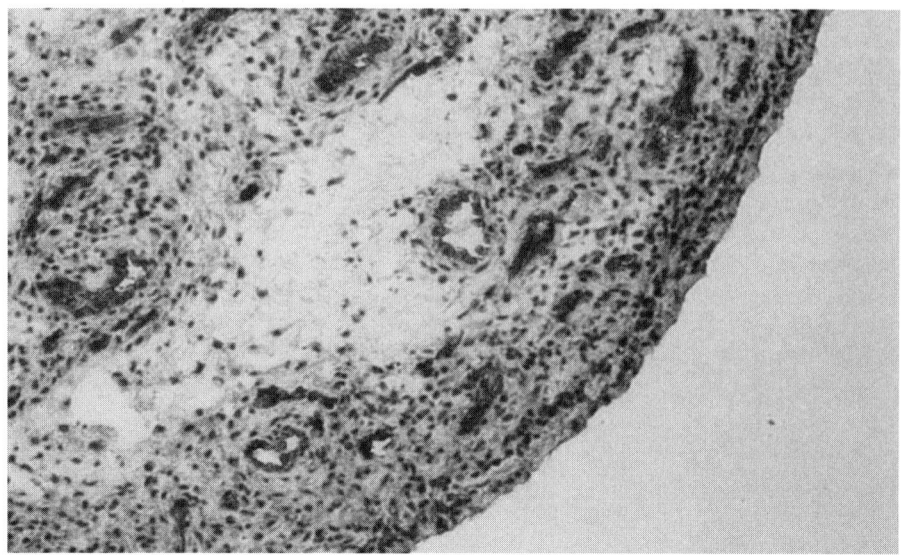

Fig. 4. CD31 staining in early PsA synovial tissue section (×25). (Color illustration in insert following p. 268.)

5. Following incubation sections are washed in PBS for 10 min and then incubated with a biotin-conjugated secondary antibody for 30 min. Sections are washed again in PBS before final incubation with avidin–biotin peroxidase complex for 30 min.
6. Following final incubation sections are incubated with DAB solution for 1 to 10 min until brown staining is visible. Sections are then rinsed in H_2O to stop chromogenic reaction and counterstained in haematoxylin for 1 min (see **Note 9**).
7. Sections are rinsed in H_2O to remove excess counterstain.
8. Sections are then dehydrated through alcohols and xylene, soaked in 95% ethanol three times for 10 s each, then xylene 3 times for 10 s.
9. Mount coverslip using a permanent mounting medium (DPX) and assess blood vessel staining by light microscopy (see **Fig. 4**) (see **Notes 10** and **11**).

3.3.2. Quantification of Blood Vessels Within the Synovial Section

Only synovial sections in which the lining layer was clearly evident were included for evaluation. Semiquantitative analysis is used to assess the number of blood vessels within the synovial section. This analysis employs an objective graticule (approx 1 cm^2), with each graticule area equal to one high powered field (HPF). Sections are examined under ×400 magnification on a binocular microscope, and the number of positively stained vessels per each HPF from each tissue section was assessed, and a mean value derived for each biopsy.

The microscopic assessment can then be correlated with the macroscopic assessment VAS of vascularity and blood vessel morphology as outlined in **Subheading 1.**

3.3.3. Blood Vessel Maturity: Dual Immunoflourescent Staining

Blood vessel maturity is a very important aspect of the analysis. Several studies in tumor biology have demonstrated that the success of blood vessel therapeutic targeting may depend on the maturity of the blood vessels. Benjamin et al. demonstrated that inhibition of both vascular endothelial growth factor (VEGF) and Ang2 in tumor tissue resulted in not only inhibition of new vessel growth but also regression, by selective EC apoptosis, of existing, immature vessels where pericyte recruitment and stabilization had not yet occurred *(27)*. Understanding the cellular and molecular basis of vessel formation and maturation in inflammation is central to elucidating their pathogenic role, and will lay the ground work for developing effective proangiogenic and antiangiogenic therapies. Dual immunofluorescent staining can be used to examine co-localisation of EC with intact pericyte recruitment around the vessels.

1. The procedure is carried out in a darkened humidified chamber at room temperature.
2. Synovial tissue section are prepared for staining procedure as described above.
3. Sections are fixed in 1% paraformadehyde for 20 min and washed in PBS for 15 min.
4. Block sections with normal serum for 1 h and then wash in PBS and incubate with PBS for 15 min.
5. Incubate sections for 1 h with anti-rabbit factor VIII polyclonal antibody and isotyped matched IgG control (*see* **Note 12**).
6. Sections are washed 3 times and then incubated for 30 min at room temperature and goat anti-rabbit IgG-fluorescein isothiocyanate (FITC).
7. Wash sections with PBS three more times.
8. Block sections with normal goat serum for 1 h.
9. Wash sections in PBS three times.
10. Incubate sections with anti-mouse αSMA monoclonal antibody (MAb) for 1 h at room temperature.
11. Wash three times in PBS.
12. Incubate sections with goat anti-mouse IgG-tetramethyl rhodamine isothiocyanate (TRITC) for 30 min at room temperature.
13. Sections are washed three times in PBS.
14. After washing, slides were counterstained in 4',6-diamidino-2-phenylindole DAPI containing antifade.
15. Store in dark at 4°C and assess using a fluorescent microscope.

3.4. Ex Vivo Assessment of the Joint Environment

To assess whether the macroscopic and microscopic assessment of the joint reflect their angiogenic capacity ex vivo, synovial biopsy explants and synovial fluid supernatants obtained from the joint at arthroscopy (*see* above) are assessed for their ability to regulate angiogenesis ex vivo. The angiogenic capacity of the explants and SF supernatants will be assessed in well-established in vitro angiogenic assays. The methods described below outline explant culture, migration assays, transendothelial migration assays, and matrigel tubule formation.

1. Explant biopsies are obtained from patients undergoing arthroscopic examination (*see* above). Synovial tissue is visibly red, inflamed, and often villous and can be taken from the site of increased vascularity under direct visualisation which is the advantage of arthroscopy.
2. Sections of biopsy are subsequently examined histologically (as described in **Subheading 2.**) to assess the synovial architecture and cellular infiltration
3. The synovial tissue is cut into approx 5 mm^3 pieces and placed in 48-well plates containing RPMI 1640 supplemented medium containing 10% FCS, penicillin (100 U/mL), and fungizone and incubated for 24, 48, and 72 h at 37°C in 5 % CO_2.
4. Spontaneous production of proinflammatory mediators are released into the medium reflecting that of the joint environment, this is termed "conditioned medium."
5. Medium is removed at 24, 48, and 72 h and stored at –70°C in cryovials.
6. The conditioned medium will then be used in a range of in vitro angiogenic assays.

3.4.2. Ex Vivo Assessment of Cell Migration

Synovial explants were incubated for 24, 48, and 72 h as outlined above in RPMI medium.

An innocyte quantitative cell migration assay (Oncogene Research Products) is used to assess the angiogenic capacity of the joint environment. This assay is designed to quantitatively determine the effects of various chemoattractant molecules on cell motility. The 8-µm pore-size of the membrane ensures that nonspecific random migration is minimized. Cell migration is quantitatively assessed by staining those cells that migrate through the pores, and attach to the lower side of the membrane, with a fluorescent dye. The assays should be carried out as per data sheet.

1. Endothelial cells are grown to confluence and harvested as previously described above *(28)*.
2. Cell viability and number are quantified, and cell number is adjusted to 500,000 cells/mL in serum free EBM medium.
3. 200 µL of explant supernatant or SF supernatant (1/3 dilution) is placed in the bottom chamber in triplicate. For negative control full supplemented medium was used (*see* **Note 13**).

4. The upper 96-well chamber plate is placed back on top of the lower plate chamber and 150 μL of endothelial cell suspension is added to the upper wells of the 96-well tray. Incubate in a CO_2 incubator at 37°C for up to 24 h (*see* **Note 14 and 15**). Follow kit protocol and measure the fluorescence at excitation 485 nm and emission 520 nm (*see* **Note 16**).
5. The extent of migration in response to either "conditioned medium" or "SF supernatant" can be correlated with macroscopic, microscopic, and SF angiogenic growth factor measurements from that patient. This will determine if the ex vivo environment reflect the angiogenic capacity of the joint.

3.4.3. Transendothelial Migration of Neutrophils are Performed by Standard Transwell Culture Plates

1. Microvascular endothelial cells are grown to confluence (as per data sheet).
2. Explant culture supernatants and SF supernatnants (1/3) are obtained as outlined above.
3. Human neutrophils are purified from normal donors by dextran sedimentation and Ficoll gradient centrifugation followed by hypotonic lysis of erythrocytes by previously standardized methods *(29)*.
4. Approximately 100,000 EC are grown on fibronectin–coated transculture chambers with pore size 3 μm and grown for 4 d in 5% CO_2 at 37°C (*see* **Notes 17 and 18**).
5. Approximately 0.6 mL of conditioned medium or SF supernatants (at 1/3 dilution) is added to the lower compartment.
6. Medium is used as a control.
7. In the upper chamber, 1×10^6 neutrophils in 0.1 mL of 0.5% FCS RPMI medium are added to the endothelial cells monolayer.
8. The chambers can be incubated for 2 to 24 h at 37°C in 5% CO_2.
9. Following incubation, cells in the lower compartments are counted on a haemocytometer.
10. Results of transendothelial migration are correlated with both macroscopic and microscopic assessments of the joint.

3.4.4. Endothelial Cell Tubule Formation

BD Matrigel™ Basement Membrane Matrix basement membrane matrix can be used to examine microvascular endothelial tube formation in response to conditioned medium and SF supernatants.

1. Matrigel (50 μL) is plated in 48-well culture plates slides after thawing on ice and allowed to polymerize at 37°C in 5% CO_2 humidified for 1 h (*see* **Notes 19–21**).
2. EC were removed from culture and resuspended at 4×10^4 cells/mL in full strength EGM growth medium.
3. Five hundred microlitres of cell suspension was added to each chamber in the presence of conditioned medium or SF supernatants.
4. The chambers are then incubated for 24 h at 37°C in 5% CO_2 humidified atmosphere.

5. Conditioned media or SF supernatants are aspirated off the Matrigel surface and cells were fixed in methanol.
6. Endothelial cell tubule formation is assessed using phase contrast microscopy and photographed.
7. A connecting branch between two discrete ECs is counted as one tube and requires a consistent intensity and thickness. The tube analysis was determined from 5 sequential fields (Magnification ×10) focussing on the surface of the matrigel.
8. Results of tubule formation are correlated with both macroscopic and microscopic assessments of the joint

4. Notes

1. Difficulty passing the introducer into the joint may result from a tough ligamentous capsule. A scalpul blade (no. 11) should be used to ensure the capsule is penetrated, this is preferable to using a sharp introducer. Occasionally the introducer may slide up the outside of the capsule preventing entry into the joint.
2. Pain experienced during the procedure may indicate that not enough local anaesthetic has been administered, further lignocaine can be given and some investigators introduce 0.5% bupivicaine into the joint prior to examination to reduce pain during the procedure.
3. Inadequate visualization may occur if insufficient fluid is in the joint, close the outflow tube and allow the joint to fill with fluid, if this does not solve the problem then the lens may be wet or dirty, remove the telescope and wipe tip with a clean, sterile gauze
4. Initially there may be no blood vessels apparent, the synovial lining appears pale or pure ivory, during the examination the vessels then blush as blood flow returns, this is thought to be caused by a reflex vasospasm either resulting from cold fluid infused into the joint or the introduction of the telescope, the synovium is highly innervated with sensory C nerve fibres suggesting it is very sensitive to such stimuli.
5. Before staining always allow sections to come to room temperature.
6. All reagents should be allowed to come to room temperature.
7. Always ensure that tissue sections are not allowed to dry out at any time during procedure as this can result in background staining.
8. Antibodies should be diluted in PBS to achieve desired concentration.
9. Prepare DAB solution as instructed on data sheet. Store 0.5 mL aliquots in the dark at –20°C. Prior to use thaw DAB solution and make up to 5 mL with PBS containing 5 µL of 3% H_2O_2.
10. Strong background staining:
    a. Check optimization of antibody concentration.
    b. Check the blocking serum is derived from the same species in which the secondary antibody is made. Check the blocking reagent concentration. Alter blocking incubation times.
    c. Further blocking can be carried out by incubating slides in 0.1 to 1% hydrogen peroxide staining to quench endogenous peroxidase staining.

d. Frozen section can contain relatively high levels of endogenous biotin, in which case an avidin-biotin block can be carried out.
   e. Ionic interactions can be blocked by preparing ABC reagents and primary antibody solutions with 0.5 M sodium chloride.
   f. Increase washing times and increase number of washes.
   g. Chromagen incubation time can result in high background, individual slides should be monitored to determine proper development time.
   h. Use a monoclonal rather than a polyclonal antibody.
11. Weak Staining:
    a. Always run a positive control to ensure procedure is working.
    b. If assessing a primary antibody for the first time, run a different well established antibody in the same experiment to ensure procedure is working.
    c. Check primary and secondary antibody concentrations and all solutions are correct.
    d. Alter temperature and time of primary antibody incubation (i.e., primary antibodies can be incubated overnight at 4°C).
    e. Check the fixing solutions are appropriate and that fixation time is correct.
    f. Shorten blocking steps.
    g. Check ABC and DAB/hydrogen peroxide solutions.
12. Ensure that the primary antibodies used when performing dual staining are different immunoglobulin isotypes and are raised in different species. The staining procedure should be carried out in a darkened humidified chamber. For Weak or strong staining carry out procedures described in trouble shooting section for ABC immunohistochemistry.
13. Synovial fluid supernatants are very viscous and therefore need at least 1/3 dilution.
14. For optimal results use 150 µL of cell suspension.
15. The cells that have migrated toward the lower chamber will cling to the underside of the membrane, careful handling of the bottom well should be carried out. Once cells have been detached, the 96-well plate should be wrapped in tin foil for further incubations.
16. For quantification, the cells should be transferred to a black 96-well plate which is not provided with the kit. If black 96-well plate is not available transfer to a 96-well plate wrapped in tin foil.
17. Ensure transculture plate is evenly coated with fibronectin.
18. Ensure an even monolayer of endothelial cells is carried out.
19. When plating the matrigel, the 48-well plates, pipet tips, matrigel endothelial cell cultures should all be kept on ice. If the matrigel is not kept cold it will begin to solidify and will not give an even coat of matrigel matrix across the well.
20. When plating the matrigel, rotate the 48-well plate to get even distribution of matrigel across the well.
21. BD matrigel can be used for either:
    a. Cell surface assessment of cells where cells are plated on a thin layer of BD matrigel™.

b. Three-dimensional assessment of cells where cells are plated using a thick gel method. Both methods are outlined in the BD Matrigel™ basement membrane Matrix protocol.

References

1. Folkman, J. and D'Amore, P. A. (1996) Blood vessel formation: what is its molecular basis. *Cell* **87,** 1153–1155.
2. Polverini, P. J. (1995) The pathophysiology of angiogenesis. *Crit. Rev. Oral. Biol. Med.* **6,** 230–247.
3. Folkman, J. (1997) Angiogenesis and angiogenesis inhibition: an overview. *EXS.* **79,** 1–8.
4. Hanahan, D. and Folkman, J. (1996) Patterns and emerging mechanisms of the angiogenic switch during tumorogenesis. *Cell* **86,** 353–364.
5. Cines, D. B., Pollak, E. S., Buck, C. A., Loscalzo, J., Zimmerman, G. A., and McEver, R. (1998) Endothelial Cells in Physiology and in the Pathophysiology of Vascular Disorders. *Blood* **91,** 3527–3561.
6. Koch, A. E. (1998) Angiogenesis: implications for rheumatoid arthritis. *Arthritis Rheum.* **41,** 951–962.
7. Thomas, K. A. (1996) Vascular endothelial growth factor: a potent and selective angiogenic agent. *J. Biol. Chem.* **271,** 603–606.
8. Reuterdah, C., Tingstrom, A., Terracio, L., Keiko, F., Heldin, C. H., and Rubin, K. (1991) Characterization of platelet-derived growth factor-β receptor expressing cells in the vasculature of human rheumatoid synovium. *J. Clin. Invest.* 1991. **64,** 321–329.
9. Feldman, M., Brennan, F. M., and Maini, R. N. (1996) Role of cytokines in rheumatoid arthritis. *Annu. Rev. Immunol.* **14,** 397–440.
10. Goddard, D. H., Grossman, S. L., Williams, W. V., et al. (1992) Regulation of synovial cell growth: coexpression of transforming growth factor B and basic fibroblast growth factor by cultured synovial cells. *Arthritis. Rheum.* **35,** 1296–1303.
11. Koch, A. E., Polverini, P. J., and Kundel, S. L. (1992) Interleukin-8 as a macrophage-derived mediator of angiogenesis. *Science.* **258,** 1798–1801.
12. Sgadari, C., Angiolillo, A. L., and Tosato, G. (1996) Inhibition of angiogenesis by interleukin-12 is mediated by the interferon-inducible protein 10. *Blood* **87,** 3877–3882.
13. Dumont, D. J., Fong, G. H., Puri, M. C., Gradwohl, G., Alitalo, K., and Breitman, M. L. (1995) Vascularisation of the mouse embryo: a study of flk-1, tie and vascular endothelial growth factor expression during development. *Dev Dyn.* **203,** 80–92.
14. Veale, D., Yanni, G., Rogers, S., Barnes, L., Bresnihan, B., and FitzGerald, O. (1993) Reduced synovial macrophage numbers, ELAM-1 expression, and lining layer hyperplasia in psoriatic arthritis compared to rheumatoid arthritis. *Arthritis Rheum.* **36,** 893–900.

15. Reece, R., Canete, J., Parsons, W., Emery, P., and Veale, D. J. (1999) Distinct vascular patterns in the synovitis of psoriatic, reactive and rheumatoid arthritis. *Arthritis. Rheum.* **42,** 1481–1485.
16. Canete, J. D., Pablos, J. L., Sanmarti R., et al. (2004) Antiangiogenic effects of anti-tumor necrosis factor alpha therapy with infliximab in psoriatic arthritis. *Arthritis Rheum.* **50,** 1636–1641.
17. Veale, D. J. (1999) The role of arthroscopy in early arthritis. *Clin. Exp. Rheumatol.* **17,** 37–38.
18. Jones, S. M., Dixey, J., Hall, N. D., and McHugh, N. J. (1997) Expression of the cutaneous lymphocyte antigen and its counter-receptor E-selectin in the skin and joints of patients with psoriatic arthritis. *Br. J. Rheumatol.* **36,** 748–757.
19. Braverman, I. M. and Yen, A. (1974) Microcirculation in psoriatic skin. *J. Invest. Dermatol.* **62,** 493–502.
20. Hull, S., Goodfield, M., Wood, E. J., and Cunliffe, W. J. (1989) Active and Inactive Edges of Psoriatic plaques: Identification by tracing and Investigation by Laser-doppler flowmetry and immunocytochemical techniques. *J. Invest. Dermatol.* **92,** 782–785.
21. Veale, D., Barnes, L., Rogers, S., and FitzGerald, O. (1995) Immunolocalisation of adhesion molecules in psoriatic arthritis, psoriatic and normal skin. *Br. J. Dermatol.* **132,** 32–38.
22. Griffiths, C. E. M., Voorhees, J. J., and Nickoloff, B. J. (1989) Characterization of intercellular adhesion molecule-1 and HLA-DR expression in normal and inflamed skin: modulation by recombinant gamma interferon and tumour necrosis factor. *J. Am. Acad. Dermatol.* **20,** 617–629.
23. Creamer, D., Jagger, R., Allen, M., Bicknell, R., and Barker, J. (1997) Overexpression of the angiogenic factor platelet-derived endothelial cell growth factor/thymidine phosphorylase in psoriatic epidermis. *Br. J. Dermatol.* **137,** 851–855.
24. Fearon, U., Griosios, K., Fraser, A., et al. (2003). Angiopoietins, growth factors, and vascular morphology in early arthritis. *J. Rheumatol.* **30,** 260–268.
25. Kraan, M. C., Reece, R. J., Smeets, T. J., Veale, D. J., Emery, P., and Tak, P. P. (2002). Comparison of synovial tissues from the knee joints and the small joints of rheumatoid arthritis patients: Implications for pathogenesis and evaluation of treatment. *Arthritis. Rheum.* **46,** 2034–2038.
26. Kraan, M. C., Versendaal. H., Jonker, M., et al. (1998). Asymptomatic synovitis precedes clinically manifest arthritis. *Arthritis Rheum.* **41,** 1481–1488.
27. Benjamin, L. E, Golijanin, D., Itin, A., Pode, D. and Keshet, E. (1999) Selective ablation of immature blood vessels in established human tumors follows vascular endothelial growth factor withdrawal. *J. Clin. Invest.* **103,** 159–165.
28. Amin, M. A, Volpert. O.V., Woods, J. M., Kumar, P., Harlow, L. A., and Koch, A. E. (2003) Migration inhibitory factor mediates angiogenesis via mitogen-activated protein kinase and phosphatidylinositol kinase. *Circ. Res.* **93,** 321–329.
29. Taniguchi, N. and Gutteridge, J. M. (2000) Experimental Protocols for Reactive Oxygen and Nitrogen Species. Oxford University Press, London pp. 40–41.

23

Analysis of Inflammatory Leukocyte and Endothelial Chemotactic Activity

Zoltán Szekanecz and Alisa E. Koch

Summary

The ingress of leukocytes into the synovium is a key event in inflammatory arthritis, such as rheumatoid arthritis. A number of soluble inflammatory mediators, such as chemokines, cytokines, and soluble adhesion molecules are involved in this event. It is feasible to test the chemotactic activity of mononuclear cells or endothelial cells using in vitro chemotaxis assays. The role of various chemotactic mediators in inflammatory synovitis can also be assessed using these systems by including these agents in the assay. There we describe monocyte and endothelial cell chemotaxis assays used in our laboratory.

Key Words: Rheumatoid arthritis; monocytes; endothelial cells; chemotaxis.

1. Introduction

The invasion of leukocytes into the synovial tissue (ST) is a crucial process in the pathogenesis of inflammatory synovitis, such as rheumatoid arthritis (RA) *(1,2)*. Monocyte-derived type A synovial lining cells and interstitial macrophages are key players in inflammatory synovitis, as they produce a number of mediators, are involved in arthritis-associated angiogenesis and express various cellular adhesion molecules (CAMs) involved in leukocyte recruitment *(1–4)*. Endothelial cells (ECs) line the lumina of synovial blood vessels. In arthritis, ECs interact with other cell types, components of the extracellular matrix (ECM), as well as inflammatory mediators. In addition, ECs themselves produce such mediators, and also express CAMs. Furthermore, ECs are active players in synovial neovascularization *(1,3,4)*. In summary, there is an existing regulatory network of cytokines, chemokines and CAMs that drive leukocytes into the synovium, which play an important role in the pathogenesis of arthritis *(1–4)*.

As seen above, the chemotaxis of myelo-monocytic cells and ECs are crucial events in the pathogenesis of inflammatory synovitis. The assessment of chemotaxis in the synovium in situ is hardly feasible. Chemotactic factors including chemokines or CAMs can be detected in the synovial tissue in situ using immunohistochemistry as we described before in a previous volume of the Methods in Molecular Biology series *(5)*. However, we can assess monocyte and EC chemotaxis using in vitro assays. In these assays, one can also study the role of various soluble factors in synovitis-associated myeloid and EC migration and chemotaxis. Using these methods we have studied the role of various cytokines, chemokines and soluble CAMs in monocyte or EC chemotaxis *(6–8)*. Therefore, in this chapter, we describe the use of monocyte and EC chemotaxis assays in arthritis studies.

The assay described in this chapter somewhat differs from those reported by others. As opposed to the originally described Boyden's chamber assay, the 48-well microwell chemotaxis chambers are suitable to test multiple samples at the same time. Furthermore, instead of assaying cell migration through the membrane, we rather investigate cells migrating into the membrane. The function of both inflammatory leukocytes and vascular endothelial cells may be assessed thus receiving information on both cell types playing a crucial role in inflammation.

2. Materials

2.1. Patients

1. Patients with RA, osteoarthritis, other types of arthritis or healthy controls.

2.2. Preparation of Chemotaxis Membranes

1. Polyvinyl pyrocarbonate (PVP)-free polycarbonate membranes (Poretics). Pore size: 5 µm for monocytes and 8 µm for ECs, size: 25 × 80 mm.
2. 0.5 M Glacial acetic acid (Sigma) in distilled water.
3. One percent calf skin type III gelatin solution (Sigma) in distilled water (only for ECs) (*see* **Note 1**).
4. 100-mm Petri dishes.
5. Whatman filter paper.

2.3. Monocyte Chemotaxis

1. Hanks balanced salt solution with Ca^{2+} and Mg^{2+} (HBSS).
2. Forty-eight—well microwell chemotaxis chambers,(Neuroprobe) (*see* **Note 2**).
3. PVP-free polycarbonate membranes (Poretics; 5µm pore size) (*see* **Note 3**),
4. Formyl-methionyl-leucyl-phenilalanine (fMet-Leu-Phe; fMLP) (Sigma).
5. Monocyte chemoattractant protein-1 (MCP-1; R&D Systems).
6. Any other chemoattractants (test substances) (e.g., cytokines, chemokines, soluble CAMs).

7. Peripheral blood, synovial fluid, or isolated synovial tissue monocytes in suspension, final concentration: 2.5×10^6 cells/mL in HBSS (monocytes can be isolated from RA patients or control subjects).
8. Light microscope equipped with ×40 objective.
9. Diff-Quick staining kit (Fisher).
10. Slides (75 × 25mm) and coverslips.
11. Mounting medium (e.g., Cytoseal 60, Stephens Scientific).

2.3. Endothelial Chemotaxis

1. Serum-free endothelial basal medium (EBM; Cambrex).
2. Phenol red-free EBM (Cambrex).
3. "Bullet" kit (Cambrex).
4. Forty-eight—well microwell chemotaxis chambers (Neuroprobe) (*see* **Note 4**).
5. Gelatin-coated PVP-free polycarbonate membranes (Poretics; 8 µm pore size).
6. Chemoattractants (e.g., cytokines, chemokines, soluble CAMs)..
7. Subconfluent human umbilican vein (HUVEC) or human dermal micovascular EC (HDMVEC) layer.
8. Light microscope equipped with ×40 objective.
9. Diff-Quick staining kit (Fisher).
10. Slides (75 × 25 mm) and coverslips.
11. Mounting medium (e.g., Cytoseal 60, Stephens Scientific).

3. Methods
3.1. Coating of Chemotaxis Membranes (Only for ECs)

1. Place membranes in Petri dishes.
2. Soak overnight in 0.5 *M* acetic acid solution.
3. Wash filters 3× for 1 to 2 h in distilled water.
4. Incubate filters in diluted gelatin solution for 12 to 16 h.
5. Air dry between Whatman filter pads (*see* **Note 5**).

3.2. Monocyte Chemotaxis Assay

1. Place 25 µL of the chemoattractant under study, HBSS with Ca^{2+} and Mg^{2+} (negative control) and 1×10^{-7} *M* fMLP or 50 ng/mL MCP-1 (positive controls) in the wells of the bottom chemotaxis chamber.
2. Place the gasket on the bottom chemotaxis chamber and make sure that the surface is even .
3. Carefully place the 5 µL pore size filter *shiny side down* on top of the filled wells and the gasket, so that the dull side is up facing the cells (*see* **Note 6**).
4. Cut off the upper right hand corner of the filter for orientation purpose.
5. Attach the top of the chamber on the filter.
6. Simultaneously tighten the screws that are opposite to each other (the two end screws first, then each of the two sets of diagonal screws) (*see* **Notes 7** and **8**).
7. Turn the chamber and tap it gently on the lab bench to allow air bubbles to go to the bottom of the bottom wells (*see* **Note 9**).

8. Place 40 µL of monocyte suspension at a concentration of 2.5×10^6 cells/mL into the upper chamber (a total of 2 mL cell solution is needed per chemotaxis chamber).
9. Wrap the chamber in parafilm.
10. Incubate chambers at 37°C for 1 h in incubator with 5% CO_2.
11. Remove chambers from the incubator and carefully remove the screws in the same order they were tightened.
12. Using forceps, flip membrane over gently (cell side up) so that cut is at the upper left-hand corner.
13. Fix and stain the membrane using Diff-Quick.
14. Place the membrane shiny side up and layer on a Whatman filter.
15. Cover with a larger box to prevent the membrane from blowing away.
16. Allow the membrane to dry for at least 2 h, or alternatively, let the membrane dry in the air hanging down overnight.
17. Trim the edges of the membrane with sharp scissors to fit onto microscope slide.
18. Cut membrane width in half, and cut the upper left hand corner of the second section to maintain correct orientation.
19. Mount membrane halves onto separate slides shiny side up so that the cuts at the upper left-hand corners of the membrane are in the lower left hand corners of the slides (leave enough room for the labels).
20. Label slides.
21. Clean off excess mounting medium before counting.

3.3. Endothelial Cell Chemotaxis Assay

1. Feed the ECs with complete medium the night before the assay.
2. Trypsinize cells, add bullet medium with 5 to 10% fetal bovine serum (FBS), spin and resuspend in 5 mL of serum- and bullet-free medium.
3. Remove an aliquot to count cell number and spin again.
4. Resuspend at appropriate concentration in EBM + 0.1% FBS (*see* **Note 10**).
5. Plate 25 µL of EC suspension into each well.
6. Position the membrane, making sure that the shiny side of the membrane is facing upward.
7. Mark orientation by cutting one corner of the membrane (upper left-hand corner).
8. Attach the top half of the chamber with the gasket in position (*see* **Note 11**).
9. Invert chambers and then tap chamber onto lab bench to remove air bubbles.
10. Invert the chemotaxis chamber again, place it into the incubator (37°C 5% CO_2), and incubate for 2 h.
11. After initial incubation, pipet 40 µL of test or control media into the wells on the top half of the chamber (*see* **Note 12**).
12. Place slide over wells to prevent evaporation, and return chamber to the incubator. Incubate again for 2 h.
13. Remove the chamber from the incubator.
14. Carefully remove the screws as described above.

15. Remove the top half of the chamber without touching the membrane using forceps (*see* **Note 13**).
16. Fix and stain the membrane using Diff-Quik (*see* **Subheading 3.4.**).
17. Rinse the membrane, dry on Whatman filter (*see* **Subheading 3.4.**).
18. Cut, mount the membrane on slide and coverslip (*see* **Subheading 3.4.**).

3.4. Cell Counting (Both monocyte and EC Chemotaxis Assay)

1. If necessary, clean excess mounting media from slide very carefully using a razor blade.
2. Using ×40 power light microscope, focus on the bottom of membrane.
3. Turn the fine focus adjustment forward slightly until the cells on the upper side of the membrane are in focus. (These are the cells, which fully migrated through the membrane.)
4. Count the number of these cells (*see* **Note 14**).
5. Again using the fine focus, focus up and down looking for "arms and legs" of cells migrating through the pores of the membrane. (These will appear as dark streaks "reaching" through the pore, sometimes "grabbing" the top side of the membrane.)
6. Again, count the number of these cells.
7. Count three high power fields per well, recording counts on sheets.
8. Add the three counts per well, and then calculate mean and standard error for these four counts (*see* **Note 15**).
9. Continue and complete counting row by row.
10. Perform statistics.

4. Notes

1. Autoclave and store in refrigerator. Just before use, dilute 1:100 (100 mg/mL).
2. Make sure you have the corresponding bottom and top chamber (compare the identification number).
3. No gelatin coating is needed for monocytes.
4. Make sure you have the corresponding bottom and top chamber (compare the identification number).
5. Membranes can be stored for approximately 1 mo.
6. Always make sure that the shiny side is facing down.
7. Do not overtighten the screws.
8. When the top chamber is applied, push down firmly or air bubbles may be trapped in the bottom wells.
9. Make sure that the top chamber is placed in the right way (clue: you must be able to read the identification number).
10. Human umbilical vein ECs (HUVECs): 1×10^6 cells/mL, human dermal microvascular ECs (HDMVECs): 1×10^6 cells/mL.
11. Gently press down while using light hand pressure to tighten the bolts finger-tight, but do not overtighten.

12. Load from the bottom and be careful to avoid air bubbles.
13. Shiny side always goes upward.
14. In the case of ECs, there are not always whole cells present.
15. If there are not three countable fields in a well, skip that well.

References

1. Jalkanen, S. (1989) Leukocyte-endothelial cell interaction and the control of leukocyte igration into inflamed synovium. *Springer Semin. Immunopathol.* **11,** 187–198.
2. Szekanecz, Z. and Koch, A. E. (2001) Update on synovitis. *Curr. Rheumatol. Rep.* **3,** 53–63.
3. Szekanecz, Z. and Koch, A. E. (2001) Chemokines and angiogenesis. *Curr. Opin. Rheumatol.* **13,** 202–208.
4. Szekanecz, Z. and Koch, A. E. (2004) Vascular endothelium and immune responses: implications for inflammation and angiogenesis. *Rheum. Dis. Clin. N. Am.* **30,** 97–114.
5. Szekanecz, Z. and Koch, A. E. (2003) Immunoperoxidase histochemistry for the detection of cellular adhesion molecule, cytokine and chemokine expression in the arthritic synovium. In: *Inflammation Protocols,* Winyard P. G. and Willoughby, D. A., eds., Humana Press, Totowa, NJ, pp. 283–290.
6. Koch, A. E., Polverini, P. J., Kunkel, S. L., et al. (1992) Interleukin-8 as a macrophage-derived mediator of angiogenesis. *Science* **11,** 1798–1801.
7. Volin, M. V., Woods, J. M., Amin, M. A., Connors, M. A., Harlow, L. A., and Koch, A. E. (2001) Fractalkine: a novel angiogenic chemokine in rheumatoid arthritis. *Am. J. Pathol.* **159,** 1521–1530.
8. Koch, A. E., Halloran, M. M., Haskell, C. J., Shah, M. R., and Polverini, P. J. (1995) Angiogenesis mediated by soluble forms of E-selectin and vascular cell adhesion molecule-1. *Nature* **376,** 517–519.

24

Acquisition, Culture, and Phenotyping of Synovial Fibroblasts

Sanna Rosengren, David L. Boyle, and Gary S. Firestein

Summary

The study of fibroblast-like synoviocytes (FLS) has yielded important insights into the pathogenic mechanisms of rheumatoid arthritis. FLS can be cultured from synovial tissue obtained at joint replacement surgery, synovectomy, or synovial biopsy. After collagenase digestion, adherent cells consist mainly of synovial fibroblasts and synovial macrophages. Proliferating FLS are enriched by repeated passage and comprise >95% of cells by passage 3. Because of cell senescence, use of FLS lines after passage 9 is generally not recommended. FLS in culture have a distinct phenotype with regard to morphology, ultrastructure, surface phenotype, and function. Surface markers that can be used to characterize FLS include positive staining for VCAM-1, CD44, CD55, CD90 (Thy-1), and cadherin-11, coupled with the absence of macrophage markers such as CD14 or CD68.

Key Words: Fibroblast-like synoviocytes (FLS); tissue culture; phenotype.

1. Introduction

Fibroblast-like synoviocytes (FLS) from the synovial membrane play a key role in the pathogenesis of rheumatoid arthritis (RA) through their elaboration of proteases, cytokines, and low-molecular-weight inflammatory mediators *(1)*. Studies of in vitro cultured FLS offer insights into mechanisms of signal transduction, mediator release, proliferation, and apoptosis, and have suggested potential therapeutic targets for the treatment of RA.

Synovial tissue excised in the course of routine joint replacement surgery, synovectomy, or synovial biopsy (either blind needle biopsy or arthroscopic procedures) provides a readily accessible source of FLS, which can be enriched by one of several different protocols:

1. Synovial tissue can be enzymatically digested to yield single cells, which are allowed to adhere to tissue culture dishes. Adherent cells from such synovial digests contain at least two readily distinguishable cell types that resemble the main cell populations found in synovial intimal lining (i.e., macrophage-like synoviocytes and FLS) *(2)*. Whereas FLS proliferate in vitro, synovial macrophages are terminally differentiated and do not expand in culture. Hence, a relatively homogeneous population of synovial fibroblasts is available after three to four passages. Although it is often assumed that these cells are derived from the intimal lining, cultures could also contain sublining fibroblasts. Typically, RA and osteoarthritis (OA) FLS cultures are routinely established using digestion of synovial tissue with collagenase and repeated passage. Detailed methods for this procedure are described in **Subheading 3.1.** Synovial tissue from donors without arthritis is occasionally obtained at surgery for sports injuries or at the time of autopsy, and these serve as valuable controls.
2. Negative selection of first-passage adherent synoviocytes using magnetic beads conjugated with antibodies to CD14, a reliable surface marker for synovial macrophages *(3)*, allows enrichment of cells with a typical FLS phenotype *(4)*. Immediately following this procedure, a significantly higher percentage of cells express vascular cell adhesion molecule-1 (VCAM-1) and major histocompatibility complex class II (MHC-II) than conventionally isolated FLS, whereas levels of prolyl-4-hydroxylase, vimentin and procollagens I and II are lower. Upon further culture, negatively-selected FLS acquire a phenotype that is similar to conventionally isolated FLS *(4)*. Few data are available on the functional characteristics of these cells compared with specimens derived by prolonged passage of dispersed cells.
3. Synovial tissue fragments can be placed in tissue culture to allow outgrowth of FLS, which can then be isolated and further expanded *(5)*. In some cases, the yield can improve by the application of pressure using a coverslip on the synovial fragment; this increases the yield as a result of the enhanced cellular contact with the substrate *(6)*. Like conventionally passaged cells, FLS cultured from explants produce hyaluronan as well as collagens type I and III *(7)* and constitutively express intercellular adhesion molecule-1 (ICAM-1) and CD58 *(8)*.

A limited number of FLS-like adherent cells may also be cultured from RA synovial fluid *(9)*. The resulting cell population expresses lower levels of surface proteins such as ICAM-1, galectin 3, and CD55 compared with synovial tissue-derived FLS. Synovial fluid adherent cells appear to have an especially aggressive phenotype when co-cultured with cartilage explants and are reminiscent of a previously described synovial lineage known as pannocytes *(10)*. Unfortunately, cultures of viable mesenchymal cells are not obtained from all synovial fluids, and the high number of mononuclear cells in the fluid may necessitate more extensive passage and culture to deplete contaminating macrophages.

2. Materials
2.1. Establishment of FLS Cell Lines

1. Synovial tissue. Tissue is routinely obtained at the time of knee or hip replacement surgery on patients with RA or OA (*see* **Note 1**). Similarly, specimens can be obtained at the time of synovectomy or synovial biopsy. Following excision, the tissue is immediately stored in a sterile screw-cap jar at +4°C. The tissue is best retrieved within 1 h and FLS can be immediately isolated. However, viable cell lines from tissue excised up to 2 d earlier have been obtained. There have been no systematic studies to determine if delayed acquisition alters the phenotype or function of isolated FLS.
2. Sterile phosphate-buffered saline (PBS) (pH 7.4).
3. Collagenase, type VIII (from Clostridium histolyticum, crude prepation, Sigma Chemical Co.), stored as powder at –20°C.
4. RPMI cell culture medium.
5. Complete Dulbecco's modification of Eagle's medium (DMEM): DMEM supplemented with glucose 4.5 g/L, L-glutamine 2 mM, penicillin 100 U/mL, streptomycin 100 µg/mL, gentamycin 50 µg/mL, and 10% fetal calf serum (FCS), (*see* **Note 2**).
6. Sterilized surgical scissors and forceps.
7. Sterile cell strainers designed to fit on 50-mL Falcon centrifuge tube, and 70 µm nylon mesh (Becton Dickinson).

2.2. Passage of FLS

1. Trypsin-ethylene diamine tetraacetic acid (EDTA): 0.25% trypsin, 0.38 g EDTA-4 Na/L in Hanks balanced salt solution (HBSS) without Ca^{2+} or Mg^{2+}.
2. Complete DMEM (*see* **Subheading 2.1.**, item 5).

3. Methods

Sterile tissue culture technique is essential for isolation and propagation of cultured FLS. Note that any human tissue is a potential source of infectious pathogens, and adequate protection for laboratory personnel should be routine. Perform all steps in a cell culture laminar flow hood suitable for safe work with low-grade biohazard materials, and discard materials and solutions that contact tissue or cells according to local regulatory guidelines.

3.1. Establish FLS cell lines

1. Mix 50 mL of RPMI with 25 mg of collagen, type VIII and filter sterilize through 0.22 µm filter into a 250-mL flask.
2. If tissue is contaminated with blood, rinse it with cold phosphate buffered saline (PBS) placed in a 150-mm petri dish.
3. Remove tissue to a fresh 150-mm petri dish and inspect tissue, which will be a mixture of synovium, fat, fibrous membranes, and possibly cartilage and/or bone

fragments. Synovium, present on the inner aspect of the joint capsule, appears pink or tan, sometimes villous, and is easily distinguishable from yellow/white fat and fibrous tissue.
4. Identify synovial tissue, excise with surgical scissors, and place in a fresh Petri dish.
5. Mince excised synovium with sterile scissors until individual fragments are less than 1 mm^3. Place minced tissue in 250-mL flask containing RPMI and collagenase from **step 1** (*see* **Note 3**).
6. Incubate flask for 90 min at 37°C, gently agitating bottle every 15 min. Alternatively, a shaking water bath can be used.
7. Pipet mixture through a 70 μm mesh cell strainer into a 50-mL conical tube. Avoid any fat, which often floats on the surface, as well as larger undigested pieces. Rinse strainer with 5 to 10 mL complete DMEM medium until tube is full. There is often a considerable amount of proteinaceous debris at this stage, which remains in the filter.
8. Pellet cells in the filtrate by centrifuge at 250g for 10 min.
9. Remove the supernatant and wash cell pellet by adding 15 mL complete DMEM and recentrifuge as above. The number of viable cells should be counted in a hemocytometer in the presence of trypan blue.
10. Resuspend pellet at 1×10^6 cells per mL in 15 mL complete DMEM, and add to T75 flask already containing 5 mL complete DMEM (*see* **Note 4**). Incubate in standard cell culture incubator at 37°C and 5% CO_2. In some cases, proteinaceous debris is still present in the cultures.
11. The next day, aspirate the nonadherent cells and wash the adherent cells with complete DMEM to yield passage 0 (P0) cells. Then add 15 to 20 mL of complete medium and culture at 37°C and 5% CO_2. Replace the medium twice a week.
12. Typically, the adherent layer is 90% confluent about 10 to 14 d following isolation, but the time can be highly variable. If the adherent cells appear too sparse following removal of the non-adherent population, concentrate them by trypsinization (*see* **Subheading 3.2.**) and reculturing in a smaller culture vessel.

3.2. Maintenance of FLS Cultures

1. FLS lines are usually passaged every 10 to 14 d, or when they reach 90% confluence. The average doubling time for FLS is about 1 wk, although it can be faster in the first few passages. The cells generally do not form foci, and ultimately stop proliferating as a result of contact inhibition.
2. To passage the cells, aspirate the supernatant, briefly rinse with 5 mL of trypsin-EDTA to remove serum, and trypsinize culture with 10 mL of trypsin-EDTA preheated to 37°C for about 5 min or until cells detach from the plastic.
3. Add 10 mL of complete DMEM and transfer the entire content of flask to a centrifuge tube.
4. Pellet cells by centrifuge at 250g for 5 min. Resuspend the cells in 10 mL of complete DMEM and recentrifuge. Resuspend pellet and divide among three fresh T75 flasks. The total culture volume in each flask should be 15 to 20 mL.

5. As a general rule, FLS should be used for experimentation between P3 and P9. Earlier passages contain appreciable numbers of contaminating cells, whereas the growth rate of cells beyond P10 slows as a result of senescence.
6. For experiments, FLS are cultured in petri dishes, six-well plates or any other vessel as needed. Cells can generally be used 1 d after passage. Alternatively, in some experiments it might be advantageous to synchronize cells or suppress proliferation. This can be accomplished by decreasing the serum content to 1% FCS. Concentrations of 0.1 to 1% can also be used to reduce activation of signaling pathways and gene expression. Upstream MAP kinase activities are especially resistant to serum starvation and might require up to 48 h culture at 0.1% FCS to suppress constitutive phosphorylation *(11)*.

3.3. Phenotyping of FLS

It should be emphasized that FLS cultures, even a few passages after isolation, do not necessarily constitute a homogeneous population of cells. Within the synovial membrane, fibroblasts in the intimal lining might differ from cells in deeper layers, whereas in culture the two subpopulations may be indistinguishable *(12)*. The heterogeneous nature of FLS in culture complicates the identification of a definite phenotype; historically, such cells were defined by the absence of macrophage markers such as CD14 and CD68, and by being incapable of ingesting latex beads, rather than by positive markers *(13)*. Even so, FLS possess typical morphology, ultrastructure, and function. Staining for some protein markers, albeit not expressed universally, can be used to confirm the identity of a particular cell line.

3.3.1. Morphology

FLS in culture appear elongated by phase light microscopy, sometimes oval or polygonal, with a few branched cytoplasmic processes (*see* **Fig. 1**). Occasional cells have a dendritic or stellate morphology. In contrast, macrophages appear rounder and smaller. When viewed by electron microscopy, FLS contain abundant rough endoplasmic reticulum and evidence of an active secretory machinery, but lack digestive vacuoles *(14)*.

3.3.2. Function

FLS proliferate in culture as adherent cells (although they can sometimes be coaxed to grow under anchorage free conditions *[15])*, and the rate of proliferation is increased by platelet-derived growth factor (PDGF) and transforming growth factor-β (TGF-β). More modest proliferative effects are observed after exposure to tumor necrosis factor (TNF)-α (TNF-α) and interleukin (IL)-1β (IL-1β) *(16)*. These cytokines also enhance FLS expression of prostaglandins *(17)*, IL-6 *(18)*, IL-8 *(19)*, granulocyte-macrophage colony-stimulating factor (GM-CSF) *(20)*, matrix metalloproteinase (MMP)-1 (MMP-1)

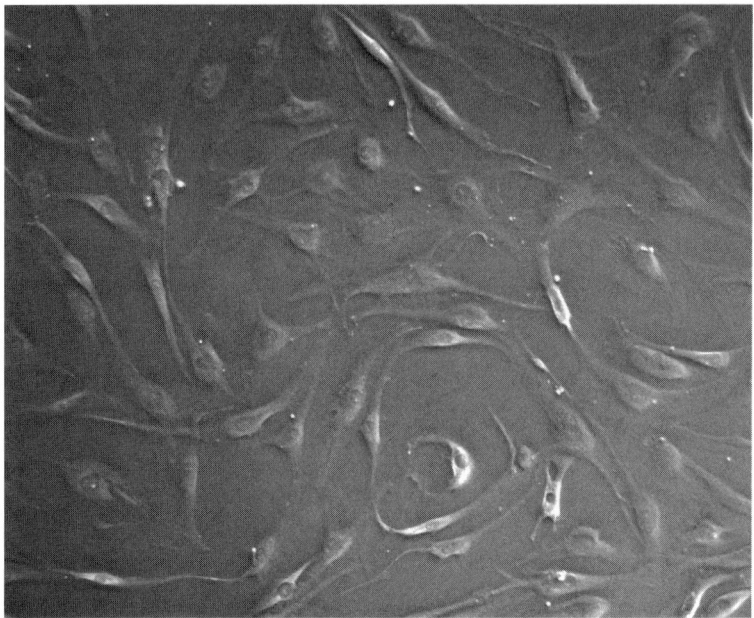

Fig. 1. Appearance of FLS (passage 4) growing in cell culture. Note the elongated appearance.

(21), MMP-3 *(22)*, and many other mediators. Comprehensive reviews of FLS function in vitro have been published *(23,24)*. At earlier passages, FLS derived from RA synovial joints can also constitutively release cytokines, growth factors, and MMPs *(25)*, especially when they are isolated by outgrowth from synovial tissue fragments instead of by enzymatic digestion *(22)*. In contrast to macrophages, FLS do not ingest latex beads *(26)*, nor do they contain nonspecific esterase activity.

3.3.3. Surface and Intracellular Protein Expression

Within the synovial lining, FLS express high levels of surface VCAM-1 *(27)* and contain the hyaluronan synthesis-related enzyme uridine diphosphoglucose dehydrogenase (UDPGD) *(28)*. CD55 (decay accelerating factor) delineate the synovial lining FLS *(29)*. In culture, however, the percentage of FLS expressing CD55 can be as low as 35% *(9)*. Similarly, VCAM-1 is usually only expressed on 15 to 25% cells in culture *(27)*, and UDPGD can be identified on cytospin preparations following collagenase digests in only a fraction of FLS *(30)*. These studies suggest that the cultures either contain de-differentiated FLS or a second population of sublining synovial fibroblasts. A thorough study of surface and intracellular antigen expression on cultured FLS was performed by flow cytometry *(31)* in which CD13 (aminopeptidase-N), CD44 (Hermes antigen, hyaluronan receptor) and CD59 (membrane attack complex inhibitor) were shown to be highly expressed on cultured FLS. Both CD13 *(32)* and CD44 *(33)* are expressed on synovial macrophages, however, and CD59 would also be expected on synovial macrophages based on its presence on myelomonocytic cells, thus making these molecules unsuitable for the selective identification of FLS. A better candidate might be CD90 (Thy-1) antigen, which is expressed specifically on FLS in the synovial membrane *(4,34)* and is present on greater than 70% of FLS in culture *(4,35)*. Staining for CD90 was recently used to distinguish FLS from synovial macrophages *(36)*. As an alternative preferential marker for FLS intracellular prolyl-4-hydroxylase, an enzyme involved in collagen fibril synthesis, may be detected by immunohistochemistry *(31)* or by flow cytometry *(4)* in permeabilized FLS, and is expressed in greater than 95% of cells. More recently, cadherin–11 was identified as a surface adhesion molecule on intimal lining FLS that contributes to homotypic aggregation *(37)*. This protein is also expressed on cultured FLS and could be a specific marker for this lineage. To harvest FLS for flow cytometry, cells are incubated on ice for 5 min with 1 mM EDTA in PBS followed by gentle pipeting to collect detached cells. Trypsin-EDTA can be used if the selected marker is resistant to proteases. If a fixative is desired, 1% formaldehyde in PBS is recommended. Standard flow cytometry protocols can then be employed *(38)*. For additional protocols please refer to selected chapters in the Immunobiology section.

4. Notes

1. Close collaboration with the personnel in the surgical suite is essential for the acquisition of high quality tissue. From the knee, the suprapatellar pouch, as well as the medial and lateral gutters, yield tissue with easily distinguishable luminal and capsular sides. Knee tissue also has the advantage of containing less contaminating blood due to the application of a tourniquet during surgery.
2. Testing of new batches of FCS to ensure adequate support of FLS growth at 10% serum is recommended. Higher serum concentrations (20%) do not accelerate growth; however, at 5% serum the growth rate is significantly reduced.
3. Several alternative digestion protocols have been explored, including DNase, trypsin, and hyaluronidase, either alone or in combination with collagenase. The cell yield of the enzymatically dispersed material is not improved by these additional steps.

4. FLS can be successfully cultured from synovial tissue obtained at arthroscopic synovectomy of joints as small as the metacarpophalangeal joint. In this case the number of resulting cells is usually lower and a reduced size culture vessel, such as a T25 flask or a small petri dish, is recommended for the initial passage.

Acknowledgements

The authors would like to thank Li Yang and Suzanne Beal for expert technical assistance.

References

1. Yamanishi, Y. and Firestein, G. S. (2001) Pathogenesis of rheumatoid arthritis: the role of synoviocytes. *Rheum. Dis. Clin. North Am.* **27,** 355–371.
2. Barland, P., Novikoff, A. B., and Hamerman, D. (1962) Electron microscopy of the synovial membrane. *J. Cell. Biol.* **14,** 207–220.
3. Athanasou, N. A. and Quinn, J. (1991) Immunocytochemical analysis of human synovial lining cells: phenotypic relation to other marrow derived cells. *Ann. Rheum. Dis.* **50,** 311–315.
4. Zimmermann, T., Kunisch, E., Pfeiffer, R., et al. (2001) Isolation and characterization of rheumatoid arthritis synovial fibroblasts from primary culture—primary culture cells markedly differ from fourth-passage cells. *Arthritis Res.* **3,** 72–76.
5. Castor, C. W. and Muirden, K. D. (1964) Collagen formation in monolayer cultures of human fibroblasts. The effects of hydrocortisone. *Lab. Invest.* **13,** 560–574.
6. Ogura, N., Tobe, M., Sakamaki, H., et al. (2002) Interleukin-1 beta induces interleukin-6 mRNA expression and protein production in synovial cells from human temporomandibular joint. *J. Oral Pathol. Med.* **31,** 353–360.
7. Vuorio, E. (1977) Rheumatoid disease in cultured human synovial cells. A biochemical study on glycosaminoglycans, proteins and plasma membranes of synovial fibroblasts in culture. *Scand. J. Clin. Lab. Invest. Suppl.* 1–72.
8. Chin, J. E., Winterrowd, G. E., Krzesicki, R. F., and Sanders, M. E. (1990) Role of cytokines in inflammatory synovitis. The coordinate regulation of intercellular adhesion molecule 1 and HLA class I and class II antigens in rheumatoid synovial fibroblasts. *Arthritis Rheum.* **33,** 1776–1786.
9. Neidhart, M., Seemayer, C. A., Hummel, K. M., Michel, B. A., Gay, R. E., and Gay, S. (2003) Functional characterization of adherent synovial fluid cells in rheumatoid arthritis: destructive potential in vitro and in vivo. *Arthritis Rheum.* **48,** 1873–1880.
10. Zvaifler, N. J. and Firestein, G. S. (1994) Pannus and pannocytes. Alternative models of joint destruction in rheumatoid arthritis. *Arthritis Rheum.* **37,** 783–789.
11. Sundarrajan, M., Boyle, D. L., Chabaud-Riou, M., Hammaker, D., and Firestein, G. S. (2003) Expression of the MAPK kinases MKK-4 and MKK-7 in rheuma-

toid arthritis and their role as key regulators of JNK. *Arthritis Rheum.* **48,** 2450–2460.
12. Edwards, J. C. (2000) Fibroblast biology. Development and differentiation of synovial fibroblasts in arthritis. *Arthritis Res.* **2,** 344–347.
13. Burmester, G. R., Jahn, B., Rohwer, P., Zacher, J., Winchester, R. J., and Kalden, J. R. (1987) Differential expression of Ia antigens by rheumatoid synovial lining cells. *J. Clin. Invest.* **80,** 595–604.
14. Mapp, P. I. and Revell, P. A. (1988) Ultrastructural characterisation of macrophages (type A cells) in the synovial lining. *Rheumatol. Int.* **8,** 171–176.
15. Aupperle, K. R., Boyle, D. L., Hendrix, M., et al. (1998) Regulation of synoviocyte proliferation, apoptosis, and invasion by the p53 tumor suppressor gene. *Am. J. Pathol.* **152,** 1091–1098.
16. Alvaro-Gracia, J. M., Zvaifler, N. J., and Firestein, G. S. (1990) Cytokines in chronic inflammatory arthritis. V. Mutual antagonism between interferon-gamma and tumor necrosis factor-alpha on HLA-DR expression, proliferation, collagenase production, and granulocyte macrophage colony-stimulating factor production by rheumatoid arthritis synoviocytes. *J. Clin. Invest.* **86,** 1790–1798.
17. Nakajima, H., Hiyama, Y., Tsukada, W., Warabi, H., Uchida, S., and Hirose, S. (1990) Effects of interferon gamma on cultured synovial cells from patients with rheumatoid arthritis: inhibition of cell growth, prostaglandin E2, and collagenase release. *Ann. Rheum. Dis.* **49,** 512–516.
18. Tan, P. L., Farmiloe, S., Yeoman, S., and Watson, J. D. (1990) Expression of the interleukin 6 gene in rheumatoid synovial fibroblasts. *J. Rheumatol.* **17,** 1608–1612.
19. Rathanaswami, P., Hachicha, M., Wong, W. L., Schall, T. J., and McColl, S. R. (1993) Synergistic effect of interleukin-1 beta and tumor necrosis factor alpha on interleukin-8 gene expression in synovial fibroblasts. Evidence that interleukin-8 is the major neutrophil-activating chemokine released in response to monokine activation. *Arthritis Rheum.* **36,** 1295–1304.
20. Alvaro-Gracia, J. M., Zvaifler, N. J., Brown, C. B., Kaushansky, K., and Firestein, G. S. (1991) Cytokines in chronic inflammatory arthritis. VI. Analysis of the synovial cells involved in granulocyte-macrophage colony-stimulating factor production and gene expression in rheumatoid arthritis and its regulation by IL-1 and tumor necrosis factor-alpha. *J. Immunol.* **146,** 3365–3371.
21. Dayer, J. M., Beutler, B., and Cerami, A. (1985) Cachectin/tumor necrosis factor stimulates collagenase and prostaglandin E2 production by human synovial cells and dermal fibroblasts. *J. Exp. Med.* **162,** 2163–2168.
22. MacNaul, K. L., Chartrain, N., Lark, M., Tocci, M. J., and Hutchinson, N. I. (1990) Discoordinate expression of stromelysin, collagenase, and tissue inhibitor of metalloproteinases-1 in rheumatoid human synovial fibroblasts. Synergistic effects of interleukin-1 and tumor necrosis factor-alpha on stromelysin expression. *J. Biol. Chem.* **265,** 17,238–17,245.
23. Firestein, G. S. (1998) Rheumatoid synovitis and pannus. In *Rheumatology* (Klippel, J. H., and Dieppe, P. A., eds.) pp. 5.13.11–15.13.24, Mosby International, London.

24. Gulko, P. S., Seki, T., and Winchester, R. (2000) The role of fibroblast-like synoviocytes in rheumatoid arthritis. In: *Rheumatoid Arthritis: Frontiers in Pathogenesis and Treatment,* Firestein, G. S., Panayi, G. S. and Wollheim, F. A., eds., Oxford University Press, Oxford, UK, pp. 113–135.
25. Bucala, R., Ritchlin, C., Winchester, R., and Cerami, A. (1991) Constitutive production of inflammatory and mitogenic cytokines by rheumatoid synovial fibroblasts. *J. Exp. Med.* **173,** 569–574.
26. Burmester, G. R., Dimitriu-Bona, A., Waters, S. J., and Winchester, R. J. (1983) Identification of three major synovial lining cell populations by monoclonal antibodies directed to Ia antigens and antigens associated with monocytes/macrophages and fibroblasts. *Scand. J. Immunol.* **17,** 69–82.
27. Morales-Ducret, J., Wayner, E., Elices, M. J., Alvaro-Gracia, J. M., Zvaifler, N. J., and Firestein, G. S. (1992) Alpha 4/beta 1 integrin (VLA-4) ligands in arthritis. Vascular cell adhesion molecule-1 expression in synovium and on fibroblast-like synoviocytes. *J. Immunol.* **149,** 1424–1431.
28. Pitsillides, A. A., Wilkinson, L. S., Mehdizadeh, S., Bayliss, M. T., and Edwards, J. C. (1993) Uridine diphosphoglucose dehydrogenase activity in normal and rheumatoid synovium: the description of a specialized synovial lining cell. *Int. J. Exp. Patho.* **74,** 27–34.
29. Stevens, C. R., Mapp, P. I., and Revell, P. A. (1990) A monoclonal antibody (Mab 67) marks type B synoviocytes. *Rheumatol. Int.* **10,** 103–106.
30. Wilkinson, L. S., Pitsillides, A. A., Worrall, J. G., and Edwards, J. C. (1992) Light microscopic characterization of the fibroblast-like synovial intimal cell (synoviocyte). *Arthritis Rheum.* **35,** 1179–1184.
31. Schwachula, A., Riemann, D., Kehlen, A., and Langner, J. (1994) Characterization of the immunophenotype and functional properties of fibroblast-like synoviocytes in comparison to skin fibroblasts and umbilical vein endothelial cells. *Immunobiology* **190,** 67–92.
32. Koch, A. E., Burrows, J. C., Skoutelis, A., et al. (1991) Monoclonal antibodies detect monocyte/macrophage activation and differentiation antigens and identify functionally distinct subpopulations of human rheumatoid synovial tissue macrophages. *Am. J. Pathol.* **138,** 165–173.
33. Johnson, B. A., Haines, G. K., Harlow, L. A., and Koch, A. E. (1993) Adhesion molecule expression in human synovial tissue. *Arthritis Rheum.* **36,** 137–146.
34. Palmer, D. G., Selvendran, Y., Allen, C., Revell, P. A., and Hogg, N. (1985) Features of synovial membrane identified with monoclonal antibodies. *Clin. Exp. Immunol.* **59,** 529–538.
35. Seemayer, C. A., Kuchen, S., Kuenzler, P., et al. (2003) Cartilage destruction mediated by synovial fibroblasts does not depend on proliferation in rheumatoid arthritis. *Am. J. Pathol.* **162,** 1549–1557.
36. Lories, R. J., Derese, I., Ceuppens, J. L., and Luyten, F. P. (2003) Bone morphogenetic proteins 2 and 6, expressed in arthritic synovium, are regulated by proinflammatory cytokines and differentially modulate fibroblast-like synoviocyte apoptosis. *Arthritis Rheum.* **48,** 2807–2818.

37. Valencia, X., Higgins, J. M. G., Simmons, B., and Brenner, M. B. (1998) Identification of cadherin-11 in type B synoviocytes derived from rheumatoid arthritis patients. *Arthritis Rheum.* **41(Suppl),** S190.
38. Hawley, T. S. and Hawley, R. G. (2004) *Flow Cytometry Protocols,* vol. 263, Humana Press, Totowa, NJ.

25

Genotyping of Synovial Fibroblasts

cDNA Array in Combination with RAP-PCR in Arthritis

Elena Neumann, Martin Judex, Steffen Gay, and Ulf Müller-Ladner

Summary

Evaluation of differentially regulated genes is essential for the development of novel therapeutic approaches in multifactorial diseases such as rheumatoid arthritis (RA). RA synovial fibroblasts (RASF) are key players in inflammation and cartilage destruction. Therefore, RASF are important cellular targets for the analysis of gene expression profiles. Such analyses may include a comparison of SF from nontreated RA patients with those from treated RA patients, and may also be used to evaluate which genes and intracellular processes can be modulated through genetic modification, gene transfer, and drug treatment for the identification of the pathways driving the destructive behavior of these cells. This chapter reports the combination of RNA arbitrarily primed polymerase chain reaction (RAP-PCR) and cDNA array with defined genes for a highly sensitive analysis of gene expression profiles in RASF using small amounts of total RNA. RNA can be extracted from cultured SF, isolated, and analyzed using RAP-PCR with different arbitrary primers for first- and second-strand synthesis to generate a radioactive labeled probe which can be used for cDNA array hybridization. Visualization and evaluation of gene expression can be performed by phosphorimaging in combination with an array-specific software analyzing system followed by statistic evaluation of the generated data. In summary, the combination of RAP-PCR combined with cDNA arrays is a sensitive method to identify differentially expressed genes in RASF with high specificity, especially for low abundant mRNAs.

Key Words: cDNA array; AtlasImage™ software; RAP-PCR; fingerprint; rheumatoid arthritis (RA); osteoarthritis.

1. Introduction

In rheumatoid arthritis (RA), there is increasing evidence that T-cell independent pathways, including alterations in the expression of growth factors, inflammatory mediators, and the production of matrix degrading enzymes lead to progressive destruction of the affected joints *(3,10)*. According to this scenario, transformed activated synovial fibroblasts at sites of invasion into articular cartilage and bone are regarded as key players *(10,11)*. Thus, the identification of differentially regulated genes in RASF should enhance our understanding of the molecular basis for this aggressive fibroblast phenotype. In addition, the identification of pathways operative specifically in RASF and not in SF of other origin (e.g., osteoarthritis [OA] or normal dermal fibroblasts) would make it possible to devise strategies to inhibit progressive growth of these cells without interfering with physiologic matrix remodeling. Moreover, to achieve a better understanding of the pathogenesis and the course of RA, it is desirable to facilitate the molecular analysis of the effects of anti-inflammatory and disease-modifying drugs following stimulation or transduction of RASF when compared with nontreated RASF of the same origin.

The cDNA expression array technique is a powerful tool to analyze gene expression profiles of different organisms, tissues, and cells, or of alterations of gene expression following distinct stimuli such as stimulation of cells, viral or nonviral transduction or drug treatment *(12,19)*. cDNA arrays can capture labeled homologous cDNAs from solution, and differential expression of a large number of genes can be measured easily in parallel in a single experiment using this method *(4,15)*. Moreover, after detection of a differentially regulated gene on a cDNA array, the sequence information of this gene is immediately available.

In the past, various strategies besides cDNA arrays such as subtractive hybridization or differential display, have been developed to examine tissue specific differences in gene expression. However, for the majority of these methods large amounts of RNA are required. Therefore, the disadvantages of these methods are that only the most abundant transcripts are visualized *(2)*. This problem is of substantial importance with regard to analysis of RASF gene expression, as in contrast to fast-growing tumor cells, because in RA only limited numbers of synovial cells can be obtained. In addition, synovial fibroblasts cannot be grown for long periods of time without changes in proliferation and gene expression pattern. This also limits the amount of available RNA.

By increasing the amount of cDNA using amplification methods such as Smart cDNA synthesis *(16)* or RNA arbitrarily primed PCR (RAP-PCR) *(21)*, less abundant messages can also be analyzed by cDNA array. The differential display approach for RAP-PCR has been demonstrated to be both efficient and reliable for numerous experimental settings *(5,7,12,20–22)*. However, the standard approaches for RAP-PCR require the product to be gel purified and sequenced before verification of differential expression can be performed. Furthermore, and with RAP-PCR, only a limited number of genes can be identified in one experiment.

The combination of cDNA array and RAP-PCR combines the advantages and excludes several disadvantages of each technique and provides insight into the molecular processes occurring in RASF. Therefore, we established a novel approach by combining RAP-PCR *(5)* with differential hybridization of cDNA arrays and defining genes for differential screening using very small amounts of RNA (*see* **Fig. 1**). This approach makes it possible to visualize low abundant RNAs with high efficiency and sensivity and to screen a variety of genes in one experiment. Based on our experience, it can also be stated that cDNA arrays can easily be adapted to other types of cells including cells from OA joints, from other arthritides, to relatively acellular cartilage specimens, as well as to premalignant cells and malignant cells *(7,14)*.

Because of the increasing number of genes spotted onto a cDNA array membrane, evaluation of cDNA arrays without using specific software has become difficult to perform. Here, we present the use and validation of the AtlasImage™ 2.0 software which facilitates evaluation of gene expression by using the same software settings for all subsequent array comparisons performed. Following the protocol outlined below, cDNA array allow for the identification of defined genes in distinct cell populations of different RA patients. Other possibilities are the comparison of RASF to other SF cultures (e.g., OA or normal dermal fibroblasts), or the evaluation of alterations in gene expression in two different settings such as stimulated cell populations and controls, or between gene transfer- or drug-treated cells of the same origin *(12)*. After RNA isolation, subsequent RAP-PCR *(17,21)* amplifies a subpopulation of the expressed RNA sequences isolated from the cultured RASF, which can be hybridized onto cDNA expression arrays to analyze the expression of thousands of genes in one experiment. The verification of the genes identified must be followed up using other methods of detecting on RNA and protein expression, such as *in situ* hybridization and immunohistochemistry in synovial tissue.

2. Materials
2.1. Preparation of Tissue Sections for Cell Culture
1. Obtain synovial tissues from synovial biopsies of patients with RA and patients with OA undergoing joint surgery, who meet the criteria of the American College of Rheumatology *(1)*.
2. Dispase II (Boehringer Mannheim, Mannheim, Germany).

2.2. Culture of Synovial Fibroblasts
1. Dulbecco's modified Eagle's medium (DMEM) (Biochrom, Berlin, Germany).
2. Fetal calf serum (FCS) (Gibco Life Technologies, Grand Island, NY).
3. Penicillin and streptomycin (PAA Laboratories GmbH, Linz, Austria or equivalent).

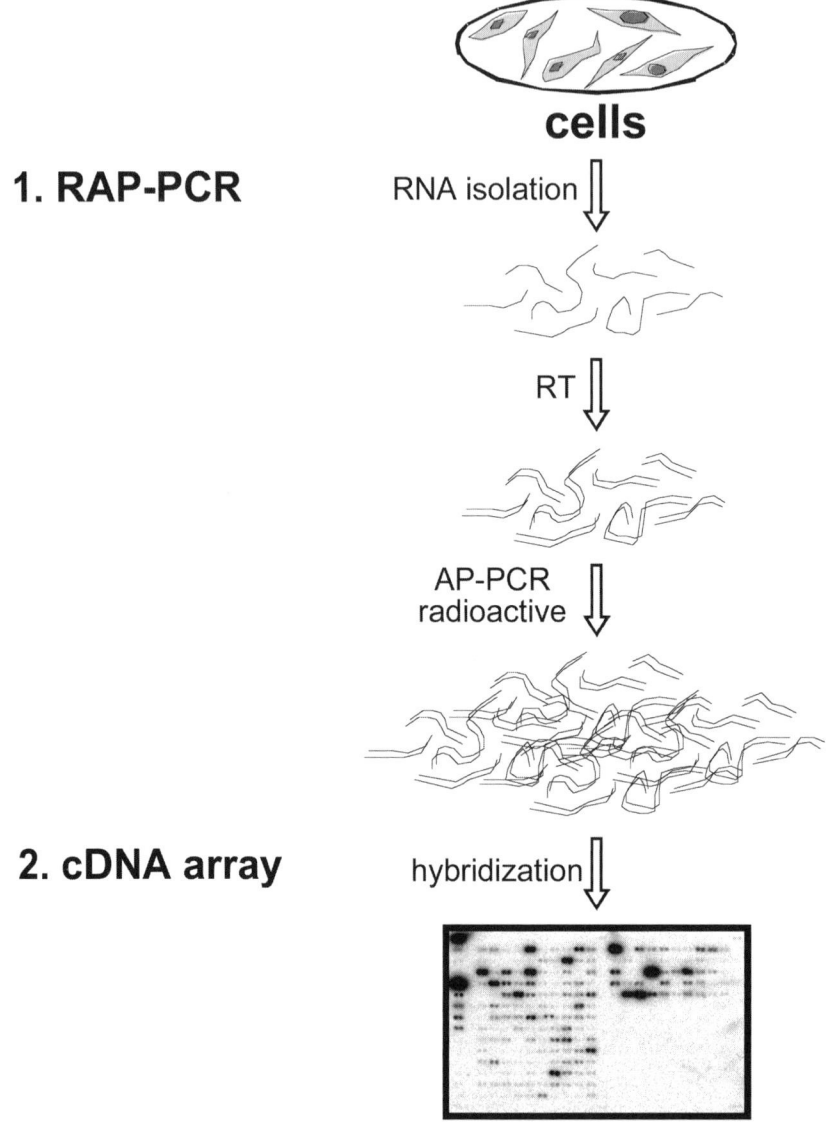

Fig. 1. Schematic diagram of cDNA array in combination with RAP-PCR. After total RNA extraction, a nonradioactive reverse transcription was performed with the first arbitrary primer followed by a radioactive AP-PCR with the second arbitrary primer. The cDNA was then hybridized onto cDNA array membranes and evaluated using the AtlasImage™ software.

Genotyping by cDNA Array and RAP-PCR

4. Phosphate buffered saline (PBS) (PAA, Linz, Austria or equivalent).
5. Trypsin/ethylene diamine tetraacetic (EDTA) (PAA Laboratories GmbH, Linz, Austria or equivalent).

2.3. RNA extraction and RAP-PCR

1. RNase Zap (Ambion, TX).
2. RNeasy spin column purification kit (Qiagen, Hilden, Germany) or other RNA isolation kit (*see* **Note 1**).
3. Ribogreen RNA quantification kit (Molecular Probes, Leiden, Netherlands).
4. Arbitrary oligonucleotide primers (10–12mer) (*see* **Note 2**) *(17)*.
5. MMuLV reverse transcriptase (e.g., Promega, Madison, WI).
6. 10X reverse transcription (RT) buffer: 500 mM Tris-HCl (pH 8.3), 750 mM KCl, 30 mM MgCl$_2$, and 200 mM dithiothreitol (DTT).
7. 100 mM dNTP mix.
8. 100 mM dNTP mix without ATP.
9. 10X RAP-PCR buffer: 100 mM Tris-HCl (pH 8.3), 100 mM KCl, and 40 mM MgCl$_2$.
10. AmpliTaq® DNA polymerase Stoffel fragment (Perkin Elmer, Norwalk, CT) and [α-^{32}P] dATP (3000 Ci/mmol).
11. Thermocycler (Model 9700, Applied Biosystems, Foster City, CA or equivalent).

2.4. Gel Electrophoresis

1. Agarose and sequencing gel equipment.
2. Gel dryer (Model 583, BioRad, Hercules, CA or equivalent).
3. Gel loading buffer: 0.25% bromophenol blue, 0.25% xylene cyanol FF, 30% glycerol.
4. Tris-borate-EDTA buffer (TBE): 90 mM Tris-borate and 2 mM EDTA.
5. 8 M Urea, 5% polyacrylamide sequencing gel, prepared with TBE buffer.
6. Running buffer: 1X TBE.
7. BioMax X-Ray film (Kodak, Stuttgart, Germany) or equivalent.

2.5. cDNA Array and Array Evaluation

1. Atlas™ human cDNA expression array membranes; the number of the spotted genes differ between the different array membranes (Clontech, Palo Alto, CA).
2. Atlas NucleoSpin® Extraction Kit (Clontech, Palo Alto, CA).
3. Hybridization solution (ExpressHyb™ Hybridization Solution, Clontech, Palo Alto, CA or equivalent).
4. Fragmented denatured salmon sperm DNA.
5. 10X denaturing solution: 1 M NaOH and 10 mM EDTA .
6. Sheared human genomic DNA (Clontech, Palo Alto, CA).
7. 2X neutralizing solution: 1 M phosphate buffer (pH 7.0).
8. 20X SSC (Sigma, Deisenhofen, Germany or equivalent).
9. Wash solution 1: 2X SSC and 2% sodium dodecyl sulfate (SDS).
10. Wash solution 2: 0.1X SSC and 0.5% SDS.

11. Phosphorimager-Screen (Molecular Dynamics, Sunnyvale, CA or equivalent).
12. Phosphorimager (STORM, Molecular Dynamics or equivalent).
13. Ambis software (ImageQuant, Molecular Dynamics or equivalent).
14. AtlasImage™ 2.0 software or AtlasImage™ 2.7 software (Clontech, Palo Alto, CA).
15. Statistic software (SPSS®11.0 for windows or equivalent).

3. Methods
3.1. Tissue Preparation and Cell Culture

1. Rinse fresh rheumatoid synovial tissue samples in cold buffered saline to remove excess fibrin and clotted blood.
2. Remove adipose tissue as much as possible.
3. Dissect the tissue specimens into individual segments no larger than 0.5 to 2 mm^2 using a sterile scalpel.
4. Digest the tissue for 1 h at 37°C in Dispase II as outlined by the manufacturer.
5. Centrifuge cells for 10 min at 1200 rpm.
6. Transfer the cell suspension into 75-cm^2 cell culture flasks and culture them at 37°C in 10% CO_2 for 24 h. Then replace medium to remove nonadherent cells.
7. Culture cells in DMEM containing 10% heat inactivated FCS, 100 U/mL penicillin, and streptomycin at 37°C in 10% CO_2 (*11*).
8. Change culture medium every 3 to 4 d or until cell confluency reaches between 80 and 90% (*see* **Note 3**).
9. At 80 to 90% confluency, wash cells with PBS and add 5 mL trypsin/EDTA (0.5% in PBS) for 10 min at 37°C.
10. Centrifuge cell suspension for 10 min at 1200 rpm for 10 min.
11. Resuspend the cells in medium and place 1/3 of the cell suspension in a 75-cm^2 cell culture flask and culture them under standard conditions (*see* **Note 4**).

3.2. Isolation of RNA

1. Clean working place with ethanol and RNase Zap® to remove any residual RNases.
2. At 80 to 90% confluency, harvest the cells and extract total RNA by silica gel binding using the RNeasy spin column purification kit or any other method providing high quality RNA, as described by the manufacturer. For better results, and if larger amounts of RNA are available, use an mRNA extraction system.
3. Treat total RNA on the spin column with DNase I at room temperature for 40 min to remove the remaining genomic DNA (*see* **Note 5**).
4. Perform the extraction according to the instructions of the manufacturer, and elute the RNA with 30 µL of RNase-free water (*see* **Note 6**).
5. Apply the eluate to the column a second time and spin through again to increase yield.
6. Measure the RNA concentrations using the Ribogreen RNA quantification kit, adjust the concentration to 50 or 100 ng/µL and store at –70°C.
7. If necessary, concentrate the RNA using a Microcon YM system for RNA (Millipore, Billerica, MA), as described by the manufacturer.

3.3. RAP-PCR

3.3.1. Principle

RAP-PCR amplifies arbitrarily a subset of the cellular RNA *(6,7)*. Unlike methods using oligo dT priming, it is not directed against the predominantly untranslated and more variable 3'-end and therefore less informative segment of the RNA, but against the translated mRNA segment. Between 50 and 100 fragments can be visualized by a single RAP-PCR approach, including relatively rare messages that happen to match with the arbitrary primers. The number of genes can be increased by testing several arbitrary primer combinations to achieve the most effective amplification method for the cell/tissue type of interest *(14)*. RAP-PCR is based on two steps:

- Transcription of total RNA into cDNA using reverse transcriptase (RT) with the first arbitrary primer, and
- Amplification of a subset of the cDNA (arbitrarily primed PCR, or AP-PCR) with the second arbitrary primer.

Short primers (10–12mers) of an arbitrarily selected sequence and a low annealing temperature of only 35°C are used for RAP-PCR, allowing annealing of the primers to regions on the target nucleic acid when matching at least five or six bases at the 3'-end, a situation that on average takes place approx every few hundred base pairs *(6,8,9,13,18,20,21)*.

3.3.2. RAP-PCR Fingerprinting

1. Mix RNA (approx 50–150 ng in 1–3 µL RNase free water) with room temperature solution to a final volume of 10 µL containing 1X RT buffer, 0.2 mM dNTPs, 2 µM primer, 10 U RNase inhibitor, and 20 U MMuLV reverse transcriptase (*see* **Note 2**).
2. When starting with fresh RNAs, a reverse transcriptase-free reaction should be included in all RAP-PCR experiments to exclude DNA contamination.
3. Apply the following reaction conditions:
   a. 5 min ramp from 25 to 37°C (primer annealing); 60 min at 37°C (reverse transcription); and 15 min at 68° (inactivation of enzyme).
4. Add 40 µL DNase free water to the RT-reaction. Use 10 µL cDNA solution out of the 50 µL for the AP-PCR.
5. Perform AP-PCR (second strand synthesis) using the second arbitrary PCR primer (*see* **Note 2**) by mixing 10 µL cDNA with polymerase chain reaction (PCR) solution for final concentrations of 1X PCR buffer, 0.2 mM dNTPs, 4 µM primer, 2.5 U AmpliTaq® DNA polymerase Stoffel fragment, and 2.8 µL [α-^{32}P]dATP (3000 Ci/mmol) in a final volume of 20 µL.
6. Cycling conditions are as follows: 5 min at 94° (initial denaturation); 30 cycles of 94°C for 30 s, 35°C for 30 s, and 72°C for 60 s; and 7 min at 72° (final extension).

3.3.3. Gel Electrophoresis

It is highly recommended to control the successful amplification of the reaction prior to cDNA array performance.

1. Mix an aliquot of the PCR reaction (maximum of 1 µL of the cDNA) 1:1 with loading buffer, denature it for 3 min at 94°C, put on ice immediately, and load onto an 8 M urea, 5% polyacrylamide sequencing gel prepared with TBE buffer.
2. Perform electrophoresis at 45°C and 100 W for about 90 min until the bromophenol blue band reaches the lower third of the gel.
3. Peel the gel away from the glass support using a 3 MM Whatman filter paper of about the same size as the gel and cover it with cling film.
4. Dry the gel under vacuum at 80°C for 1 to 2 h and expose it to a BioMax X-Ray film (Kodak, Stuttgart, Germany) at room temperature for 12 to 72 h depending on the intensity of radiation of the amplified fragments (*see* **Fig. 2**).

3.4. cDNA Array

All hybridizations using the combined method should be performed in duplicates to confirm viability of the hybridization in two independent experiments (*see* **Note 7**).

3.4.1. Probe Purification

1. Purify the ^{32}P-labeled RAP-PCR probe from unincorporated ^{32}P-labeled nucleotides and small cDNA fragments (<0.1 kb) by column chromatography using the Atlas NucleoSpin® Extraction Kit and elute the purified probe in 100 µL DNase free water as outlined by the manufacturer.
2. The fractions should be measured by scintillation counting prior to hybridization.

3.4.2. Hybridization

1. Rinse the cDNA array membranes with 2X SSC.
2. Prewarm the hybridization solution to 68°C and add 100 µg/mL fragmented denatured salmon sperm DNA.
3. Prewarm a hybridization oven with roller bottles to 68°C (*see* **Note 8**).
4. Transfer the rinsed membranes to roller bottles, add 5 mL hybridization solution to the membranes and prehybridize for a minimum of 30 min (*see* **Note 9**).
5. Dilute the labeled cDNA probe 1:10 with 10X denaturing solution and incubate at 68°C for 20 min.
6. Add 5 µL (1 µg/µL) sheared human genomic DNA and an equal volume of 2X neutralizing solution to the cDNA reaction and incubate for 10 min at 68°C.
7. Add the mixture to the membranes with the hybridization solution quickly and hybridize overnight at 68°C.

3.4.3. Wash

1. Prewarm wash solutions 1 and 2 to 68°C.
2. Wash the filters 3 times in wash solution 1 for 20 min at 68°C (*see* **Note 10**).

Fig. 2. Fingerprint gel with total RNA from one RA patient prior to hybridization to cDNA array membranes. RAP-PCR using the same amounts of RNA was performed in two independent reactions to control the reproducibility of the amplification.

3. Perform 2 washing steps with wash solution 2 at 68°C for 10 or 20 min depending on the intensity of the radiation of the membranes after washing with solution 1.
4. Wash 5 min with 2X SSC at RT.
5. Rinse the membranes with 2X SSC at room temperature, shrink-wrap the membranes in plastic without air, avoid drying of the membranes (see **Note 11**).

3.4.4. Phosphorimaging

1. Expose the membranes to a phosphorimager for approx 3 to 5 d depending on the intensity of radiation.
2. Scan the screen with a phosphorimager (see **Fig. 3**).

3.5. Evaluation of cDNA Arrays and Statistics

Use the AtlasImage™ software developed specifically for analysis of the Atlas™ cDNA Expression Arrays or an equivalent software for evaluation. The following steps are integral to the software (see **Note 12**).

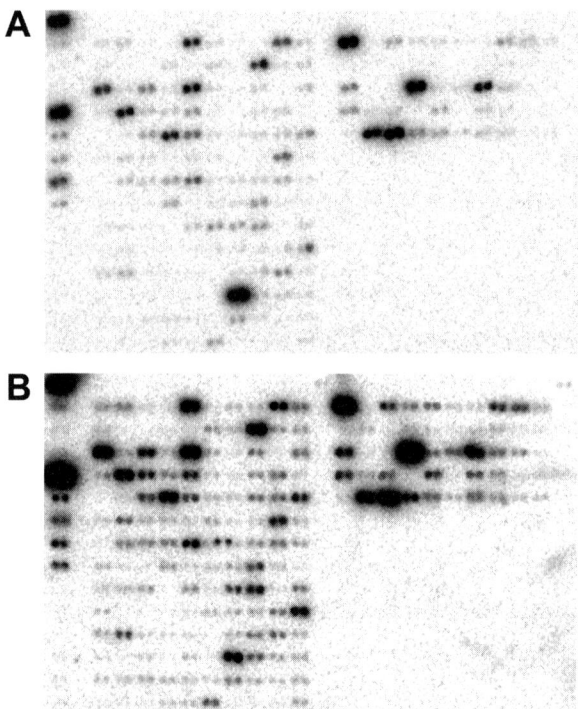

Fig. 3. Examples of cDNA arrays with and without amplification using the same total RNA of one RA patient. (**A**) cDNA array without amplification by radioactive labeled reverse transcription of the total RNA (1.5 ng total RNA). (**B**) cDNA array using the same total RNA (0.15 mg total RNA) with amplification by RAP-PCR.

cDNA array with or without amplification using the same total RNA resulted in a stable expression pattern in both approaches. In addition, low abundant RNAs could be visualized by amplification with RAP-PCR. Reprinted with permission from **ref. *13***.

3.5.1. Evaluation of cDNA Arrays Using the AtlasImage™ Software

3.5.1.1. Array Alignment

1. Adjust the arrays to the "array image view" as described by the manufacturer using two anchor spots.
2. Choose "logarithmic scale" as view option for the best optical analysis of weak exposures. Use "auto-alignment" for fine tuning of the spots.
3. Use the "auto set" function to control and manually adjust when necessary.
4. Exclude ambiguous genes (marked in red by the software) or genes affected by signal outshone from the analysis.
5. Save aligned arrays for comparison.

3.5.1.2. Array Comparison

1. Compare two arrays as described by the manufacturer.
2. Set the normalization to "global normalization." Using general normalization, signal values of all genes on the array were used for the calculation of the "normalization coefficient." In general, hybridization intensity varies for each array, therefore, global normalization using the sum method (default method) allows for the correction of hybridization efficiency (*see* **Note 13** and **Fig. 4**) *(14)*.

3.5.2. Statistics

1. For statistical evaluation of multiple array comparisons use the Lavene-statistics (values above 0.05) followed by *t*-test (parametric) or by Mann-Whitney test (non-parametric).
2. Add a significance level correction (Bonferroni adjustment) according to the number of compared genes: $p = 0.05/n_{comparisons}$ (*see* **Note 14**).

3.6. Confirmation of Results

It is strongly recommended to confirm the results by additional quantitative or semiquantitative methods on RNA, and if possible, on protein level, with methods such as reverse transcriptase (RT)-PCR, real-time PCR, or western-blot for cells, and with *in situ* hybridization and/or immunohistochemistry in tissues (*see* **Note 15**).

4. Notes

1. RNA should be kept at very low temperatures as long as possible during the procedure (use dry ice for transportation). Because of slow but continuous RNA degradation even at −20°C, RNA should be aliquoted to avoid freeze-thaw-cycles.
2. Of the different arbitrary primers that were tested, for our purposes the combination of OPN23 (5'-CAG GGG CAC C-3') as RT primer and OPN21 (5'-ACCAGGGGCA-3') for the AP-PCR step resulted in the most stable and reproducible results in the fingerprints and in the cDNA arrays performed. However, other arbitrarily chosen 10–12mer primers may work as well or even better for samples of different origin.
3. Fibroblasts should not be cultured until 100% confluency because of contact inhibition of the cells, nor under 50% of confluency because these conditions may alter gene expression.
4. Passaging of RASF to passages higher than 4 to 5 can lead to changes in the geneexpression pattern. Therefore, only low culture passages should be used for cDNA array analysis to exclude such effects.
5. Although RNA fingerprints are relatively insensitive to DNA contamination, no chances should be taken here; a DNase digest should always be performed.
6. Addition of carrier RNA was tested but showed no improvement in yield and was thus discontinued.

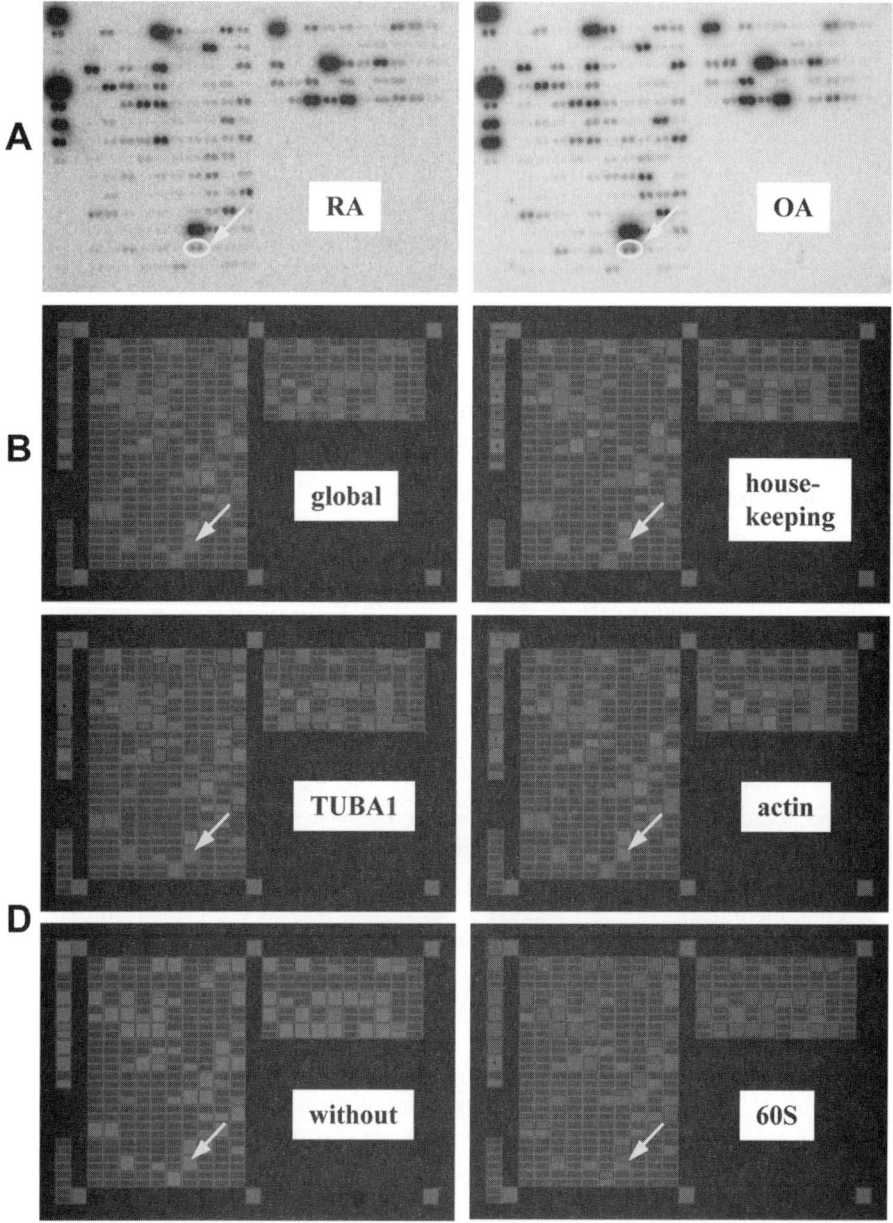

Fig. 4. Comparison views using different normalization settings of two cDNA arrays. (**A**) cDNA array 1 and 2. (**B**) Global normalization of array 1 in comparison to array 2. (**B**) User-defined normalization using all housekeeping genes to compare array 1 to array 2. (**C**) User-defined normalization using β-actin. (Color illustration in insert following p. 268.)

7. It is highly recommended to perform array analysis in duplicate to exclude hybridization events such as nonrinsed regions of the membranes or mechanical damage of a membrane which lead to unspecific signals. In addition, when using amplification methods, false-positive results must be excluded.
8. Lower temperatures than the hybridization temperature lead to nonspecific binding of cDNA to the membrane-bound cDNA. Therefore, prewarming of bottles and hybridization solution should be performed. In addition, every step should be performed at room temperature, such as adding the cDNA probe to the membranes, and as fast as possible to avoid nonspecific bindings.
9. Prehybridization avoids nonspecific binding to the cDNA array membranes. Longer prehybridization does not reduce the specificity. In contrast, when nonspecific background hybridization is a problem, elongation of the prehybridization period may reduce the background.
10. The first 1 to 2 washing steps should be performed in the roller bottles. Subsequent wash steps can be performed with all membranes together in a washing-box in a prewarmed water bath at 68°C.
11. Do not let the membranes run dry. Drying the membranes will exclude the reuse of membranes by stripping. The bound cDNA cannot be removed after drying.
12. Software settings are specific for the AtlasImage™ software and may differ from softwares from other manufacturers. Nevertheless, it is recommended not to use one single housekeeping gene for normalization.
13. As control, use the "user-defined normalization," adjusted manually to all clearly expressed housekeeping genes in the arrays for normalization as described by the manufacturer. This normalization allows normalizing towards single or groups of housekeeping genes. It is preferable to chose more than a single housekeeping gene for normalization. Compare the results with the global normalization. "Global normalization" as well as "user defined normalization" referring to all housekeeping genes most likely results in almost equal comparison views, reflecting the visual impression of differentially regulated genes *(14)*. For more than two comparisons, select one array as basis for all comparisons using the same signal threshold and identical normalization settings for each comparison.
14. Example for the Bonferroni adjustment: For 100 comparisons a random significance of $p < 0.05$ in 5 of 100 comparisons may occur (α-Factor). Therefore, the significance level needs to be adjusted: $p = 0.05/n_{comparisons}$. The correct significance level for 100 comparisons in this example is $p < 0.0005$).
15. As outlined in **Note 15**, a high risk of false positive events in cDNA array experiment needs to be taken into account. Genes of special interest, which shall be investigated further should therefore be evaluated by additional methods.

Fig. 4. *(continued) Red*: Upregulation of array 2 vs array 1; *blue*: Downregulation of array 2 vs array 1; *green*: equal expression; *gray*: one or both arrays below signal threshold (background level). Upper half of each gene "box:" ratio between array 2 and array 1. Lower half of a gene "box:" difference of array 2 vs array 1. Reprinted with permission from **ref. 14**.

Acknowledgments

The authors want to thank Birgit Riepl, Wibke Ballhorn, and Gabi Flossmann for excellent technical assistance and Rainer Straub for statistical advice. The experiments were supported by grants of the German Research Society (DFG Mu 1383/1–3 and Mu 1383/3–4) and the Swiss National Fond (SNF-3200-64142.00).

References

1. Arnett, F. C., Edworthy, S. M., Bloch, D. A., et al. (1988) The American Rheumatism Association 1987 revised criteria for the classification of rheumatoid arthritis. *Arthritis Rheum.* **31,** 315–324.
2. Boll, W., Fujisawa, J., Niemi, J., and Weissmann, C. (1986) A new approach to high sensitivity differential hybridization. *Gene* **50,** 41–53.
3. Gay, S., Gay, R. E., and Koopman, W. J. (1993) Molecular and cellular mechanisms of joint destruction in rheumatoid arthritis: two cellular mechanisms explain joint destruction? *Ann. Rheum. Dis.* **52,** 39–47.
4. Hellmann, G. M., Fields, W. R., and Doolittle, D. J. (2001) Gene expression profiling of cultured human bronchial epithelial and lung carcinoma cells. *Toxicol. Sci.* **61,** 154–163.
5. Judex, M., Neumann, E., Lechner, S., et al. (2003) Laser-mediated microdissection facilitates analysis of area-specific gene expression in rheumatoid synovium. *Arthritis Rheum.* **48,** 97–102.
6. Kullmann, F., Judex, M., Ballhorn, W., et al. (1999) Kinesin-like protein CENP-E is upregulated in rheumatoid synovial fibroblasts. *Arthritis Res.* **1,** 71–80.
7. Lechner, S., Müller-Ladner, U., Neumann, E., profiles in laser-microdissected cell populations. *Lab. Invest.* **81,** 1233–1242.
8. Mathieu-Daude, F., Cheng, R., Welsh, J., and McClelland, M. (1996) Screening of differentially amplified cDNA products from RNA arbitrarily primed PCR fingerprints using single strand conformation polymorphism (SSCP) gels. *Nucleic Acids Res.* **24,** 1504–1507.
9. McClelland, M., Mathieu-Daude, F., and Welsh, J. (1995) RNA fingerprinting and differential display using arbitrarily primed PCR. *Trends Genet.* **11,** 242–246.
10. Müller-Ladner, U., Gay, R. E., and Gay, S. (1998) Molecular biology of cartilage and bone destruction. *Curr. Opin. Rheumatol.* **10,** 212–219.
11. Müller-Ladner, U., Kriegsmann, J., Franklin, B. N., et al. (1996) Synovial fibroblasts of patients with rheumatoid arthritis attach to and invade normal human cartilage when engrafted into SCID mice. *Am. J. Pathol.* **149,** 1607–1615.
12. Neumann, E., Judex, M., Kullmann, F., et al. (2002) Inhibition of cartilage destruction by double gene transfer of IL-1Ra and IL-10 involves the activin pathway. *Gene Ther.* **9,** 1508–1519.
13. Neumann, E., Kullmann, F., Judex, M., et al. (2002) Identification of differentially expressed genes in rheumatoid arthritis by a combination of complementary DNA array and RNA arbitrarily primed-polymerase chain reaction. *Arthritis Rheum* **46,** 52–63.

14. Neumann, E., Lechner, S., Tarner, I. H., et al. (2003) Evaluation of differentially expressed genes by a combination of cDNA array and RAP-PCR using the AtlasImage 2.0 software. *J. Autoimmun.* **21,** 161–166.
15. Sehgal, A., Boynton, A. L., Young, R. F., et al. (1998) Application of the differential hybridization of Atlas Human expression arrays technique in the identification of differentially expressed genes in human glioblastoma multiforme tumor tissue. *J. Surg. Oncol.* **67,** 234–241.
16. Spirin, K. S., Ljubimov, A. V., Castellon, R., et al. (1999) Analysis of gene expression in human bullous keratopathy corneas containing limiting amounts of RNA. *Invest. Ophthalmol. Vis. Sci.* **40,** 3108–3115.
17. Welsh, J., Jung, B., Mathieu-Daude, F., and McClelland, M. Non-stoichiometric reduced complexity probes for cDNA arrays. (1998) *Nucleic Acids Res.* **26,** 3883–3891.
18. Trenkle, T., Welsh, J., and McClelland, M. (1999) Differential display probes for cDNA arrays. *Biotechniques* **27,** 554–560, 562–564.
19. Usui, H., Ichikawa, T., Miyazaki, Y., Nagai, S., and Kumanishi, T. (1996) Isolation of cDNA clones of the rat mRNAs expressed preferentially in the prenatal stages of brain development. *Brain Res. Dev. Brain Res.* **97,** 185–193.
20. Vogt, T. M., Welsh, J., Stolz, W., et al. (1997) RNA fingerprinting displays UVB-specific disruption of transcriptional control in human melanocytes. *Cancer Res.* **57,** 3554–3561.
21. Welsh, J., Chada, K., Dalal, S. S., Cheng, R., Ralph, D., and McClelland, M. (1992) Arbitrarily primed PCR fingerprinting of RNA. *Nucleic Acids Res.* **20,** 4965–4970.
22. Welsh, J. and McClelland, M. (1990) Fingerprinting genomes using PCR with arbitrary primers. *Nucleic Acids Res.* **18,** 7213–7218.

26

Gene Transfer to Synovial Fibroblast

Methods and Evaluation in the SCID Mouse Model

Ingmar Meinecke, Edita Rutkauskaite, Antje Cinski,
Ulf Müller-Ladner, Steffen Gay, and Thomas Pap

Summary

The use of gene transfer techniques has become of utmost importance both for the analysis of molecular pathways of rheumatic joint destruction and for the evaluation of novel therapeutic concepts to treat rheumatic diseases. However, gene transfer into synovial fibroblasts faces several challenges, which result mainly from the lack of specific surface markers and the low-proliferation rate of these cells. This chapter describes both nonviral and viral strategies of transferring gene constructs into synovial fibroblasts. It focuses on the use of lipofection for the gene transfer of siRNA to synovial fibroblasts and the use of AMAXA-nucleofection for the nonviral transfer of gene expression constructs. In addition, retro- and lentiviral strategies of gene transfer are introduced. Finally, the SCID mouse in vivo model of rheumatoid joint destruction is described as a means of evaluating the effects of gene transfer on the invasiveness of synovial fibroblasts.

Key Words: Rheumatoid arthritis (RA); synovial fibroblasts; gene transfer; electroporation; nucleofection; lipofection; retrovirus; lentivirus; siRNA; animal models; SCID mouse model; matrix degradation.

1. Introduction

Numerous ways have been described to transfer genes into primary fibroblasts, but none of them is free of problems. Generally, gene transfer methods can be divided into nonviral and viral methods. Nonviral methods include the transfer of naked DNA as well as the use chemical compounds (e.g., liposomes) or physical procedures (e.g., electroporation) to facilitate entry of gene constructs into cells. Although simple to use, these techniques have major disadvantages, most of which are linked to the nonspecific and thus potentially harmful way of how gene constructs enter the cells. As a consequence, transfection rates are often dissatisfying or a considerable proportion of cells may die after transfection. These problems are further enhanced when synovial fibroblasts need to be transfected, because these cells have low-proliferation rates *(1–3)*. The transfer of short inhibitory molecules (e.g., small interfering [si]RNA) constitutes an exception as these constructs exert their specific effects in the cytoplasm and do not require nuclear transport. Therefore, delivery of siRNA into synovial fibroblasts can be achieved successfully by nonviral methods. In contrast, delivery of gene expression constructs (e.g., antisense RNA-expression constructs or full-length gene expression constructs) is often troublesome when nonviral methods are used. Recently, a novel nonviral method has been described that uses a modified electroporation technique to deliver gene constructs directly into the nucleus of cells. This technique has been termed nucleofection (AMAXA Biosystems).

This chapter describes the use of lipofection for the gene transfer of siRNA to synovial cells and the use of AMAXA-nucleofection for the gene transfer of gene expression constructs into synovial fibroblasts. Of note, nonviral techniques will result only in the transient transfection of synovial fibroblasts, thus allowing for the analysis of short-term effects in these cells. The investigation of long-term effects such as invasiveness requires the use of viral gene transfer methods together with the application of viruses that are capable of integrating the gene construct into the fibroblast genome. So far, retroviruses (particularly Moloney Murine Leukaemia Virus [MMLV] based systems) have been used most widely to obtain stably transduced synovial fibroblasts. Numerous studies have shown that long-term transgene expression can be achieved when MMLV-derived retroviruses are used to transduce synovial fibroblasts. However, these systems have some important limitations. As retroviruses infect only dividing cells, initial transduction efficacy is usually low. Consequently, multiple rounds of infection need to be applied, and fibroblasts need to be selected for transgene expression following transduction (e.g., by growing the transduced cells in antibiotic-containing medium over at least 7–10 d if the retroviral vector carries an antibiotic-resistance gene). However, this procedure may take several weeks and result in the (potentially biased) selection of more rapidly proliferating fibroblast populations. Lentiviral gene transfer systems have become a most interesting alternative as lentiviruses show both a high transduction efficacy and stable transgene expression. Because lentiviral systems are more difficult to handle (which results from the common use of a two-plasmid strategy and lack of broadly available, stable packaging cell lines) this chapter describes both the use of an easy to handle retroviral system and a lentiviral strategy.

Apart from the question of how to deliver genes or gene constructs to synovial fibroblasts, evaluation of gene transfer effects constitutes a major challenge. This is particularly true for the assessment of long-term behavioral aspects such as invasiveness, which require the maintenance of synovial fibroblasts under conditions close to the in vivo situation. While a number of in vitro methods have been tested to achieve this goal, there are few well established in vivo models to study the effects of gene transfer strategies in the invasiveness of rheumatoid arthritis (RA) synovial fibroblasts. Among these, the severe combined immune deficient (SCID) mouse in vivo model has been used widely. In this model, synovial fibroblasts are coimplanted with normal human articular cartilage into SCID mice and kept there for at least 60 d. As these immune deficient mice do not reject the implants, the model allows one to study the behavior of (parental and genetically altered) synovial fibroblasts towards the coimplanted cartilage. It has been shown that RA synovial fibroblasts maintain their activated phenotype under such conditions and progressively destroy the cartilage. In contrast, normal synovial fibroblast or OA-SF do not exhibit such an aggressive behavior. This makes the SCID mouse model a very valuable tool to investigate the effects of (stable) gene transfer on the aggressive behaviour of rheumatoid arthritis synovial fibroblasts *(4–9)*.

2. Materials
2.1. Nonviral Transfection of Synovial Fibroblasts
2.1.1. Gene Transfer of siRNA Using Lipofection

1. Dulbecco's modifed Eagle's medium (DMEM), containing 4.5 g/L glucose with HEPES, penicillin, streptomycin, and amphotericin.
2. Fibroblast culture (60–80% confluent).
3. Fetal calf serum (FCS) (Biochrom AG seromed, Berlin, Germany).
4. 20 µM siRNA (1 µg total siRNA for 24-well, 5 µg total siRNA for 6-well).
5. RNAiFect reagent (Qiagen).
6. 6- or 24-Well plates.

2.1.2. Gene Transfer of a Gene Expression Construct Using Electroporation With the Nucleofectorô System (Amaxa Biosystems)

1. pEGFP-C1 expression vector (BD Biosciences, Pharmingen, Germany).
2. Nucleofector and Human Dermal Fibroblast Nucleofectorô Kit (Amaxa Biosystems).
3. DMEM 4.5 g/L glucose with HEPES, penicillin, streptomycin, and amphotericin.
4. Fibroblasts in culture (5×10^5 cells at 80% confluence per transfection).
5. EndoFree Plasmid Maxi Kit (Qiagen).
6. Six-well plates.
7. FCS.
8. 1X Trypsin/ethylene diamine tetraacetic acid (EDTA).

2.2. Viral Transduction of Synovial Fibroblasts

2.2.1. Retroviral Transduction Using the pLXIN-Retroviral Vector and the PT67 Packaging Cell Line

1. pLXIN vector (Clontech Laboratories, Heidelberg, Germany).
2. RetroPackô PT67 packaging cells (Clontech).
3. G418 (800 µg/mL), (Clontech).
4. HpaI restriction enzyme (4 U) and buffer (Roche Diagnostics).
5. RNAse free water.
6. 20 mg/mL Glycogen (Roche Diagnostics).
7. 7.5 M Ammonium acetate (Roche Diagnostics).
8. 100% Ethanol.
9. DNA Ligation Kit (Qiagen).
10. Purified DNA of the gene of interest (1.5 µg total, blunt end).
11. TOP10 competent *Escherichia coli* (Invitrogen, Karlsruhe, Germany).
12. Luria-Bertani (LB)-Broth (20 g/L) (Sigma).
13. Antibiotics: Ampicillin (50 µg/mL) (Sigma).
14. QIAGEN Plasmid Midi Kit (Qiagen).
15. Tris-EDTA (TE)-buffer : 10 mM Tris-HCl and 1 mM EDTA (pH 8.0) (Sigma).
16. RNAiFect- solution (Qiagen).
17. NIH3T3 cells (Clontech).
18. pLEIN as control vector (Clontech).
19. LB-agar (Sigma).
20. 10X Phosphate buffered saline (PBS) (pH 7.4).
21. 1X Trypsin-EDTA.
22. Polybrene (hexadimethrine bromide) : 4 mg/mL stock (Sigma-Aldrich).
23. Multifuge 3S-R (Heraeus, Germany).
24. 24- and 96-Well plates.
25. 0.45 µm Filter (Sartorius).

2.2.2. Lentiviral Gene Transfer of an Expression Construct Using the pLenti6/UbC/V5-DEST Gateway Vector System

1. Lentiviral expression system pLenti6/UbC/V5-DEST Gateway vector (Invitrogen, California).
2. Pfx-Polymerase (0.4 µL of 2.5 U/µL, Invitrogen, Karlsruhe, Germany).
3. QIAquick Gel Extraction Kit (Qiagen).
4. Polymerase chain reaction (PCR) product of gene of interest (blunt end), the forward primer must begin with the sequence CACC at the 5'-end of the primer.
5. One Shot® TOP10 *E. coli* and SOC medium (Invitrogen).
6. LB-agar and LB-broth (Sigma).
7. Antibiotics: kanamycin (50 µmg/mL), Blasticidin (50 µg/mL), Ampicillin 100 µg/mL), (Sigma).
8. EndoFree Plasmid Maxi Kit (Qiagen).
9. LR Clonase enzyme mix (Invitrogen, California).

10. 5X LR Clonase™ Reaction buffer (Invitrogen, California).
11. Proteinase K solution (2 µg/µL), (Invitrogen, California).
12. pENTR Directional TOPO® Cloning Kit (Invitrogen, California).
13. TE-buffer: 10 mM Tris-HCl, 1 mM EDTA (pH 8.0) (Sigma).
14. 293FT cells (Invitrogen, Karlsruhe, Germany).
15. Agaroseplates (Sigma).
16. DMEM containing 4.5 g/L glucose (PAA Laboratories GmbH, Cölbe, Germany) with HEPES, penicillin, streptomycin and amphotericin.
17. Crystal violet (Sigma).
18. PCR thermocycler.

2.3. The SCID Mouse Coimplantation Model

1. Four-wk-old female SCID mice (Charles River, Sulzfeld, Germany).
2. Normal human articular cartilage (e.g., from patients undergoing joint replacement surgery).
3. Inert collagen sponge (e.g., Gelfoam; Pharmacia & Upjohn).
4. Instruments for the surgical procedure (scalpel, pincers, scissors, syringes, needles, 5.0 prolene suture material),
5. 0.014 mg/g Xylocaine (Lidocain hydrochloride; Astra Pharmaceutica) and 0.09 mg/g of Ketalar (ketamine hydrochloride; Parke-Davis).
6. 5.0 Prolene suture material.
7. 4% Buffered formalin and paraffin.

3. Methods
3.1. Nonviral Transfection of Synovial Fibroblasts
3.1.1. Gene Transfer of siRNA Using Lipofection

The general principle of various chemical methods to transfect synovial fibroblasts is the formation of complexes between positively charged chemical compounds and negatively charged DNA or RNA. These complexes are incorporated into cells through endocytosis. Lipid formulations, where cationic lipids bind negatively charged DNA or RNA through electrostatic interactions, are used most widely for the purpose of gene transfer. These DNA-lipid-complexes adsorb onto cell surfaces, then fuse with the cell membrane or are incorporated by nonspecific endocytosis. The efficacy of the methods largely depends on size and composition of the DNA-lipid complexes; there are a number of different reagents on the market. For the delivery of siRNA into synovial fibroblasts, we have used the RNAiFect transfection reagent (Qiagen), the procedure and pitfalls of which are described below.

1. Seed the fibroblasts into multiple well plates 24 h before transfection (standard culture conditions: DMEM with 10% FCS and antibiotics supplementation, humidified atmosphere with 5% CO_2 at 37°C). The confluence of cells should be

between 60 and 80% on the day of transfection. It is recommended to use at least 5×10^4 synovial fibroblasts (24-well plate); for a 6-well format, 2×10^5 cells should be used.
2. On the day of transfection, dilute the siRNA (1 µg total siRNA in case of a 24-well format and 5 µg total siRNA in case of a 6-well format) in complete DMEM to reach a final volume of 100 µL. Add the RNAiFect reagent (6 µL RNAiFect-solution in case of a 24-well format and 15 µL RNAiFect-solution in case of a 6-well format), vortex for 10 s and incubate for 15 min at room temperature (*see* **Note 1**) .
3. During incubation, aspirate the medium from the cell culture and add a defined volume of complete DMEM (300 µL in case of a 24-well format and 1900 µL in case of a 6-well format) to the plate.
4. Slowly transfer the mixture of siRNA, DMEM, and RNAiFect onto the cells and mix it carefully by shaking the plate.
5. Incubate the cells under normal cell culture conditions (37°C, 5%CO_2) for 48 to 72 h (*see* **Note 2**), then remove the medium and use the cells for further experiments.

3.1.2. Gene Transfer of a Gene Expression Construct Using Electroporation With the Nucleofector™ System

Electroporation is another method to transiently transfect cells. It uses pulsed electric fields to temporarily open microscopic pores into cell membranes. These pores allow DNA molecules to enter the cells and reseal spontaneously when the pulses are sequenced properly. Here, we describe the transfer of a green fluorescence protein (GFP)-expression vector (*see* **Fig. 1**) that may stand for any gene of interest and at the same time allows for the direct monitoring of transgene expression by fluorescence microscopy. Several systems have been established that differ in terms of efficacy of transfection and the operability of the equipment. In our laboratory, we have successfully used a novel electroporation technique termed nucleofection, which delivers gene constructs directly into the nucleus of cells. This technique has been demonstrated to transfect effectively a number of different cells *(10–12)*. For some cells, such as dendritic cells, AMAXA nucleofection has been described as the only successful way of nonviral gene delivery *(10)*. Because the electrical parameters used by the device and the compositions of the buffers are proprietary, no real adjustments can be made in the transfection process. One can only choose between a number of preset programs designed to ensure either a high "survival rate" or a "high transfection rate." In our laboratory we use program "U23" which transfects fibroblast-like cells well and with a sufficient survival rate. For synovial fibroblasts, we recommend the following protocol:

1. Prepare 2 mL prewarmed DMEM (37°C) in a 6-well plate and 500 µL prewarmed DMEM (37°C) in a tube.

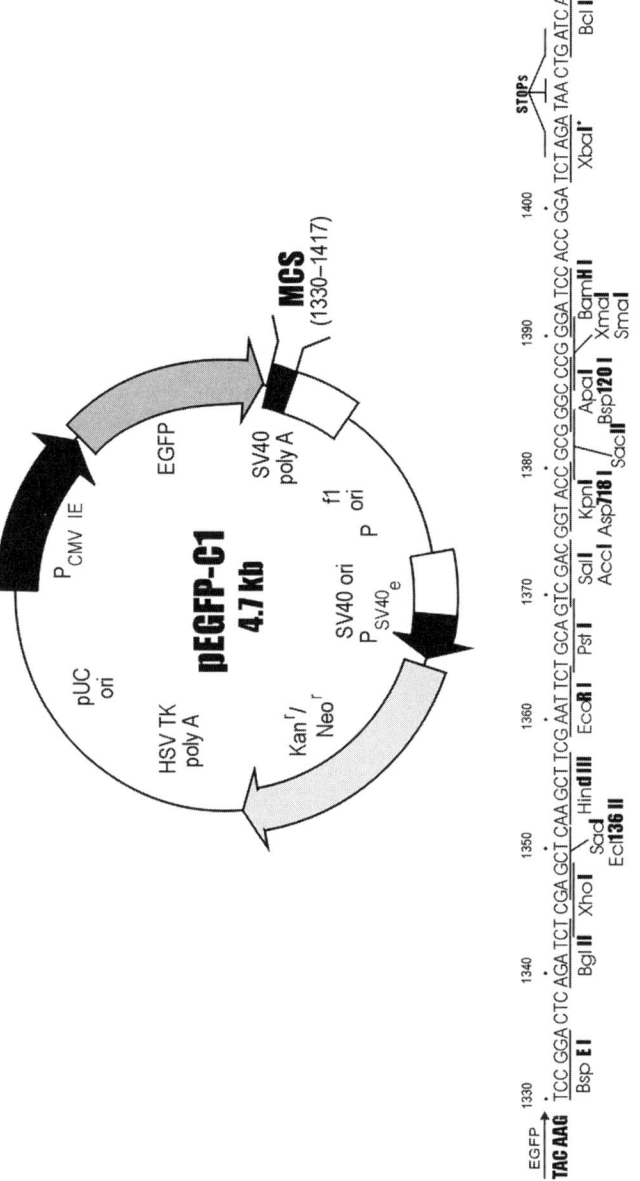

Fig. 1. Vector map of pEGFP-C1.

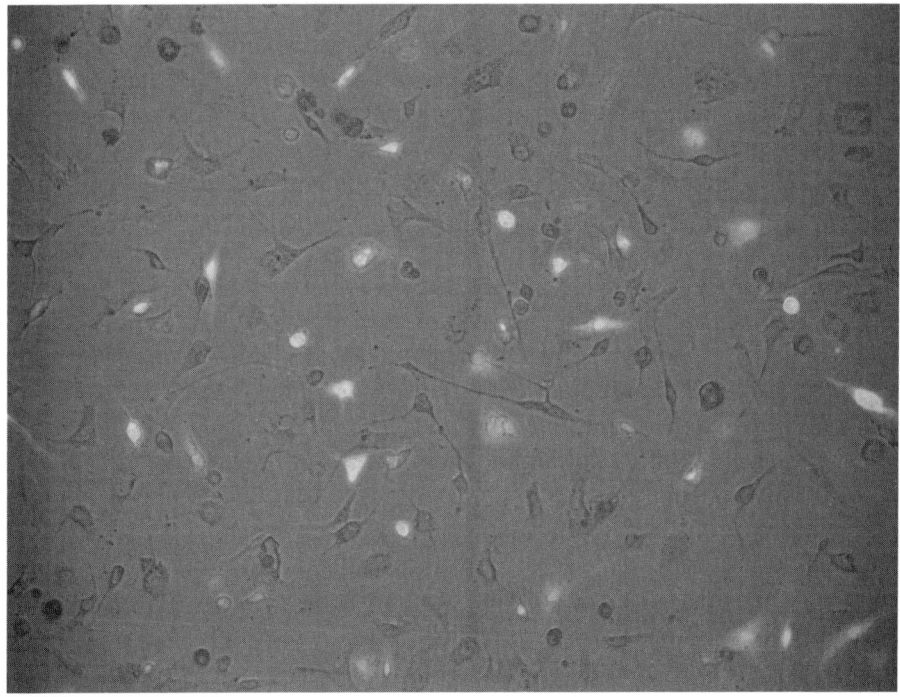

Fig. 2. RA synovial fibroblasts 1 d after transfection with pEGFP-C1-vector by AMAXA- nucleofection.

2. Trypsinise and count the synovial fibroblasts from the culture flask.
3. Transfer 5×10^5 cells into a 1.5-mL tube and centrifuge at $200g$ for 10min.
4. Remove the supernatant completely and add 100 µL prewarmed Nucleofector™ solution.
5. Add 2 µg DNA and resuspend the cells using a new 100 µL pipet tip.
6. Transfer this mix into the cuvet and close it with the cap.
7. Place the cuvet in the Nucleofector™ and press the start-button.
8. When the display shows "OK," remove the cuvet from the Nucleofector™, add 500 µL prewarmed DMEM, and mix carefully.
9. Transfer the cells from the cuvet into the prepared 6-well plate using the pipet from the kit and place the plate in the incubator at 37°C with 5% CO_2.
10. Press the start-button again to reset the Nucleofector and repeat **steps 8–12** if you have more than one sample.

The transfection efficiency strongly depends on the DNA quality, but between 40 and 60% transfected cells should be achieved (*see* **Fig. 2**) (*see* **Notes 3–5**). This can be analysed by counting the cells under a fluorescence microscopy or by fluorescence activated cell sorting (FACS).

3.2. Viral Transduction of Synovial Fibroblasts

Viral delivery systems have been demonstrated to be an effective tool for the long-term expression of gene constructs *(13,14)*. Retroviral systems are used most widely, because retroviruses integrate into the host genome and can achieve long lasting expression of the transgene. However, one prominent disadvantage of retroviral systems, is their inability to infect nondividing cells. This is a particular problem for the transduction of synovial fibroblasts, because these cells have a low proliferation rate *(2,3,15)*. To overcome this disadvantage, a number of retroviral vectors hav been developed that allow one to select successfully transduced cells by adding antibiotics such as neomycin (G418). Because of the ease of generation and use, retroviral gene transfer is still preferred by many investigators when stable gene expression is needed.

3.2.1. Retroviral Transduction Using the pLXIN- Retroviral Vector and the PT67 Packaging Cell Line

This bicistronic retroviral vector contains elements derived from the MMLV and the Moloney Murine sarcoma virus (MMSV). **Figure 3** shows the structure of the pLXIN vector. The vector contains an internal ribosome entry site (IRES), which is located between the multiple cloning site (MCS) and the neomycin resistance gene (Neo^r). Consequently, the 5' viral LTR promoter in this vector is responsible for the expression of bicistronic message of the gene of interest and the neomycin resistance (Neo^r) gene. This vector can be propagated in bacteria and selected within ampicillin. To avoid replication of retroviruses after gene transfer, retroviral vectors lack important genes that are required for retroviral replication (including those needed for the production of virus envelopes). As a consequence, infectious viral particles for gene transfer must be generated from specific packaging cells that contain these genes and, upon transfection with the retroviral vector, are capable of producing infectious but replication deficient retroviral particles.

Here, the packaging cell line RetroPacko PT67 Cell Line was used to generate infectious, replication-incompetent retroviral particles after transfection with pLXIN. The PT67 packaging cell line derived from the NIH3T3-fibroblast cell line and expresses the 10A1 viral envelope. Virus particles generated from these cells can be used to infect a broad range of mammalian cells. This is because the virus is able to enter cells via two different surface molecules, the amphotropic retrovirus receptor (Ram-1) and the GALV receptor. Stably transfected PT67 cells can be selected using neomycin (G418).

Retroviral gene transfer using the pLXIN vector consists of three steps: (1) cloning of the gene of interest into the retroviral pLXIN vector (*see* **Subheadings 3.2.1.1–3.2.1.2.**), (2) transfection of PT67 packaging cells and harvesting the viral particles (*see* **Subheading 3.2.1.3.**) (3) determination of the virus titer

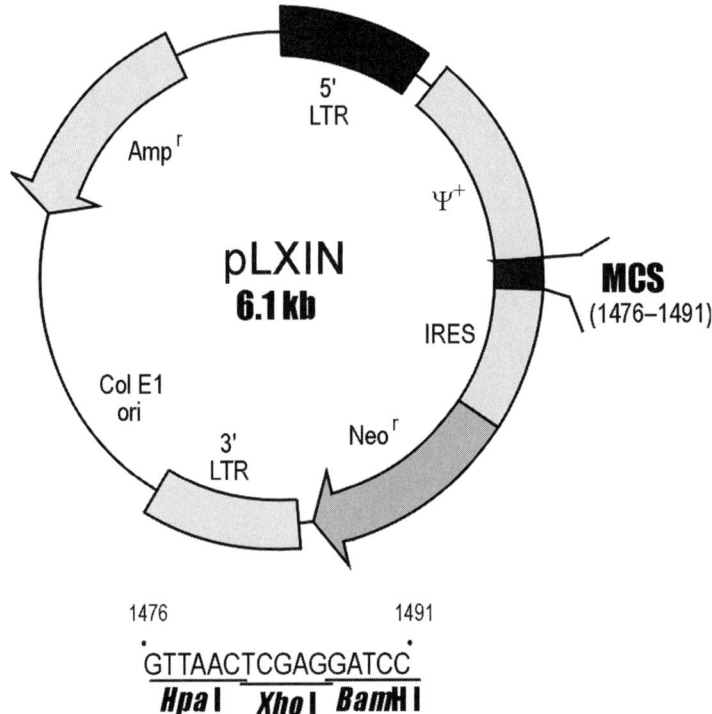

Fig. 3. Vector map of the pLXIN retroviral vector (BD Biosience).

(*see* **Subheading 3.1.2.4.**), and (4) transduction and selection of human synovial fibroblasts (*see* **Subheading 3.2.1.5.**). The suggested procedure assumes that the gene of interest is available as a blunt end DNA fragment (as obtained either by PCR using Pfu polymerase or blunt-end restriction).

3.2.1.1. PREPARATION OF THE PLXIN VECTOR

1. First, the pLXIN- vector is linearised using the HpaI restriction enzyme that creates blunt-ends. Prepare an incubation sample of pLXIN vector with a total volume of 50 μmL and incubate for 1 h at 37°C for enzymatic digestion and for 10 min at 65°C to stop the reaction:
   10 mL pLXIN (500 ng/ mL pLXIN).
   5 mL Appropriate restriction enzyme buffer.
   1 mmL HpaI (4U).
   34 mL RNAse free water.
2. Carry out a precipitation step to concentrate DNA (for pLXIN and the gene of interest) and to remove buffers containing undesirable enzyme activities. The following components are added per 50 μL sample:

2 µL Glycogen (20 mg/mL).
   10 mL 7.5 M Ammonium acetate
   125 mL 100% Ethanol (ice-cold).
3. Centrifuge for 20 min at 10,000g at 4°C.
4. Remove the supernatant with a pipette and air-dry the pellet for 5 min.
5. Wash the pellet (200 µL 70% ethanol) and centrifuge for 20 min at 10,000g at 4°C.
6. Discard the supernatant carefully and air-dry the pellet for 5 min.
7. Resuspend the pellet in 10 µL RNase free water.

3.2.1.2. Ligation of the Gene of Interest Into the pLXIN Vector and Propagation in *E. coli*

This procedure can be performed using the DNA Ligation Kit (Qiagen). The (blunt end) ligation is carried out using a vector-insert-ratio of 2:1

1. Add the following components into a 1.5-mL tube (*see* **Notes 6** and **7**): 6 µL of the pLXIN (equivalent to 3 µg of pLXIN), 0.7 µL of the sample (equivalent to 1.5 µg of DNA of interest), 1 µL of 10X ligase buffer, 1 µL of 10 mM rATP (pH 7.5), 0.5 µL of T4 DNA ligase (4 U/µL), and 0.8 µL RNAse free water. Incubate the sample at 12°C for 4 h.
2. Next, propagate the ligated vector in *E. coli* using a standard transformation protocol. For this purpose, the following is needed: transfer 3 µL of the plasmid to 100 µL bacteria, incubated on ice for 30min, incubate at 42°C for 90 s (heat shock step), add 500 µL LB-medium, incubate for 15 min at 37°C, and plate 50 µL of the sample onto an agar plate containing 50 µg/mL Ampicillin. Keep at 37°C overnight (12 h).
3. Extract the DNA (e.g., by using the QIAGEN Plasmid Midi Kit (Qiagen) and precipitated.

3.2.1.3. Transfection of PT67 Packaging Cells

To obtain replication deficient retroviral particles, PT67 cells are transfected by lipofection. The transfection procedure is as follows:

1. The day before transfection, seed 6×10^5 PT67 cells per 60-mm dish in 5 mL DMEM. Incubate the cells under the normal conditions (generally 37°C and 5% CO_2).
2. On the day of transfection, dilute 5 µg DNA (in TE buffer) in cell growth medium containing no FCS and antibiotics to a total volume of 150 µL.
3. Add 30 µL of SuperFect Transfection Reagent to the DNA solution. Mix by pipetting.
4. Incubate the samples for 5 to 10 min at room temperature, which will allow the formation of transfection-complexes.
5. Aspirate the growth medium from the dishes, and wash the cells once with 4 mL phosphate buffered saline (PBS).

6. Add 1 mL cell growth medium (containing FCS and antibiotics) to the reaction tube containing the transfection complexes. Mix the solution by pipetting and immediately transfer onto the cells in the 60-mm dishes.
7. Incubate the cells with the transfection complex for 3 h under normal growth conditions.
8. After 3 h, remove the medium and wash the cells once with 4 mL PBS.
9. Add fresh cell growth medium (containing serum and antibiotics) and incubate the cells for 24 h.
10. After 24 h trypsinise the cells by adding 1 mL trypsin, wash with PBS and centrifuge at 700g for 10 min at 20°C. Add the medium containing G418 (800 µg/mL) and the cell growth medium to the pellet at the ratio 1:15. Seed the cells into 24-well plates.
11. Start selection with G418 for at least 7 d to get the stable clones.

3.2.1.4. DETERMINATION OF THE VIRAL TITER

The viral titer produced by stable virus-producing packaging cell lines is determined by limiting dilution.

1. Seed the packaging cells in 25-cm^2 cell culture flask.
2. Collect the viral particle containing medium from the packaging cells after 48 h.
3. Add polybrene to a final concentration of 8 µg/mL and filter the medium through a 0.45 µm filter to remove packaging cells.
4. Prepare six 10-fold serial dilutions. To dilute the virus supernatant, use fresh medium (DMEM).
5. Infect target cells (e.g., NIH3T3, plated one day before in 96-well plates, 5×10^3 cells per well, in 200 µL of medium) by adding virus-containing medium to the wells.
6. Replace the fresh medium after 24 h.
7. Start the selection with G418 within 24 h. After selection, the presence of 1 colony in the 10^8 dilution of pL(gene of interest)IN and in 10^7 of the control vector pLEIN indicates a viral titer of 1×10^8 and 1×10^7, respectively.

3.2.1.5. TRANSDUCTION OF FIBROBLASTS

After obtaining clones of packaging cells that produce high virus titers, the filtered supernatants from these packaging cells are used for transduction. Transduction is usually carried out in 24-well culture plates. On day 1, synovial fibroblasts (4–6 passage) are plated at a density of 1×10^4 cells per well. After the cells reach 50 to 60% confluence, they are incubated with the retroviral supernatants twice for 8 h, with a 24 h interval. All transductions are carried out in the presence of polybrene (8 µg/mL).

Successfully transduced cells are selected with G418 (800 µg/mL). For further experiments, pL(gene of interest)IN-transduced, mock (pLEIN)-transduced, and non-transduced cells are used.

Gene Transfer to Synovial Fibroblast

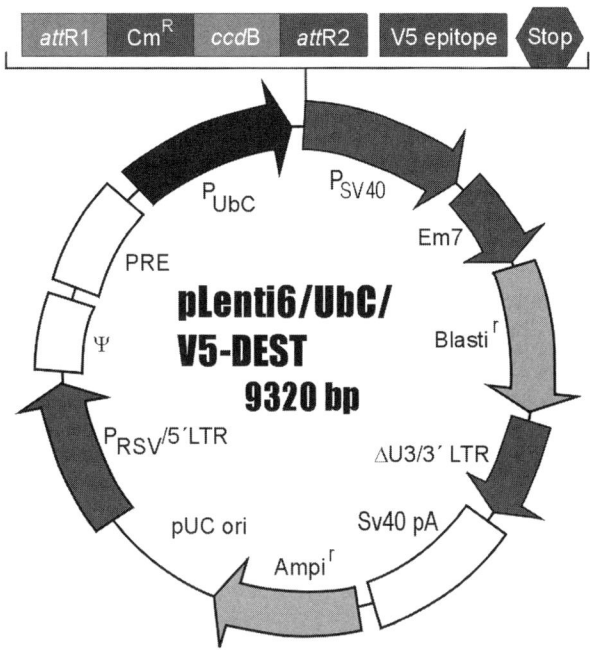

Fig. 4. Vector map of thee pLenti6/UbC/V5-DEST Gateway vector (Invitrogen).

3.2.2. Lentiviral Gene Transfer Using the pLenti6/UbC/V5-DEST Gateway Vector

Lentiviruses offer a number of advantages over onco-retroviruses and have gained increasing interest for gene transfer into slowly dividing cells such as synovial fibroblasts *(16,17)*. Whereas most lentiviral systems are still being developed by research laboratories and distributed in the frame of collaborations, the pLenti6 /UbC/V5-DEST-vector (Invitrogen, Germany) is an example of a commercially available lentiviral expression system.

The pLenti6/UbC/V5-DEST Gateway vector is derived from HIV-1 (*see* **Fig. 4**). This expression system can transduce both dividing and nondividing cells and provides stable, long-term expression of a gene of interest. Handling of the system, however, is somewhat more complex than the pLXIN- based retroviral system. This is because the lentiviral vector is generated through recombination of the pENTR TOPO® vector with the pLenti-DEST vector, and as a result of rapid cell death of the 293FT packaging cells following transfection with the packaging mix (*see* **Notes 8** and **9**), no stable packaging cells are available. As a consequence, generation of replication deficient lentiviruses from this system, requires a number of additional steps.

First, a blunt end polymerase chaing reaction (PCR) product of the gene of interest is required, that has some special requirements. The forward primer must contain the sequence CACC at the 5'-end of the primer to facilitate the recombination of PCR product into the pENTR/D-TOPO® vector. The reverse primer must not contain the sequence GTGG (this is the complementary sequence to the overhang at the 5'-end of pENTR/D-TOPO®), as this would increase the chance that the open reading frame is cloned in the opposite orientation. The stop codon in the gene of interest should be removed from the reverse primer if the product is to be fused to a C-terminal tag. We have performed the PCR by using the Pfx-polymerase, which provides a higher fidelity than the Pfu DNA polymerase.

3.2.2.1. Recombination of pENTR/D-TOPO™ Vector With Blunt End PCR Product and Transformation of Chemically Competent One Shot™ TOP10 *E. coli*

We have used the pENTR Directional TOPO™ Cloning Kit for cloning the PCR product into the pENTR/D-TOPO™ vector. For highest recombination efficiency, the molar ratio of the PCR product and the TOPO™ vector should be 5:1.

1. Add the following components and mix them gently: 10 to 20 ng (in 4 µL) of a fresh PCR product, 1 µL of salt solution from the kit, and 1 µL TOPO® vector.
2. Incubate the reaction mix for 10 min at room temperature.
3. Add 2 µL of the reaction mix to the tube containing the One Shot™ TOP10 *E. coli* and mix gently without pipetting up and down.
4. Incubate the tube on ice for 15 min.
5. Incubate the reaction for 30 s at 42°C (heat shock) and put the tube on ice immediately.
6. Add 250 µL SOC medium at room temperature.
7. Shake the tube horizontally at 37°C for 30 min.
8. Preheat an LB agar plate containing 50 µg/mL kanamycin.
9. Plate 100 µL of the transformation mix on an LB agar plate and incubate at 37°C for 16 h.
10. Pick 5 colonies and grow the bacteria in 30 mL LB medium containing 50 µg/mL kanamycin.
11. Isolate plasmid DNA using the EndoFree Plasmid Maxi Kit (Qiagen) and dissolve it in RNAse free water.
12. Make sure the gene of interest is cloned into the plasmid (PCR and/or sequencing)

3.2.2.2. L- and R-Attachment Site (LR-) Recombination of the pENTR/D-TOPO® Vector With pLenti6/UbC/V5-DEST Gateway Vector

1. Mix the following components in a 1.5-mL tube and resuspend: 1 µL pENTR/D-TOPO™ vector with blunt end PCR product, 2 µL pLenti6/UbC/V5-Dest Gateway™ Vector, 4 µL LR Clonase™ reaction buffer, 9 µL TE buffer (pH 8.0), and

Gene Transfer to Synovial Fibroblast

4 μL LR Clonase™ enzyme mix. Incubate the mix for 5 h at 25°C and add 2 μL of the proteinase K solution and incubate for 10 min at 37°C.

The product of this reaction can be used to transform a suitable *E. coli*.

3.2.2.3. GENERATING A LENTIVIRAL STOCK BY USING 293FT CELL LINE

Generation of lentiviral supernatant stock using the 293FT cell line cotransfection of 293FT cells with the lentiviral vector and helper plasmids ("packaging mix") will result in the production of replication deficient lentiviral particles by the 293FT packaging cells. However, as transfection with the packaging mix results in the death of these cells after about 72 h, it is necessary to test all generated lentiviral stock for virus titer.

1. One day before transfection, transfer 5×10^6 293FT cells into a 10 cm plate with 10 mL medium.
2. On the following day, remove the medium from the 293FT cells and replace with 5 mL medium containing serum but no antibiotics.
3. Prepare the DNA-Lipofectamine™2000 complex as follows: Add 9 μg of the optimized packaging mix and 3 μg of pLenti expression plasmid DNA to 1.5 mL medium without serum and mix. Add 36 μL of Lipofectamine™2000 to 1.5 mL medium without serum and incubate for 5 min at room temperature. Mix the diluted DNA and the Lipofectamine™2000 and incubate for 20 min at room temperature.
4. Pipet the DNA-Lipofectamine complex onto the cell plate dropwise and gently mix by shaking.
5. Incubate the cells for 12 to 18 h at 3°C in a CO_2 incubator.
6. Following incubation, remove and replace the supernatant with complete medium.
7. Collect the virus-containing supernatant after 48 to 72 h in a 15-mL sterile tube and centrifuge at 700*g* force for 15 min at 4°C.
8. This supernatant should be filtrated through a 0.45 μm low protein binding filter and viral stocks be stored in 1-mL aliquots at –80°C.

The viral supernatants should be titered using synovial fibroblasts to obtain reproducible and cell specific expression results. The procedure is similar to that described above for retroviral transduction.

1. Trypsinise the synovial fibroblasts and count the cells.
2. Seed 1×10^5 cells into each well of a 6-well plate.
3. On the following day, prepare 10-fold serial dilutions raging from 10^{-2} to 10^{-6} into complete culture medium to a final volume of 1 mL.
4. Add polybrene to each dilution to a final concentration of 6 μg/mL and mix gently.
5. Remove the cell culture medium from the fibroblasts in the 6-well plates and add the mix from point 4 to the wells.
6. On the following day remove the mix and replace complete medium.

7. One day later, remove the medium again and replace with blasticidin (final concentration of 1.6 µg/mL for synovial fibroblasts) containing medium.
8. After 10 to 12 d, add 1 mL crystal violet solution to each well for 10 min at room temperature.
9. Wash the cells and count blue-stained colonies to determine the viral titer of the stock.

3.3. The SCID Mouse Coimplantation Model

The SCID mouse coimplantation model allows one to investigate in vivo interactions of synovial fibroblasts with normal articular cartilage in the absence of human inflammatory cells. By using synovial fibroblasts, in which signaling or effector pathways are altered through stable transduction with gene constructs, it offers the opportunity to investigate the specific contribution of these pathways and molecules to fibroblast-mediated cartilage destruction.

3.3.1. Preparation of Fibroblasts and Cartilage

Synovial fibroblasts are harvested after 4 to 6 passages and should be tested for mycoplasma infection before implantation. It is also recommended to characterize the cells as fibroblasts by flow cytometry. For the coimplantation of synovial fibroblasts and normal human cartilage into the SCID mice, we use the sponge technique (*see* **Fig. 5**). Before implantation, fibroblasts are trypsinized, washed with PBS, resuspended in sterile culture medium, and soaked in small pieces of an inert collagen sponge. The articular cartilage is also cut into small pieces and then inserted into a cavity of the gel sponge. This step guarantees close contact of fibroblasts and cartilage. The detailed protocol is as follows:

1. Trypsinize the synovial fibroblasts (0.25% trypsin, containing 0.02% EDTA) and count the cells.
2. Resuspend 10^5 synovial fibroblast in 200 µL sterile filtered cell culture medium (containing FCS and antibiotics) in a 15-mL Falcon tube.
3. Cut small (approx 0.5 mm–1 mm) pieces of sponge under sterile conditions in a petri dish.
4. Add 2 to 5 pieces of the sponge into the 15-mL Falcon tube, so that they get fully soaked with the cell containing medium.
5. Store the sponges with the cells at room temperature until used.

3.3.2. SCID Mouse Coimplantation Procedure

Implantation of fibroblasts and cartilage is carried out under sterile conditions. This is not only required because of the potential leakiness (*see* **Note 10**) but also because of the increased susceptibility of these animals to bacterial infection. The implants are most commonly placed under the renal capsule of the SCID mice and kept there for 60 d. The detailed protocol is as follows:

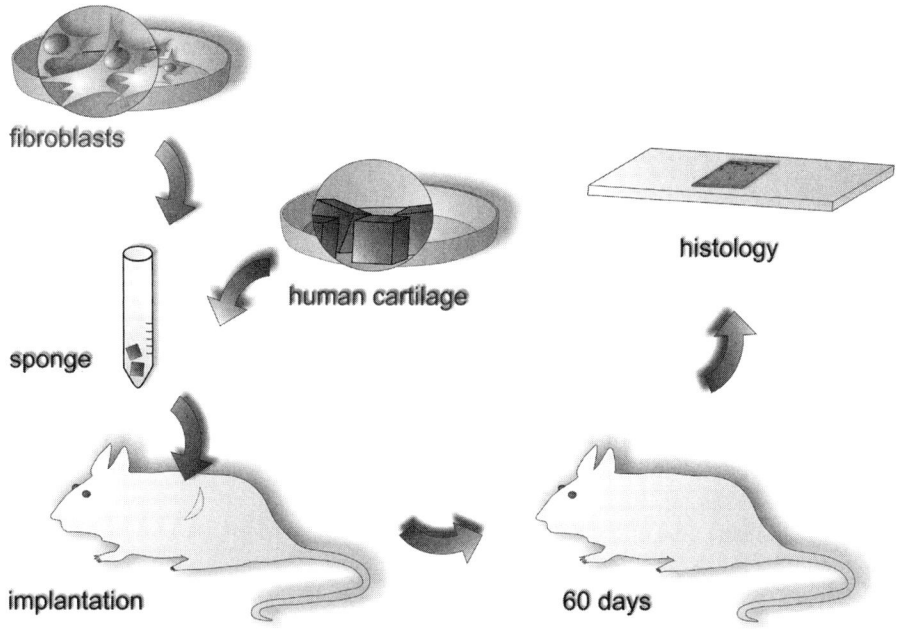

Fig. 5. The SCID-mouse co-implantation model of rheumatoid cartilage destruction (for details *see* **Subheading 3.3.**).

1. Mice are anesthetized intraperitoneally with 0.014 mg/g Xylocaine (lidocaine hydrochloride) and 0.09 mg/g Ketalar (ketamine hydrochloride) in an isotonic solution.
2. The animals are then placed onto their right side, and a 1-cm incision is made on the left flank.
3. Following exteriorsation of the left kidney, a small incision of the renal capsule is made, and one piece of the sponge is placed together with the cartilage directly under the capsule using a small forceps (*see* **Notes 11** and **12**).
4. The peritoneal layer and the skin are closed using 5.0 Prolene suture material.
5. The animals are placed onto a warming plate that keeps their body temperature for about 20 min and then transferred into their cages.

3.3.3. Sacrifice and Histological Assessment of Invasion

After 60 d, mice are sacrificed and the implants are removed. Tissues are fixed in 4% buffered formalin and embedded in paraffin. 4 µm sections are cut and stained using standard hematoxylin &eosin staining. Histologic evaluation of the H&E stained section is performed using a semiquantitative score for invasion under the microscope. In this score, the depth of invasion is evaluated by counting the number of cell layers invading the cartilage. The scores used ranged from 0 = no or minimal invasion, 1 = visible invasion (2–5 cell layers), 2 = invasion (6–10 cell layers), 3 = deep invasion (>10 cell layers), and 4 = overall deep invasion.

4. Notes

1. Usually, different ratios of siRNA and the lipid complexes need to be tested for optimal transfection efficacy and best survival. Thus, the RNAiFect protocol proposes 3 different dilutions of siRNA with RNAiFect, namely 1:3, 1:6, and 1:9. We have transfected synovial fibroblasts with a number of different dilutions and have obtained the best results with a dilution ratio of 1:6.
2. The duration of transfection is limited mainly by potential nonspecific (toxic) effects of the lipid formulations. For synovial fibroblasts, we recommend a transfection of no longer than 48 h. Alternatively, (i.e., if transfection efficacy is too low), multiple rounds of transfection can be applied with 12 h intervals, in which the lipids containing supernatant is replaced by normal medium.
3. Above all, the time line of the experiment must be prepared very carefully. Every delay in the course of the experiment can increase cell mortality and decrease the rate of transfection. For multiple transfections, it is recommended not to keep the fibroblasts in the Nucleofector^{TM} solution for longer than 15 min.
4. Preparation of cells is also critical. Six days before nucleofection, the cells should be passaged the last time. The transfection can be carried out if cells show about 80% confluence.
5. We used the EndoFree Plasmid Kit for the DNA isolation, because the transfection rate depends on DNA quality and concentration. The concentration should be from 1 to 5 µg/µL and the A260/280 ratio after purification control should be at least 1.8.
6. The ratio of vector and insert for the ligation of the gene of interest into the pLXIN vector depends on the ends of the DNA (blunt end or sticky end). For ligation of inserts with 'sticky' ends the vector-insert ratio should be 1:5–1:10.
7. The efficiency of transduction of synovial fibroblasts can be increased by performing the transfection multiple times with 12 to 24 h intervals. Extension of transfection period is not recommended, because polybrene is toxic for the transfected cells and prolonged exposure may result in cell death.
8. The packaging mix is the most critical part of lentivirus production. pLP1 encodes for gag/pol, pLP2 encodes for a regulatory protein REV- and pLP/VSVG – encodes for the envelope protein VSVG. The key to HIV-1 splicing is Rev (supplied in the pLP2 vector). When one cotransfects the packaging mix (containing pLP2) into 293FT cells, Rev accumulates and stimulates the nuclear export of unspliced and singly spliced RNA into the cytoplasm for encapsidation. The unspliced RNA becomes the viral genome after encapsidation. If pLP2 vector is left out (i.e., in the absence of Rev), one would get mostly multispliced RNA and viral titers would drop significantly. We typically get 2 (sometimes 3) 24-h harvests from 1 transfected plate (approx 30 mL of virus per 10-cm plate). After that, the cells are "wiped-out." The VSV-G causes the cells to fuse, preventing further cell division and leading to cell death.

9. Based on our experience, we recommend a "multiplicity of infection" (MOI) of 1 or higher for transduction.
10. The age of the SCID mice is an important issue. In about 15% of young mice functional antigen receptor rearrangement occurs through "illegitimate" recombination. Thus, a few T- and B-cell clones survive and expand in vivo after antigen exposure. This phenomenon is called "leakiness" and increases in frequency with age especially in mice that are not kept under strict germ free conditions. At the age of 12 to 14 mo, the majority of SCID mice become leaky. Therefore, the use of young animals at an age of 6 to 8 wk is recommended.
11. The implanted cartilage should be taken from healthy patients undergoing joint replacement surgery after severe trauma. The cartilage should not be older than 24 ho (i.e., availability of the cartilage is a major determinant of the operation procedure). The cartilage is also cut into small (approx 1-mm^3 pieces) and stored in a petri dish containing normal medium.
12. Whereas implantation under the kidney capsule has been associated with good vascularization, little scar formation and, thus, reproducible results, several groups prefer implantation under the skin. The implantation technique is considerably easier and it has been demonstrated that the results are also reliable. In addition, several implants can be used in one mouse.

References

1. Kinne, R. W., Emmrich, F., Bail, H., et al. (1995) Expression of activation markers of the rheumatoid arthritis synovial membrane: comment on the article by Qu et al. *Arthritis Rheum.* **38,** 1346–1348.
2. Mohr, W., Hummler, N., Pelster, B., and Wessinghage, D. (1986) Proliferation of pannus tissue cells in rheumatoid arthritis. *Rheumatol. Int.* **6,** 127–132.
3. Nykanen, P., Bergroth, V., Raunio, P., Nordstrom, D., and Konttinen, Y. T. (1986) Phenotypic characterization of 3H-thymidine incorporating cells in rheumatoid arthritis synovial membrane. *Rheumatol. Int.* **6,** 269–271.
4. Baier, A., Meineckel, I., Gay, S., and Pap, T. (2003) Apoptosis in rheumatoid arthritis. *Curr. Opin. Rheumatol.* **15,** 274–279.
5. Franz, J. K., Pap, T., Hummel, K. M., et al. (2000) Expression of sentrin, a novel antiapoptotic molecule, at sites of synovial invasion in rheumatoid arthritis. *Arthritis Rheum.* **43,** 599–607.
6. Geiler, T., Kriegsmann, J., Keyszer, G. M., Gay, R. E., and Gay, S. (1994) A new model for rheumatoid arthritis generated by engraftment of rheumatoid synovial tissue and normal human cartilage into SCID mice. *Arthritis Rheum.* **37,** 1664–1671.
7. Muller-Ladner, U., Kriegsmann, J., Franklin, B. N., et al. (1996) Synovial fibroblasts of patients with rheumatoid arthritis attach to and invade normal human cartilage when engrafted into SCID mice. *Am. J. Pathol.* **149,** 1607–1615.
8. Pap, T., Muller-Ladner, U., Gay, R. E., and Gay, S. (2000) Fibroblast biology. Role of synovial fibroblasts in the pathogenesis of rheumatoid arthritis. *Arthritis Res.* **2,** 361–367.

9. Pap, T., Aupperle, K. R., Gay, S., Firestein, G. S., and Gay, R. E. (2001) Invasiveness of synovial fibroblasts is regulated by p53 in the SCID mouse in vivo model of cartilage invasion. *Arthritis Rheum.* **44,** 676–681.
10. Lenz, P., Bacot, S. M., Frazier-Jessen, M. R., and Feldman, G. M. (2003) Nucleoporation of dendritic cells: efficient gene transfer by electroporation into human monocyte-derived dendritic cells. *FEBS Lett.* **538,** 149–154.
11. Hamm, A., Krott, N., Breibach, I., Blindt, R., and Bosserhoff, A. K. (2002) Efficient transfection method for primary cells. *Tissue Eng* **8,** 235–245.
12. Maasho, K., Marusina, A., Reynolds, N. M., Coligan, J. E., and Borrego, F. (2004) Efficient gene transfer into the human natural killer cell line, NKL, using the Amaxa nucleofection system. *J. Immunol. Methods* **284,** 133–140.
13. Gouze, J. N., Stoddart, M. J., Gouze, E., Palmer, G. D., Ghivizzani, S. C., Grodzinsky, A. J., and Evans, C. H. (2004) In vitro gene transfer to chondrocytes and synovial fibroblasts by adenoviral vectors. *Methods Mol. Med.* **100,** 147–164.
14. Tonini, T., Claudio, P. P., Giordano, A., and Romano, G. (2004) Transient production of retroviral- and lentiviral-based vectors for the transduction of Mammalian cells. *Methods Mol. Biol.* **285,** 141–148.
15. Aicher, W. K., Heer, A. H., Trabandt, A., et al. (1994) Over-expression of zinc-finger transcription factor Z-225/Egr-1 in synoviocytes from rheumatoid arthritis patients. *J. Immunol.* **152,** 5940–5948.
16. Naldini, L. and Verma, I. M. (2000) Lentiviral vectors. *Adv. Virus Res.* **55,** 599–609.
17. Naldini, L. (1998) Lentiviruses as gene transfer agents for delivery to non-dividing cells. *Curr. Opin. Biotechnol.* **9,** 457–463.

27

In Vitro Matrigel Fibroblast Invasion Assay

Tanja C. A. Tolboom and Tom W. J. Huizinga

Summary

Rheumatoid arthritis is characterized by inflammation of the joints and degradation and invasion by fibroblast-like synoviocytes (FLS) of the cartilage. To assess the invasiveness of FLS an in vitro invasion assay was developed. In this invasion assay the FLS grow through an artificial matrix composed mainly of collagen IV. First, the walls of transwells are coated with paraffin to avoid meniscus formation. Subsequently, the bottoms of the transwells (on top of the membrane) are coated with a thin layer of matrigel. On top of this matrigel fibroblast-like synoviocytes are seeded at a density of 100,000 cells per milliliter. The cells are cultured in serum free medium in the inner compartment inside the transwell. To the outer compartment outside the transwell IMDM with 10% fetal calf serum and 10% NHS is added. The cells are incubated for 3 d at 37°C and 5% CO_2. After 3 d the cells are fixed with 2% glutaraldehyde in phosphate buffered saline and stained with 1% crystal violet in water. The matrix on the inside of the transwells and the cells that have not grown through the matrix and the membrane are removed. The cells that have grown through the matrix and through the membrane under the transwell can be visualized by light microscopy and counted.

Key Words: Fibroblast-like synoviocytes; rheumatoid arthritis (RA); invasion; matrigel; in vitro.

1. Introduction

Rheumatoid arthritis (RA) is characterized by inflammation of the joint, hyperplasia of the synovium, and degradation of cartilage and bone. The degradation of cartilage is thought to be mediated by fibroblast-like synoviocytes (FLS) from the synovium. These FLS show characteristics of transformation and invade the cartilage *(1)*. To study this invasion into the cartilage a high-throughput assay is needed. Müller-Ladner et al. have developed a system where synovium of RA patients together with normal human cartilage is implanted under the renal capsule of SCID mice *(2)*. After 60 d the mice are sacrificed, synovium and cartilage are recovered and invasion can be assessed by immunohistochemistry.

The major drawbacks of this system are the use of animals, the cost, the time investment and relatively low-throughput. Therefore, an in vitro system to measure invasiveness of FLS was developed. This system is adapted from a system used in cancer research to measure the metastatic potential of tumors (3).

In this assay, transwells with a membrane in the bottom are placed in normal wells. These transwells are coated with Matrigel basement membrane matrix. This is a solubilized basement membrane preparation extracted from the Engelbreth-Holm-Swarm (EHS) mouse sarcoma, which is a tumour rich in extracellular matrix proteins. Its major component is laminin, followed by collagen IV, heparan sulfate proteoglycans, entactin, and nidogen (4). Furthermore, it contains several growth factors that occur naturally in the EHS tumor, like transforming growth factor (TGF)-β, fibroblast growth factor and tissue plasminogen activator (5). On top of this matrigel the cells are seeded in serum-free medium (in the inner compartment). In the outer compartment medium with fetal calf serum (FCS) and normal human serum is added. The cells are incubated for three days and during this time they can grow through the matrigel and the membrane (*see* **Fig. 1**).

2. Materials

1. Transwells 8 μm pore width, 6.5-mm diameter (Costar, Cambridge NY, USA).
2. Sterile 24-well plate (to place transwells in).
3. Sterile paraffin.
4. Toothpicks.
5. Matrigel (matrigel basement membrane matrix) (Becton Dickinson). Store at –20°C. Always keep on ice, because it solidifies above 4°C.
6. Iscove's Modified Dulbecco's medium (IMDM) (Biowhittaker).
7. 100X Glutamax I: L-analyl-L-glutamine in 0:85% NaCl (GibcoBRL).
8. 100X Penicillin and streptomycin: 10,000 U and 10,000 μg/mL respectively (Boehringer Mannheim).
9. FCS (GibcoBRL).
10. Normal human serum (NHS).
11. 0.25% Trypsin in Gibco solution A (GibcoBRL).
12. Sterile phosphate buffered saline (PBS).
13. 70% Ethanol.
14. 2% Glutaraldehyde in PBS.
15. 1% Crystal violet in water.

3. Methods

The methods described below outline coating of sides of transwells with paraffin, coating of transwells with matrigel, seeding of matrigel coated

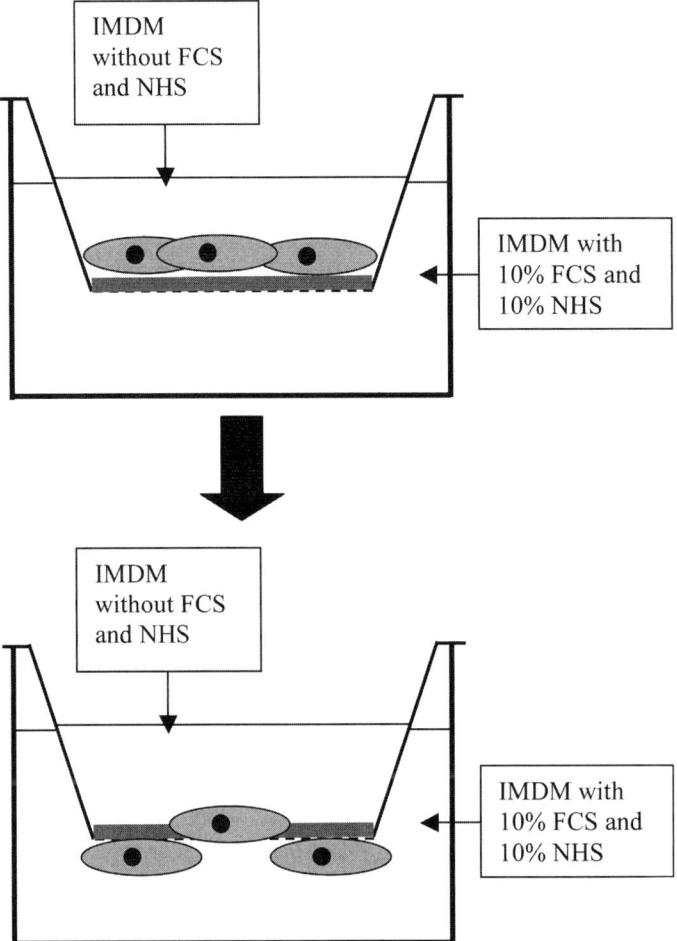

Fig. 1. Schematic view of cells growing through Matrigel and membrane of transwell insert.

transwells, fixing and staining of the cells, and measuring the invasiveness of the cells.

3.1. Coating Sides of Transwells With Paraffin

1. Autoclave the toothpicks to sterilize them.
2. Warm the solid paraffin up in between 60 and 70°C until all paraffin has melted.
3. Coat the tips of the sterile toothpicks with paraffin by putting the toothpicks in the molten paraffin. Be sure that the tip of the toothpick is completely covered in paraffin (see **Note 1**).

4. Let the paraffin on the toothpick cool down again before putting it aside. Be sure to keep the paraffin and toothpick sterile. Use sterile tweezers to handle the toothpicks and keep the tweezers sterile by putting them in 70% ethanol. Coating of the walls of the transwells with paraffin is important to avoid meniscus formation.
5. Rub the inside walls of the transwells carefully with the paraffin coated toothpicks. This leaves a thin layer on the inside of the transwells. Be sure that also the walls just above the membrane is coated, but do not touch the membrane, because it breaks easily (*see* **Note 2**).
6. When the transwells are coated with paraffin, add 100 µL IMDM inside the wells and incubate for 30 min in an incubator at 37°C 5% CO_2. This is to moisten the membrane.

3.2. Coating of Transwells With Matrigel

1. Thaw the matrigel on ice (*see* **Note 3**). Dilute the matrigel in ice cold IMDM to a concentration of 0.375 mg/mL using ice cold pipettete tips (keep them in the freezer). Pipet the matrigel slowly up and down for five or six times and mix it with the IMDM (*see* **Note 4**). The matrigel is now ready for use.
2. Aspirate the IMDM from the transwells thoroughly, but be careful not to go through the membrane (*see* **Note 5**).
3. Add 100 µL of the diluted matrigel to the inside of each transwell. Use ice cold pipet tips and use a fresh one for every well. Pipet the diluted matrigel slowly up and down five or six times before adding to the transwell. Avoid air bubbles because such bubbles gives holes in the matrigel layer leading to spurious results.
4. Keep the plate without the lid overnight in a full functioning laminar flow cabinet to let the matrigel solidify and the remnants of IMDM dampen. This leaves a thin layer of matrigel on the bottom of the transwells (*see* **Note 6**).

3.3. Seeding of Cells on Matrigel-Coated Transwells

1. Warm PBS, trypsin, IMDM with 10% FCS, IMDM without 10% FCS, FCS and NHS to 37°C.
2. Wash transwells by adding 100 µL of IMDM on the inside of the transwell. Incubate for at least 1 h, in order to remove any proteins or other chemicals dissolved in the IMDM used to dilute matrigel in.
3. Make IMDM 10% FCS and 10% fresh NHS. Filter the medium through a filter with a pore size of 0.22 µm. Keep it at 37°C.
4. Wash cells with PBS, add 3 mL 0.25% trypsin (T75 flask) and incubate for 5 to 7 min in an incubator at 37°C and 5% CO_2. Knock the flask gently against your hand palm to be sure all cells are detached. Check under the microscope that all cells have detached (*see* **Note 7**).
5. Add 7 mL IMDM with 10% FCS, rinse flask, and bring cells over in 50-mL tube. Rinse flask with 5 mL IMDM 10% FCS.
6. Centrifuge cells for 10 min at 1500 rpm at 20°C. Aspirate supernatant and resuspend cells in 5 mL IMDM. Count cells under a light microscope using a counting

Invasion Assay

chamber (Bürker). Dilute the cells 1:1 in trypanblue, to exclude dead cells and count 50 small squares in duplicate. Multiply the mean number of cells by 10,000 and then by the total volume of cell suspension (5 mL) (*see* **Note 8**).

8. Dilute the cell suspension to a concentration of 100,000 cells/mL and add 200 µL of the diluted cellsuspension to the inner compartment of the transwell (on top of the matrigel). Add 900 µL of the freshly made IMDM of 10% FCS and 10% NHS to the outer compartment of the transwell (*see* **Note 9**).
9. Incubate the cells for 3 d in an incubator at 37°C and 5% CO_2.

3.4. Fixing and Staining of the Cells

1. Place the transwells in clean empty wells. Aspirate the medium from the inside of the transwell as well as any large droplets on the outside (underneath) the transwells. Add 1 mL 2% glutaraldehyde in PBS to the transwells; apply a few drops in the inner compartment and the rest in the outer compartment. Incubate for 30 min at room temperature.
2. Aspirate the glutaraldehyde from the inner and outer compartments. Wash the inside and the outside of the transwells once with PBS (approx 2 mL/well applied to the inside of the transwell will flow over through the gaps to the outside of the well).
3. Add 1 mL 1% crystal violet to an empty well, aspirate PBS from the inner and outer compartment of the transwell and put the transwell in the well with crystal violet. Incubate again for 30 min at room temperature.
4. Place the transwells in an empty well. The crystal violet can be reused. Wash the transwells thrice with PBS (*see* **Note 10**). When the crystalviolet stain has been removed by washing, aspirate the PBS and remove the matrigel layer on the inside of the transwells by gently rubbing the upper side of the membrane with a cotton bud.

3.5. Measuring the Invasiveness of the Cells

1. Cells that have grown through the matrigel and through the filter can be seen through a light microscope (*see* **Fig. 2** for examples). To assess the invasiveness of each cell preparation, the cells have to be counted. This can be done by an image analyzer. However, this can give problems with the pores of the filter.
2. Another method to count the cells is to count nine visions of the ocular (*see* **Fig. 3**). In this way almost all cells can be counts (*see* **Note 11**).
3. A third method to count the cells, is to use a counting grid and count ten - fifteen grids at random and correct the mean of these numbers for the surface size of the membrane.

4. Notes

1. If the tip of the toothpick is not completely covered with paraffin it is difficult to coat the walls of the transwells just above the membrane with paraffin. It is important to coat just above the membrane because of meniscus formation when fluid levels are low (100 µL) should be avoided. Also be sure that no big droplets

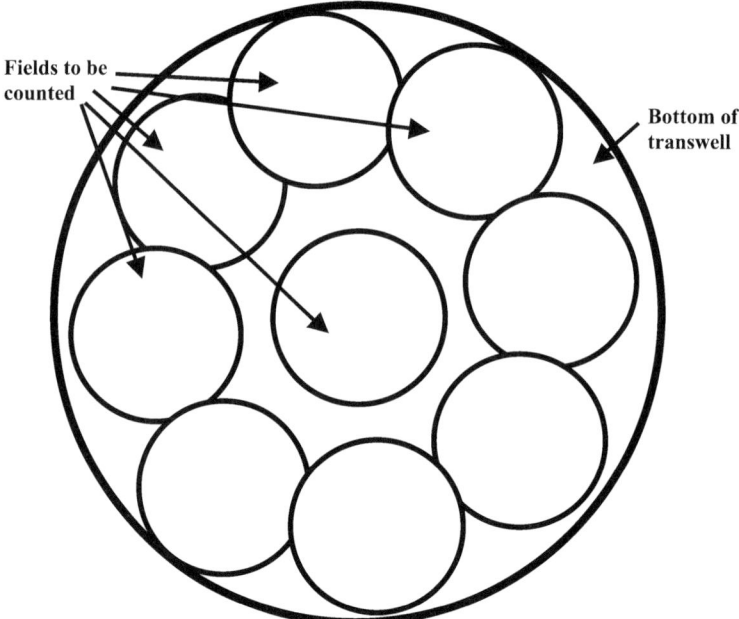

Fig. 2. Bottom of transwell divided into nine ocular views in which the cells can be counted.

are hanging under the tip of the toothpick. These drops can break when the toothpick is knocked against the wall of the transwell and it is more difficult to coat just above the membrane.
2. To avoid rupturing of the membrane, it is important to work carefully when handling the transwells. During the coating of the transwells with paraffin hold the transwells between 2 or 3 fingers of one hand. Be sure hands are sterile or wear gloves that are sterilized with 70% ethanol. Handle the toothpick with the other hand. Rest your elbows on the working bench of the laminar flow cabinet, with your hands at eye height. When you add medium (or any other liquid) to the inside of the transwells, it should stay on top of the membrane. If the fluid leaks through the membrane into the outer compartment, the membrane has ruptured. This sometimes shows immediately, but may only become apparent after 15 to 30 min. These transwells should be discarded.
3. Matrigel solidifies above 4°C, so be sure that the matrigel is kept on ice at all times. Do not thaw in the refrigerator as this tends to cause fluctuations in temperature. To avoid repeated freeze-thaw cycles, pippette the matrigel out into small aliquots of about 200 to 300 µL. Thaw the flask with matrigel overnight on ice in the refrigerator. Color variations may occur in frozen or thawed vials of matrigel, but this does not affect product efficacy. The color may range from straw yellow to dark red. This results from the interaction of carbon dioxide with the bicarbonate buffer and phenol red.

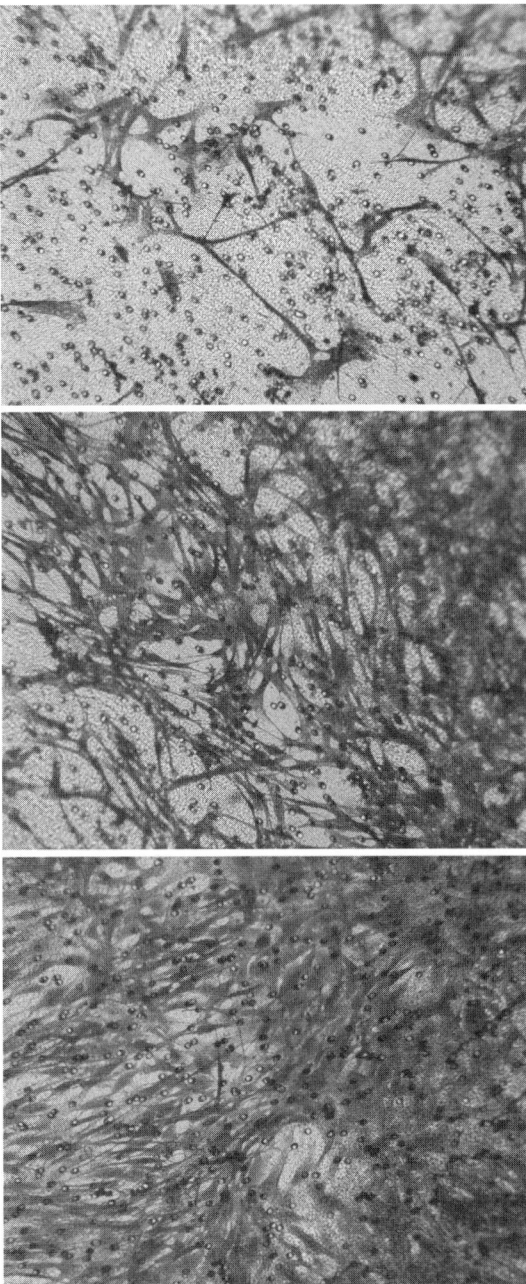

Fig. 3. Example of invaded cells. Upper panel: hardly invasive. Middle panel: intermediately invasive. Lower panel: very invasive. The small circles scattered in the photographs are the pores in the membrane. The cells were fixed with 2% glutaraldehyde in PBS and stained with 1% crystalviolet.

4. Swirl the vial with matrigel when it is thawed to be sure that material is evenly dispersed. Be sure to pipet slowly, because high concentrations of matrigel are viscous the solution may be sucked rather slowly into the pipette tip. Also, swirl the vial with diluted matrigel just before use to evenly disperse material. Be sure to keep matrigel cold at all times.
5. When the medium is aspirated be sure that the transwell is not sucked up, this can rupture the membrane. If it is possible, use a crystal tip (pipet tip 0.5–10 µL) to decrease suction. Always apply pressure to the edge of the transwell to avoid the transwells being sucked up.
6. Be sure that the matrix is dry before adding medium. If it takes too much time for the matrix to dry, turn on the heater in the room or use another safe source as heater keeping the plate sterile at all times. Do not incubate in an incubator, because the air inside is moist and this inhibits the evaporation of the fluid.
7. If the cells have not completely detached, incubate for longer periods at 37°C, but avoid death of the cells. Do not incubate longer than 15 min. If the cells are still not detached after prolonged incubation use a cell scraper to detach the cells. Be sure that the trypsin has not been thawed for too long, as this influences enzymatic activity.
8. If another dilution is used or other parameters are changed the following equation should be used to obtain the number of cells: number of cells in suspension = number of cells counted per 25 squares the dilution 10^4 volume of suspension.
9. The transwells have holes in the upper side of the walls. These can be used to fill the outer compartment, so the transwells do not have to be removed to fill the outer compartment.
10. To avoid staining of the aspiration hose and other equipment do not aspirate the transwells after the first wash, but empty the transwells thoroughly on a tissue and put the transwells in a clean well in a new plate. Discard the old plate with the PBS from the first wash. In the new plate the wells can washed two or three times according to how much crystal violet is removed after each wash.
11. When only a small amount of cells have grown through the matrix and the membrane, it is easy to count all the cells. However when many cells have grown through the matrix and the membrane and cells are (almost) confluent under the membrane counting cells may be more difficult. As a rule of thumb, we assessed that when a field under the microscope is confluent, it contains about 1100 cells. The maximum number of cells under the filter is about 10,000. When one is more experienced in counting the cells, it is easier to see how many cells are in one field. It is also possible to assign scores to the invasiveness, this avoids counting the cells.

References

1. Firestein, G. S. (1996) Invasive fibroblast-like synoviocytes in rheumatoid arthritis. Passive responders or transformed aggressors? *Arthritis Rheum.* **39**, 1781–1790.

2. Muller-Ladner, U., Kriegsmann, J., Franklin, B. N., et al. (1996) Synovial fibroblasts of patients with rheumatoid arthritis attach to and invade normal human cartilage when engrafted into SCID mice. *Am. J. Pathol.* **149,** 1607–1615.
3. Repesh, L. A. (1989) A new in vitro assay for quantitating tumor cell invasion. *Invasion Metastasis* **9,** 192–208.
4. Kleinman, H. K., McGarvey, M. L., Liotta, L. A., et al. (1982) Isolation and characterization of type IV procollagen, laminin, and heparan sulfate proteoglycan from the EHS sarcoma. *Biochemistry* **21,** 6188–6193.
5. McGuire, P. G. and Seeds, N. W. (1989) The interaction of plasminogen activator with a reconstituted basement membrane matrix and extracellular macromolecules produced by cultured epithelial cells. *J. Cell Biochem.* **40,** 215–227.

28

Culture and Analysis of Circulating Fibrocytes

Timothy E. Quan and Richard Bucala

Summary

Fibrocytes circulate in the peripheral blood, produce collagen and other matrix proteins, and express cell surface markers indicative of a hematopoetic origin distinguishing them from fibroblasts. Circulating fibrocytes were first identified in 1994 in a model system of wound repair, and defined by their growth characteristics and unique surface phenotype. The methods currently employed for the isolation, growth, and characterization of peripheral blood fibrocytes rely on the entry of "fibroblast-like" cells into wound chambers, or the derivation of "fibroblast-like" cells from the buffy coat of peripheral blood obtained from different mammalian species. In this protocol, we culture fibrocytes from the mononuclear cells of peripheral blood and harvest the cultured cells for flow cytometry analysis.

Key Words: Antigen presenting cell; fibrocyte; fibrosis; $CD34^+$; $CD45^+$; collagen; culture.

1. Introduction

Circulating fibrocytes were first identified in 1994 as a result of studies in a model system of wound repair, and defined by their growth characteristics and unique surface phenotype *(1,2)*. Fibrocytes produce collagen and other matrix proteins, but they also circulate in the peripheral blood and express cell surface markers indicative of a hematopoetic origin. This new leukocyte subpopulation hence was termed "fibrocytes," which combines the Greek "kytos" referring to cell, and "fibro", which is from the Latin denoting fiber (*see* **Figs. 1** and **2**).

In older literature, references to so-called "blood-borne fibroblasts" and "fibroblast-like cells" exist and likely represent the first observations of cells with the current, molecular-defined features of circulating fibrocytes. Indeed, the potential derivation of matrix-producing cells from the peripheral blood was discussed for almost 150 yr beginning with the work of Paget in 1863, Conheim in 1867, Fischer in 1925, and Maximov in 1928 *(3,4)*. It is likely that

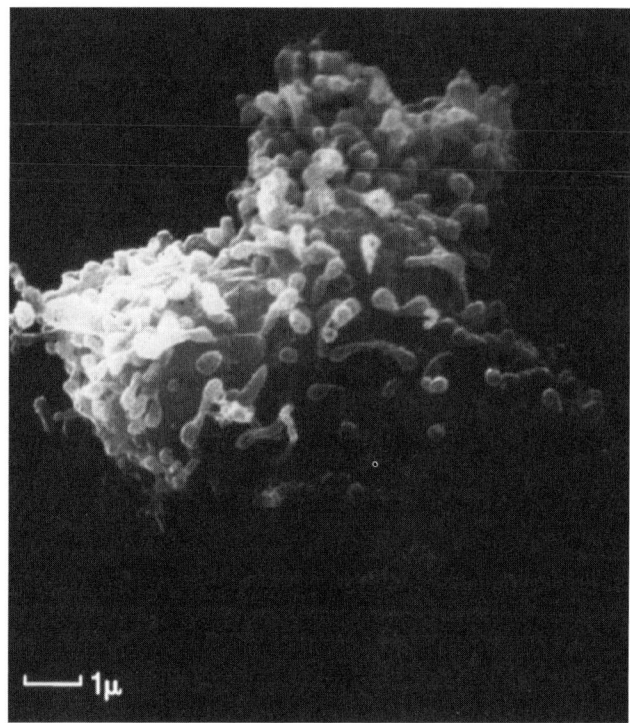

Fig. 1. Scanning electron microscopy of a fibrocyte. (Reprinted with permission from **ref. 2**.)

in experimental studies of wound repair that go back to the 1940s, the cells in the circulating blood that appeared capable of producing connective tissue were indeed "fibrocytes" (4). In one such study, Stirling and Kakkar used careful cannulation and diffusion chamber techniques to demonstrate that collagen-producing cells were not contaminants dislodged from the blood vessel wall, but indeed were derived from circulating blood elements (4). Fibrocytes are collagen producing, marrow derived, and found in the peripheral circulation, typical features found in tissues of mesenchymal origin.

Cells of mesenchymal origin are the precursors to numerous constituents of structural, supportive, blood, cardiac, and bone tissue. In structural and supportive tissue, these cells typically have an irregular star (stellate), or spindle (fusiform) shape and delicate branching cytoplasmic extensions that form an interlacing network throughout the tissue. It is generally considered that mesenchymal cells can mature into tissue fibroblasts (5). Fibroblasts play a major role in wound repair and although morphologically similar to fibrocyte are *not* of hematopoietic origin. For a number of decades, there was an active debate

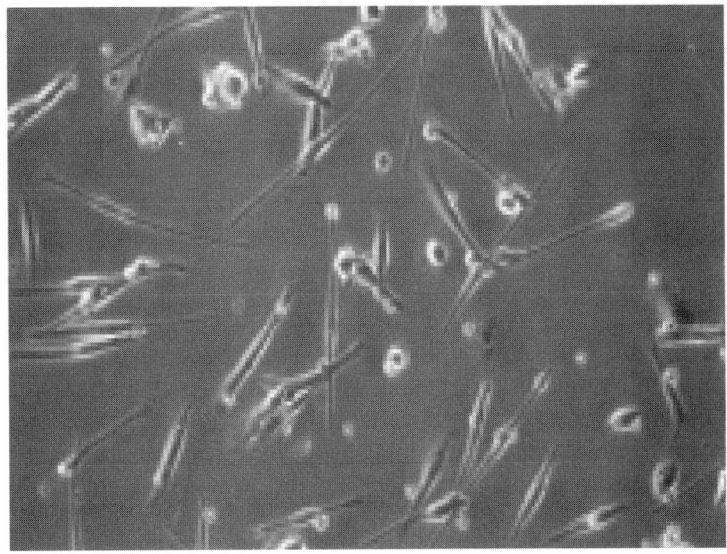

Fig. 2. Light microscopic view of a fibrocyte culture. (Reprinted with permission from **ref. 1**.)

regarding the extent to which connective tissue scar was the result of an ingrowth of adjacent mesenchymal or fibroblast-like elements vs the hematogenous entry of circulating, fibroblast precursors *(6)*.

Since the original description of circulating fibrocytes in 1994, our knowledge of this unique cell population has grown steadily. Whereas initially described in the context of wound repair, fibrocytes have since been found to participate in granuloma formation, antigen presentation, and various fibrosing disorders. Fibrocytes produce matrix proteins such as vimentin, collagens I and III, and they participate in the remodeling response by secreting matrix metalloproteinases. Fibrocytes also are a rich source of inflammatory cytokines, growth factors, and chemokines that provide important intercellular signals within the context of the local tissue environment. Moreover, fibrocytes express the immunological markers typical of an antigen-presenting cell (APC), and they are fully functional for the presentation of antigen to naïve T-cells. Fibrocytes can further differentiate, and they may represent a systemic source of the contractile myofibroblast that appears in many fibrotic lesions. Clinically, there is evidence that patients with hypertrophic scars such as keloids, and those affected by scleroderma and other fibrosing disorders have fibrocytes in their lesions. Recently, a new disease entity called nephrogenic fibrosing dermopathy (NFD) has been described, and the fibrocyte may play an important etiopathogenic role in disease development. Nephrogenic fibrosing dermopathy occurs in patients with renal insufficiency and leads to thickening and hardening of the skin, especially of the extremities. Ongoing research is focusing on the molecular signals that influence fibrocyte migration, proliferation, and function in the context of normal physiology and pathology.

2. Materials

1. 23 g 3/4 12" tubing with multiple sample Luer Adapter (Becton Dickinson).
2. Sodium heparin blood collection tubes (Becton Dickinson)
3. Dulbecco's modified Eagle's medium (DMEM) Media (Gibco).
4. Heat inactivated fetal calf serum(Gibco).
5. Phosphate buffered saline (PBS) (Gibco).
6. 0.5% Trypsin-ehtylene diamine tetraacetic acid (EDTA) (Gibco).
7. Cell scraper (Becton Dickinson Labware).
8. Human fibronectin cellware 6-well plate (Becton Dickinson Labware).
9. Falcon culture slide (cat. no. 354101) 1 chamber polystyrene vessel tissue culture treated glass slide (Becton Dickinson Labware).
10. Fibronectin F2006 (Sigma).
11. Ficoll-paque (Pharmacia).
12. FACScalibur flow cytometer (Becton Dickinson).
13. Flow Jo software (Tree Star Inc,).
14. 4-mL Facs tubes (Becton Dickinson Labware).
15. Anti-human CD34-PE, anti-human CD45-PE, anti-mouse IgG-APC (Becton Dickinson).
16. Anti-collagen I AB 765 (Chemicon).

3. Methods

The methods currently employed for the isolation, growth, and characterization of peripheral blood fibrocytes are similar to the ones described in older literature *(1)*. The present techniques rely on the entry of "fibroblast-like" cells into wound chambers, or the derivation of "fibroblast-like" cells from the buffy coat of peripheral blood obtained from different mammalian species. We use the latter to harvest cells for flow cytometry.

3.1. Collection of Peripheral Blood Monocytes

1. Peripheral blood is drawn from the ante-cubital vein of a blood donor. Universal precautions with protective eyewear and gloves is used by the phlebotomist.
2. The arm of the subject is hyper-extended with the arm below the level of the heart.
3. A tourniquette is placed over the upper arm tight enough to occlude venous return of blood.
4. The area is cleaned with isopropanol wipe.

5. A 23-Gage butterfly needle (bevel up) is inserted into the middle of the antecubital vein following the course of the vein (*see* **Note 1**).
6. The other end of the vacutainer butterly system is placed into the vacuum sodium heparin tube. Thirty milliliters of blood is collected.
7. The tubes are immediately rotated to mix the heparin and blood.
8. The blood can sit for 24 h at room temperature on its side if not immediately used.
9. Once blood draw is complete, the tourniquette is released. The needle is quickly removed. Direct pressure with gauze is applied for 2 to 3 min over the entry site.

3.2. Culturing and Harvesting of Fibrocytes

1. Fibronectin is dissolved at a concentration of 10 µg/mL in PBS. The solution is filter sterilized.
2. One ml of solution is used to cover the bottom of 1 mL chamber slides, TI-75 flask or 2 mL wells of 6-well plate.
3. Wells are coated with fibronectin and subsequently incubated for 1 h at 37°C or overnight at 4°C (*see* **Note 2**). All labware is kept sterile at 4°C.
4. The heparinized blood sample is diluted 3:1 with PBS.
5. Fifteen milliliters of Ficoll Hypaque is slowly layered underneath the diluted blood in a Falcon 50-mL polypropylene tubes (*see* **Fig. 3**. and **Note 3**).
6. Centrifugation is performed at 400g at room temperature with the break off.
7. The leukocyte-rich buffy coat is then removed, washed twice in 3 to 4X volume PBS (*see* **Note 3**).
8. The peripheral blood mononuclear cells (PBMC) are then resuspended in DMEM supplemented with 10%, heat-inactivated FBS (*see* **Note 4**).
9. The PBMCs are counted on a hemocytometer and the resulting mononuclear cells then are plated onto fibronectin coated tissue culture wells or flasks and grown in DMEM supplemented with 10% heat-inactivated FCS (*see* **Note 4, step 4**).
10. The 2 mL of cells are plated out at a minimum concentration of 2.5 million cells/well in a 2 mL 6-well plate.
11. After 2 d, the nonadherent cells (largely T-cells) are aspirated off, and the remaining adherent cells cultivated for 14 d. Over time, the contaminating monocytes die off, and fibrocytes appear as clusters of stellate, elongated, or spindle-shaped cells that show long cellular processes. After a prolonged period of culture, fibrocytes then begin contracting into a fusiform shape (*see* **Fig. 4**).
12. Fibrocytes are very adherent. Ice cold 0.05% EDTA in PBS was used by Chesney et al. to detach the cells from the plastic surface. We have found that trypsin-EDTA 0.05% aids in the detachment of the cells. Incubate the plated cells just covering the surface for 1.5 min at 37° C trypsin–EDTA .05% buffer.
13. Horizontal shear force is applied to the cells by slapping the side of the tissue culture flask.
14. Immediately media with 10% FCS is added (*see* **Note 5**). Cells are then lightly scraped and harvested.
15. They are immediately washed in PBS.

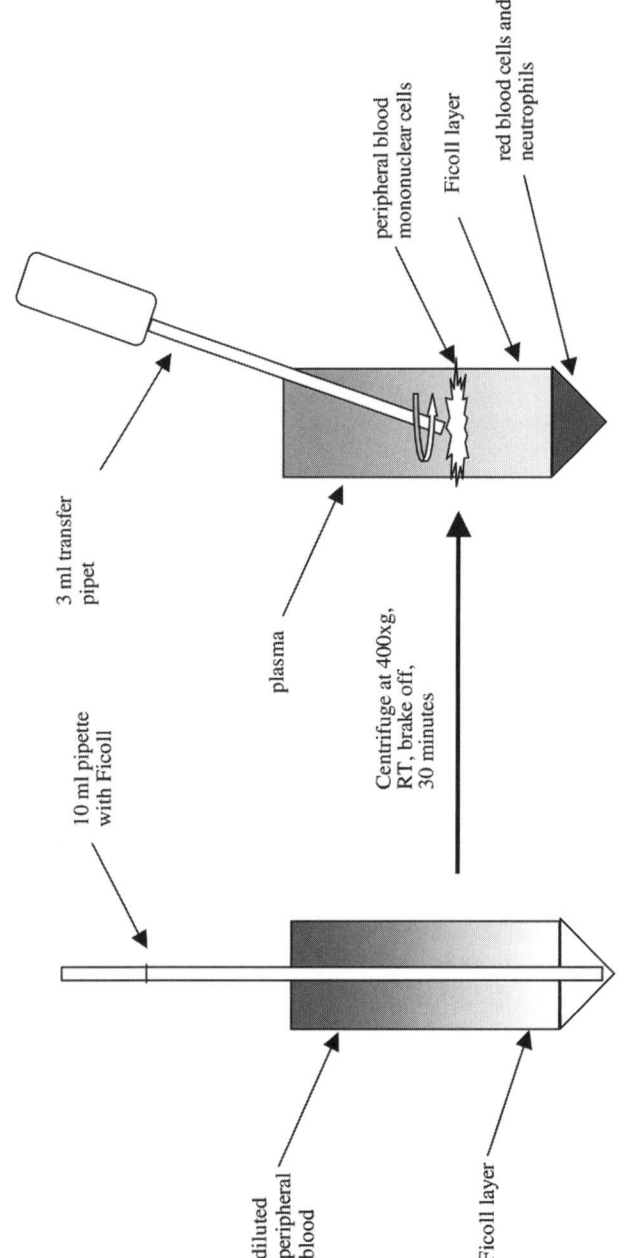

Fig. 3. Underlaying of Ficoll and removal of buffy coat.

Culture and Analysis of Circulating Fibrocytes

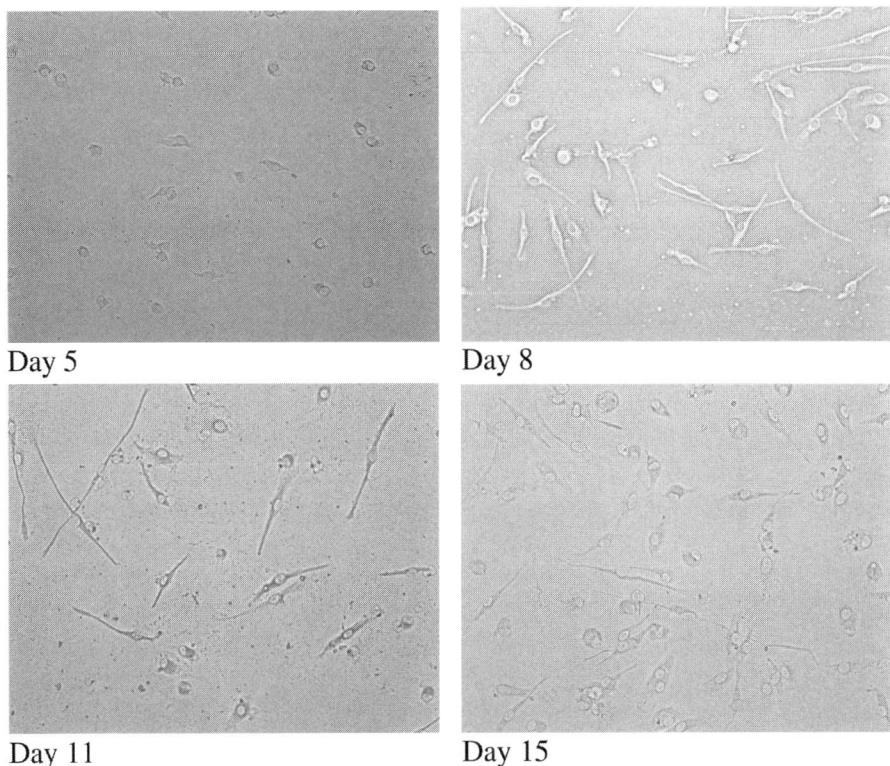

Fig. 4. Fibrocytes grown for 15 d from healthy control. Light microscopy of fibrocytes over 5, 8, 11, and 15 d. Note the change in morphology.

3.3. Characterization of Fibrocytes

1. Harvested cells are used for flow cytometric characterization. Approximately 0.5×10^6 cells are spun down in 4-mL Facs tubes.
2. The liquid is poured off and the remaining (approx 180 μL) is mixed with the fluorochrome-labeled antibodies as described below.
3. The cells are washed in PBS, and 2 μL of mouse anti-collagen AB are incubated with the cells for 30 min. The cells are washed in PBS.
4. Secondary goat anti-mouse AB linked to PE is then added for 30 min. The cells are washed twice with 3 mL of PBS.
5. One microliter of CD34PE or 1 μL of CD45-PE is added to stain surface markers (Becton Dickinson Pharmingen). The cells are incubated for 30 min at 4°C.
6. Cells are washed and then fixed in 200 μL of 2% paraformaldehyde.
7. The cells are analyzed on a Facscalibur flow cytometer and quantified by Flojo Software (Treestar) (*see* **Note 7**).

3.4. Future Directions

Studies performed over the last ten years have provided strong support for the fibrocyte as a collagen-producing cell of the peripheral blood. The accumulated evidence also indicates that circulating fibrocytes play an important role in the inflammatory and proliferative phases of wound repair. This is evidenced by the early migration of fibrocytes into wounds, and by their regulated production of both matrix proteins and growth regulating cytokines. Fibrocytes also have been identified in granulomas, hypertrophic scars, and in the bronchial lesions of asthma. Our current molecular definition of fibrocytes by flow cytometry and expression studies reinforces many decades of observations, some made as far back as the mid-nineteenth century, that have supported a blood-borne source for a connective tissue cell population *(3,4)*.

The recent emergence of NFD has served to focus much attention on the potential pathologic role of fibrocytes in this disease. Whether NFD is a disease of aberrant fibrocyte activation, trafficking, or synthetic function remains unknown, and there is at present very little information about the precise etiopathogenesis of this disorder. Conceivably, there may exist exogenous stimuli for fibrocyte activation that lead to the pathologic features of NFD. Identifying such stimuli would be of obvious interest in NFD and other fibrosing conditions such as scleroderma, asthma, and hypertrophic scarring.

Future studies will need to better define the fundamental properties of fibrocytes over time in different inflammatory and tissue settings. The reduction in the cell surface expression of the CD34 antigen both in vitro and in several in vivo contexts may reflect an important phenotypic transition of the fibrocyte that may be related to differentiation, α-smooth muscle actin expression, or other phenomena specific to a particular tissue microenvironment. Cytokines such as TGF-β, interleukin-1, and SLC influence fibrocyte function, proliferative potential, differentiation into myofibroblasts, and trafficking properties. Whether the fibrocyte explains the origin of the myofibroblast, which features prominently in many pathologic fibroses, is also of great interest in better understanding many chronic diseases.

The precise role of fibrocytes in antigen presentation and in adaptive immunity, and their capacity to support angiogenesis are additional functional properties that will require further investigation in vivo. The ability to expand fibrocytes ex vivo for therapeutic readministration may prove to be of clinical utility, as has been suggested by one clinical study of fibrocytes for immunotherapy in metastatic cancer. Finally, closer examination of the molecular signals that control fibrocyte activation and trafficking may prove instructive for unraveling the etiopathogenesis of a number of different fibrosing conditions, such as NFD, scleroderma, asthma, and other diseases in which fibrosis contributes to pathological manifestations.

4. Notes

1. It is important to quickly break through the epidermis where most pain occurs. Then slowly enter the vein stopping after flash back and advancement by 1 to 2 mm more.
2. Fibronectin pre-coating of the culture surfaces increases the yield and better sustains the growth of fibrocytes. However, the use of this matrix also makes harvesting cells from the plate much more difficult.
3. Because the Ficoll is a density-gradient, it is very important to add the Ficoll slowly so that it has minimal mixing with the diluted blood above.
4. This step is extremely tricky for the novice (*see* **Fig. 3**). One must carefully remove the white layer of buffy-coated cells that lies between the plasma and the Ficoll. It is important to remove as little Ficoll as possible with the maximum number of cells. We use a 3-mL Falcon transfer pipet. While gently squeezing the bulb with thumb and index finger, slowly aspirate the white buffy coat with the tip in a deliberate even circular motion. An average yield is 1 to 2 million PBMCs/mL of starting blood from a healthy control). It is important to wash cells immediately as Ficoll can be toxic to cells.
5. Trypsin can be toxic to cells and one may need to change the timing of the trypsin digestion. FCS is used to stop the trypsin digestion by providing excess protein substrate found in the FCS. The number of dead cells should be estimated by trypan blue exclusion. Briefly, equal amount of trypan blue and washed cells are mixed together and counted on a hemocytometer. Cells that are dark blue are dead and are excluded.
6. Flow cytometric studies confirmed that peripheral blood fibrocytes express the CD34 cell surface antigen. CD34 is a 110-kD integral membrane glycoprotein that was initially believed to be expressed exclusively on hematopoietic stem cells, including various myeloid and lymphoid progenitor cells *(2,7)*. We now know that embryonic fibroblasts, endothelial cells, and bone marrow stromal cells also express CD34 *(7,8)*. Studies in different laboratories have affirmed the utility of CD34 as a marker for identifying fibrocytes, although it has become apparent that fibrocyte expression of CD34 decreases over time, both in culture, and under certain conditions in vivo *(2,9–13)* Additional cell surface markers that are indicative of the hematologic origin of fibrocytes include CD11b, CD45, HLA-DR, CD71, CD80, and CD86 (**Table 1**). It is likely that the *in situ* environment influences the durability of fibrocyte CD34 expression. For example, fibrocytes derived from wound chambers, which likely represent an inflammatory milieu, maintain CD34$^+$ expression over time *(2)*. More recent studies have utilized additional markers to identify fibrocytes, such as Type I procollagen *(14)*. The minimum criteria of the coexpression of collagen production and uniquely hematologic markers is likely sufficient to describe fibrocytes in different pathologic lesions. Other commonly used markers for fibrocytes include the pan-leukocyte antigen CD45, and HLA-DR, which reflects the antigen presenting ability of these cells *(2,15)*. Other markers of connective tissue matrix production, such as vimentin, and prolyl 4-hydroxy-

Table 1
Cell Markers of Fibrocytes

Markers	Function	Other cell types
α-Smooth muscle actin	Contractile element	Myofibroblasts
Vimentin	Intermediate filament	Fibroblasts
Fibronectin	Pericellular matrix	Mesenchymal cells
Collagen I	Extracellular matrix	Fibroblasts, smooth muscle, connective tissue
Collagen III		
Type I procollagen	Secreted unprocessed precursor of collagen I	Fibroblasts
Prolyl 4-hydroxylase	Collagen hydroxylation	Fibroblasts
CD11a (LFA-1)	αL subunit of integrin LFA-1, adhesion molecule	Lymphocytes, granulocytes, monocytes, granulocytes
CD11b (MAC-1)	αM subunit of integrin CR3, adhesion molecule	Monocytes, granulocytes, NK cells
CD13 (Aminopeptidase N)	Pan-myeloid antigen	Myelomonocytic cells
CD34	Hematopoietic stem cell, capillary endothelial marker	hematopoetic progenitors, embryonic fibroblasts
CD18	β2 integrin	Myeloid and DC
CD45RO	Leukocyte Common Antigen	Leukocytes
CD54 (ICAM-1)	Intracellular adhesion molecule binds LFA-1 and Mac-1	Hematopoetic and nonhematopoetic cells
CD58 (LFA-3)	Adhesion molecule, binds CD2	Hematopoetic and nonhematopoetic cells
CD71	Transferrin receptor	Macrophages, activated cells
CD86 (B7.2)	Costimulatory molecule binds CD28	Antigen presenting cells
CD80 (B7.1)	Costimulatory molecule binds CD28	Antigen presenting cells
MHC II DP, DQ, DR	Major histocompatibility molecule for antigen presentation	Antigen presenting cells
CXCR4 (CD184)	Receptor for stromal derived cell factor (SDF)	Immature CD34+ hematopoetic cells
CCR3	Receptor for Secondary Lymphoid Chemokine (SLC)	lymphocytes
CCR5 (CD195)	Receptor for a CC type cytokine	Promyelocytic cells
CCR7 (CDw197)	Mediates lymphocyte function, receptor for MIP-3β	Activated lymphocytes

lase, also have been employed to identify fibrocytes in hypertrophic scars and keloids *(12)*.

Acknowledgments

These studies were supported by the Scleroderma Foundation (RB).

References

1. Chesney, J. and Bucala, R. (2001) Peripheral Blood Fibrocytes. In: *In Human Cell Culture*, Vol. 5. B. O. Pallsson, M. R. Koller, and J. Masters, eds. Kluwer Academic Publishers, London, p. 209–219.
2. Bucala, R., Spiegel, L. A., Chesney, J., Hogan, M., and Cerami, A. (1994) Circulating fibrocytes define a new leukocyte subpopulation that mediates tissue repair. *Mol Med* **1,** 71–81.
3. Paget, J. (1863) Lectures on Surgical Pathology. In *Royal College of Surgeons of England*. Longmans, London.
4. Stirling, G. A. and Kakkar, V. V. (1969) Cells in the circulating blood capable of producing connective tissue. *Br. J. Exp. Pathol.* **50,** 51–55.
5. Burkitt, Y. and Heath, H. (1997) Wheater's Functional Histology, Vol. 1. Churchill Livingston, London, p. 404.
6. Dunphy, J. (1963) The Fibroblast-A unique Ally for the Surgeon. *New Engl. J. Med.* **268,** 1367–1377.
7. Brown, J., Greaves, M. F., and Molgaard., H. V. (1991) The gene encoding the stem cell antigen, CD34, is conserved in mouse and expressed in haemopoietic progenitor cell lines, brain, and embryonic fibroblasts. *Int. Immunol.* **3,** 175–184.
8. Fina, L., Molgaard, H. V., Robertson, D., et al. (1990) Expression of the CD34 gene in vascular endothelial cells. *Blood* **75,** 2417–2426.
9. Barth, P. J., Ebrahimsade, S., Ramaswamy, A., and Moll, R. (2002) CD34+ fibrocytes in invasive ductal carcinoma, ductal carcinoma in situ, and benign breast lesions. *Virchows Arch.* **440,** 298–303.
10. Hirohata, S., Yanagida, T., Nagai, T., et al. (2001) Induction of fibroblast-like cells from CD34(+) progenitor cells of the bone marrow in rheumatoid arthritis. *J. Leukoc. Biol.* **70,** 413–421.
11. Barth, P. J., Ebrahimsade, S., Hellinger, A., Moll, R., and Ramaswamy, A. (2002) CD34+ fibrocytes in neoplastic and inflammatory pancreatic lesions. *Virchows Arch.* **440,** 128–133.
12. Aiba, S. and Tagami, H. (1997) Inverse correlation between CD34 expression and proline-4-hydroxylase immunoreactivity on spindle cells noted in hypertrophic scars and keloids. *J. Cutan. Pathol.* **24,** 65–69.
13. Aiba, S., Tabata, N., Ohtani, H., and Tagami, H. (1994) CD34+ spindle-shaped cells selectively disappear from the skin lesion of scleroderma. *Arch. Dermatol.* **130,** 593–597.
14. Schmidt, M., Sun, G., Stacey, M. A., Mori, L. and Mattoli, S. (2003) Identification of circulating fibrocytes as precursors of bronchial myofibroblasts in asthma. *J. Immunol.* **171,** 380–389.

15. Chesney, J., Bacher, M., Bender, A., and Bucala, R. (1997) The peripheral blood fibrocyte is a potent antigen-presenting cell capable of priming naive T cells in situ. *Proc. Natl. Acad. Sci. USA* **94,** 6307–6312.
16. Kirchmann, T. T., Prieto, V. G., and Smoller, B. R. (1994) CD34 staining pattern distinguishes basal cell carcinoma from trichoepithelioma. *Arch. Dermatol.* **130,** 589–592.

Index

A

Actin, ring formation detection in osteoclasts, 295
ADAMTS, *see* Aggrecanase
Aggrecan, degradation fragment analysis from cartilage explants,
  dimethyl methylene blue assay,
    advantages, 205
    limitations, 205
    staining, 203
    standard curve, 204, 205
  explant culture, 202, 203, 208
  materials, 202, 208
  overview, 201
  uronic acid analysis,
    advantages, 207, 208
    incubation conditions, 206, 208
    limitations, 206, 207
    sensitivity, 206
Aggrecanase,
  catalytic mechanism, 168
  immunohistochemical analysis,
    blocking and antibody incubation, 172
    cartilage explant studies, 177, 178
    chondroitinase ABC treatment effects, 173–177
    controls, 179
    counterstaining, 173
    detection, 173
    hydrogen peroxide treatment effects, 172–177
    materials, 170, 171
    overview, 168, 170
    pretreatment, 172
    processing, 171, 172
    product development, 173
    specimen preparation, 171, 178, 179
  types in cartilage, 168
Angiogenesis,
  assessment in joint,
    arthroscopy,
      large joint, 347, 354
      materials, 344
      small joints, 347, 348
    ex vivo assessment,
      endothelial cell tubule formation, 353–356
      explant culture, 352
      innocyte quantitative cell migration assay, 352–354
      materials, 346
      neutrophil migration in transwell culture plates, 353, 355
    macroscopic assessment, 344, 348
    microscopy,
      dual immunofluorescent staining for vessel maturity analysis, 345, 346
      section preparation, 345, 348, 349
      vascularity analysis, 349–351, 354, 355
  inflammatory arthritis pathogenesis, 344
Apoptosis,
  articular cartilage injury, 185

435

chondrocyte assays,
   fluorescent dye staining, 192–196
   materials, 188–190, 192
   overview, 186–188
   TUNEL assay, 196
morphology, 183, 184
signaling, 184
Arthroscopy,
   angiogenesis assessment,
      large joint, 347, 354
      materials, 344
      small joints, 347, 348
   historical perspective, 27–29
   indications in rheumatology, 31, 32
   instrumentation, 33
   scoring of biopsies, 30, 31
   SFA system, 32
   synovial examination,
      biopsy under direct vision,
         collection, 39
         forceps, 39
         insertion and triangulation, 39
         site selection, 39
         specimen handling, 40
      knee,
         biopsy, 38
         joint cannulation, 36
         lavage, 36, 37, 41
         orientation, 37
         systematic examination, 37, 38
         termination and post-procedure care, 38, 39, 41
      materials, 34, 35
      patient preparation and consent, 35
      portal selection and local anesthesia, 36
      set-up, 35
      skin disinfection and draping, 36
      white balance setting, 36
   synovitis, 29, 30, 33
   training, 40
AtlasImage™ software, synovial fibroblast DNA microarray analysis, 385–387, 389

B

Biopsy, *see* Arthroscopy
Bone histomorphometry,
   animal models of arthritis, 269, 270
   arthritis study techniques, 280–282
   decalcified paraffin-embedded paw sections,
      preparation,
         decalcification and confirmation, 273, 281
         fixation, 273, 281
         paraffin embedding, 273
         paw preparation, 272, 281
         sectioning, 274, 281
      processing, 274
   materials, 270, 271
   undecalcified plastic-embedded paw sections,
      preparation,
         fixation, 276
         plastic embedding, 276, 281
         sectioning, 277, 282
      processing,
         Goldner-Trichome stain, 277, 278
         tartrate-resistant acid phosphatase stain, 278, 280
         toluidine blue stain, 278
         Von Kossa stain, 278
Bone marrow edema, magnetic resonance imaging, 10

C

Cartilage histomorphometry,
   counting particles, 157–161, 163, 165
   quantitative analysis overview, 147, 148
   reference volume, 163, 164
   sampling strategy,
      random sampling , 149
      systematic sampling, 149, 150, 152, 153, 165
   surface area, 154, 155, 157
   surface density, 154, 155, 157
   volume fraction of a tissue component, 153, 154

Cartilage, magnetic resonance imaging, 10, 11
Cathepsin K, immunostaining on osteoclasts, 293, 294
Cell adhesion, see Leukocyte adhesion
Chemotaxis, see Leukocyte recruitment
Chondrocyte,
  apoptosis assays,
    fluorescent dye staining, 192–196
    materials, 188–190, 192
    overview, 186–188
    TUNEL assay, 196
  death role in osteoarthritis, 185
  functions, 227
Collagenase,
  assays,
    bovine nasal cartilage assay,
      cartilage preparation, 218, 219
      dimethyl methylene blue assay for proteoglycan content, 220, 221, 224
      hydroxyproline assay, 219, 220
      incubation conditions, 219
    diffuse fibril assay,
      activation of promatrix metalloproteinases, 218, 224
      calculations, 216, 217
      incubation conditions, 216
      microtiter plate assay, 217, 223, 224
      radiolabeling of substrate, 216, 223
      substrate preparation, 215, 223
      tissue inhibitor of matrix metalloproteinase assay, 218
    materials, 212–215
    overview, 211, 212
  function, 211
  immunohistochemistry,
    cryosectioning, 222, 224, 225
    fixation, 221, 223
    immunostaining, 222, 223
    mounting and freezing, 221
    slide coating, 222

Computed tomography (CT), synovial joint imaging, 18
CT, see Computed tomography

D

DAPI, chondrocyte death detection, 192–194, 196
Digital image analysis,
  file formats, 130, 137
  hardware,
    analog-to-digital converter, 126, 137
    cameras, 123–126, 135
    filter sets, 142, 143
    microscope, 126, 127, 137, 139
    objectives, 127, 137, 141
  image acquisition, 128, 129, 137
  integrated optical density calculation, 134
  masking, 130, 131
  overview, 121–123
  segmentation, 131–134
  signal-to-noise optimization, 129, 130, 137
  software, 127, 137
  specifications in commercial systems, 139, 140
Dimethyl methylene blue,
  bovine nasal cartilage assay of collagenase activity, 220, 221, 224
  glycosaminoglycan degradation analysis,
    advantages, 205
    limitations, 205
    staining, 203
    standard curve, 204, 205
DNA microarray, synovial fibroblast genotyping in rheumatoid arthritis,
  AtlasImage™ software analysis, 385–387, 389
  confirmation of results, 387, 389
  hybridization, 384, 389
  overview, 378, 379

phosphorimaging, 385
probe purification, 384
RNA isolation, 382, 387
washing, 384, 385, 389

E

EC, see Endothelial cell
Electroporation, synovial fibroblast gene transfer,
   materials, 395
   Nucleofector™ system, 398, 400, 410
ELISA, see Enzyme-linked immunosorbent assay
Endothelial cell (EC), characterization,
   activation,
      cytokine treatment, 316
      enzyme-linked immunosorbent assay, 316, 317, 319
      flow cytometry, 317–319
      flow cytometry of surface antigens, 315, 317–319
      low-density lipoprotein uptake, 313, 314
      materials, 307
      microtubule formation, 313
      morphology, 313
      UEA-1 staining, 315, 319
      von Willebrand factor staining, 314, 315
   chemotaxis assay,
      cell counting, 363, 364
      endothelial cell chemotaxis assay, 362–364
      materials, 360, 361, 363
      membrane coating, 361, 363
      monocyte chemotaxis assay, 361–363
   isolation,
      human cells,
         dermal microvascular cells, 307, 308, 318
         human umbilical vein cells, 309, 310, 318, 319, 326
         materials, 306, 307
         mouse cardiac endothelial cells, 311, 312, 319
         pig arterial endothelial cells, 310, 311, 319
   leukocyte interactions, see Leukocyte adhesion; Leukocyte recruitment
   rationale for study, 305, 306
   storage in liquid nitrogen, 307, 312, 313
   tubule formation assay in angiogenesis, 353–356
Enthesitis, magnetic resonance imaging, 9, 10
Enthesopathy, ultrasonography, 16
Enzyme-linked immunosorbent assay (ELISA),
   endothelial cell activation assay, 316, 317, 319
   tissue inhibitor of matrix metalloproteinase, 264
Erosions,
   magnetic resonance imaging, 10
   ultrasonography, 17
Ethidium homodimer, chondrocyte death detection, 194, 196

F

Fibroblast, synovial,
   gene transfer and evaluation in severe combined immunodeficient mouse,
      electroporation,
         materials, 395
         Nucleofector™ system, 398, 400, 410
      lipofection,
         materials, 395
         small interfering RNA, 397, 398, 410
      overview of vectors, 393, 394
      severe combined immunodeficient mouse,
         coimplantation, 408, 409, 411
         fibroblast and cartilage preparation, 408

Index 439

histological assessment of
invasion, 409, 410
viral transduction,
lentiviral transduction, 405–408,
410, 411
materials, 396, 397
retroviral transduction, 401–404,
410
genotyping in rheumatoid arthritis,
cell culture, 379, 382, 387
DNA microarray analysis,
AtlasImage™ software, 385–387,
389
confirmation of results, 387, 389
hybridization, 384, 389
overview, 378, 379
phosphorimaging, 385
probe purification, 384
washing, 384, 385, 389
materials, 379, 381, 387
rationale, 378
RNA arbitrarily primed
polymerase chain reaction,
fingerprinting, 383, 387
gel electrophoresis, 384
overview, 378, 379, 383
RNA isolation, 382, 387
Fibroblast-like synoviocyte (FLS),
enrichment and culture,
adherent cell preparation, 368
dissection, 367, 368, 372
maintenance culture, 368, 369
materials, 367, 372
overview, 366
function, 365
marker expression and flow
cytometry, 371
Matrigel invasion assay,
cell seeding, 416, 417, 420
data analysis, 417, 420
fixation and staining, 417, 420
materials, 414
overview, 413, 414
paraffin coating, 415–418

transwell coating with Matrigel,
416, 418, 420
morphology, 369
phenotyping, 369–371
proliferation response, 369–371
Fibrocyte,
definition, 423
fibroblast comparison, 424, 425
function, 425
history of study, 423, 424
isolation and culture,
culture and harvesting, 427, 428, 430
materials, 426
peripheral blood monocyte
collection, 426–428
markers and flow cytometry, 428,
430, 431
nephrogenic fibrosing dermopathy
role, 425, 426
prospects for study, 432, 433
Flow chamber, see Leukocyte adhesion
Flow cytometry,
endothelial cell surface antigens,
315, 317–319
fibroblast-like synoviocytes, 371
fibrocytes, 428, 430, 431
mononuclear cell isolation from
rheumatoid arthritis synovium,
114, 116, 117
FLS, see Fibroblast-like synoviocyte

G

Gelatinases,
assays, overview, 228
developmental expression, 229, 230
immunohistochemistry,
cartilage preparation, 233
detection, 234, 236
immunostaining, 234–236
materials, 232, 233
overview, 228–231
sectioning, 234, 235
specimen preparation, 234, 235
substrate specificity, 231

types, 227
zymography,
  gel electrophoresis, 234–236
  materials, 233, 235
  overview, 231
  staining, 235, 236
Glycosaminoglycan, see Aggrecan
Goldner-Trichome stain, undecalcified plastic-embedded paw section staining, 277, 278

H

Histology, synovium,
  digital image analysis, see Digital image analysis
  inflammation, 47, 48, 105
  normal tissue, 47, 105
Histomorphometry, see Bone histomorphometry; Cartilage histomorphometry

I

IC_{50}, comparison with inhibition constant, 260
IL-1, see Interleukin-1
IL-10, see Interleukin-10
Immunohistochemistry, synovium,
  aggrecanase proteinases and products,
    blocking and antibody incubation, 172
    cartilage explant studies, 177, 178
    chondroitinase ABC treatment effects, 173–177
    controls, 179
    counterstaining, 173
    detection, 173
    hydrogen peroxide treatment effects, 172–177
    materials, 170, 171
    overview, 168, 170
    pretreatment, 172
    processing, 171, 172
    product development, 173
    specimen preparation, 171, 178, 179
  angiogenesis vessel maturity analysis, 345, 346
  antibodies, 48, 49
  collagenase,
    cryosectioning, 222, 224, 225
    fixation, 221, 223
    immunostaining, 222, 223
    mounting and freezing, 221
    slide coating, 222
  detection, 50
  digital image analysis, see Digital image analysis
  dual-labeled immunofluorescence staining, 55
  gelatinases,
    cartilage preparation, 233
    detection, 234, 236
    immunostaining, 234–236
    materials, 232, 233
    overview, 228–231
    sectioning, 234, 235
    specimen preparation, 234, 235
  kits, 49, 50
  materials, 51, 52, 57
  quantitative analysis, 56, 57
  reagent preparation, 52, 53, 57, 58
  sample sources, 48
  slide preparation,
    coating, 53, 58
    sectioning of synovial biopsies, 53, 54, 58, 59
  T-cell immunoperoxidase staining, 54, 55, 59
  troubleshooting,
    background staining, 60, 61
    weak or absent staining, 61, 62
In situ hybridization (ISH), synovium,
  advantages, 65, 66
  applications, 65, 66
  historical perspective, 65
  hybridization, 71, 72
  immunodetection, 72–74
  materials, 68–70

prehybridization, 71, 73
pretreatment, 71
principles, 66–68
RNase,
   contamination prevention, 70, 73
   treatment, 72
section preparation, 70, 71, 73
washing, 72
Integrated optical density (IOD), digital image analysis, 134
Interleukin-1 (IL-1), joint disease expression, 106–108
Interleukin-10 (IL-10), joint disease expression, 107, 108
Intravital microscopy, leukocyte recruitment,
   carotid artery cannulation, 337, 340
   endothelial adhesion molecule staining, 339, 340
   interaction analysis, 338–341
   knee joint preparation, 337, 340
   labeling of leukocytes, 336, 337
   materials, 334–336
   statistical analysis, 340
   synovial endothelium activation, 336, 340
   synovial vessel visualization, 337
IOD, *see* Integrated optical density
ISH, *see* *In situ* hybridization

L

Laser mediated microdissection (LMM), synovium,
   differentially expressed gene cloning, sequencing, and confirmation, 102
   materials, 92, 93
   microdissection, 96, 98
   overview, 92, 95
   RNA arbitrarily primed polymerase chain reaction,
      band excision and elution of fragments, 101
      fingerprinting, 100, 103
      gel electrophoresis, 100, 101
      principles, 98, 100
   RNA extraction, 98, 103
   section preparation, 94, 95, 102
   single strand conformation polymorphism analysis, 101, 103
   slide preparation, 94, 102
   tissue preparation and embedding, 94, 102
LDL, *see* Low-density lipoprotein
Lentivirus, synovial fibroblast gene transfer, 405–408, 410, 411
Leukocyte adhesion, flow assays,
   human umbilical vein endothelial cell isolation, 326
   materials, 324, 325
   motion analysis, 328, 329
   neutrophil isolation and culture, 326, 327, 329
   overview, 323, 324
   parallel plate flow chamber adhesion assay, 327–329
   slide preparation, 326, 329
Leukocyte recruitment,
   angiogenesis and neutrophil migration in transwell culture plates, 353, 355
   chemotaxis assay,
      cell counting, 363, 364
      endothelial cell chemotaxis assay, 362–364
      materials, 360, 361, 363
      membrane coating, 361, 363
      monocyte chemotaxis assay, 361–363
   endothelium recruitment, 323, 324, 359, 360
   intravital microscopy,
      carotid artery cannulation, 337, 340
      endothelial adhesion molecule staining, 339, 340
      interaction analysis, 338–341
      knee joint preparation, 337, 340
      labeling of leukocytes, 336, 337

materials, 334–336
statistical analysis, 340
synovial endothelium activation, 336, 340
synovial vessel visualization, 337
modeling, 333, 334
rheumatoid arthritis pathogenesis, 333, 334
Lipofection, synovial fibroblast gene transfer,
materials, 395
small interfering RNA, 397, 398, 410
Live/Dead™ kit, chondrocyte death detection, 194—196
LMM, *see* Laser-mediated microdissection
Low-density lipoprotein (LDL), endothelial cell uptake assay, 313, 314

M

Macrophage, *see* Mononuclear cells; Osteoclast
Magnetic cell separation, mononuclear cell isolation from rheumatoid arthritis synovium, 113, 116
Magnetic resonance imaging (MRI),
contraindications, 8, 9
contrast agents, 6, 8
display, 5, 6
equipment, 4, 5
pulse sequences, 6
synovial joint imaging,
advantages, 4
animal models of arthritis, 11
bone marrow edema, 10
cartilage, 10, 11
enthesitis, 9, 10
erosions, 10
prospects, 11, 122
synovitis, 9
Masking, digital image analysis, 130, 131

Matrigel invasion assay, *see* Fibroblast-like synoviocyte
Matrix metalloproteinases, *see* Collagenase; Gelatinases; Membrane-type 1 matrix metalloproteinase; Tissue inhibitor of matrix metalloproteinase
Membrane-type 1 matrix metalloproteinase (MT1-MMP),
collagen degradation assay,
collagen layer preparation, 243, 244
incubation conditions and gel electrophoresis, 244, 245
materials, 242
overview, 240
immunofluorescence microscopy of cell layers, 241, 243
inhibitors, 239
promatrix metalloproteinase-2 activation assay,
incubation conditions and gel electrophoresis, 242, 244
materials, 241
overview, 240
substrates, 239
zymography,
culture and visualization, 243, 244
gelatin preparation, 242, 244
materials, 241
overview, 240
slide chamber preparation, 242–244
Mononuclear cells, preparation from rheumatoid arthritis synovium,
culture, 114, 117
flow cytometry, 114, 116, 117
immunomagnetic cell isolation, 113, 116
materials, 108–111, 116
overview, 108
synovial fluid sample preparation, 111, 112
synovial tissue sample preparation, 112, 116

Index   443

Morphological analysis, *see* Bone
    histomorphometry; Cartilage
    histomorphometry
MRI, *see* Magnetic resonance imaging
MT1-MMP, *see* Membrane-type 1
    matrix metalloproteinase

N

Necrosis,
    articular cartilage injury, 185
    features, 183
Neutrophil, *see* Leukocyte adhesion;
    Leukocyte recruitment
Nucleofector™ system, *see*
    Electroporation

O

Osteoarthritis, chondrocyte death role, 185
Osteoclast,
    characterization,
        actin ring formation, 295
        cathepsin K immunostaining, 293, 294
        pit formation assay, 294, 295, 299
        tartrate-resistant acid phosphatase staining, 292, 293, 299
    culture system,
        bone marrow culture, 287, 288, 297, 298
        bone marrow-derived macrophage culture, 288, 298, 299
        collagen gel culture, 291, 292
        inducing factors, 286, 295, 297
        materials, 286, 287, 297
        osteoclast formation in cocultures, 290, 291, 299
        overview, 285, 286
        RAW264.7 cell culture, 290, 299
        purification of derived cells, 292

P

PCR, *see* Polymerase chain reaction
PET, *see* Positron emission tomography

Polymerase chain reaction (PCR),
    RNA arbitrarily primed polymerase chain reaction after laser mediated microdissection of synovium,
        band excision and elution of fragments, 101
        fingerprinting, 100, 103
        gel electrophoresis, 100, 101
        principles, 98, 100
    suppressive subtractive hybridization, 86–89
    synovial fibroblast genotyping with RNA arbitrarily primed polymerase chain reaction,
        fingerprinting, 383, 387
        gel electrophoresis, 384
        overview, 378, 379, 383
        RNA isolation, 382, 387
Positron emission tomography (PET), synovial joint imaging, 18
Propidium iodide, chondrocyte death detection, 192–194, 196

R

RAP-PCR, *see* Polymerase chain reaction
Reference volume, cartilage histomorphometry, 163, 164
Retrovirus, synovial fibroblast gene transfer, 401–404, 410
Rheumatoid arthritis,
    angiogenesis, *see* Angiogenesis
    bone histomorphometry, *see* Bone histomorphometry
    leukocyte recruitment in pathogenesis, 333, 334
    mononuclear cell preparation, *see* Mononuclear cells
    synovial fibroblasts, *see* Fibroblast, synovial
RNA interference, synovial fibroblast gene transfer, 397, 398, 410

S

SCID mouse, *see* Severe combined immunodeficient mouse
Segmentation, digital image analysis, 131–134
Severe combined immunodeficient (SCID) mouse, synovial fibroblast gene transfer analysis,
  coimplantation, 408, 409, 411
  fibroblast and cartilage preparation, 408
  histological assessment of invasion, 409, 410
Single-photon emission computed tomography (SPECT), synovial joint imaging, 18
Single-strand conformation polymorphism (SSCP), laser mediated microdissection sample analysis, 101, 103
SPECT, *see* Single-photon emission computed tomography
SSCP, *see* Single-strand conformation polymorphism
SSH, *see* Suppressive subtractive hybridization
Subtractive hybridization, *see* Suppressive subtractive hybridization
Suppressive subtractive hybridization (SSH),
  adaptor ligation, 84, 85, 88
  complementary DNA synthesis,
    first strand, 82, 83
    second strand, 83, 87, 88
  hybridizations, 85, 86
  materials, 81, 82
  polymerase chain reaction, 86–89
  principles, 79, 81
  reverse transcription with SMART technology, 78
  RNA isolation and quality control, 78, 87

  *Rsa*I digestion, 84, 88
  subtractive hybridization overview, 77, 87
Synovial fibroblasts, *see* Fibroblast, synovial
Synovitis,
  arthroscopy, see Arthroscopy
  magnetic resonance imaging, 9
  ultrasonography, 15, 16, 33, 34

T

Tartrate-resistant acid phosphatase (TRAP), undecalcified plastic-embedded paw section staining, 278, 280
T-cell, immunoperoxidase staining, 54, 55, 59
Tenosynovitis, ultrasonography, 16, 17
TIMP, *see* Tissue inhibitor of matrix metalloproteinase
Tissue inhibitor of matrix metalloproteinase (TIMP),
  assays,
    fluorogenic substrate assay for titration against matrix metalloproteinases,
      principles, 254, 255
      titration, 255, 256, 264, 265
    kinetic constant determination,
      equilibrium titration, 258, 265, 266
      IC_{50} utility, 260
      inhibition constant, 259, 260
      on-rate constant, 259, 260
      principles, 256, 258
    materials, 252–254, 264–266
    reverse zymography, 260–262, 266
  collagenase assay, 218
  enzyme-linked immunosorbent assay, 264
  inactivation with reductive alkylation, 262, 263
  types and properties, 251, 252
  Western blot, 263

TNF-α, *see* Tumor necrosis factor-α
Toluidine blue, undecalcified plastic-
    embedded paw section
    staining, 278
TRAP, *see* Tartrate-resistant acid
    phosphatase
Tumor necrosis factor-α (TNF-α),
    joint disease expression, 106–108
    therapeutic targeting, 108
TUNEL assay, chondrocyte apoptosis
    detection, 196

U

UEA-1, endothelial cell staining, 315,
    319
Ultrasonography,
    equipment and facilities, 13, 14
    principles, 14, 15
    synovial joint imaging,
        advantages, 4
        enthesopathy, 16
        erosions, 17
        prospects, 17, 18
        synovitis, 15, 16
        tenosynovitis, 16, 17, 33, 34
    training, 12, 13
Uronic acid, glycosaminoglycan
    degradation analysis,
    advantages, 207, 208
    incubation conditions, 206, 208
    limitations, 206, 207

sensitivity, 206

V

Von Kossa, undecalcified plastic-
    embedded paw section
    staining, 278
von Willebrand factor, endothelial cell
    staining, 314, 315

W

Western blot, tissue inhibitor of matrix
    metalloproteinase, 263

Z

Zymography,
    gelatinases,
        gel electrophoresis, 234–236
        materials, 233, 235
        overview, 231
        staining, 235, 236
    membrane-type 1 matrix
        metalloproteinase,
        culture and visualization, 243, 244
        gelatin preparation, 242, 244
        materials, 241
        overview, 240
        slide chamber preparation, 242–244
    tissue inhibitor of matrix
        metalloproteinase reverse
        zymography, 260–262, 266